P9-AGV-160

NANOTECHNOLOGY APPLICATIONS TO TELECOMMUNICATIONS AND NETWORKING

NANOTECHNOLOGY APPLICATIONS TO TELECOMMUNICATIONS AND NETWORKING

Daniel Minoli
Managing Director
Leading-Edge Networks Incorporated

A JOHN WILEY & SONS, INC., PUBLICATION

Published by John Wiley & Sons, Inc., Hoboken, New Jersey.
Published simultaneously in Canada.

Library of Congress Cataloging-in-Publication Data:

Minoli, Daniel, 1952–
Nanotechnology applications to telecommuications and networking / by Daniel Minoli.
p. cm.
Includes bibliographical references and index.
ISBN-13 978-0-471-71639-6
ISBN-10 0-471-71639-1 (cloth)
1. Telecommunication—Equipment and supplies. 2. Computer networks—Equipment and supplies. 3. Nanotechnology. I. Title.

TK5103.M474 2005
621.382–dc22 2004063825

Printed in the United States of America
10 9 8 7 6 5 4 3 2 1

For Anna

And for my Father and Mother

CONTENTS

PREFACE

This is believed to be the first book that takes a view of nanotechnology from a *telecommunications and networking* perspective. Nanotechnology refers to the manipulation of materials at the atomic or molecular level. Nanotechnology is getting a lot of attention of late not only in academic settings and in laboratories around the world, but also in government and venture capitalists' initiatives. There now is a major drive to commercialize the technology by all sorts of firms, ranging from start-ups to Fortune 100 companies.

At the start of the decade, Charles Vest, the president of MIT, observed: "We are just beginning to understand how to use nanotechnology to build devices and machines that imitate the elegance and economy of nature. The gathering nanotechnology revolution will eventually make possible a huge leap in computing power, vastly stronger yet much lighter materials, advances in medical technologies, as well as devices and processes with much lower energy and environmental costs."

Nanotechnology is a nanometer-level *bottom-up*[1] assembly approach that allows developers to engineer particles at the molecular level, building them up to the "right size," with engineered functional properties. A nanometer is one billionth of a meter (a meter being about 3 ft). Bottom-up process technology provides a control mechanism over development of particles with respect to their size, shape, morphology, and surface conditions. Because of the challenges involved in working at this microscopic scale of a few nanometers, research and engineering efforts involving manipulation of components as "large" as 100 nm are typically included in the field of nanotechnology. Atoms are typically between one-tenth and one-half of a nanometer wide.

Research and development topics in nanotechnology range from molecular manipulation to nanomachines (microscopic devices that can themselves carry out tasks at the atomic or sub-atomic level). While nanomachines represent futuristic initiatives with relatively little current (commercial) achievement, nanomaterials, nanomaterial processing, nanophotonics, and nanoelectronics are already resulting (or will do so in the next 3–5 years) in usable technologies.

In this book we focus on developments and technologies that have the potential to be used (or are already being used) in communication and networking environments. Such applications include faster and smaller non-silicon-based processors, faster and smaller switches (particularly optical switches), and MEMSs (microelectromechanical

[1]In the nanotechnology field the term *bottom-up* is preferred to the (perhaps) more common English-language term *bottoms-up*.

systems). MEMS are microscale systems (~100 μm) that include both mechanical and electrical devices integrated on a single die or chip. MOEMS are microoptical-electromechanical systems consisting of MEMS devices with integral optical components such as mirrors, lenses, filters, laser diodes, emitters or other optics. A MEMS system may include microfluidic elements, integral microelectronics or ICs, "lab-on-chip" systems, actuators, micromotors, or sensors. Efforts are already underway to create nanoscale MEMSs, also known as NEMSs.

In Chapter 1 we review the basic concepts of nanotechnology and applications. In Chapters 2 and 3 we cover supportive topics such as physics and chemistry basics (e.g., electron, atoms, atomic structures, molecules, bonded structures); electrical properties (e.g., insulators, semiconductors, conductors); and chemical bonds and reactions. Chapter 2 also provides a basic introduction to transistors, in support of the discussion to follow in Chapter 6. It turns out that while classical Newtonian mechanics can predict with precision the motions of masses ranging in size from microscopic particles to stars, it cannot predict the behavior of the particles in the atomic domain; at these dimensions quantum theory (physics) comes into play. Hence, as a spin-off of Chapters 2 and 3, in Appendices D and E we discuss some of the basic scientific principles that support quantum theory; the reader who may find these two appendices somewhat demanding may chose to skip this material and move on to the chapters that follow, which are generally self-contained. In Chapter 4 we look at nanomaterials and nanomaterial processing: Individual nanoparticles and nanostructures (e.g., nanotubes, nanowires) are discussed. Nanophotonics is discussed in Chapter 5 (e.g., nanocrystals, nanocrystal fibers). Nanoelectronics (e.g., metal nanoclusters, semiconducting nanoclusters, nanocrystals, quantum dots) is covered in Chapter 6. Both Chapters 5 and 6 provide a discussion of near-term and longer-term applications in the field of computers, telecommunications, and networking. An extensive glossary is also included. Appendix F discusses nanoinstrumentation, while Appendix G provides detailed information on quantum computing.

This book is intended as an introduction to the field of nanotechnology for telecommunications vendors, researchers, and students who want to start thinking about the potential opportunities afforded by these emerging scientific developments and approaches for the next-generation networks to be deployed 5–10 years in the future. Advanced planning is a valuable and effective exercise. When the author first joined Bell Telephone Laboratories in 1978, he was involved in planning networks 5–10 years into the future. While, recently, advanced planning and strategic development have suffered at the hand of the "next-quarter" mentality, it is indeed advantageous to plan 10 years out, only if for the reason that it takes about 10–15 years to grow a carrier (such as a CLEC, a hotspot provider, a 3G wireless operator) to turn a profit from a cold start.

As noted, this book is intended as an introduction to the field. We hope it will serve as motivation, by raising interest, to continue the line of investigation and research into the field. We have made every effort to make it relatively self-contained by discussing the introductory fundamental principles involved, and by providing an extensive glossary. Most professionals outside the field of basic sciences probably have forgotten freshman college physics and chemistry. The most basic take-aways from these courses are summarized in the book, to facilitate the discussion of nanotechnology applications.

The reader is encouraged, after reading this text, to seek out additional books that go into greater detail. Each chapter included here can be supported by an entire book just covering each individual chapter-level topic.

Finally, it should be noted that nanotechnology is a highly active burgeoning field at this time, with (hundreds of) thousands of articles, publications, lectures, seminars, and books available. Given this plethora of research, this book is based liberally on industry sources. In this context, we have made every effort to acknowledge the source of the material we cover and provide appropriate credit thereof; we hope, with said diligence, that any unwitting omissions are strictly minimal and/or essentially inconsequential. Hence, while the actual synthesis of the topic(s) as presented here is original, the intrinsic material itself is based on the 750+ references that we cite and utilize throughout the body of the text.

Acknowledgement

I would like to thank Mr. Emile A. Minoli for contributions in Chapters 2 and 3.

The cover page shows Daniel Minoli (center front) with a slide rule next to an AM radio the student trio built based on discrete electronic components. Students Melvin Lee (left front) and Steven Lightburn (right front) part of the student trio are with Mr. Tepper (middle front), electronics teacher in a Technical Electronics Laboratory in Hight School in Brooklyn, NY in the fall of 1970. Two second-row students are unidentified. As this textbook shows, electronics and electronics density has come a long way in the past 35 years, and will continue to do so under the thrust of nanotechnology.

DANIEL MINOLI

ABOUT THE AUTHOR

Daniel Minoli has many years of telecom, networking, and information technology (IT) experience for end-users, carriers, academia, and venture capitalists, including work at ARPA think tanks, Bell Telephone Laboratories, ITT, Prudential Securities, Bell Communications Research (Bellcore/Telcordia), AT&T, NYU, Rutgers University, Stevens Institute of Technology, and Societe General de Financiament de Quebec (1975–2001). Recently, he also played a founding role in the launching of two networking companies through the high-tech incubator Leading Edge Networks Inc., which he ran in the early 2000s: Global Wireless Services, a provider of secure broadband hotspot mobile Internet and hotspot VoIP services to high-end marinas; and InfoPort Communications Group, an optical and gigabit Ethernet metropolitan carrier supporting data center/SAN/channel extension and grid computing network access services (2001–2003). In the recent past, Mr. Minoli was involved (on behalf of a venture capitalist considering a $15 million investment) in nanotechnology-based systems using quantum cascade lasers (QCLs) for 10-μm-transmission free space optics communication systems.

An author of a number of technical references on IT, telecommunications, and data communications, he has also written columns for *ComputerWorld*, *NetworkWorld*, and *Network Computing* (1985–2005). He has taught at New York University (Information Technology Institute), Rutgers University, Stevens Institute of Technology, Carnegie Mellon University, and Monmouth University (1984–2003). Also, he was a Technology Analyst At-Large, for Gartner/DataPro (1985–2001); based on extensive hands-on work at financial firms and carriers, he tracked technologies and wrote around 50 distinct CTO/CIO-level technical/architectural scans in the area of telephony and data communications systems, including topics on security, disaster recovery, IT outsourcing, network management, LANs, WANs (ATM and MPLS), wireless (LAN and public hotspot), VoIP, network design/economics, carrier networks (such as metro Ethernet and CWDM/DWDM), and e-commerce. Over the years he has advised venture capitalists for investments of $150 million in a dozen high-tech companies and has acted as expert witness in a (won) $11 billion lawsuit regarding a wireless air-to-ground communication system.

CHAPTER 1

Nanotechnology and Its Business Applications

1.1 INTRODUCTION AND SCOPE

1.1.1 Introduction to the Nanoscale

Nanotechnology is receiving a lot of attention of late across the globe. The term *nano* originates etymologically from the Greek, and it means "dwarf." The term indicates physical dimensions that are in the range of one-billionth (10^{-9}) of a meter. This scale is called colloquially *nanometer scale*, or also *nanoscale*. One nanometer is approximately the length of two hydrogen atoms. Nanotechnology relates to the design, creation, and utilization of materials whose constituent structures exist at the nanoscale; these constituent structures can, by convention, be up to 100 nm in size.[1–3] Nanotechnology is a growing field that explores electrical, optical, and magnetic activity as well as structural behavior at the molecular and submolecular level. One of the practical applications of nanotechnology (but certainly not the only one) is the science of constructing computer chips and other devices using nanoscale building elements. This book is a basic practical survey of this field with an eye on computing and telecom applications.

The nanoscale dimension is important because quantum mechanical (non-Newtonian) properties of electronics, photons, and atoms are evident at this scale. Nanoscale structures permit the control of fundamental properties of materials without changing the materials' chemical status. Nanostructure, such as nanophotonic devices, nanowires, carbon nanotubes, plasmonics devices, among others, are planned to be

[1]Measures are relatives; hence, one can talk about something being 1000 nanometers (nm), or 1 microm (μm), of 10,000 Angstroms (Å). A micron is a unit of measurement representing one-millionth of a meter and is equivalent to a micrometer. An angstrom is a unit of measurement indicating one-tenth of a nanometer, or one ten-billionth of a meter (often used in physics and/optics to measure atoms and wavelengths of light).

[2]Atoms are typically between 0.1 and 0.5 nm wide.

[3]For comparison, a human hair is between 100,000 and 200,000 nm in diameter and a virus is typically 100 nm wide.

Nanotechnology Applications to Telecommunications and Networking, By Daniel Minoli

incorporated into telecommunication components and into microprocessors in the next few years, leading to more powerful communication systems and computers—these nanostructures are discussed in the chapters that follow. Nanotechnology is seen as a high-profile emerging area of science and technology. Proponents prognosticate that, in the next few years, nanotechnology will have a major impact on society. Recently, Charles Vest [1], the president of MIT, observed: "The gathering nanotechnology revolution will eventually make possible a huge leap in computing power, vastly stronger yet much lighter materials, advances in medical technologies, as well as devices and processes with much lower energy and environmental costs." There already are an estimated 20,000 researchers worldwide working in nanotechnology today.

In the sections that follow in this chapter we preliminarily answer questions such as: What is nanotechnology? What are the applications of nanotechnology? What is the market potential for nanotechnology? What are the global research activities in nanotechnology? Why would a practitioner (the likely reader of this book), need to care? We then position the reader for the balance of the book, which looks at the nanotechnology topic from an application, and, more specifically, from a telecom- and networking-perspective angle.

While many definitions for nanotechnology exist, the National Nanotechnology Initiative (NNI[4]), calls an area of research, development, and engineering "nanotechnology" only if it involves all of the following [2]:

1. Research and technology development at the atomic, molecular, or macromolecular levels, in the length scale of approximately 1- to 100-nm range
2. Creating and using structures, devices, and systems that have novel properties and functions because of their small and/or intermediate size
3. Ability to control or manipulate matter on the atomic scale

Hence, nanotechnology can be defined as the ability to work at the molecular level, atom by atom, to create large structures with fundamentally new properties and functions. Nanotechnology can be described as the precision-creation and precision-manipulation of atomic-scale matter [3]; hence, it is also referred to as *precision molecular engineering*. Nanotechnology is the application of nanoscience to control processes on the nanometer scale, that is, between 0.1 and 100 nm [4]. The field is also known as *molecular engineering* or *molecular nanotechnology (MNT)*. MNT deals with the control of the structure of matter based on atom-by-atom and/or molecule-by-molecule engineering; also, it deals with the products and processes of molecular manufacturing [5]. The term *engineered nanoparticles* describes particles that do not occur naturally; humans have been putting together different materials throughout time, and now with nanotechnology they are doing so at the nanoscale.

[4]The National Nanotechnology Initiative (NNI) is a U.S. government-funded R&D and commercialization initiative for nanoscience and nanotechnology. The 21st Century Nanotechnology Research and Development Act of 2003 put into law programs and activities supported by the initiative.

As it might be inferred, nanotechnology is highly interdisciplinary as a field, and it requires knowledge drawn from a variety of scientific and engineering arenas: Designing at the nanoscale is working in a world where physics, chemistry, electrical engineering, mechanical engineering, and even biology become unified into an integrated field. "Building blocks" for nanomaterials include carbon-based components and organics, semiconductors, metals, and metal oxides; nanomaterials are the infrastructure, or building blocks, for nanotechnology.

The term *nanotechnology* was introduced by Nori Taniguchi in 1974 at the Tokyo International Conference on Production Engineering. He used the word to describe ultrafine machining: the processing of a material to nanoscale precision. This work was focused on studying the mechanisms of machining hard and brittle materials such as quartz crystals, silicon, and alumina ceramics by ultrasonic machining. Years earlier, in a lecture at the annual meeting of the American Physical Society in 1959 (There's Plenty of Room at the Bottom) American Physicist and Nobel Laureate Richard Feynman argued (although he did not coin or use the word *nanotechnology*) that the scanning electron microscope could be improved in resolution and stability, so that one would be able to "see" atoms. Feynman proceeded to predict the ability to arrange atoms the way a researcher would want them, within the bounds of chemical stability, in order to build tiny structures that in turn would lead to molecular or atomic synthesis of materials [6]. Based on Feynman's idea, K. E. Drexler advanced the idea of "molecular nanotechnology" in 1986 in the book *Engines of Creation,* where he postulated the concept of using nanoscale molecular structures to act in a machinelike manner to guide and activate the synthesis of larger molecules. Drexler proposed the use of a large number (billions) of roboticlike machines called "assemblers" (or nanobots) that would form the basis of a molecular manufacturing technology capable of building literally anything atom by atom and molecule by molecule. Quite a bit of work has been done in the field since the publication of the book, although the concept of nanobots is still speculative.[5]

At this time, an engineering discipline has already grown out of the pure and applied science; however, nanoscience still remains somewhat of a maturing field. Nanotechnology can be identified precisely with the concept of "molecular manufacturing" (molecular nanotechnology) introduced above or with a broader definition that also includes laterally related subdisciplines [7]. This text will encompass both perspectives; the context should make clear which of the definitions we are using. The nanoscale is where physical and biological systems approach a comparable dimensional scale. A basic "difference" between systems biology and nanotechnology is the goal of the science: systems biology aims to uncover the fundamental operation of the cell in an effort to predict the exact response to specific stimuli and genetic variations (has scientific discovery focus); nanotechnology, on the other hand, does not attempt to be so precise but is chiefly concerned with useful design

[5]The possibility of building tiny motors on the scale of a molecule appears to have been brought one step closer of late: researchers recently have described how they were able—using light or electrical stimulation—to cause a molecule to rotate on an axis in a controlled fashion, similar to the action of a motor [8].

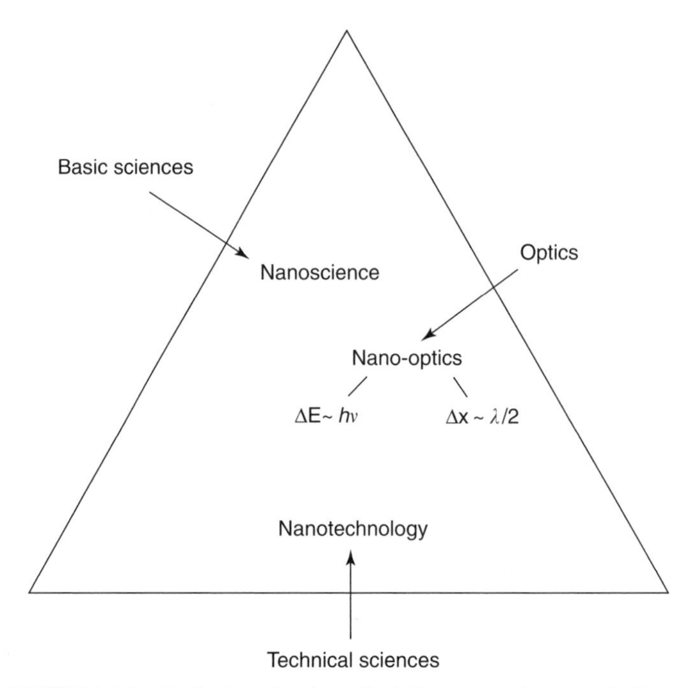

FIGURE 1.1 Evolution of various disciplines toward nanoscale focus.

(has engineering design focus) [9]. Figure 1.1 depicts the current evolution of various disciplines toward a nanoscale focus.

Figure 1.2 places "nano" in the continuum of scales, while Figure 1.3 depicts the size of certain natural and manmade objects (Table 1.1, loosely based on [10] depicts additional substances, entities, and materials). A nanometer is about the width of four silicon atoms (with a radius of 0.13 nm) or two hydrogen atoms (radius of 0.21 nm); also see Figure 1.4. Figure 1.5 depicts an actual nanostructure. For comparison purposes, the core of a single-mode fiber is 10,000 nm in diameter, and a 10-nm nanowire is 1000 times smaller than (the core of) a fiber. The nanoscale exists at a boundary between the "classical world" and the "quantum mechanical world"; therefore, realization of nanotechnology promises to afford revolutionary new capabilities. In this context, the following quote is noteworthy [11]:

> When the ultimate feature sizes of nanoscale objects are approximately a nanometer or so, one is dealing with dimensions an order of magnitude larger than the scale exploited by chemists for over a century. Synthetic chemists have manipulated the constituents, bonding, and stereochemistry of vast numbers of molecules on the angstrom scale, and physical and analytical chemists have examined the properties of these molecules. So what is so special about the nanoscale? There are many answers to this question, possibly as many as there are people who call themselves nanoscientists or nanotechnologists. *A particularly intriguing feature of the nanoscale is that this is the scale on which*

Prefix	Factor	Symbol	Prefix	Factor	Symbol
deci	10^{-1}	d	yotta	10^{24}	Y
centi	10^{-2}	c	zetta	10^{21}	Z
milli	10^{-3}	m	exa	10^{18}	E
micro	10^{-6}	μ	peta	10^{15}	P
nano ←	10^{-9}	n	tera	10^{12}	T
pico	10^{-12}	p	giga	10^{9}	G
femto	10^{-15}	f	mega	10^{6}	M
atto	10^{-18}	a	kilo	10^{3}	k
zepto	10^{-21}	z	hecto	10^{2}	h
yocto	10^{-24}	y	deka	10^{1}	da

FIGURE 1.2 Putting *nano* in context.

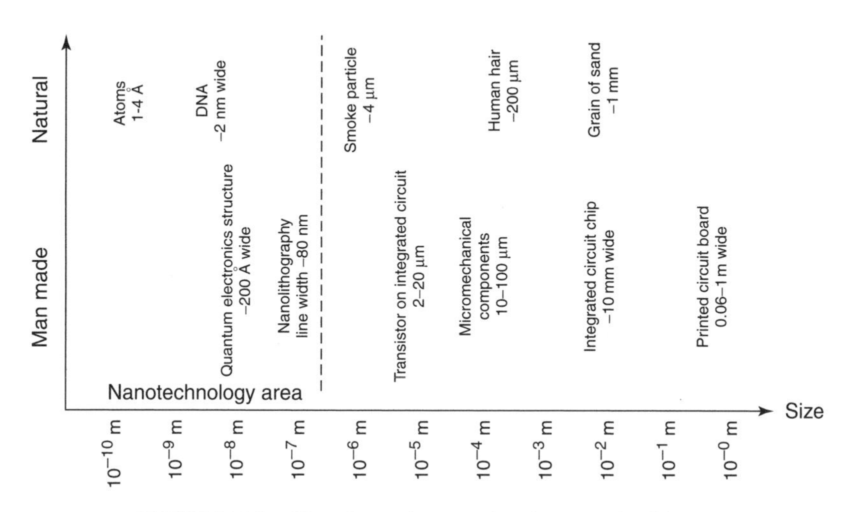

FIGURE 1.3 Size of certain natural and manmade objects.

biological systems build their structural components, such as microtubules, microfilaments, and chromatin. The associations maintaining these and the associations of other cellular components seem relatively simple when examined by high-resolution structural methods, such as crystallography or Nuclear Magnetic Resonance—shape complementarity, charge neutralization, hydrogen bonding, and hydrophobic interactions. A key property of biological nanostructures is molecular recognition, leading to self-assembly and the templating of atomic and molecular structures. Those who wish to create defined nanostructures would like to develop systems that emulate this behavior.

TABLE 1.1 Scale of Some Substances and Entities

The Planck length (the smallest measurement of length that has meaning)	1.616×10^{-35} m
One fermi (aka a femtometer: a unit suitable to express the size of atomic nuclei)	1×10^{-15} m
Diameter of proton	1.66×10^{-15} m
Classical diameter of neutron	2.2×10^{-15} m
Diameter of the nucleus of a helium atom	3.8×10^{-15} m
Classical diameter of an electron	5.636×10^{-15} m
Diameter of the nucleus of an aluminum atom	7.2×10^{-15} m
Diameter of the nucleus of a gold atom	1.4×10^{-14} m
Wavelength of γ rays	1×10^{-12} m
Diameter of flourine ion	3.8×10^{-11} m
Most likely distance from electron to nucleus in a hydrogen atom (bohr radius)	5.29×10^{-11} m
Distance between bonded hydrogen atoms	7.41×10^{-11} m
One angstrom	1×10^{-10} m
Van der Waals radius of hydrogen atoms (max distance between atoms that are not bonded)	1.2×10^{-10} m
Resolution (size of smallest visible object) of a transmission electron microscope	2×10^{-10} m
Distance between bonded iron atoms	2.48×10^{-10} m
Van der Waals radius of potassium atoms (max distance between atoms that are not bonded)	2.75×10^{-10} m
Diameter of water molecule	3×10^{-10} m
Distance between base pairs in a DNA molecule	3.4×10^{-10} m
Diameter of xenon ion	3.8×10^{-10} m
Distance between bonded cesium atoms	5.31×10^{-10} m
One nanometer	1×10^{-9} m
Size of glucose molecule	1.5×10^{-9} m
Diameter of DNA helix	2×10^{-9} m
Diameter of insulin molecule	5×10^{-9} m
Diameter of a hemoglobin molecule	6×10^{-9} m
Thickness of cell wall (Gram-negative bacteria)	1×10^{-8} m
Size of typical virus	7.5×10^{-8} m
Thickness of gold leaf	1.25×10^{-7} m
Diameter of smallest bacteria	2×10^{-7} m
Resolution (size of smallest visible object) of an optical microscope	2×10^{-7} m
Length of the smallest transistor in a Pentium 3 chip	2.6×10^{-7} m
Wavelength of violet light	4.1×10^{-7} m
Wavelength of red light	6.8×10^{-7} m
One micrometer (micron)	1×10^{-6} m
Size of typical bacterium	1×10^{-6} m
Diameter of average human cell nucleus	1.7×10^{-6} m
Thickness of typical red blood cell	2.4×10^{-6} m
Length of the smallest transistor in an Intel 286 chip	3×10^{-6} m
Diameter of typical capillary	4×10^{-6} m
Length of the smallest transistor in an Intel 8086 chip	6×10^{-6} m
Diameter of a single yeast organism	7×10^{-6} m

TABLE 1.1 (*Continued*)

Diameter of a single yeast organism	7×10^{-6} m
Diameter of typical red blood cell	8.4×10^{-6} m
Diameter of average cell in human body	1×10^{-5} m
Size of a grain of talcum powder	1×10^{-5} m
Length of the smallest transistor in the first 6502 chips	1.6×10^{-5} m
Length of the smallest transistor in an Intel 4004 (the first microprocessor)	2×10^{-5} m
Diameter of a small grain of sand	2.0×10^{-5} m
Diameter of a typical human hair	2.5×10^{-5} m
Thickness of typical sheet of paper	8.38×10^{-5} m
Optical resolution: minimum size of object that can resolved by unaided eye	1×10^{-4} m
Size of a grain (crystal) of salt	1×10^{-4} m
Diameter of a period printed at end of typical sentence	3×10^{-4} m
Diameter of the most common type of optical fiber (including cladding)	3.7×10^{-4} m
Size of largest known bacterium	7.5×10^{-4} m
Diameter of the head of the average pin	1.7×10^{-3} m
Diameter of a large grain of sand	2×10^{-3} m

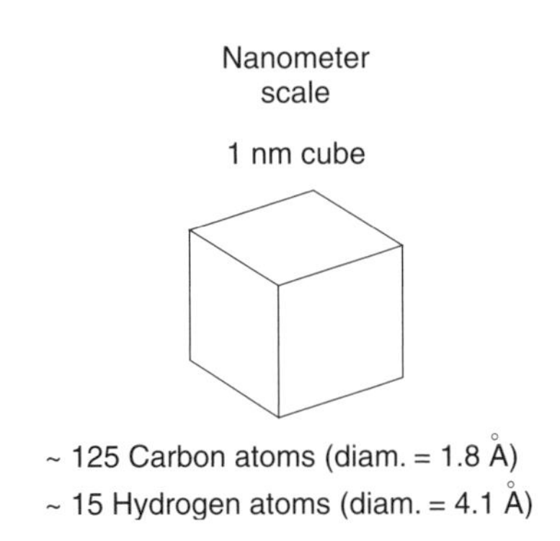

FIGURE 1.4 What one gets at the nanometer scale.

1.1.2 Plethora of Potential Applications

Nanotechnology is an enabling and potentially disruptive technology that can address requirements in a large number of industries. Developments in nanoscale science and engineering promise to impact, if not revolutionize, many fields and lead to a new technological base and infrastructure that can have major impact on telecom, computing, and information technology (in the form of optical networking/nanophotonics, nanocomputing/nanoelectronics, and nanostorage); health care and biotechnology; environment; energy; transportation; and space exploration, among

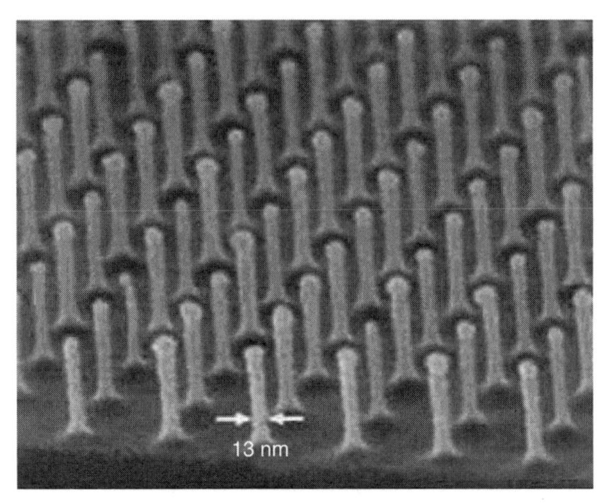

100-nm period posts in Si

FIGURE 1.5 Example of nanostructure. This scanning electron micrograph shows a grid of 13-nm-diameter posts on 100-nm centers that have been etched into the material. Light emission from nanostructured silicon may have applications in optical communications, displays, and various other uses. (Courtesy: Research Laboratory of Electronics NanoStructures Laboratory, Massachusetts Institute of Technology).

others [12]. Nanotechnology will enable manufacturers to produce computer chips and sensors that are considerably smaller, faster, more energy efficient, and cheaper to manufacture than their present-day counterparts. Specifically, nanotechnology is now giving rise to many new applications such as quantum computing, surface and materials modification, novel separations, sensing technologies, diagnostics, and human biomedical replacements.

The technology will also open up completely new areas of research because, as already stated, matter behaves differently at this physical scale [13]. Interfacing materials with biology is widely believed to be the exciting new frontier for nanotechnology [14]. For example, the National Aeronautics and Space Administration (NASA) foresees a *zone of convergence* between biotechnology, nanotechnology, and information technology; consequently, NASA, is funding basic nanoscience, as well as work on nanostructured materials, nanoelectronics, and research into sensors [15]. As another example, the U.S. Army is funding *soldier nanotechnologies* to develop products to substantially reduce the weight that soldiers must carry while increasing physical protection.

Nanomaterials give impetus to new applications of the (nano)technology because they exhibit novel optical, electric, and/or magnetic properties. The first generation of nanotechnology (late 1990s–early 2000s) focused on performance enhancements to existing micromaterials; the second generation of nanotechnology (slated for 2006–2007) will start employing nanomaterials in much more significant and radical ways. Industry observers assert that nanotechnological advances are essential if one is to continue the revolution in computer hardware beyond about the next decade; furthermore, nanotechnology will also allow us to fabricate an entire new generation of products that are cleaner, stronger, lighter, and more precise[6] [7]. Nanomaterials with structural features at the nanoscale can be found in the form of clusters, thin films, multilayers, and nanocrystalline materials often expressed by the dimensionality of 0, 1, 2 and 3; the materials of interest include metals, amorphous and crystalline alloys, semiconductors, oxides, nitride and carbide ceramics in the form of clusters, thin films, multilayers, and bulk nanocrystalline materials [16].

All products are manufactured from atoms, however, interestingly, the properties of those products depend on how those atoms are arranged. For example, by rearranging the atoms in coal (carbon), one can make diamonds. It should be noted that current manufacturing techniques are very rudimentary at the atomic/molecular level: casting, grinding, milling, and even lithography move atoms in bulk rather than in a "choreographed" or "highly controlled" fashion. On the other hand, with nanotechnology one is able to assemble the fundamental building blocks of nature (atoms, molecules, etc.), within the constraints of the laws of physics, but in ways that may not occur naturally or in ways to create some existing structure but by synthesizing it out of cheaper forms or constituent elements. Nanomaterials often have properties dramatically different from their bulk-scale counterparts; for example, nanocrystalline copper is five times harder than ordinary copper with its micrometer-sized crystalline structure [17]. A goal of nanotechnology is to close the size gap between the smallest lithographically fabricated structures and chemically synthesized large molecules [18].

As scientists and engineers continue to push forward the limits of computer chip manufacturing, they have entered into the nanometer realm in recent years without much public fanfare: The first transistor gates under 100 nm went into production in 2000, and microprocessor chips that were coming to market at press time had gates 45 nm wide [19]. A Pentium 4 chip contains in the range of 50 million transistors. However, as the physical laws related to today's telecom chipsets, computer memory, and processor fabrication reach their limits, new approaches such as single-electron technology (nanoelectronics) or plasmonics (nanophotonics) are needed. The invention of the scanning tunneling microscope, the discovery of the fullerene family of molecules, the development of materials with size-dependent properties, and the ability to encode with and manipulate biological molecules such as deoxyribonucleic acid (DNA), are a few of the crucial developments that have advanced nanotechnology in the recent past [20]. A gamut of products featuring the unique properties of nanoscale materials are already available to consumers and industry at this time. For example, most computer hard drives contain giant magnetoresistance (GMR) heads that, through nanothin layers of magnetic materials, allow for a significant increase in storage capacity. Other electronic applications include nonvolatile magnetic memory, automotive sensors, landmine detectors, and solid-state compasses. Some other

[6]It is worth noting that the National Science Foundation has estimated that 2 million workers will be needed to support nanotechnology industries worldwide within 15 years.

nanotechnology uses that are already in the marketplace include (also see Table 1.2 [2] and Table 1.3 [21]):

- Burn and wound dressings
- Water filtration
- Catalysis
- A dental-bonding agent
- Coatings for easier cleaning glass
- Bumpers and catalytic converters on cars
- Protective and glare-reducing coatings for eyeglasses and cars
- Sunscreens and cosmetics
- Longer-lasting tennis balls

TABLE 1.2 Recent Achievements in Nanotechnology (Partial List)

Use of the bright fluorescence of semiconductor nanocrystals for dynamic angiography in capillaries hundreds of micrometers below the skin of living mice—about twice the depth of conventional angiographic materials—and obtained with one-fifth the irradiation power.
Nanoelectromechanical sensors that can detect and identify a single molecule of a chemical warfare agent—an essential step toward realizing practical field sensors.
Nanotube-based fibers requiring 3 times the energy to break of the strongest silk fibers and 15 times that of Kevlar fiber.
Nanocomposite energetic materials for propellants and explosives that have over twice the energy output of typical high explosives.
Prototype data storage devices based on molecular electronics with data densities over 100 times that of today's highest density commercial devices.
Field demonstration that iron nanoparticles can remove up to 96% of a major contaminant (trichloroethylene) from groundwater at an industrial site.

TABLE 1.3 Short-Term Commercially Viable Nanotechnology Products

Sector	Examples of Products Generating Revenues in 2005 and Beyond
Building materials	Scratchless long endurance treatments for vinyl, roofing, furniture, etc.; self-cleaning windows
Communications and computers	Nanodrives and memory, enhanced displays and electronic paper, copiers and printers; sensors as inputs for security and monitoring systems
Military and aerospace	Materials and coatings for hardening products; sensors
Chemicals	Advanced catalysts and additives
Pharmaceuticals and medical devices	Better targeted pharmaceuticals and cosmetics; drug delivery systems; nanobiotechnology products
Energy	Filters, additives, and catalysts for hydrocarbon-based fuels; photovoltaics

- Light-weight, stronger tennis racquets
- Stain-free clothing and mattresses
- Ink

Telecommunications- and computing-specific applications include, among others:

- Nanoelectronics, nanophotonics, nanomaterials, new chipsets
- Optical transmission [e.g., in the emerging optical transport network (OTN)]
- Optical switching [e.g., in the emerging automatically switched optical network (ASON)]
- Microelectromechanical systems (MEMS) and microoptical-electromechanical systems (MOEMS) applications [e.g., tunable optical components and modules, optical switches, fiber-optic networks, electromagnetic radio frequency (RF) MEMS switch; sensor; actuators; information storage systems including magnetic recording, optical recording, and other recording devices, e.g., rigid disk, flexible disk, tape and card drives; processing systems including copiers, printers, scanners, and digital cameras]
- Speech recognition/pattern recognition/imaging
- Advanced computing (e.g., quantum computing, pervasive computing, ubiquitous computing, autonomic computing, utility computing, grid computing, molecular computing, massively parallel computing, and amorphous computing)
- Storage
- "Terascale integration" microprocessors
- Quantum cryptography
- Nanosensors and nanoactuators

Focusing on electronics and photonics, note that the micrometer (10^{-6} m) range is representative of typical computing technology of the late 1990s–early 2000s: random-access memory (RAM), read-only memory (ROM), and microprocessors have feature sizes on the order of micrometers. The entire advancement of processor technology and (optical) communication is essentially the effort to shrink circuits from micrometers down to fractions of a micrometer (e.g., 0.1 μm or less). Silicon can be machined into slabs 0.3–0.1 μm wide (this is smaller than the wavelength of deep violet light). This is what one could do at press time with conventional processing technology. Somewhere between 0.5 and 0.1 μm some of the basic laws (such as Ohm's law) begin to break down, and the rules of quantum theory begin to become important if not overriding [5].

Consider for illustrative purposes one example of nanoelectronic (nano)structures, specifically nanowires. Nanowires are electrical conductors that function like wires but exist at the nanoscale. Nanowires can be used to manufacture faster computer chips, higher-density memory, and smaller lasers. Nanowires are molecular structures with characteristic electrical or optical properties. They are one of the key components to be used for the creation of "molecular electronics chips." These wires have been manufactured in the 40- to 80-nm-diameter range. Nanowires are relatively easy to produce, and they can be assembled in grids to become the basis of nanoscale logic circuits. Nanowires can have a number of (very) different shapes: They often are thin and short "threads" but also have other shapes.

Nanotubes are the ultimate form of nanowires. *Carbon nanotube* is a generic term referring to molecular structures with cylindrical shapes that are based on the carbon atom; there are several other kinds of nanotubes based on noncarbon atoms. Single-wall nanotubes are 1–2 nm in diameter. Nanotubes have interesting electrical properties. A carbon nanotube (Fig. 1.6*b*) (discovered in 1991 by the Japanese researcher Sumio Iijima, Meijo University) is an assembly of carbon (graphite) atoms with extraordinary properties. The carbon nanotube is a single molecule of graphite shaped in a cylindrical sheet (a hexagonal lattice of carbon). Each end of the cylinder is terminated by a hemispherical cap. A nanotube's length can be in the millimeter range (this being millions of times greater than its diameter). Carbon nanotubes have many possible applications, given that they are 100 times stronger than steel (and 6 times lighter), they are good conductors, and they can resist very high temperatures. These important advances provide a foundation to build the nanoelectronic devices and chips of the future.

Carbon nanotubes are based on fullerenes. Fullerene is a third form of carbon (the other two being the diamond form and the graphite form); it is a molecular form of pure carbon that has a cagelike structure. Fullerenes are closed, convex cage molecules containing only hexagonal and pentagonal faces. This class of carbon molecules was

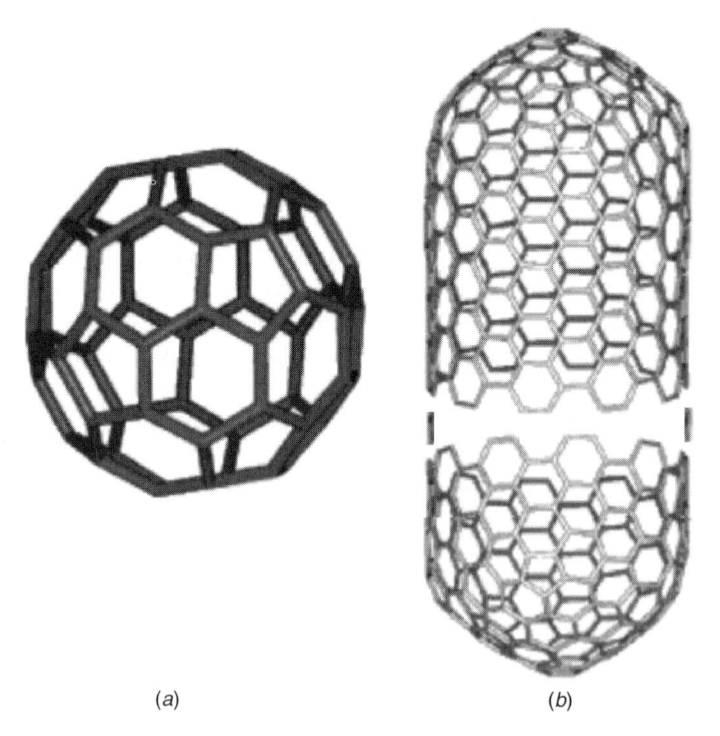

FIGURE 1.6 Fullerene: basic carbon nanomaterial structure: (*a*) buckyballs and (*b*) carbon nanotubes.

discovered by Richard Smalley in 1985. The fullerene structure can be spherical or tubular in shape, as shown in Figure 1.6 (this form of carbon is named in honor of the architect Buckminster Fuller, who designed the geodesic structures that the fullerene resembles[7]). Fullerenes are formed when vaporized carbon condenses in an atmosphere of inert gas. Fullerenes enjoy extraordinary properties, such as superconducting. *Buckyballs* are the most famous fullerene molecules (pictorially they are close to the shape of an European soccer ball.) The buckminsterfullerene (buckyball C_{60}) (Fig. 1.6*b*) is a nanostructure composed of 60 atoms of carbon, organized in a perfectly symmetric closed cage; much larger fullerenes also exist as seen in Figure 1.6.

1.1.3 Challenges and Opportunities

In 2004 the semiconductor industry reliably crossed the 100-nm fabrication barrier, and manufacturers were able to place 100 million transistors on a chip, but from 2005 onward, major challenges were expected to begin to materialize, according to observers. Continued improvements in lithography have resulted in integrated circuits (ICs) with linewidths that are less than 1 μm (1000 nm): This work is often called "nanotechnology," especially if/when the 100-nm barrier is crossed.[8] However, the challenge arises when scientists seek to create structures less than 100 nm in *two* or *three* dimensions [22]. Submicron lithography is a useful technique, but it is equally clear that conventional lithography will not permit the building of semiconductor devices in which individual dopant atoms are located at specific lattice sites: Many of the exponentially improving trends in computer hardware capability have remained operative for the last 50 years, and there is fairly widespread belief that these trends are likely to continue for a number of years, but thereafter conventional lithography will start to reach its limits [7].

There are challenges in the area of the gate dielectric, gate electrodes, substrate and device structure, and device interconnects [23]. Specifically, (i) there are the power implications of Moore's law[9]; (ii) two major gaps in the EDA (electronic design automation) chain, at the architectural and the physical levels; and, (iii) the deep-submicron physical effects that jeopardize the separation of design and manufacturing [24]. To continue to follow (and/or exceed) the performance goals of Moore's law, one needs to develop new manufacturing techniques and approaches that will let one build computer systems with "mole quantities" of logic elements that are molecular in both size and precision and are interconnected in complex patterns, in an inexpensive manner [7].

[7]American architect Richard Buckminster Fuller designed a dome presenting this kind of symmetric pattern for the 1967 Montreal World Exhibition.

[8]For example, in early 2004 Intel announced the first lot of chips based on the 65-nm process: It announced the first fully functional 4-Mbit SRAM chips (static random access memory). Intel was planning mass production for 2005. In early 2002, Intel demonstrated prototypes of first SRAM chips based on the 90-nm process. At that time this was a technological breakthrough; but at press time the 90 nm is well in reach, with many chip makers releasing such chips [24a].

[9]Gordon Moore made his well-known observation (now known as Moore's law) in 1965, just a few years after the first ICs were developed. In his original paper Moore observed an exponential growth in the number of transistors per integrated circuit and predicted that this trend would continue. Through technology advances, Moore's law, the doubling of transistors every couple of years, has been maintained and still holds true today. Observers (such as Intel) expect that it will continue at least through the end of this decade [25].

The silicon transistor, as embodied in the complementary metal oxide semiconductor (CMOS), is the dominant technology and will likely remain so for the foreseeable future; only a breakdown in Moore's law provides a chance for other technologies, including nanotechnology, to compete. However, such a break is more likely to be the result of economics rather than technological problems, according to some [26]. Nanoelectronics and nanophotonics are of particular interest in this context. Contemporary nanotechnology research is concerned, at the macrolevel, with two avenues of research: (i) the development of new manufacturing techniques and (ii) the development of new devices, for example, single-electron transistors, nanowires, and photonic bandgap devices (to mention only a few).

Manufacturing techniques for growing and fabricating structures with dimensions as small as a few nanometers using electron beam lithography, dry etching, and molecular beam epitaxial growth are under development. Novel techniques of manufacturing nanometer-scale structures by stamping are also under development. Recent accomplishments have included the first demonstration of 3-nm electron beam lithography and the invention of new low-damage dry etch processes for selective and unselective patterning of the Ga(Al, In)As and InP systems [27].

In reference to *new devices*, research work in this arena goes on in earnest. For example, a baseline 2001 paper for nanoelectronic circuit design demonstrated that all of the important logic functions for building complex circuits can be built from a bottom-up assembly process of chemically synthesized nanowires [28] and/or nanotubes. As far back as 2001, a team at Nanosys Incorporated (Cambridge, MA) arranged nanowires into a simple crossbar architecture that allowed communication among nanowires; the team constructed logic circuits from silicon and gallium nitride nanowires. A team at UCLA demonstrated more recently that a simple 16-bit memory circuit could be built from semiconducting crossbars that took advantage of chemical transistor switches made from organically synthesized molecules [29]. Advancements like these and other nanotechnology-driven developments will play an important role in the future of telecommunications.

Major opportunities exist for the development of new usable technologies during the next few years. As stated earlier, nanostructure, such as nanophotonic devices, nanowires, carbon nanotubes, plasmonics devices, among others, are being developed to the point where these devices can be incorporated into telecommunication components and into microprocessors, leading to powerful new communication systems and computers. These opportunities will be described throughout this text.

To provide a balance to this discussion, note that some see nanotechnology (just) as a new label for chemistry, materials science, and applied physics as the industry starts working at the molecular level. Others see nanotechnology as being hyped as the next "dot.com" and call for a need to recognize the opportunities and discount the hype. Yet others make the case that because near-term applications will be largely invisible in existing products—offering higher strength, safety, sensitivity, accuracy, and overall performance—the nanotechnology phenomenon is an incremental one, not revolutionary [30]. Also, despite much recent publicity concerning potential applications of new inorganic materials in nanotechnology and optoelectronics, a number of chemists believe that self-organizing *organic polymers* hold the greatest promise for future important discoveries and applications [31]; the previously discovered polymers comprise only a small set from a large array of possible chain molecules.

1.1.4 Technology Scope

As implied by the discussion above, significant breakthroughs have taken place during the past two decades in a wide range of issues related to nanoscale science and engineering. Progress in molecular nanotechnology is being made on several fronts, producing breakthroughs in molecular manipulation for chemical bond formation, molecular electronics, and the harnessing of biomolecular motors [32]. Nanomaterials and nanoscience concepts have evolved rapidly of late, and at this point in time nanoscience concepts are becoming broadly understood. Nanotechnology is now an interdisciplinary science that spans topics such as microengineering, precision machinery, nanoelectronics, nanophotonics, nanomaterials/nanostructures, and bio/biomedical nanotechnology.

The three major nanotechnology areas of current emphasis are: (i) nanomaterials, (ii) nanobiotechnology, and (iii) nanoelectronics/photonics. A more granular view of subfields include the items depicted in Table 1.4, which also provides a sense of the concentration of worldwide research (in terms of studies published from 1996 to 2000—data generalized from [14].) Table 1.5 (inspired partially by [33]) depicts the many areas and subdisciplines of nanotechnology (this expanding the more coarse view provided in Table 1.4). A number of these subareas (but by no means all) are discussed in this book.

As hinted at in the opening paragraphs, nanotechnology relies on quantum theory, specifically, on quantum mechanics. Quantum theory, a branch of physics, is based on the quantum principle, that is, that energy is emitted not as a continuous

TABLE 1.4 Concentration of Research (in Terms of Articles Published from 1996 to 2000)

	Number of Papers	% of Papers
Nanoparticles	41,918	(18.4%)
Quantum computing	41,795	(18.4%)
Semiconductor nanostructures	22,308	(9.8%)
Catalysis	18,758	(8.2%)
Self-assembly	18,214	(8.0%)
Nanomedicine	16,176	(7.1%)
Fullerenes	13,230	(5.8%)
Nanocomposites and coatings	12,237	(5.4%)
Quantum dots	11,863	(5.2%)
Biosensors	9,921	(4.4%)
Nanotubes	9,024	(4.0%)
Dendrimers and supramolecular chemistry	6,628	(2.9%)
Energy storage and distribution	2,883	(1.3%)
Soft lithography (nanoimprinting)	2,500	(1.1%)
	227,455	100.0%

TABLE 1.5 Taxonomy of Areas and Subdisciplines in the Nanotechnology Field

Nanostructured materials	Nanofunctional materials
	Nanoparticles
	Carbon nanotubes
	Carbon nanotube "peapod"
	Nanodiamond
	Nanowires
	Nanorods
	Nanostructured polymer
	Nanoscale manipulation of polymers
	Nanostructured coatings
	Nanocatalysis
	Nanocrystals
	Nanocrystals in Si-based semiconductors
	Nanocrystalline materials and nanocomposites
	Thin-film photonic crystals
	Biomolecules
Nanomaterials synthesis and assembly	Nanoparticles
	Carbon nanotube
	Nanomachining
	Nanodeposition
	Sol–gel methods
	Ball-milling
	Nanocomposites
Nanofabrication methods	"Top-down" approaches: lithography (E-beam, extreme ultraviolet); Dip-Pen nanolithography
	"Bottom-up" approaches: selective growth; self-assembly; scanning tip manipulation
Nanomanipulation	Scanning probe microscope-based nanomanipulator
	Nanotweezer
Nanolithography	Scanning probe microscope (SPM)
	Dip-pen nanolithography
	Extreme ultraviolet (EUV)
	Electron beam nanolithography/X-ray
	Focused ion beam
	Light coupling nanolithography
	Imprint nanolithography
Nanosensors	Nanotube and nanowire sensors
	Nanocomposite sensor
Quantum behaviors and scaling limit of CMOS	Moore's law/scaling and limits of CMOS
	Quantum theory/mechanics
	Wave interference, quantum mechanics, tunneling, diffraction
	Quantum dots
	Quantum wires
	Quantum wells
	Quantum corrals

TABLE 1.5 (*Continued*)

Nanoelectronics	Silicon nanoelectronics
	Molecular nanoelectronics, carbon nanoelectronics
	DNA nanoelectronics
	Neuromorphic nanoelectronics
	Ballistic magnetoresistance (BMR) and nanocontacts
	Single electronics
	Josephson arrays
	RTD (resonant tunneling diode)-based devices
	Spintronics
	Molecular nanoelectronics
Nanophotonics	Optical metal nanoshells
	Photonic bandgap structures
	Photonic crystal structures
	Nanooptics
	Nanocavities
	Photonic crystal waveguide
	Atom optics and nanofocusing
Nanomechanics	Atomic force microscope (AFM)
	Nanoresonators
	Nanocantilevers
	Nanomechanical transistor
	Nanoacoustics
	Nanofluidics
	Nanoindentation
	Nanorobots, nanoelectromechanicals (NEMS) (also MEMS) and AFM nanomanipulator
Nanomagnetics	Nanoscale magnets
	Magnetic nanoparticles
	Giant magnetoresistance (GMR)
	Spintronics
	Magnetic nanosensor for ultra-high-density magnetic storage
Nanochemistry	Self-assembly nanochemistry (self-assembly is the construction principle that nature uses to create functionally)
	Nanocatalysts, batteries, fuel cells
	Nanoelectrochemical lithography
Nanobiotechnology	Molecular motors
	Biomolecular electronics
	mtDNA (mitochondrial DNA)/nanotechnology interplay
	Molecular motors
	Micromanipulation techniques, self-assembly, gene chips
	Nanobiomedicine
	Nanobiosensors
	Self-assembled biomolecular structures
	Bio-MEMS
	Bioelectronics
	DNA nanoelectronics
Nanoinstrumentation	Nanometrology and tools for nanoscale materials/structures: SPM, TEM, etc.

quantity but in discrete discontinuous units. Quantum theory is the science of all complex elements of atomic and molecular spectra and the interaction of radiation and matter [34]. Quantum physics/mechanics principles will be covered in this text. Table 1.6 identifies some key terms of interest in nanotechnology; other terms are provided in the Glossary at the end of the book. Related to the Glossary, we have made every effort to include as many of the terms used in this text as possible. Hence, unfamiliar terms should, in most instances, be defined in the glossary.

1.1.5 Commercialization Scope

Commercial R&D work is now being focused on nanotechnology in order to translate the pure science discoveries into usable products. While there is extensive academic and institutional interest and activity, there is also rapidly expanding commercial activity. Significant nanotechnology research work has been undertaken in recent years at the six national Nanoscale Science and Engineering Centers, located at Columbia University (New York), Cornell University (Ithaca, NY), Rensselaer Polytechnic Institute (Troy, NY), Harvard University (Cambridge, MA), Northwestern University (Evanston, IL), and, Rice University (Houston). In the past few years there has been a lot of coverage on nanotechnology in scientific journals, at conferences, in university programs, in market research reports, and even in the financial and business press (in the United States, press-time network TV advertisements from NEC also extolled the virtues of nanotechnology). For example, in 2002 Merrill Lynch published the first nanotechnology equity report. *Science Magazine* named nanotechnology the 2001 Breakthrough of the Year,[10] and quantum dot nanocrystals ("tiny" 5- to 10-nm semiconductor nanocrystals that glow in various colors when excited by laser light and used to tag biological molecules) were named by *Science Magazine* as Breakthrough of the Year #5 in 2003 [35]. High-tech companies such as, but not limited to, NEC have highlighted its nanotechnology research in its corporate ads. There are now hundreds of labs, companies, and academic institutions involved in this work (ranging at the corporate level, to name a few, from IBM, Intel, NEC, and HP to Veeco Instruments, Perkin-Elmer, and FEI Corp). As of the early 2000s there were more than 100 startups developing nanotechnology-based products that will be marketable in the 2005–2007 timeframe. Figure 1.7 (based on data from [14]) shows that some countries are focusing more research (as a percentage of the total scientific publications) on nanotechnology than other countries. This book is a step in the direction of advocating practical attention to this field, specifically from a computing and telecom perspective.

As noted in the previous section, at a macrolevel, commercially focused research falls into six functional categories, as follows: (i) nanomaterials and nanomaterials processing, (ii) nanophotonics, (iii) nanoelectronics, (iv) nanoinstrumentation, (v) nanobiotechnology, and (vi) software. This is generally how this book is organized (with the exception of nanobiotechnology, which is not covered here.) As a point of reference, in 2002 there were around 50 companies focused on nanomaterials and

[10]*Science Magazine* cited work of the team at Nanosys Incorporated (Cambridge, MA) that arranged nanowires into a simple crossbar architecture that allowed communication among nanowires.

TABLE 1.6 Glossary of Key Nanotechnology-Related Terms Assembled from Various Scientific Sources

Term	Definition
Buckminsterfullerene (aka Buckyball, C_{60})	Most famous of the fullerenes, it is a nanostructure composed of 60 atoms of carbon arranged in a perfectly symmetric closed cage. Discovered in 1985 by Richard Smalley, Harold Kroto, and Robert Curl for which they won the 1996 Nobel Prize in chemistry [36].
Carbon nanotube	Cylinder-shaped structure resembling a rolled-up sheet of graphite (carbon) that can be a conductor or semiconductor depending on the alignment of its carbon atoms. It is 100 times stronger than steel of the same weight, although due to high fabrication costs, widespread commercial use is still distant [36].
Fullerene	Third form of carbon, after diamond and graphite. Can be spherical or tubular in shape [36].
Nanotechnology	Creation and utilization of materials, devices, and systems through the control of matter on the nanometer-length scale, that is, at the level of atoms, molecules, and supramolecular structures [37].
Quanta	a. Fundamental units of energy. b. Light can carry energy only in specific amounts, proportional to the frequency, as though it came in packets. The term *quanta* was given to these discrete packets of electromagnetic energy by Max Planck [38]. c. Smallest physical units into which something can be partitioned, according to the laws of quantum mechanics. For example, photons are the quanta of the electromagnetic field [38]. d. Each particle is surrounded by a field for each of the kinds of charges it carries, such as an electromagnetic field if it has electric charge. In the quantum theory, the field is described as made up of particles that are the quanta of the field. More loosely, the smallest amount of something that can exist [38].
Quantum chemistry	Application of quantum mechanics to the study of chemical phenomena.
Quantum device	Semiconductor device whose operation is based on quantum effects [36, 39].
Quantum dot (QD)	Nanometer-scale "boxes" for selectively holding or releasing electrons; the size of the box can be from 30 to 1000 nm [40, 41]. Something (usually a semiconductor island) capable of confining a single (or a few) electron and in which the electrons occupy discrete energy states just as they would in an atom [42]. QDs are grouping of atoms so small that the addition or removal of an electron will change its properties in a significant way [36]. QDs are small devices fabricated in semiconductor materials that contain a tiny droplet of free electrons; the size and shape of these structures and, hence, the number of electrons they contain, can be precisely controlled; a QD can have from a single electron to a collection of several thousands [43, 44].

TABLE 1.6 *(Continued)*

Quantum effect	Properties of transistors and wires become altered at the nanoscale level, so that they can no longer be characterized by classical electronic circuit theory. Quantum effects, such as the quantization of electronic charge and the interfering wave properties of electrons as they propagate through transistors and wires, need to be take into account [36].
Quantum electronics	Name used for those parts of quantum optics that have practical device applications [36, 38].
Quantum optics	Science concerned with the applications of the quantum theory of optics; i.e., optics defined in terms of the quanta of radiant energy or photons [34].
Quantum physics	Physics based upon the quantum principle that energy is emitted not as a continuum but in discrete units [38].
Quantum theory	a. General term describing quantum physics. b. Theory that seeks to explain that the action of forces is a result of the exchange of subatomic particles [38]. c. Theory used to describe physical systems that are very small, of atomic dimensions or less. A feature of the theory is that certain quantities (e.g., energy, angular momentum, light) can only exist in certain discrete amounts, called quanta. d. Initially, the theory developed by Planck that radiating bodies emit energy not in a continuous stream but in discrete units called quanta, the energy of which is directly proportional to the frequency. Now, all aspects of quantum mechanics. e. Quantum theory provides the rules with which to calculate how matter behaves. Once scientists specify what system they want to describe and what the interactions among the particles of the system are, then the equations of the quantum theory are solved to learn the properties of the system.
Quantum well (QW)	In a diode laser, a region between layers of gallium arsenide and aluminum gallium arsenide, where the density of electrons is very high, resulting in increased lasing efficiency and reduced generation of heat [34]. Semiconductor heterostructure fabricated to implement quantum effects in electronic and photonic applications; typically an ultrathin layer of narrower bandgap semiconductor is sandwiched between two layers of larger bandgap semiconductor; electrons and holes are free to move in the direction perpendicular to the crystal growth direction but not in the direction of crystal growth, hence, they are "confined" [36, 39].
Quantum wire	Narrow channel created by cleaving a crystal made of alternating layers of gallium arsenide and aluminum gallium arsenide, and adding additional layers on the cleaved end face, at right angles to the first, resulting in an efficient diode laser [34].

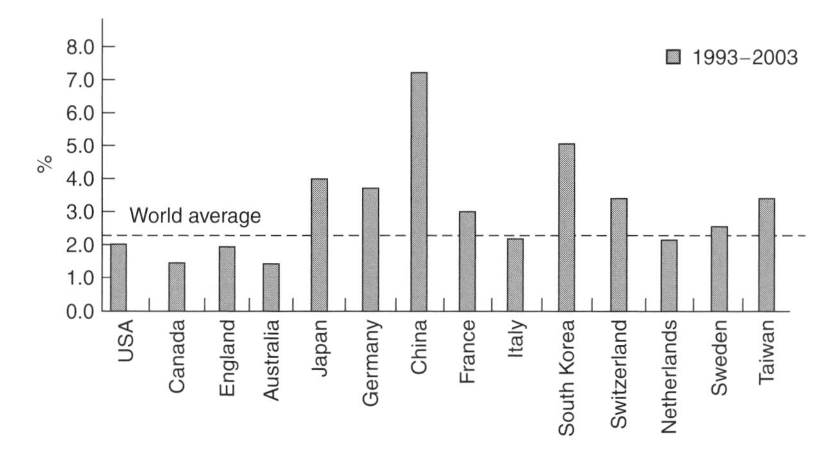

FIGURE 1.7 Percentage of all science publications related to nanotechnology.

nanomaterials processing; these companies were in the process of developing the materials and methods to manipulate and manufacture nanomaterials-based products. From a commercial perspective, nanophotonics, nanoelectronics, and nanobiotechnology (a hybrid discipline that combines biology and nanoelectronics) also hold near-term promise. Many fundamental biological functions are carried out by molecular ensembles (e.g., single enzymes, transcription complex, ribosome) that have the sizes in the range of 1–100 nm. To understand the functions of these ensembles, one has to describe their movements, shapes, and so on at this scale. Nanotechnology developments and tools (e.g., scanning probe microscopy, micromanipulating techniques, etc.), hence, are applicable to the field of biology.

Nanoelectronics includes electronic (and optoelectronic) devices where individual assemblies of nanoscale components operate as active device elements. Applications of nanoelectronics include memory, logic, passive optical components, field emission devices, and flat-panel display and light-emitting diodes. As another example, there is research underway into ultrasmall magnetism, an atomic property that results from electronic "spin." Nanospintronics is a new area of nanoscience that could help engineers build faster and more powerful computers and electronic devices in the future. Spintronics (short for spin-based electronics) seeks to isolate the spin from the electron's charge [45]; spintronics-based devices are devices that rely on an electron's spin to perform their functions. The spin is a fundamental property of an electron and is the basis of the magnetic bits on a computer hard drive. Current computer drives store information with tiny magnets, but these drives need millions of the devices to do the job. An improvement might be to use nanoscale magnets; the use of these nanoscale magnets is expected to lead to a storage drive one or two orders of magnitude smaller than current-day technology. Nanobiotechnology companies focus on developing a gamut of biological diagnostic tools, for example, arrays of tiny sensors that can detect specific biological molecules and/or individual components of DNA.

The nanotechnology developments and trends highlighted so far may be of interest to telecommunications vendors, researchers, and students who want to start thinking about the potential opportunities afforded by these emerging scientific developments and approaches for the next-generation networks to be deployed 5–10 years in the future; advanced planning is a valuable and effective exercise. For example, nanophotonics companies are developing highly integrated optical-communications components using nanooptic and nanomanufacturing technologies (nanomanufacturing as applied to optical components allows rapid prototyping, performance improvements, smaller forms, and lower costs [12].) MEMSs are microscale systems (~100 μm) (but not nanoscale yet) that include both mechanical and electrical devices integrated on a single die or chip. MOEMS consist of MEMS devices with integral optical components such as mirrors, lenses, filters, laser diodes, emitters or other optics. A MEMS system may include microfluidic elements, integral microelectronics or ICs, "lab-on-chip" systems, actuators, micromotors, or sensors. Lab-on-chip (also known as nanolab) is a chip that uses (i) microfluidics to transport individual cells across the device, (ii) nanowire sensors ~10 nm in diameter to identify genes and proteins in the cell, and, (iii) nanomechanical sensors to detect protein and gene interactions.

These and other nanoscale advances will be critical to the computing and telecom industry in the coming decade; hence, the focus of this book.

1.1.6 Opportunities of the Technology and the 21st Century Nanotechnology Research and Development Act of 2003

National Science Foundation (NSF[11]) representatives have been quoted as saying: "Some call [nanotechnology] the next industrial revolution, anticipating an economic bonanza that dollar for dollar, and job for job, will outstrip the introduction of electricity, the automobile or the new information and communications technologies.... The expectations raised by nanotechnology have inspired governments worldwide to increase support for nanotechnology research and education, and sparked international competition to bring nanotechnology from the bench to the boardroom" [46]. The NSF has recently being advocating for closer cooperation between the scientific and engineering research communities on nanotechnology development; this cooperation would create what has been dubbed "nanotransformations" in scientific and social fields.

From a current market point of view, there was an expectation that about $1 billion would be invested in nanotech by the venture capital (VC) community in 2003 (about 2% of the total VC pool of money), and most of that investment will be in software (e.g., computer-based development/modeling tools). Nanomaterials are estimated to have a $150 million market in the short term [47]. In addition to this private investment, there is government-based funding.

In late 2003, President George W. Bush signed the 21st Century Nanotechnology Research and Development Act into law. The $3.7 billion appropriation, which earlier had been approved by Congress, was to be divided among eight government agencies. Nanotechnology "has the potential to be the making of a revolution because it can be an enabling technology, fundamentally changing the way many items are designed and manufactured," said Rep. Mike Honda, D-Calif. "And we've all probably heard the

[11]The NSF invests over $3.3 billion per year in about 20,000 research and education projects in science and engineering.

National Science Foundation prediction that the worldwide market for nanotechnology products and services could reach $1 trillion by 2015." The bill authorizes the president to create a permanent National Nanotechnology Research Program (NNRP) to replace the expiring National Nanotechnology Initiative. The NNRP, according to the bill, is a "coordinated interagency program that will support long-term nanoscale research and development leading to potential breakthroughs in areas such as materials and manufacturing, nanoelectronics, medicine and health care, environment, energy, chemicals, biotechnology, agriculture, information technology, and national and homeland security." The act encourages the development of networked facilities linking academic institutions, national labs and industry[12] [48].

The act identifies a list of "grand challenges" as the long-term guiding principles for individual research groups and for the national Nanoscale Science and Engineering Centers. Those grand challenges include [48][13]: (i) To design materials (nanoelectronics, optoelectronics, and magnetics) that are stronger, lighter, harder, safer, and self-repairing. Health care applications are specifically cited and so are nanoscale processes, environmental solutions, energy management, and energy conservation. (ii) To foster the development of economical, efficient, and safe transportation, including the development of microspacecraft that can overcome Earth's gravity field when blasting off and survive the rigors of space flight in a manner that is cheaper and more environmentally friendly than current technology. (iii) To foster the development of biologically oriented nanodevices for detection and mitigation of biologically based threats to humans.

These funding initiatives will be a major impetus to the science, the engineering, and the applications of nanotechnology. Hence, this practical book comes at an opportune time. We focus on applied technology and product development.

1.2 PRESENT COURSE OF INVESTIGATION

We have so far implied that at the nanoscale ordinary matter often displays surprising properties that can be exploited to increase computer speed and memory capacity and to manufacture materials that are stronger, lighter, and "smarter" by orders of magnitude. The underlying principles for nanoscale devices are significantly different than ordinary semiconductor techniques because the systems are so small that quantum effects govern their behavior. Recent developments in surface microscopy, silicon

[12]The president's 2005 budget provides about $1 billion for the multiagency National Nanotechnology Initiative (NNI), a doubling over levels in 2001, the first year of the initiative. The 2005 budget request is designed to support the NNI activities of 10 federal agencies in order to advance understanding of nanoscale phenomena [2].

[13]The act also directs the establishment of a National Nanotechnology Coordination Office to handle day-to-day technical and administrative support and act as the point of contact on all federal nanotechnology activities for government organizations, academia, industry, professional societies, state nanotechnology programs, and others wishing to exchange technical and programmatic information. The act also seeks to "establish a new center for societal, ethical, educational, legal and work force issues related to nanotechnology... to encourage, conduct, coordinate, commission, collect and disseminate research." The act also authorizes public hearings and expert advisory panels, as well as an American Nanotechnology Preparedness Center that will study nanotechnology's potential societal and ethical impact [48].

fabrication, physical chemistry, and computational engineering have come together to help scientists better understand, fabricate, and manipulate structures at this level. The ability to construct matter and molecules one atom at a time, coupled with new methods to fabricate novel materials and devices, has made the field of nanoscience an interesting discipline for both scientists and engineers [22].

Furthermore, nanotechnology operates at the dimension where the living and non-living worlds meet [46]. Quantum dots (grouping of atoms so small that the addition or removal of an electron alters the properties of the grouping in a significant way), nanowires, and related nanoscale structures are now key topics in contemporary semiconductor research; this research is aimed at downsizing chip components to the molecular scale. These just-named factors and the issues listed earlier in the chapter make this body of research a very interesting topic of investigation. As noted, in this text we focus on telecom and networking applications of nanotechnology.

In the chapters that follow, we discuss at a high level the basic science (physics and chemistry) behind nanotechnology. In Chapters 2 (physics) and 3 (chemistry) we cover topics such as: physics and chemistry basics (e.g., electron, atoms, atomic structures, molecules, bonded structures), electrical properties (e.g., insulators, semiconductors, conductors), and chemical bonds and reactions. Chapter 2 also provides a basic introduction to transistors, in support of the discussion to follow in Chapter 6. It turns out that while classical Newtonian mechanics can predict with precision the motions of masses ranging in size from microscopic particles to very large stars, it cannot predict the behavior of the particles in the atomic domain; quantum theory comes into play instead. Hence, taking off from the coverage of Chapters 2 and 3, in Appendices D and E we discuss some of the basic scientific principles that support this field; the reader who may find these two appendices somewhat demanding may chose to skip this material and move on to the chapters that follow; which are generally self-contained.

In Chapter 4 we look at nanomaterials and nanomaterial processing: Individual nanoparticles and nanostructure (e.g., nanotubes, nanowires) are discussed. Nanophotonics is discussed in Chapter 5 (e.g., nanocrystals, nanocrystal fibers). Nanoelectronics (e.g., metal nanoclusters, semiconducting nanoclusters, nanocrystals, quantum dots) is covered in Chapter 6. Both Chapters 5 and 6 provide a discussion of near-term and longer-term applications to the field of computers, telecommunications, and networking. Appendix F discusses nanoinstrumentation, while Appendices A, B, and C provide other background material. Appendix G provides information on quantum computing.

The goal of this book is to provide a self-contained, reasonably pedagogical introduction to the field for professionals wanting to obtain an entry-level view of this rapidly evolving field. It is not the purpose of this book to be a research monogram for the in-field scientist, nor to describe ultrarecent research breakthroughs. The treatment is *not* intended to be exhaustive: We only survey some of the more well-established/well-developed areas. We believe that we have provided for the prospective reader a reasonable mix of technical (introductory) material and a sense of the applications to enable the practitioner to get started in this field.

CHAPTER 2

Basic Nanotechnology Science—Physics

As we saw in Chapter 1, nanoscience (nanotechnology) is the field of study that explores activity at the molecular and submolecular level and deals with the precise manipulation (manufacturing) of materials at this atomic and/or molecular level. Advocates see nanotechnology as the "next industrial revolution." In Chapters 2 and 3 and Appendices D and E we discuss some of the basic scientific principles that support this field. Appendices A, B, and C provide some supplementary information. The reader who may find these chapters and/or appendices somewhat demanding may choose to only read selected subsections or to move on to the chapters that follow, which are generally self-contained.

Physics and chemistry have always dealt at the molecular and/or atomic level, but not in the true sense of material "fabrication" per se. These fields have looked at the issues either from a science point of view ("how things work") or from an (exploitable) phenomenon point of view (e.g., how to make use of the fact that when a photon hits a detector a current flows). By now the reader will have internalized that nanotechnology relates to the design, creation, and utilization of materials whose constituent structures exist at the nanoscale (10^{-9} m). Hence, an understanding of nanotechnology requires a (fundamental) understanding of the physics and chemistry that operates at this molecular/atomic level. These disciplines provide the understanding and a prediction model of the properties of matter at this scale. In turn, the disciplines of physics and chemistry have to advance in the next few years at the technological–application level, so that atom-level nanostructures can be fabricated reliably and cost effectively—new techniques and/or understandings may need to evolve within the context of physics and chemistry to harvest in full the benefit of nanotechnology. With molecular manufacturing one can "essentially" get every atom in the right place, thereby making almost any structure consistent with the laws of physics that one can specify in molecular detail [7].

A comprehensive description of the basic sciences requires a complex mathematical framework; in these chapters that follow and in the appendices, however, we only look at some basic concepts with, *relatively speaking*, light mathematical

Nanotechnology Applications to Telecommunications and Networking, By Daniel Minoli

machinery. In this chapter (physics) and in Chapter 3 (chemistry) we cover topics such as physics and chemistry basics (e.g., electron, atoms, atomic structures, molecules, bonded structures), electrical properties (e.g., insulators, semiconductors, conductors), and chemical bonds and reactions. Quantum theory comes into play at the nanoscale; hence, in Appendices D and E we discuss some of the basic scientific principles that support this field. Application of the sciences to nanotechnology per se, for example, individual nanoparticles (e.g., metal nanoclusters, semiconducting nanoclusters, nanocrystals) and nanostructures (e.g., quantum wells, quantum wires, quantum dots), is provided starting in Chapter 4.

2.1 APPROACH AND SCOPE

In this section we look at some basic ideas to start the discussion. These issues are then treated in more detail in the sections and chapters that follow. The focus of nanotechnology is the technical and scientific ability to work at the molecular level, atom by atom, to create large structures with fundamentally new molecular organization. Functionality, behavior, and performance of structural features in the range of 10^{-9}–10^{-7} m (1–100 nm) demonstrate important changes compared to isolated molecules of about 1 nm (10^{-9} m) or of bulk materials [2, 33].

It turns out that the machinery required to understand nanotechnology (at least at the theoretical level) is quantum theory (also known as quantum mechanics or quantum physics). Some call quantum physics "just the physics of the incredibly small" [49]; quantum theory can be used to describe the behavior of electrons, atoms, molecules, and photons, among other particles. While empirical methods may suffice in some cases, particularly at the higher range of the dimensions (e.g., 100 nm) and/or for practical chemistry applications, quantum theory is needed for a thorough comprehension of the nanotechnology field. In general, quantum theory will be needed at the atomic level (e.g., 0.5 nm and smaller.) Quantum theory differs from classical Newtonian physics, the latter being the science governing the motions of macroscopic entities. Quantum theory aims at explaining the behavior of matter and, with extensions, it aims at explaining the interaction of matter with light. Quantum theory describes the laws of physics that apply on very small (atomic) scales: The essential feature is that energy (charge), momentum, and angular momentum come in discrete amounts called *quanta* [50].

The discovery and formulation of the fundamental concepts of atomic/quantum physics took place in the early part of the twentieth century (1900–1930) by such people as Planck, Einstein, de Broglie, Schrödinger, and Heisenberg, among others. Additional refinements were brought in the 50 years that followed. Because of what sounds like an esoteric name, people are either intimidated or believe that it is something new. It is true that the equations involved may be difficult to solve in closed form, but with supercomputers one does not always have to have a closed-form formula to obtain an answer or predict an outcome under some new set of parameters. Also, while it may be "startling" to some that some well-accepted rules for environment x do not apply to environment y, we do not believe that this is a far-fetched logical concept.

A priori, the rules that apply at $1 + 10^{30}$ K of temperature do not necessarily apply at $1 - 10^{-30}$ K of temperature; the rules at 10^{30} atm of pressures do not necessarily apply at 10^{-30} atm of pressures; and the rules at play at 10^{28} m from a nucleus are not necessarily the same rules at play at 10^{-28} m from a nucleus. And so on.

Our approach in this chapter 2, Chapter 3, and Appendices D and E is to provide a mix of the practical empirical science as well as some of the formalism. We cover the pragmatics and tangible results. We also briefly allude to the complex science, but do so only at a high level.

2.2 BASIC SCIENCE

2.2.1 Atoms

Atoms are fundamental particles of matter: They are the smallest particle of an element that can take part in a chemical reaction. The hypothesis that matter is composed of particles dates to the sixth-century BC: the idea of "atomicity" was studied by Greek scholars of antiquity such as Leucippus, Democritus, and Epicurus. Of course, most of the scientific understanding was developed in the twentieth century (see Appendix A). An element is a simple substance that cannot be resolved into simpler substances by normal chemical means[1] [51].

Particle physics[2] is a branch of physics that studies the elementary constituents of matter (subatomic particles) and radiations; it also studies the interactions between them. Subatomic particles include atomic building blocks such as electrons, protons, and neutrons (protons and neutrons are composite particles, comprised of quarks) and particles produced by radiative and scattering processes, such as photons, neutrinos, and muons.

At a simplified level, atoms consist of a positively charged *nucleus* with negatively charged *electrons "orbiting" around it*. See Figures 2.1 and 2.2. The electron is present in all atoms. The diameters of nuclei fall in the range of 1×10^{-14} and 1×10^{-15} m (i.e., a million times smaller than the nanoscale we are focusing on in this book); the diameter of an atom is typically of the *order of magnitude* of 1×10^{-10} m (i.e., 10 times smaller than the nanoscale of 10^{-9} m—a cube that is $1 \times 1 \times 1$ nm could contain up to 125 such atoms of diameter 2×10^{-10} m). An electron is an elementary particle[3] of an atom having a low negative charge and mass ($e = 1.602192 \times 10^{-19}$ C and $m = 9.109381 \times 10^{-28}$ g). The electron has a mass of 1/1836 the mass of a hydrogen

[1]Because of the existence of isotopes of elements, an element cannot be regarded as a substance that has identical atoms but is regarded as one that has the same atomic number. Isotopes are defined as two or more nuclides having an identical nuclear charge (i.e., same atomic number) but differing atomic mass; such substances have almost identical chemical properties but differing physical properties, and each is said to be an isotope of the element of a given atomic number; the difference in mass in accounted for by the differing number of neutrons in the nucleus [51].

[2]Following convention, we use "elementary particles" to refer to entities such as electrons and photons, with the understanding that these "particles" also exhibit what can be considered "wavelike" properties. Some recent models attempt to dispense with the "particle" and "wave duality" view altogether, e.g., string theory.

[3]Elementary particles are also known as subnuclear particles or subatomic particles.

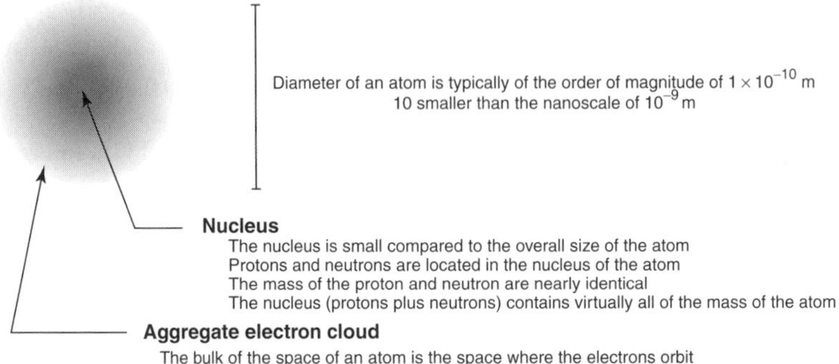

FIGURE 2.1 Simplified view of an atom.

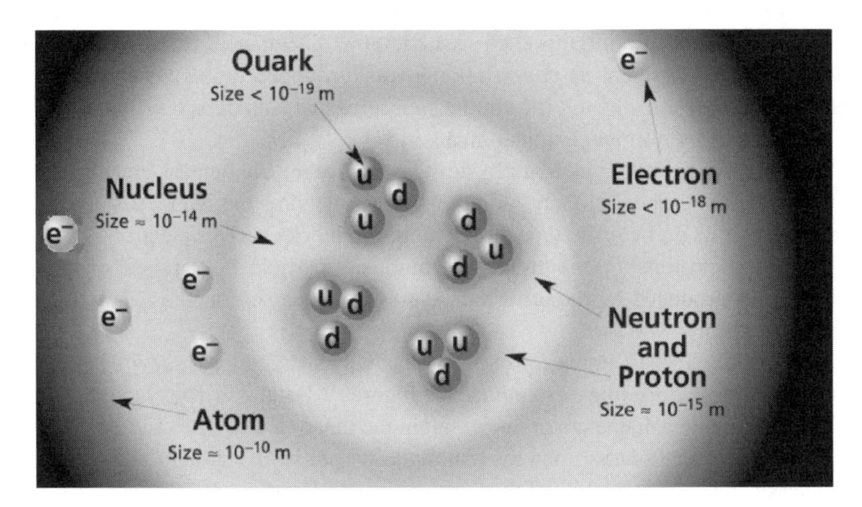

FIGURE 2.2 Another view an atom.

atom. The electrons occupy a volume of space through their orbital motion many times larger than that occupied by the nucleus (said volume is of the order of the nanoscale).

2.2.2 Key Subatomic Particles

As noted, atoms are comprised of a number of different subatomic particles. In terms of practical engineering applications (at the present time) the following are of interest: *protons*, which have a rest mass $m = 1.6726231 \times 10^{-24}$ g, a diameter of the order of 1×10^{-15} m, and a positive charge; *neutrons*, which have approximately the same mass as the proton and a diameter of the order of 1×10^{-15} m but have no charge; *electrons* (already discussed), which are small (about 1/2000 mass of proton), have a diameter of the order of 1×10^{-18} m, and have negative charge; and *photons*, the carrier particle of electromagnetic interactions, moving at the speed of light (299, 792,460 m/s). The nucleus, which comprises most of the mass of an atom (but encompasses only a small part of atom's volume), contains *protons* and *neutrons.* Protons are present in all atoms; a proton is about the same mass as a hydrogen atom and carries positive charge equal in magnitude but opposite in sign to an electron. Neutrons are electrically neutral and have about the same mass as a proton.

Table 2.1 describes some of these subatomic particles, along with other key concepts [50, 51]. Table 2.2 lists a few terms from classical physics that should be known [654]; other terms are provided in the Glossary at the end of the book.

The electron cloud around the nucleus typically has a radius of 0.05–2 nm (0.5–20 Å). Table 2.3 depicts the radius of the elements in angstroms. Notice that Ne (neon) has the smallest radius (0.51 Å), while Fr (francium) has the largest radius (2.7 Å). As seen later in the chapter, electrons move not in spherical orbits but in a variety of "fuzzy orbits" (some of these orbits, however, are spherical); furthermore, we can only determine the probability of finding the electron(s) at some distance from the nucleus, not a deterministic certainty.

As seen in Figure 2.2, the diameter of an atomic nucleus is about 10^{-4} Å (1×10^{-14}), making it about 0.01% the diameter of the atom. The nucleus of an atom is dense. For example, a nucleus containing one neutron and one proton has the following parameters:

- Mass of nucleus = (approximately) 3.32×10^{-24} g.
- Diameter of nucleus = (approximately) 1×10^{-4} Å $= 1 \times 10^{-14}$ m.
- Volume of nucleus = $(\frac{4}{3}) \times (\pi) \times (\text{radius of nucleus})^3 = 5.24 \times 10^{-43}\,\text{m}^3$.
- Ratio of mass/volume = $3.32 \times 10^{-24}\,\text{g}/5.24 \times 10^{-43}\,\text{m}^3 = 6.34 \times 10^{18}\,\text{g/m}^3$.

As a point of reference, the hydrogen *atom* has a radius of 0.208×10^{-9} m (0.28 nm) while the radius of an electron is 0.2817×10^{-14} m (radius of the nucleus is about 100,000 times smaller than the radius of the entire atom). The radius of *atomic nuclei* falls in the range of 0.5×10^{-15}–5×10^{-15} m. This means that the hydrogen atom has a radius about 73,837 the radius of an electron and that, for hydrogen, the nucleus is approximately but not exactly the size of the electron. Figures 2.3 and 2.4 depict these distances diagrammatically.

We want to give the reader a sense of scales involved. If we were to think of Earth, with its 3987-mile radius, as being the electron, then the center of the nucleus could be as far away as 294,352,858 miles out (= $3987 \times 73{,}837$). Now, the mean distance of Earth from the sun is 93,505,864 miles (149,600,000 km).[4] Hence, if Earth were an electron, the center of the nucleus could be as far as (about) 3 times further out than our distance from the sun. (This model is for scale purposes only and should not be construed as in any way implying a planetary configuration of the atom.)

[4]For comparison Mars' distance from the sun (maximum) is 154,700,000 miles; Jupiter's distance from the sun (maximum) is 507,000,000 miles.

TABLE 2.1 Basic Entities/Particles of Interest

Angstrom (Å)	Unit of measurement for atomic distances equal to 1×10^{-10} m, or 10 nm. Most atoms are between 1 and 5 Å in diameter.
Anion	A negatively charged ion. Nonmetals typically form anions.
Atom	**The smallest particle of an element that can take part in a chemical reaction**.
Atomic mass unit (amu)	The units of mass used to describe atomic particles. An atomic mass unit is equal to 1.66054×10^{-24} g.
Baryon	A hadron made from three quarks. They may also contain additional quark–antiquark pairs. For example, the proton (uud) and the neutron (udd) are both baryons.
Cation	A positively charged ion. Metals typically form cations.
Electron (e)	**An elementary particle having a negative** charge $e = 1.602192 \times 10^{-19}$ C and a mass $m = 9.109381 \times 10^{-28}$ g ($= 5.486 \times 10^{-4}$ amu). It is the most common lepton, with electric charge −1.
Element	**Simple substance that cannot be resolved into simpler substances by normal chemical means**.
Fermion	Any particle that has odd-half-integer ($\frac{1}{2}, \frac{3}{2}, \ldots$) intrinsic angular momentum (spin), measured in units of h (Planck's constant). Many of the properties of ordinary matter arise because of this rule. Electrons, protons, and neutrons are all fermions, as are all the fundamental matter particles, both quarks and leptons. As a consequence of the peculiar angular momentum, fermions obey a rule called the Pauli exclusion principle, which states that no two fermions can exist in the same state at the same place and time.
Fundamental particle	A particle with no internal substructure. In the standard model the quarks, leptons, photons, gluons, $W^{\pm}$ bosons, and Z^0 bosons are fundamental and all other objects are made from these.
Gluons (g)	The carrier particle of strong interactions.[a]
Hadron	A particle made of strongly interacting constituents such as quarks and/or gluons. (Hadrons include the mesons and baryons.) Such particles participate in residual strong interactions.[b]
Ion	An electrically charged atom, molecule, or group of atoms or molecules.
Ionic bonding	The electrostatic attraction between oppositely charged ions.
Isotopes	Two or more nuclides having an identical nuclear charge (i.e., same atomic number) but different atomic mass.
Lepton (K)	A fundamental fermion that does not participate in strong interactions. A meson containing a *strange* quark and an anti-*up* (or an anti-*down*) quark or an anti-*strange* quark and an *up* (or *down*) quark. The electrically charged leptons are the *electron* (e), the *muon* (μ), the *tau* (τ), and their antiparticles. Electrically neutral leptons are called *neutrinos* (ν).

Meson

A hadron made from an even number of quark constituents. The basic structure of most mesons is one quark and one antiquark.

Neutron (n)

An elementary particle, having zero charge and a rest mass of 1.6749286 × 10^{-24} g (939.6 MeV/c^2), that is a constituent of the atomic nucleus. The mass equates to 1.0087 amu. It is a baryon with electric charge zero; it is a fermion with a basic structure of two *down* quarks and one *up* quark (held together by gluons). The neutral component of an atomic nucleus is made from neutrons. Different isotopes of the same element are distinguished by having different numbers of neutrons in their nucleus.

Nucleon

A proton or a neutron; that is, one of the particles that makes up a nucleus.

Nucleus

A collection of neutrons and protons that forms the core of an atom (plural: *nuclei*).

Particle

A subatomic "object" with a definite mass and charge.

Particle physics

(also called high-energy physics): a branch of physics that studies the elementary constituents of matter and radiations and the interactions between them. "Elementary particle" refers to a particle of which other, larger particles are composed. For example, atoms are made up of smaller particles such as electrons, protons, and neutrons; the proton and neutron are, in turn, composed of more elementary particles known as quarks.

Phonon

In the lattice vibrations of a crystal, the phonon is a quantum of thermal energy (given by hf, where h is the Planck constant and f the vibrational frequency and $h = 6.6260755 \times 10^{-34}$ J · s).

Photon (γ)

The quantum of electromagnetic radiation. The carrier particle of electromagnetic interactions. Photons move at the speed of light: 299,792,460 m/s.

Proton (p)

A positively charged elementary particle that forms the nucleus of the hydrogen atom and is a constituent particle of all nuclei. Rest mass $m = 1.6726231 \times 10^{-24}$ g (=1.0073 amu). The proton has a charge of +1 electron charge (or $+1.602 \times 10^{-19}$ C). The proton is the most common hadron, a baryon with electric charge (+1) equal and opposite to that of the electron (−1). Protons have a basic structure of two *up* quarks and one *down* quark (bound together by gluons). The nucleus of a hydrogen atom is a proton. A nucleus with electric charge Z contains Z protons; therefore the number of protons is what distinguishes the different chemical elements

TABLE 2.1 (*Continued*)

Quarks (q)	A fundamental fermion that has strong interactions. Names *up, charm, top, down, strange*, and *bottom* are used to characterize different types of quarks. Quarks have electric charge of either $\frac{2}{3}$ (*up, charm, top*) or $-\frac{1}{3}$ (*down, strange, bottom*) in units where the proton charge is 1.
Standard model	A model for the theory of fundamental particles and their interactions. The model contains 24 fundamental particles that are the constituents of matter: 12 species of elementary fermions ("matter particles") and 12 species of elementary bosons ("radiation particles") plus their corresponding antiparticles. It describes the strong, weak, and electromagnetic fundamental forces using mediating bosons known as "gauge bosons." The species of gauge bosons are the photon, W^- and W^+ and Z bosons, and the gluons. The model predicts the existence of a type of boson known as the Higgs boson, but these are yet to be discovered. While it is widely tested and is currently accepted as correct by particle physicists, the standard model is currently perceived to be a provisional theory (until a more com prehensive theory is developed), also because it appears that there may be some elementary particles that are not properly described by the model (such as graviton—the hypothetical particle that carries gravitational force).
W± bosons	A carrier particle of the weak interactions; it is involved in all electric-charge-changing weak processes.[c]
Z^0 bosons	A carrier particle of weak interactions; it is involved in all weak processes that do not change flavor.

[a]*Strong interactions:* The interaction responsible for binding quarks, antiquarks, and gluons to make hadrons.
[b]*Residual strong interactions:* Interaction between objects that do not carry a charge but do contain constituents that have charge. Residual strong interactions provide the nuclear binding force.
[c]*Weak interaction:* The interaction responsible for all processes where "flavor" changes, hence for the instability of heavy quarks and leptons and particles that contain them. Flavor is the name used for the different quark types (*up, down, strange, charm, bottom, top*) and for the different lepton types (*electron, muon, tau*). Hence, flavor is the quantum number that distinguishes the different quark/lepton types. Each flavor of quark and charged lepton has a different mass.
Note: Items in bold are the most basic for an understanding of nanotechnology.

TABLE 2.2 Basic Physics Terms of Interest

Amplitude	For a wave or vibration, the maximum displacement on either side of the equilibrium (midpoint) position
Kinetic energy	Energy of motion, described by the relationship Kinetic energy $= \frac{1}{2} \times$ Mass $\times$ Speed^2
Momentum	Product of an object's mass and its velocity
Node	Point of zero amplitude in a standing wave. Antinodes are points of maximum amplitude.
Potential energy	Stored energy that an object possesses by virtue of its position with respect to other objects; for example, gravitational potential energy by virtue of the position of one mass relative to other(s)
Wavelength	Distance between successive crests, troughs, or identical parts of a wave
Wavelength–momentum relation	Wavelength = Planck's constant/momentum, namely, $\lambda = h$/momentum

TABLE 2.3 Atomic Radius in Angstroms (1 Å = 0.1 nm)

H 2.08																	He NA
Li 1.55	Be 1.12											B 0.98	C 0.91	N 0.92	O 0.65	F 0.57	Ne 0.51
Na 1.9	Mg 1.6											Al 1.43	Si 1.32	P 1.28	S 1.27	Cl 0.97	Ar 0.88
K 2.35	Ca 1.97	Sc 1.62	Ti 1.45	V 1.34	Cr 1.3	Mn 1.35	Fe 1.26	Co 1.25	Ni 1.24	Cu 1.28	Zn 1.38	Ga 1.41	Ge 1.37	As 1.39	Se 1.4	Br 1.12	Kr 1.03
Rb 2.48	Sr 2.15	Y 1.78	Zr 1.6	Nb 1.46	Mo 1.39	Tc 1.36	Ru 1.34	Rh 1.34	Pd 1.37	Ag 1.44	Cd 1.71	In 1.66	Sn 1.62	Sb 1.59	Te 1.42	I 1.32	Xe 1.24
Cs 2.67	Ba 2.22	La 1.38	Hf 1.67	Ta 1.49	W 1.41	Re 1.37	Os 1.35	Ir 1.36	Pt 1.39	Au 1.46	Hg 1.6	Tl 1.71	Rb 1.75	Bi 1.7	Po 1.67	At 1.45	Rn 1.34
Fr 2.7	Ra 2.33	Ac 1.88															
Lanthanides				Ce 1.81	Pr 1.82	Nd 1.82	Pm	Sm 1.81	Eu 1.99	Gd 1.8	Tb 1.8	Dv 1.8	Ho 1.79	Er 1.78	Tm 1.77	Yb 1.94	Lu 1.75
Actinides				Th 1.8	Pa 1.61	U 1.38	Np 1.3	Pu 1.51	Am 1.84	Cm NA	Bk NA	Cf NA	Es NA	Fm NA	Md NA	No NA	Lr NA

Nanotechnology is currently focused at the molecular level; hence, one is able to "zoom" out one "layer" from the subatomic particles listed in Table 2.1. However, to understand molecules one needs to understand the atomic structure, particularly the structure of the set of electrons that comprise the atom. The number of electrons in a stable atom is equal to the number of protons; therefore, the overall electrical charge in an atom is zero. All atoms of an element have the same number of protons in the nucleus. It follows that since the net charge on an atom is zero, all atoms of an element must have an equal number of electrons. Although the number of neutrons

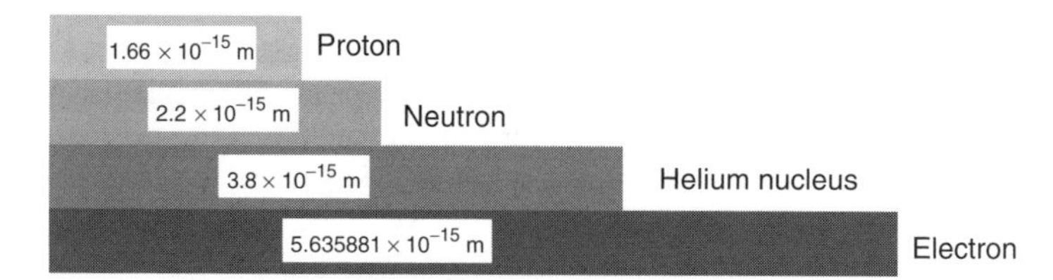

FIGURE 2.3 Scale for a few atomic entities.

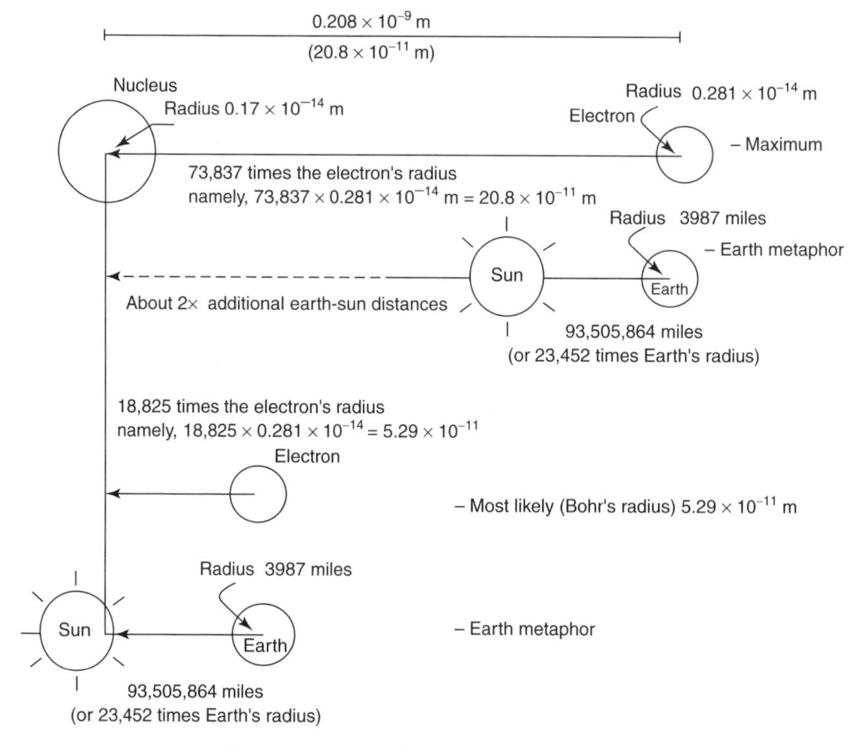

FIGURE 2.4 Some atomic distances and some metaphors.

typically is equal to the number of protons, the number of neutrons can vary to some degree: atoms that differ only in the number of neutrons are called isotopes; it follows that given that the neutron has a mass of about 1.0087 amu, different isotopes have different masses.

In general scientific terms, the understanding that matter is comprised of atoms goes back (in modern times) to Dalton, who, in 1803, stated that the atom is an "indivisible, indestructible, tiny ball." By the middle of the nineteenth century evidence was accumulating that the atom is itself composed of smaller particles. J. J. Thompson (1897) measured the charge-to-mass ratio for a stream of electrons (using a cathode ray tube apparatus) at 1.76×10^8 C/g. Robert Millikan (1909) measured the charge on a single electron (the well-known Milliken oil drop experiment), obtaining 1.60×10^{-19} C. It follows that the mass of a single electron could then be determined to be 9.10×10^{-28} g (this being derived as 1 g/1.76×10^{8} C multiplied by 1.60×10^{-19} C). As noted in Table 2.1, the currently accepted value for the mass of the electron is 9.109381×10^{-28} g. Atomic physics progressed rapidly at the beginning of the last century, due in large part, to optical spectroscopy; quantization and spin were discovered through optical studies. With the introduction of the laser, physicists learned how to manipulate atomic wave functions by applying coherent optical fields [52] (wave functions are discussed in Appendix D; refer to Appendix A for some additional historical background).

The discovery of the "secondary" subatomic particles listed in Table 2.1 took place during the middle of the twentieth century. All the particles observed in the past 100 years have now been catalogued in a theory called the standard model. The standard model is a theoretical framework whose basic idea is that all the visible matter in the universe can be described in terms of the elementary particles *leptons* and *quarks* and the forces acting between them [53]. Leptons are a class of pointlike fundamental particles showing no internal structure and no involvement with the strong forces. Electrons and neutrinos are among the particles classified as leptons. The strong force (nuclear strong force) is one of the four fundamental forces: the gravitational force, the electromagnetic force, the nuclear strong force, and the nuclear weak force; the strong force approximately one hundred times stronger than the electromagnetic force. A quark is a hypothetical fundamental particle having charges whose magnitudes are one-third or two-thirds of the electron charge and from which the elementary particles may in theory be constructed. At the present time, ongoing experimental projects in particle physics are expected to permit a completion of the standard model, but a unified theory of all forces known to physics is not yet in sight [53].

The standard model defines fundamental subatomic particles (e.g., quarks and several other particles) and describes how all other atomic entities are constituted from these fundamental particles. The model contains 24 fundamental particles (plus corresponding antiparticles—47 species of elementary particles including particles and antiparticles) that are the constituents of matter. Some of these species can combine to form composite particles, accounting for the hundreds of other species of particles discovered since the 1960s. For example, protons and neutrons are composed of quarks. The standard model has been found to agree with almost[5] all the experiments conducted to the present time. However, most particle physicists believe that it is an incomplete description and that a more fundamental theory awaits future discovery. Another approach utilizes string theory (see Appendix D.3).

Table 2.4 provides a summary of atomic particles (also see Figure 2.5 [54]). One (but not the only) question of interest is whether the particle is affected by the strong interaction or not: If it is affected by the strong interaction (term defined in Table 2.1), it is called a *hadron*; if not, it is called a *lepton* [55]. Particle physics follows the

[5]In the recent past, measurements of neutrino mass have provided the first experimental deviations from the standard model.

TABLE 2.4 Atomic Particles—A Current View

Nonelementary Particles

Molecule. Consists of atoms (a molecule is comprised of two or more chemically bonded atoms; the atoms may be of the same type of element or they may be different)

- *Atom*. Consists of a nucleus containing both protons (baryon variety of hadron) and neutrons (baryon variety of hadron) with an orbiting electron (lepton) cloud

Hadrons. Composed of quarks and gluons of three types:

- *Baryons*. Three quarks:
  (a) *Proton*. Part of the atomic nucleus
  (b) *Neutron*. Part of the atomic nucleus
- *"Exotic baryons"*:
  (a) *Pentaquark*. Made of five quarks
- *Mesons*, *quark*, and *antiquark*:
  (a) *Pion*
  (b) *Kaon*

Elementary Particles

Fermions. Of these particles there also exist an antiparticle

- *Leptons*. Six types:
  (a) *Electron*. The antiparticle of the electron is called a positron
  (b) *Muon*
  (c) *Tauon*
  (d) *Electron-*, *muon-*, and *tauon-neutrino*
- *Quarks*. Six types (*up*, *down*, *strange*, *charm*, *top*, and *bottom*). Quarks form protons and neutrons.

Bosons. These particles are their own antiparticle. They carry the fundamental forces of nature.

- *Graviton*. Transmits gravity (not yet observed)
- *Higgs boson*. Gives particles mass (not yet observed)
- *Photon*. Transmits electromagnetic force
- *W and Z bosons*. Transmits weak nuclear force
- *Gluons*. Transmits strong nuclear force

principles of quantum theory (e.g., see [56, 57, 58]). Subatomic particles exhibit what is peceived (by some) as wave–particle duality. This means they display particlelike behavior under certain experimental conditions and what can be considered wavelike behavior in others. At the theoretic level, these particles are described neither as waves nor as particles, but as state vectors in an abstract Hilbert space (this is covered in Appendix D; also see Appendix B for a short tutorial on the pure mathematical concept of Hilbert spaces).

2.2.3 Atomic Structure

In physics and chemistry, one is interested to determine *how* the electrons are arranged when they bind to nuclei to form atoms and molecules. This arrangement

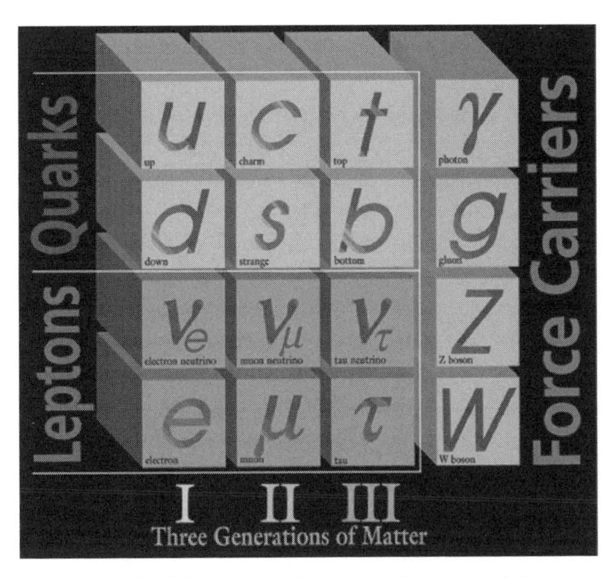

FIGURE 2.5 Quarks, leptons, and other particles.

of electrons is termed the *electronic structure* of the atom or molecule. There is a relationship between the electronic structure of an atom and its physical properties. The forces that bind the atom (aside from the nuclei themselves) are electrostatic: The positively charged nucleus attracts the negatively charged electrons. There also are magnetic forces that arise from the motions of the charged particles; these magnetic forces, however, are smaller in magnitude than the electrostatic forces and are not responsible for binding matter (these magnetic forces do, however, give rise to a number of important phenomena). (These forces, known as bonding forces, are discussed in more detail in Chapter 3.)

The electron cloud is the space where the *electrons* orbit the nucleus at very high speed; the cloud comprises a large part of an atom's volume. The cloud is subdivided into *energy levels* (also called shells.) Electrons occupy the lowest energy levels. For energy level n the maximum number of electrons is $2n^2$. For $n = 1$, the maximum number of electrons is 2, for $n = 2$ the number is 8; for $n = 3$ the maximum is 18, and for $n = 4$ the maximum is 32. Table 2.5 shows the *electron shell configuration* for the first few elements (see Appendix C for a more inclusive list).

The electronic structure is altered during a chemical reaction; however, only the number and arrangement of the electrons are changed. During a chemical reaction the nucleus remains unaltered. This mechanism is responsible for retaining the atom's chemical identity during the chemical reaction. It follows that for the purpose of understanding the chemical properties and behavior of atoms, the nucleus of a given element may be regarded as a point charge of constant magnitude, giving rise to a central field of force that binds the electrons to the atom [59, 60].

The atomic number is the number of protons in the atom. As noted, the number of electrons is equal to the number of protons. The mass number is the total number

TABLE 2.5 Electron Configuration

		K	L		M			N				O				P				Q	
Number	Symbol	1s	2s	2p	3s	3p	3d	4s	4p	4d	4f	5s	5p	5d	5f	6s	6p	6d	6f	7s	7p
Period 1																					
1	**H**	1																			
2	**He**	2																			
Period 2																					
3	**Li**	2	1																		
4	**Be**	2	2																		
5	**B**	2	2	1																	
6	**C**	2	2	2																	
7	**N**	2	2	3																	
8	**O**	2	2	4																	
9	**F**	2	2	5																	
10	**Ne**	2	2	6																	
Period 3																					
11	**Na**	2	2	6	1																
12	**Mg**	2	2	6	2																
13	**Al**	2	2	6	2	1															
14	**Si**	2	2	6	2	2															
15	**P**	2	2	6	2	3															
16	**S**	2	2	6	2	4															
17	**Cl**	2	2	6	2	5															
18	**Ar**	2	2	6	2	6															

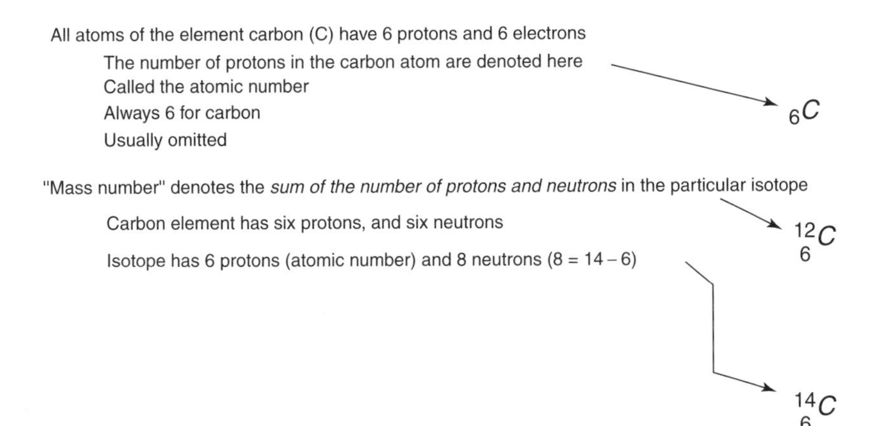

FIGURE 2.6 Nomenclature.

of particles in the nucleus, namely, number of protons in the nucleus plus the number of neutrons in the nucleus. The notation used is (by way of an example) $^{12}_{6}C$, where the top number is the mass number in the carbon element and the bottom number is the atomic number (see Figure 2.6). An element consists of atoms *all having the same atomic numbers.* Atoms that have the same atomic number but different mass numbers are called isotopes (i.e., they belong to the same element but have different number of neutrons). Elements are known by common names as well as by their abbreviations, these consisting of one or two letters with the first one capitalized. Some examples are carbon, C; oxygen, O; sulfur, S; aluminum, Al; copper, Cu (from *cuprum*); iron, Fe (from *ferrum*); lead, Pb (from *plumbum*); and mercury, Hg (from *hydrargyrum*). Names are maintained by the International Union of Pure and Applied Chemists (IUPAC).[6]

2.2.4 Substances and Elements

Next we look at matter and forms of matter. Matter is an *aggregate* of atoms. Matter is defined as anything that has mass and occupies space. It can be manifest in three physical states, as depicted in Table 2.6: gas (vapor), liquid, and solid.[7]

[6]Since elements above 109 have not yet been proven to exist, they have only generic names and symbols based on those by the IUPAC. The name is based on the digits in the element's atomic number. Simply replace each digit with the expression shown next, then end the name with the usual *-ium* suffix. The corresponding chemical symbol is the first letter of each of the three main syllables. The symbols are 0 = nil; 1 = un; 2 = bi; 3 = tri; 4 = quad; 5 = pent; 6 = hex; 7 = sept; 8 = oct; 9 = enn. Example: Element 125: 1 = *un*, 2 = *bi*, 5 = *pent*, so one has *un-bi-pent-ium* or unbipentium, Ubp.

[7]Liquid crystals can be considered a fourth phase of matter, a state qualitatively different from the ordinary three phases, gas, liquid, and solid. Liquid crystals flow like a liquid, but there is order in at least one dimension in the arrangement of the molecules [31]. Some physicists consider plasma as yet another state of matter. Plasma is a mixture of ions and electrons (such as in an electrical discharge).

TABLE 2.6 States of Matter

Gas	Form of matter without fixed shape or volume. Shape: This form of matter conforms to the shape of its container. Volume: Can be *compressed* or *expanded* (up to a certain limiting point), to encompass different volumes.
Liquid	Form of matter where there is a fairly definite volume but there is no specific *shape*. Shape: This form of matter conforms to the shape of its container. Volume: Liquids can be compressed, but only to a limited degree.
Solid	Form of matter where there is definite shape and volume. This form of matter is rigid. Volume: Solids can be compressed, but only to a very limited degree.

A pure substance has a fixed composition and has distinct properties (note that most "everyday" matter is not a pure substance but a mixture of substances). Pure substances have a set of properties that are fairly unique to each specific substance, particularly if the substance is a pure element. These characteristics allow us to distinguish a substance from other substances. These properties fall into two general categories: *physical properties* and *chemical properties*. Physical properties are properties that can be measured without altering the basic identity of the substance. These include but are not limited to weight, temperature, electrical resistance, and so on. Chemical properties are properties that describe the manner in which a substance may change or "react" to form other substances. Some of the key physical properties of elements (of possible interest to nanotechnology), making them unique, are defined in Table 2.7 [61, 62, 63, 64, 65, 66, 67]. Also see Appendix C for some numerical values and the Glossary for additional terms.

For illustrative purposes, Table 2.8 provides some of these parameters for three elements.

Substances can undergo a series of changes in properties, and these changes may be classified as either physical changes or chemical changes. Physical changes are those changes where a substance changes its physical state or structure but not its basic identity. Examples include going from solid to liquid, heating up a substance, electrically charging a substance, and/or magnetizing a substance. Chemical changes (also known as chemical reactions) are changes where a substance is transformed into a substance with different chemical properties. It has been demonstrated empirically (and also theoretically) that all chemical and physical differences between elements are due to the differences in the number of protons, electrons, and neutrons in the atom.

A mixture is a combination of two or more substances where each substance retains its own chemical and physical identity and/or properties. Because in mixtures the individual components retain their physical and chemical properties, it is possible to reversibly separate the components based on their properties. For example, we can separate salt from water by removing (evaporating) the water. Mixtures can be heterogeneous or homogeneous. Heterogeneous mixtures are not uniform throughout the resulting aggregate and may have areas of different appearance and properties. Homogeneous mixtures (also called solutions) are uniform throughout the

TABLE 2.7 Key Physical Properties of Elements

Atomic number	The number of positively charged protons in the nucleus of an atom. Atomic number Z is a characteristic property of an element, equal to the number of protons present in the nucleus of an atom. In neutral species, it is also equal to the number of electrons present in the atom.
Atomic radius	Atomic radius is usually referred to as one-half of equilibrium internuclear distance between two adjacent atoms (which may either be bonded covalently or be present in a closely packed crystal lattice) of an element.
Atomic volume	The atomic (or molecular) volume V_m is the average volume per $10^3 N_0$ of atoms in the structure, where N_0 is Avogadro's number (6.022×10^{23}/mol). *Units*: SI: m^3/kmol; cgs: $10^6 cm^3$/kmol; Imperial: in.^{3}/kmol. For a pure element, the atomic volume is $V_m = A/\rho$, where A is the atomic weight in kg/kmol and ρ is the density in kg/m^3. For compounds the average atomic volume is $V_m = M/n\rho$, where M is the molecular weight and n is the number of atoms in the molecule. Thus, for a compound with the formula A_xB_y the atomic volume is $$V_m = \frac{xA_A + yA_B}{(x+y)\rho}$$ where A_A is the atomic weight of element A and A_B is the atomic weight of element B. For a polymer $(C_xH_yO_z)_n$ the atomic volume is $$V_m = \frac{XA_C + YA_H + ZA_O}{(X+Y+Z)\rho}$$ where A_C is the atomic weight of carbon, and so on. The atomic volume is involved in many property correlations (and thus is crucial for checking and estimating properties) and, together with the density, it gives the atomic weight.
Atomic weight	The average relative weight of the atoms of an element referred to an arbitrary standard of 16.0000 for the atomic weight of oxygen. The atomic weight scale used by chemists takes 16.0000 as the average atomic weight of oxygen atoms as they occur in nature. The scale used by physicists takes 16.00435 as the atomic weight of the most abundant oxygen isotope. Division by the factor 1.000272 converts an atomic weight on the physicists' scale to the corresponding atomic weight on the chemists' scale. See also "atomic number."
Boiling point	The temperature at which the vapor pressure of a liquid is equal to the atmospheric pressure. The normal boiling point is the boiling point at normal atmospheric pressure (101.325 kPa).
Covalent radius	Half the distance between the nuclei of two identical atoms when they are joined by a single covalent bond.
Density at 300 K	Mass per unit volume of a substance at 300 K.
Electrical conductivity	Characterizes the conduction capacity (electrical and thermal) of a substance. Electrical conductivity is expressed in siemens per unit of length. Electronic or ionic conduction is the phenomenon by which an electron or an ion moves in a material. (Sideline: Thermal conduction is the phenomenon by which, in a given medium, heat

TABLE 2.7 (*Continued*)

	flows from a high-temperature region to a lower temperature region or between two media in contact with each other. Ionic or protonic conductivity quantifies the ease with which an ion or a proton moves in a material.)
Electronegativity	Electronegativity is a parameter that describes, on a relative basis, the power of an atom or group of atoms to attract electrons from the same molecular entity.
Electronic configuration (structure)	Electronic configuration is the arrangement of electrons in an atom when it is in its ground state. All the properties of elements depend on their electronic configuration.
Energies of electrons	Measured and expressed in terms of a unit called an *electron volt* (eV), the most commonly used unit of energy, defined as the energy acquired by an electron when it is accelerated through a potential difference of 1 V. If the charge on the electron is denoted by e, then the energy change in EV is given by the charge multiplied by the voltage V, namely, $\frac{1}{2}mv^2 = \text{eV}$. One electron volt equals 1.6021×10^{-19} joules (J). Also note that 1 eV $= 1.602 \times 10^{-12}$ erg and charge on electron $= e = 4.8029 \times 10^{-10}$ esu.
First ionization potential	Ionization energy is the minimum energy required to remove an electron from an isolated atom or molecule (in its vibrational ground state) in the gaseous phase.
Heat of fusion	The energy absorbed during the change of a mole of a solid to liquid without a change in temperature. (The heat of atomization is energyneeded to decompose 1 mol of a certain substance into atoms.)
Heat of vaporization	The energy absorbed during the change of a mole of liquid to a vapor without a change in temperature.
Melting point	The temperature at which the solid and liquid phases of a substance are in equilibrium at a specified pressure (normally taken to be atmospheric unless stated otherwise).
Specific heat capacity	(Or also specific heat.) The heat capacity of a system divided by its mass. It is a property solely of the substance of which the system is composed. As with heat capacities, specific heats are commonly defined for processes occurring at either constant volume (C_v) or constant pressure (C_p). Heat capacity (also called thermal capacity) is the ratio of the energy or enthalpy absorbed (or released) by a system to the corresponding temperature rise (or fall). Heat capacities are defined for particular processes. For a constant-volume process, $$C_v = \frac{\partial U}{\partial T}$$ where U is the internal energy of a system and T is its temperature. For a constant-pressure process, $$C_p = \frac{\partial H}{\partial T}$$ where H is the system enthalpy.

TABLE 2.7 (*Continued*)

	The heating rate Q for a constant-volume process is $Q = C_v\, dT / dt$ (V = const), whereas in a constant-pressure process, $Q = C_p\, dT / dt$ (p = const).
Temperature	The thermal state of matter with respect to its ability to transfer heat to other matter. Heat is the energy that is transferred between matter by means of radiation, conduction, and/or convection. The common scales for measuring temperature are Celsius (centigrade), Fahrenheit, and Kelvin.
Thermal conductivity	Rate of heat flow divided by area and by temperature gradient.

resulting aggregate. The most common type of solution is comprised of a solid (the solute) dissolved in a liquid (the solvent).

Pure substances are composed of either elements or compounds. Elements are substances that cannot, when in pure form, be altered, reduced, or decomposed into other substances by chemical techniques. It is a substance that cannot be broken down or reduced further. They correspond to matter comprised of a single type of atom. Compounds are substances that can be altered, reduced, or decomposed into other substances by chemical techniques. In other words, they can be decomposed into two or more elements. Compounds are substances of two or more elements united chemically in specific proportions by mass. For example, pure water is composed of the elements oxygen (O) and hydrogen (H) and at the defined ratio by mass of 89% oxygen and 11% hydrogen. The specific ratio representing the elemental composition of a pure compound is always the same; known as the law of constant composition (or the law of definite proportions), this is credited to the eighteenth-century French chemist Joseph Louis Proust. Most elements are found in nature in molecular form with two or more atoms bonded together. For example, oxygen is found in its molecular form O_2 as well as O_3 (also known as ozone). Although O_2 and O_3 are both compounds of oxygen, they are different in their chemical and physical properties.

Eventually we will be interested in the following entities and how they exist and/or can be improved by nanotechnology means:

- *Conductor:* A material through which electricity can flow with relatively little resistance.
- *Insulator:* A material that does not allow electricity to flow through it.
- *Semiconductor:* A substance (such as silicon) through which electricity can flow under certain circumstances. Its conductive properties are between those of a good conductor and an insulator.
- *Superconductor:* A material through which electricity flows with basically zero resistance.

TABLE 2.8 Example of Elements and Physical Parameters

Hydrogen	1.00794 Atomic weight
1 Atomic number	1 Oxidation states
20.28 Boiling point, K	
13.81 Melting point, K	H Symbol
0.0899 Density at 300 K	
$1s^1$ Electron configuration	

Hydrogen
0.32 Covalent radius
2.08 Atomic radius
14.1 Atomic volume
13.598 Ionization potential
14.304 Specific heat capacity
hcp Crystal structure
2.1 Electronegativity
0.4581 Heat of vaporization
0.0585 Heat of fusion
0.1815 Thermal conductivity

Helium	4.0026 Atomic weight
2 Atomic number	
4.216 Boiling point, K	
0.95 Melting point, K	He Symbol
0.1785 Density at 300 K	
$1s^2$ Electron configuration	

Helium
0.93 Covalent radius
31.8 Atomic volume
24.587 Ionization potential
5.193 Specific heat capacity
hcp Crystal structure
0 Electronegativity
0.084 Heat of vaporization
0.021 Heat of fusion
0.152 Thermal conductivity

Carbon	12.011 Atomic weight
6 Atomic number	±4, 2 Oxidation states
5100 Boiling point, K	
3825 Melting point, K	C Symbol
2.26 Density at 300 K	
$1s^2 2s^2 p^2$ Electron configuration	

Carbon
0.77 Covalent radius
0.91 Atomic radius
5.3 Atomic volume
11.26 Ionization potential
0.709 Specific heat capacity
hcp Crystal structure
2.55 Electronegativity
715 Heat of vaporization
0.07 Electrical conductivity
155 Thermal conductivity

2.2.5 Nomenclature and Periodic Table

Elements are the fundamental substances from which all matter is composed. Figure 2.7 provides a useful taxonomy related to properties of matter. The current number of known and officially named elements is 109 (a tentative 110th element has been synthesized); a handful of other elements could be synthesized at higher atomic number in the future.

Elements can be organized according to a table (the periodic table of the elements) based on the atomic number, where elements share some of their physical and chemical properties. In the midnineteenth century researchers realized that elements could be grouped, or classified, according to their chemical behavior. For example, certain elements have similar characteristics: helium (He), neon (Ne), and argon (Ar) are very nonreactive gases; lithium (Li), sodium (Na), and potassium (K) are all soft, very reactive metals. When the elements are arranged in order of increasing atomic number, their chemical and physical properties exhibit a repeating, or periodic, pattern. This kind of investigation eventually (1869) resulted in the development of the periodic table. The elements in a column of the periodic table are known as a family or group.

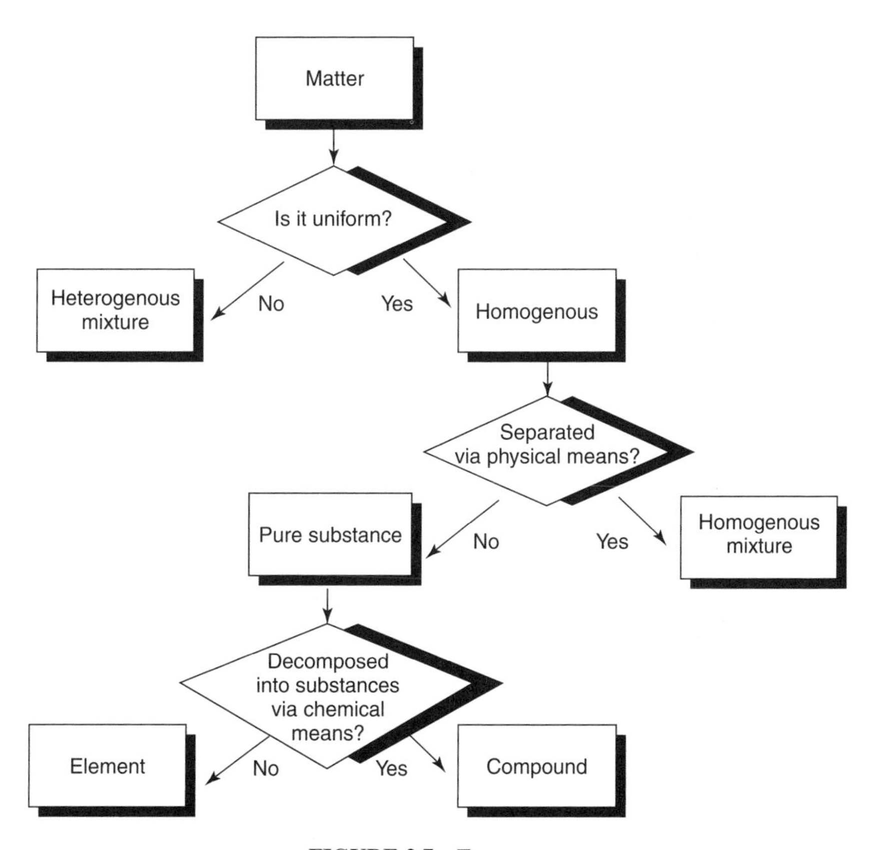

FIGURE 2.7 Taxomomy.

The periodic table (see Table 2.9) is an arrangement of elements in a geometric pattern designed to represent the periodic law by aligning elements into periods and groups. Periods are shown as horizontal rows and groups are vertical columns. Elements follow a standard arrangement in octaves (groups of eight) where the properties of each "note" (individual entries) are to some extent repeated by heavier elements within subsequent octaves; after the first two octaves, subsequent octaves are disrupted by the introduction additional blocks of elements between the second and third notes for the next three octaves; the sixth octave remains to be completed and awaits the discovery of new elements [68]. Elements with the same number of orbitals are in the same period; elements with the same number of electrons in the filling orbital have a number of properties that are similar to the properties of elements in the same group. Periods are characterized by the number of energy levels (shells) of electrons surrounding the nucleus. Elements in first period have only one shell and have a 2-electron maximum (hydrogen has 1 electron and helium has 2 electrons); elements in second period have two shells and have a 10-electron maximum; and so on [61].

TABLE 2.9 Periodic Table of the Elements by Atomic Number

	Group																	
Period	1	2	3	4	5	6	7	8	9	10	11	12	13	14	15	16	17	18
1	H 1																	He 2
2	Li 3	Be 4											B 5	C 6	N 7	O 8	F 9	Ne 10
3	Na 11	Mg 12											Al 13	Si 14	P 15	S 16	Cl 17	Ar 18
4	K 19	Ca 20	Sc 21	Ti 22	V 23	Cr 24	Mn 25	Fe 26	Co 27	Ni 28	Cu 29	Zn 30	Ga 31	Ge 32	As 33	Se 34	Br 35	Kr 36
5	Rb 37	Sr 38	Y 39	Zr 40	Nb 41	Mo 42	Tc 43	Ru 44	Rh 45	Pd 46	Ag 47	Cd 48	In 49	Sn 50	Sb 51	Te 52	I 53	Xe 54
6	Cs 55	Ba 56	La 57	Hf 72	Ta 73	W 74	Re 75	Os 76	Ir 77	Pt 78	Au 79	Hg 80	Ti 81	Pb 82	Bi 83	Po 84	At 85	Rn 56
7	Fr 87	Ra 88	Ac 89	Rf 104	Db 105	Sg 106	Bh 107	Hs 108	Mt 109	Uun 110	Uuu 111	Uub 112						
	Lanthanides				Ce 58	Pr 59	Nd 60	Pm 61	Sm 62	Eu 63	Gd 64	Tb 65	Dy 66	Ho 87	Er 88	Tm 89	Yb 70	Lu 71
	Actinides				Th 90	Pa 91	U 92	Np 93	Pu 94	Am 95	Cm 96	Bk 97	Cf 98	Es 99	Fm 100	Md 101	No 102	Lr 103

2.2.6 Making Compounds

Quantum theory shows that atoms contain electrons that are constrained to exist in locations relative to one another that are strictly prescribed (as discussed in Appendix D). These possible locations are termed atomic orbitals, with the most important ones, frontier orbitals, being on the outside. When two atoms are brought together to form a molecule, the atomic orbitals on each atom can mix to form a new set of molecular orbitals, which are termed bonding, antibonding, or nonbonding, depending on whether electrons in the molecular orbitals have a stabilizing, destabilizing, or neutral effect, respectively, on holding the molecule together. When atoms

Notes to Table 2.9

The group number is an identifier employed to describe the column of the periodic table, based on IUPAC conventions. Moving down a group, the elements all have the same valence structure but with an increasing number of shells. Groups 1–2 (except hydrogen) and 13–18 are termed main-group elements. Groups 3–11 are termed transition elements. Transition elements are those whose atoms have an incompleted subshell or whose cations have an incomplete d subshell. Main-group elements in the first two rows of the table are called typical elements. The first row of the f-block elements are called lanthanoids (or, less desirably, lanthanides). The second row of the f-block elements are called actanoids (or, less desirably, actanides) [69]. The following names for main groups in common use are as follows:

- Group 1: alkali metals (Li, Na, K, Rb, Cs, Fr)
- Group 2: alkaline earth metals (Be, Mg, Ca, Sr, Ba, Ra)
- Group 11: coinage metals (not an IUPAC approved name)
- Group 15: pnictogens (not an IUPAC approved name)
- Group 16: chalcogens (“chalk formers”) (O, S, Se, Te, Po)
- Group 17: halogens (“salt formers”) (F, Cl, Br, I, At)
- Group 18: noble gases (He, Ne, Ar, Kr, Xe, Rn)

In addition, groups may be identified by the first element in each group, so the group 16 elements are sometimes called the oxygen group.

Observation: There is considerable confusion surrounding the group labels (the above is based on the current IUPAC convention). The other two systems are less desirable since they are confusing, but they are still in common usage. The designations A and B are completely arbitrary. The first of these (A left, B right) is based upon older IUPAC recommendations and frequently used in Europe. The last set (main-group elements A, transition elements B) is still used in the United States [69, 70, 71, 72]:

IUPAC, EUROPEAN, AND AMERICAN GROUP LABELLING SCHEMES

Group	1	2	3	4	5	6	7	8	9	10	11	12	13	14	15	16	17	18
European	IA	IIA	IIIA	IVA	VA	VIA	VIIA	VIIIA	VIIIA	VIIIA	IB	IIB	IIIB	IVB	VB	VIB	VIIB	VIIIB
American	IA	IIA	IIIB	IVAB	VB	VIB	VIIB	VIIIB	VIIIB	VIIIB	IB	IIB	IIIA	IVA	VA	VIA	VIIA	VIIIA

combine to form molecules, the number of orbitals is conserved. For example, bringing together an atom with four frontier atomic orbitals with an atom containing three frontier atomic orbitals results in a molecule with seven frontier molecular orbitals, perhaps as a set of three bonding, three antibonding, and one nonbonding orbital. If three atoms were combined, more molecular orbitals would result and similarly with

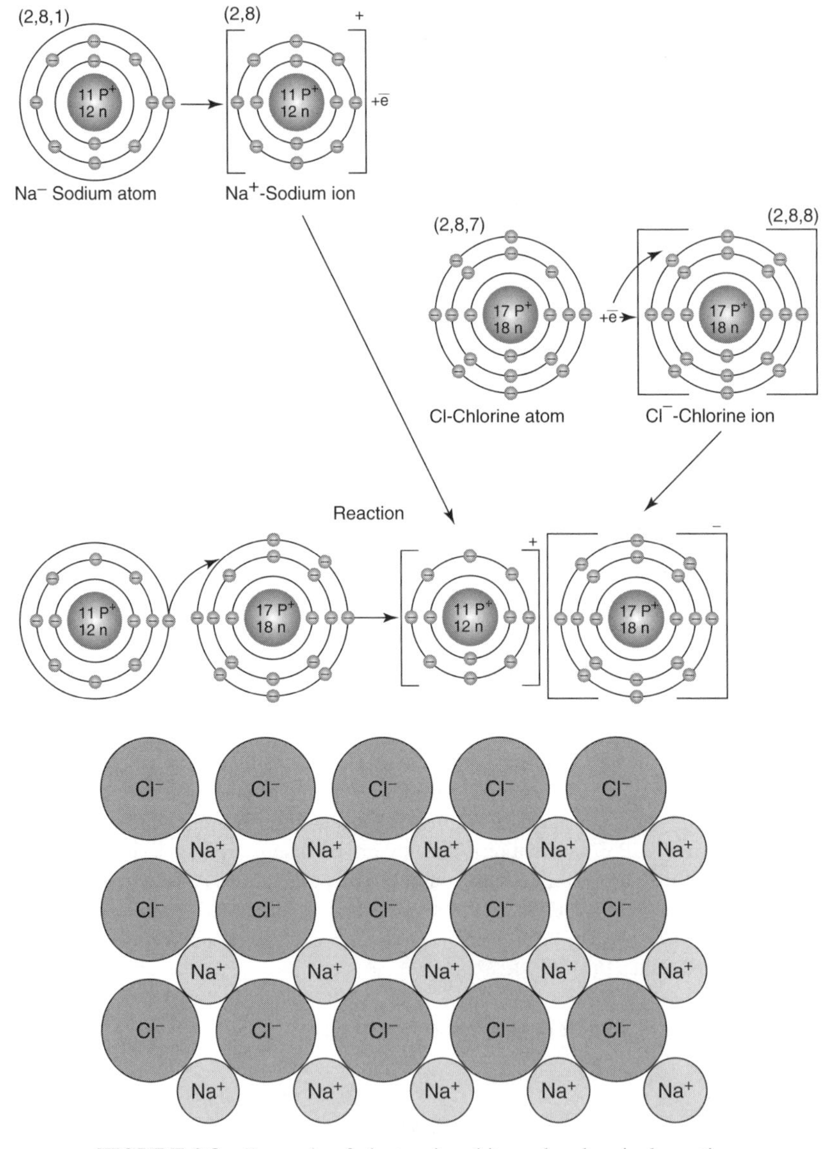

FIGURE 2.8 Example of electronic orbits and a chemical reaction.

four, five, or six atoms. When the number of atoms becomes large, the differences in energies among adjacent orbitals becomes small as more and more are packed into the same energy region [73].

Under the right conditions, groups of elements can combine through a chemical reaction to form compounds. A variety of "empirical laws" describe these processes (some of these are discussed in Chapter 3); in reality some very complex quantum theory mechanism is at play, but the outcome may be explainable and/or predictable through one of these "empirical laws." The chemical formula for water, H_2O, illustrates the method of describing compounds in atomic terms: in this particular compound there are two atoms of hydrogen and one atom of oxygen (the 1 subscript is omitted). H_2O_2 is a different compound: hydrogen peroxide. Although both compounds are composed of the same types of atoms, they are chemically different: Hydrogen peroxide is reactive (i.e., tends to participate readily in chemical reactions), while water is inert.

Figure 2.8 shows for illustrative purposes a sodium chloride (NaCl) molecule being formed via a chemical reaction. Sodium (in its neutral atom state) has two electrons in the first orbital (K shell), eight electrons in the second orbital (L shell), and one electron in the third (M shell). The M-shell electron is loosely bound and can be removed to form a positive sodium ion. Chlorine has two electrons in the K shell, eight electrons in the L shell, and seven electrons in the M shell. The Cl atom is short one electron for completely filling the last orbital. When a Na and a Cl atom come in close proximity, they easily bond together. The electron given up by a Na atom is easily accommodated in the last orbital of a Cl atom. The Na and Cl atoms are attracted to each other because of opposite electrical charges and the final reaction between Na and Cl^- is Na^+–Cl^-. This topic of chemical formulas and which are feasible and why is treated in more detail in Chapter 3 (there is a strong relationship to the electronic structure).

2.3 BASIC PROPERTIES OF CONDUCTORS, INSULATORS, AND SEMICONDUCTORS

To get a full sense of the properties of conductors, insulators, and semiconductors, one must utilize condensed-matter physics. Condensed-matter physics is a subdiscipline of physics that focuses on the various properties that describe solid and liquid substances, including their thermal, elastic, electrical, chemical, magnetic, and optical characteristics. In terms of solid matter, theoretical advances have been made in recent years to study crystalline materials whose simple repetitive geometric arrays of atoms are multiparticle systems. These systems are described in terms of quantum theory; however, because atoms in solids are coordinated over large distances, the theory must extend beyond the atomic and molecular levels. In addition, conductors such as metals contain free electrons that govern the electrical conductivity of the whole material, which is a property of the entire solid rather than its individual atoms. Crystalline and amorphous semiconductors and insulators as well as properties of the liquid state of matter (e.g., liquid crystals and quantum liquids) are also studied in condensed-matter physics. The macroscopic quantum phenomena observed in quantum liquids, such as superfluidity, is also seen in certain metallic and ceramic materials as superconductivity [22].

Metals[8] do not hold on to their electrons very tightly. An important effect of the electropositive nature of metals is that the orbitals in metals are highly diffuse and cover a large volume of space around the atoms, which allows the orbitals to interact directly with many other atoms at once to create molecular orbitals which are delocalized, or spread out over a large distance. The net effect is a continuous band of available locations and energies for the electrons within the substance. Finally, metals tend to have more orbitals available than electrons to fill them. Combining this fact with the banded nature of the available energies and the delocalization of the orbitals results in material that easily transports excess charge from location to location within the bulk substance—a conductor in other words.

The electrons in nonmetals are held more tightly to the atomic nuclei. Those electrons used for bonding are not diffuse and interact strongly only with a few neighbors. Instead the molecular orbitals and the electrons in them are constrained to exist only in small areas between individual atoms. This results in consequences for the band structure of nonmetals. Bulk nonmetals also contain many atoms and therefore many molecular orbitals that form continuous bands. Unlike metals, however, the nonmetals are characterized by multiple nonoverlapping band sets. The lower energy bands that are full of electrons are referred to as the valence band, whereas the empty higher energy bands are referred to as the conduction band. The energy difference between the valence and conduction bands, referred to as the bandgap, is usually much too large to easily promote an electron from the valence band to the conduction band or even to add an electron to the conduction band using an external electrical bias. For this reason, these materials are termed insulators.

The semimetal or metalloid materials such as aluminum and silicon are intermediate in properties between metals and nonmetals. For these materials in the bulk (with many atoms), a band structure develops into distinct bands separated by a bandgap, such as in insulators; however, the bandgap can be small. Under the right conditions (e.g., applied electrical potential, heat, or light), the conduction band becomes accessible and the material can be induced to conduct like a conductor. Such materials with intermediate properties are termed semiconductors. For semiconductors, the size of the bandgap is a critically important parameter that determines among other things the color of the material since light of wavelengths comparable to the bandgap can be absorbed by the material to promote an electron from the valence band to the conduction band.

Expanding on the observations above, a bandgap is a forbidden energy band; specifically a "band" is a closely spaced group of energy levels in atoms, a range of energies that electrons can have in a solid. Each band represents a large number of allowed quantum states. The outermost electrons of the atom forms the "valence band" of the solid. In order for electrons to move through a solid, there must exist empty quantum states with the same energy, and this can occur only in an unfilled band, the

[8]These observations are based on reference [73].

"conduction band." In general, so-called metals are good conductors because the partly filled conduction band overlaps with a filled valence band, and vacant energy states in the conduction band are thus readily available to electrons. In "insulators," the conduction band and valence band are separated by a wide forbidden band, and electrons do not have enough energy to jump from one band to another. In intrinsic "semiconductors," the forbidden gap is narrow, and at normal temperatures some electrons at the top of the valence band can move by thermal agitation into the conduction band. In a so-called doped semiconductor, the doping impurities essentially create one or more thin separate conduction bands in the forbidden band. In this context, the "gap" refers to the gap between energy bands, that is, from the upper boundary of the valence band to the lower boundary of the conduction band [31].

2.4 BASIC PROPERTIES OF SILICON AND BASICS OF TRANSISTOR OPERATION

This section provides a basic view of the operation of a transistor, which rests on the operation of semiconductors discussed earlier. The field of physics, which we briefly introduced above, naturally has a very broad (even universal) scope, but, consistent with the focus of this book, we are looking at a very specific and pragmatic set of issues. Hence, we focus parochially on microelectronics in the subsections that follow.

The development of microelectronics, starting with the transistor and then the aggregation of transistors into microprocessors, memory chips, and controllers, has ushered in the age of information technology (IT); these devices all work by streaming electrons through silicon [74]. Field-effect transistors (FETs), among the workhorse devices of microelectronics technology, are small switches in which the passage of electric current between a "source" and a "drain" is controlled by an electric field in a middle component called a "gate." Researchers exploring ways to build ultrasmall electronic system these days start with the existing semiconductor science and then move forward by utilizing extensions, expansions, or new phenomena. For example, devices made of atom-thick carbon cylinders have incorporated "carbon nanotubes" into a new kind of FET [75]; there is also interest in developing polymer FETs and semiconducting polymer blends, with submicrometer critical features in planar and vertical configurations [76, 77]. Some basic methods of traditional operation are discussed below; the topic of nanoelectronics is revisited in Chapter 6.

2.4.1 Transistors

A transistor operates on the principle of semiconducting. Silicon (Si) is the most common element used commercially to date for this purpose. A material such as silicon acts as insulator; to obtain semiconducting behaviors in silicon, impurities are added to form a compound. A semiconductor's structure is comprised of electronic bands (a direct or indirect electronic bandgap) characterized by the presence of a bandgap between an allowed and full band (or almost full band) called the valence

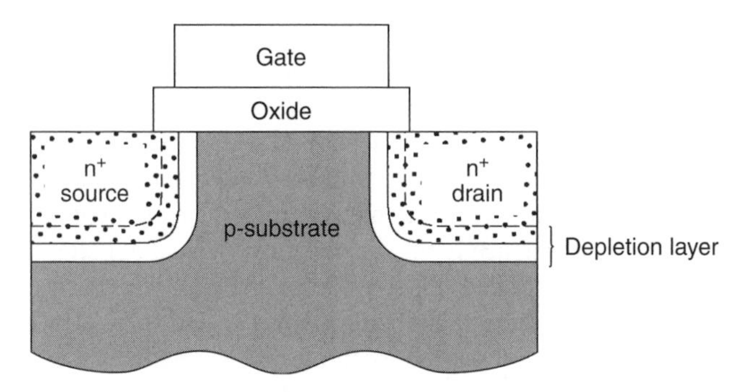

FIGURE 2.9 Field-effect transistor.

band and an empty band (or almost empty band) called the conduction band [64]. The operation of the semiconductor is expected to be at room temperature.[9]

Silicon has four outer orbital electrons that form tetrahedral bonds to four other Si atoms in a crystal; it follows, as noted, that all electrons are bound and the material acts as insulator. When a phosphorus atom with five electrons in the outer orbital (or some other group V element) is added, after four electrons are used to bond the surrounding Si atoms, there is one "leftover" electron. This orbital arrangement allows electronic conduction when a voltage is applied. The Si atom is said to form an n-type (region): The electron conduction is via the electrons which carry a negative charge.

When an atom such as boron (or some other group III element) replaces one silicon atom, the silicon is said to be a p-type (region). In this arrangement, the boron can bond to three silicon atoms in the crystal lattice; however, there is a "hole" left where the fourth bond used to be. It follows that bonding electrons from the outer orbital of other Si atoms can move to fill this hole; in turn, this leaves a hole behind where the electrons came from in the crystal. The net effect is that the Si can sustain conduction as electrons move into holes in the system.

If a p-type Si region is placed beside an n-type Si region, then a depletion layer forms where electrons and holes "annihilate" each other; also, electrons cannot flow between these two regions unless a suitable voltage is applied between them (stated in other words, the p–n junction stops electrons from flowing across the system).

These observations can be used to construct a usable transistor, such as the metal–oxide–semiconductor field effect transistor (MOSFET), depicted diagrammatically in Figure 2.9. A MOSFET is a planar device of semiconducting material whose free-carrier concentration and, hence, its conductivity are controlled by the gate. This transistor is a p-type silicon substrate with an insulator (oxide) and a metal strip (gate) on top. It is comprised of two back-to-back p–n junctions. Without a gate voltage applied, no current flows from the source n-type Si contact and the sink (drain) n-type Si contact. When a voltage is applied to the gate, electrons are attracted to the gate and they form a thin layer in the proximity of the surface of the p-type Si. It then follows

[9]The rest of Section 2.4 is loosely based on references [19] and [26].

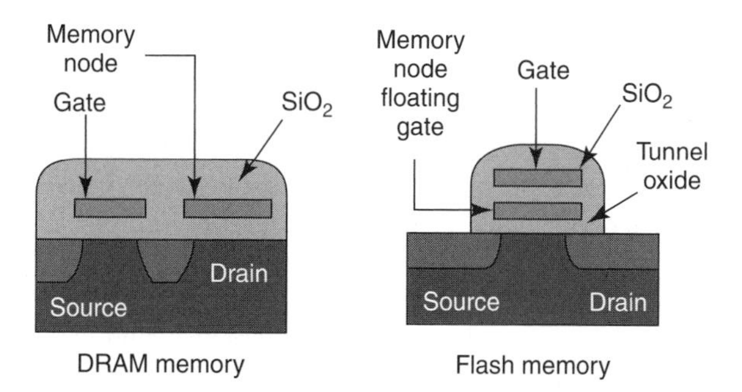

FIGURE 2.10 DRAM and flash memory.

that electrons can be transported from the source contact to the drain contact. The gate, in effect, switches the current off and on between the source and the drain. The gate operates by inducing electrical charge in the silicon below it, which provides a channel for current to flow; if the capacitance of the gate is too low, not enough charge will be present in this channel for it to conduct; hence, capacitance must be properly engineered [19].

Microprocessing unit (MPU) chips in use today employ a CMOS architecture that incorporates two MOSFET transistors; one transistor uses electrons as the conducting layer and a second transistor uses holes. The advantage of this arrangement is that power is only dissipated (aside from that resulting from any leakage currents in the transistor) when the circuit is switched, thereby substantially reducing the power consumption of the chip. Since the integration density of a chip is limited by the power dissipation, the ability to scale to smaller transistors facilitates the integration density from both the power and size requirements. Figure 2.10 depicts two examples of memory gates.

2.4.2 Manufacturing Approaches

Microprocessing units are built through the use of wafers which provide the underlying substrate upon which the multitude of transistors are placed (grown). At this time, CMOS wafers are fabricated using a top-down approach.[10] Here, deep ultraviolet photons are shone through a patterned mask made of glass (or quartz) and chrome; where the light is transmitted through the mask, it reacts with an optical resist on the wafer and this area may be dissolved in a solvent to leave the mask pattern in the resist; this pattern is then transferred to the underlying substrate by etching. By blanket deposition of insulators or metallic layers along with lithography and etching, complex circuits are patterned onto chips. The smallest feature which may be produced using this lithographic technique is given by the Rayleigh resolution criteria $k\lambda/\mathrm{NA}$, where k is a constant, λ is the wavelength of light, and NA is the

[10]"Top-down" assembly is the building of nanostructures and materials by mechanical methods (e.g., molding, machining, and laser-based tools) and by bulk technology. This topic is revisited in Chapter 4.

numerical aperture of the lens in the system. At press time, 248 nm KrF excimer laser sources were being utilized in most instances, as noted (197 nm ArF lasers were also available and/or contemplated as well as EUV). To produce smaller features, the wavelength must be reduced and the numerical aperture of the lens in the optical system must be increased. By using phase-shifting technology (i.e., parts of the mask also rotate the polarization of the light), interference effects can be achieved which produce smaller features than the Rayleigh criteria; the drawback is that phase-shifting technology is complex.

Advances in the engineering of the silicon substrate have proven useful in recent years. For example, recently manufacturers have utilized a technique called *strained silicon*: Stretching the crystal lattice by about 1% increases the mobility of electrons passing through it, enabling the transistors to operate at a faster rate (as an example Intel was using strained silicon in a Prescott processor).

The design of the transistors built out on silicon substrates themselves has also improved over time. One of the steps in the fabrication of transistors on a chip is growing a thin layer of silicon dioxide on the surface of a wafer (this is done by exposing it to oxygen and water vapor). The oxide layers insulate the gates of the discrete transistors. Semiconductor manufacturers have also looked at materials other than silicon dioxide to insulate the gate; for example, high-K materials have been used, but the manufacturing process to place these insulators on top of silicon is more complex, involving atomic-layer deposition (see Glossary).

After the insulator material is put in place, portions of it must be selectively removed using lithographic methods (lithography is used to imprint desired topological features). For a while manufacturers operated under the assumption that it is impossible to use lithography to define features smaller than the wavelength of light employed; however, as of press time 70-nm features were being routinely created using ultraviolet light with a wavelength of 248 nm, as mentioned earlier. Techniques such as but not limited to optical proximity correction and phase-shifting masks are used for these reduced-size features. When the size of the features is smaller than the wavelength of the light, the distortions that arise through optical diffraction can be calculated and compensated for. Namely, one can establish an arrangement for a given mask that, after diffraction takes place, yields the desired pattern on the silicon that the manufacturer was looking for (e.g., if a rectangle is needed, a dog-bone biscuit shape can be used; this is because if the mask had an actual rectangular-opening shape, diffraction would round the corners projected on the silicon, while if the pattern on the mask resembles a dog-bone biscuit, the outcome is a rectangle with sharp corners). This technique now allows transistors with 50-nm features to be produced using light with a wavelength of 193 nm. However, one can extend these diffraction–correction techniques only to a point, which is why manufacturers are seeking to develop the means for higher resolution patterning; the most promising approach employs lithography but with light of much shorter wavelength at the EUV range (e.g., 13 nm wavelength). This reduces the wavelengths, and in turn the size of the features that can be printed, by an order of magnitude; however, manufacturers face a number of challenges as they migrate to EUV lithography [19].

Manufacturers must also deal with the removal of exposed parts of the photoresist and with etching the material that remains uncovered in such a manner that there is no damage to adjacent areas. Also, manufacturers must be able to wash off the photoresist and the residues left over after etching (this is now done with supercritical fluids, e.g., supercritical carbon dioxide). The next step is the addition of the junctions of the transistors to serve as the current "source" and "drain." Junctions are made by infusing the silicon with trace elements that transform it from a semiconductor to a conductor. Finally, atomic-layer deposition is used to lay down an insulating layer of glass on which a pattern of lines is printed and etched; the grooves are then filled with metal to form the wires [19].

2.4.3 Manufacturing Limitations

While semiconductor technology has sustained significant (exponential-growth) advancements in Boolean gate density on a chip during the past 40 years, these advancements may be reaching an asymptotic limit in the next 10 years or so. In CMOS-based systems, as interconnects are reduced in size, the resistance increases faster than the decrease in capacitance (ability to store electrical charge/energy); therefore, the *RC* time constants increase for reduced interconnect size and the speed of the circuit correspondingly is reduced. This is just one of several problems faced with miniaturization.

At a given design rule, the scaling of the physical processes breaks down and new phenomena that are absent in larger structures start to dominate the device behavior. For example, at some point, upon decreasing the size of the MOSFET, the channel length approaches the depletion layer widths of the source and drain. This results in a degradation of the subthreshold characteristics of the device and a failure to achieve current saturation, among other challenges. One can suppress these and other short-channel effects by high doping in the channel, but this comes at the expense of reduced mobility, lower operating speed, and increased risk for avalanches at the drain [78] (second-order effects in CMOS design such as mobility degradation, corrected threshold value, corrected bulk threshold parameter, and effective channel length are due to narrow or short-channel dimensions less than 3 μm). In MOSFETs, at a given drain current, there are two independent degrees of design freedom: (i) inversion level and (ii) channel length (note that channel width, required for layout, is found from the operating drain current and selected inversion level and channel length). There are fairly conspicuous trade-offs in circuit bandwidth, transconductance, output conductance, direct-current (DC) gain, DC matching, linearity, white noise, flicker noise, and layout area resulting from the selection of inversion level and channel length [79]. As the MOSFET devices are scaled into the 100-nm-gate-length arena, it becomes important to reduce the short-channel effect (SCE). Furthermore, if one wants to advance into 100-nm area with lower threshold voltage, it is more important to reduce SCE, which is manifested as the lowering of the threshold voltage at short channel lengths for a given technology; it has been reported that the channel and drain engineering can reduce the SCE [80]. Some of the approaches studied in this context include SSR (supersteep retrograde) channel profile, halo structure, recessed-channel structures, and silicon-on-insulator (SOI) devices. In SOI devices the silicon channel can be fully depleted; moreover, SOI allows for the fabrication of 3D structures with the gate surrounding the channel. The specific gate geometry may invert the channel, under favorable conditions, and this volume inversion is believed to be advantageous for the electrical properties [78].

Besides the 248/197-nm lithographic techniques, other lithographic-oriented approaches at the research stages include EUV sources, X-ray, and electron beams (see Fig. 2.11 for a sense of the throughput of the various methods). These approaches, however, all had technical limitations at press time. For example, early electron beam lithography systems used a single focused beam with a spot of a few nanometers; the pattern is then built up by rastering the beam over the chip area, but for large complex chips this is a very slow process. Vector beam systems have been developed where a much larger standard feature (e.g., a rectangle or a repeated feature) can be exposed at one time, substantially increasing the throughput. Imprint lithography is a recent development where a master mold is imprinting into a polymer resist through pressure and heating. When the mold is removed, the polymer may be transferred to the underlying substrate using etching. This technique can potentially be applied on a wafer scale; the present limit appears to be more related to the ability to fabricate small features on the master mold rather than through the imprint technique itself. Features as small as 10 nm have been demonstrated, although for high

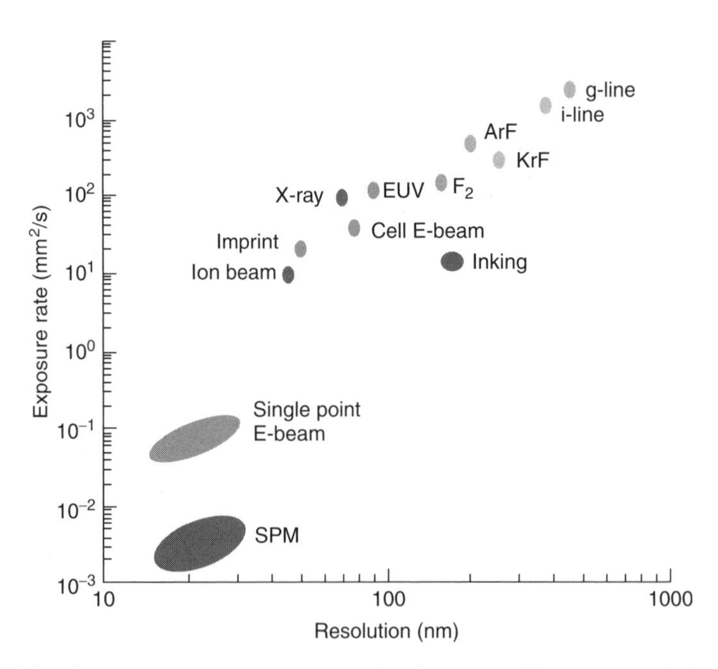

FIGURE 2.11 Throughput versus resolution for various lithographic techniques.

throughput, features are typically 50 nm or above. The major problem with such a technique is that it is difficult to align and calibrate the mold to different layers. As a first-stage lithographic process this is adequate, but for large-scale nanofabrication, many different layers are required; an additional potential problem is that the mechanical contact is prone to dirt and defects and may be damaged over time. A derivative of the imprint technique is inking; with inking, the polymer or ink is applied to the mold so that only raised features have any ink on them; the pattern is then transferred to the substrate through contact. These techniques show promise to scaling to features of tens of nanometers but may ultimately be limited by their inability to be flexible to other layers in real structures.

Lithographic approaches require very clean environments because particulates and dust can mask part of the exposed area. For high yields, clean rooms must have particulate densities that are extremely low and at sizes much smaller than the lithographic minimum feature size. These requirements imply that the cost of lithography is nontrivial, and it increases rapidly as the feature size decreases and may eventually slow down the progress of microelectronics.

Self-assembly approaches attempt to address these limiting issues. In a bottom-up technique the devices self-assemble into circuits; if successful, these techniques will be substantially cheaper than lithographic techniques. This work, however, is at a relatively early stage of research. In the late 1990s and early 2000s, a number of biological, organic, and inorganic systems have been shown to be able to self-assemble. Examples of documented self-assembly include Langmuir–Blogett films, S-bacterial layers, self-assembled monolayers, and antibody–antigen recognition. Much of this self-assembly, however, has been in the area of metallic and semiconductor dots that have some optoelectronic properties for detectors, light-emitting diodes (LEDs), and lasers. Most of these approaches are not appropriate for electronic systems because no interconnects are available. In conclusion, these techniques are still at a very early stage, and substantial research is required to get any of the techniques to a manufacturing level for computing and telecom applications.

We make some passing comments on optics as we wrap up this chapter in the context of the desire to develop integrated optoelectronic devices. Compared with groups III–V *semiconductors* (e.g., GaAs and InAs), silicon is a poor material for *optical components* due to the indirect bandgap. However, since the Si-based technology is more widespread and much cheaper than the groups III–V technology, it would be desirable if effective Si-based optical materials can be developed. An area with some potential promise is Si and SiO_2 with nanocrystalline inclusions. Silicon layers can be fabricated by molecular beam epitaxy (MBE) which are doped with In and As at levels above their solubility as well as SiO_2 layers doped with Ge. With MBE it is possible to grow layers (<1 nm) embedded in undoped Si films. The layers can be grown into a diode structure with optical properties [33].

Nanotechnology (nanophotonics and nanoelectronics) is being studied as a way to break through the design restrictions and bottlenecks. Nanophotonics is discussed in greater detail in Chapter 5; the topic of electronics, specifically nanoelectronics, is discussed in more detail in Chapter 6.

2.5 CONCLUSION

In this chapter we looked at basic topics in the area of physics, for example, electrons, atoms, atomic structures, and electrical properties, among others. Appendices A–D provide additional information; in particular, Appendix D (optional reading) provides a basic introduction to quantum theory. Chapter 3 looks at basic chemistry principles.

CHAPTER 3

Basic Nanotechnology Science—Chemistry

This chapter continues the technical treatment of nanoscience, by exploring activity at the molecular and submolecular level from the perspective of chemistry. One of the most important observation of chemistry is that there are functional groups of elements that share a characteristic set of properties. This empirical observation, and the ancillary implications, enable the practitioner to predict the behavior of chemical reactions. However, while a chemist can write down a certain "feasible" chemical formula for a compound of interest, to actually carry out the chemical reaction and affect the molecular synthesis is, often, more demanding: Certain reaction-inducing processes (e.g., addition of heat, pressure, catalysts, accelerators, etc.) are typically required. Even beyond this basic synthesis, generally speaking, nanotechnology-based products entail the assemblage of large molecules; hence, in a way, this is (and/or may require) an extension of traditional chemical procedures where the chemist is involved with compounds comprised of a handful of atoms in the basic molecule.

In this chapter we provide a short synopsis of the highlights of chemistry that the reader needs to be familiar with, at least at the superficial level. We look at basic chemistry concepts such as bonding, molecular structure, ionic properties, covalency, carbon strings (chains) and rings, and other key concepts. Appendix E extends this discussion: After some basic concepts, we look at empirical methods ("mechanical" molecular model, in particular), followed by a brief application of quantum theory methods to chemistry (the reading of Appendix E is optional.) Chemistry is a very broad field, and only the very basics are included herewith.

3.1 INTRODUCTION AND BACKGROUND

Chemistry is the discipline that studies the properties of materials and the changes that materials undergo under various circumstances. Practical chemistry, in contrast to physics, which leans toward abstract mathematical representation, relies more on symbolism and empirical approaches [68]. Chemists make their observations in the

Nanotechnology Applications to Telecommunications and Networking, By Daniel Minoli

macroscopic world and then seek to understand the fundamental properties of matter at the level of the *microscopic* world (i.e., molecules and atoms) [81]. Researchers are making progress in developing new tools to understand atomic/molecular structure; these tools include but are not limited to X-ray diffraction, linacs (linear accelerators), colliders, electron microscopes, and nuclear reactors (see Appendix F for a description of some of the tools used in physics, chemistry, and nanoscience.) Computer graphics allows chemists to calculate key variables related to molecular structures and then visualizing such structures on a computer screen.

As the reader may be aware by now, the arrangement of electrons around the nucleus is known as the *electronic structure* of the atom or molecule. The reason why certain chemicals react the way they do is a direct consequence of their atomic (electron) structure. In the previous chapter we looked at the arrangement of electrons in energy levels or "shells." Also we have noted that physicists have identified a long list of particles that make up the atomic nucleus. Chemists, however, are primarily concerned with key subatomic particles: electrons, protons, and neutrons (more interested with former and to a lesser extent with the latter) [81].

3.2 BASIC CHEMISTRY CONCEPTS

3.2.1 Physical Aspects

We start this section with a quick pass at some of the basics; then we expand further in the subsections that follow. In effect, the paragraphs that follow summarize the key takeaways from Chapter 2. For reference, Table 3.1 depicts a list of topics typically covered in a one-year college chemistry course [82].

As noted in Chapter 2, there are only about 109 kinds of known atoms, based on the number of protons, neutrons, and electrons. All matter as we know it, is comprised of combinations of these atoms. Molecular structure determines not only the appearance of materials, but also their properties. Electrons and protons have, respectively, negative and positive charges of the same magnitude, 1.6×10^{-19} C. Neutrons are electrically neutral. Protons and neutrons have the same mass, 1.67×10^{-24} g. The mass of an electron is much smaller, 9.11×10^{-28} g (this mass can be neglected in calculation of atomic mass). The atomic mass (A) is equal to mass of protons added to the mass of neutrons. The number of protons in effect determines the chemical identification of the element. The atomic number (Z) is equal to the number of protons. The number of neutrons defines the isotope number of the atom.

Atomic weight is often expressed in terms of the atomic mass unit (amu): 1 amu is defined as $\frac{1}{12}$ of the atomic mass of the most common isotope of the carbon atom; this isotope has 6 protons ($Z = 6$) and 6 neutrons ($N = 6$). Hence, by definition, the atomic mass of the ^{12}C atom is 12 amu. The mass of a proton, which is also the approximate mass of a neutron, is 1.67×10^{-24} g $= 1$ amu. The atomic weight of an element is defined as the weighted average of the atomic masses of the atom's naturally occurring isotopes. It follows that the atomic weight of carbon is 12.011 amu.

TABLE 3.1 Typical Introductory Topics in Chemistry

Basics of chemistry	Chemistry and measurement
	Atoms, molecules, and ions
	Calculations with chemical formulas and equations
	Chemical reactions, and introduction
	The gaseous state
	Thermochemistry
Atomic and molecular structure	Quantum theory of the atom
	Electron configurations and periodicity
	Ionic and covalent bonding
	Molecular geometry and chemical bonding theory
States of matter and solutions	States of matter; liquids and solids
	Solutions
Chemical reactions and equilibrium	Rates of reaction
	Chemical equilibrium
	Acids and bases
	Acid–base equilibria
	Solubility and complex-ion equilibria
	Thermodynamics and equilibrium
	Electrochemistry
Nuclear chemistry and chemistry of the elements	Nuclear chemistry
	Metallurgy and chemistry of the main-group metals
	Chemistry of the nonmetals
	Transition elements and coordination compounds
	Organic chemistry
	Biochemistry

The atomic weight is also often specified in *mass per mole*. A mole is the amount of matter that has a mass in grams equal to the atomic mass in amu of the atoms. For example, a mole of carbon has a mass of 12 g. The number of atoms in a mole is called the Avogadro number, $N_{av} = 6.023 \times 10^{23}$; $N_{av} = 1$ g/1 amu.

The number n of atoms per cm^3 for a material of density d (g/cm^3) and atomic mass M (g/mol) is defined to be $n = (N_{av} \times d)/M$. The mean distance between atoms L can be computed as follows: $L = (1/n)^{1/3}$. For example, for the graphite form of carbon: $d = 2.3$ g/cm^3, $M = 12$ g/mol; it follows that $n = 6 \times 10^{23}$ atoms/mol $\times$ 2.3 g/cm^3/12 g/mol $= 11.5 \times 10^{22}$ atoms/cm^3. By way of comparison, the diamond form of carbon has a higher density, and, hence a higher number of atoms per cm^3; here, $d = 3.5$ g/cm^3, $M = 12$ g/mol, so that $n = 6 \times 10^{23}$ atoms/mol $\times$ 3.5 g/cm^3/12 g/mol $= 17.5 \times 10^{22}$ atoms/cm^3. For a substance with $n = 6 \times 10^{22}$ atoms/cm^3 the mean distance between atoms is $L = 0.25$ nm (2.5 Å), or nanoscale levels. The electrons form a (probabilistically defined) cloud around the nucleus of radius of 0.05–2 nm.

An element is a substance composed of atoms with identical atomic number. A compound is a substance comprised of two or more atoms joined together chemically with a bond (such as a covalent or ionic bond.) An inorganic compound is a compound that does not contain carbon chemically bound to hydrogen. (*Note:* bicarbonates, carbides, carbonates, and carbon oxides are considered inorganic compounds, even though they contain carbon.) An organic compound is a compound that contains carbon chemically bound to hydrogen. These compounds often contain other elements (particularly O, N, halogens, or S). While organic compounds may exist in nature, they can also be synthesized in the laboratory. Table 3.2 provides a short list of *some* key chemistry concepts.

We have seen in Chapter 2 (also Appendix D) that electrons move "around" the positively charged nucleus in various orbits. Only certain *orbitals* or *shells* of electron probability densities are possible. The electron topology of the orbital of the electrons is determined by n, the principal quantum number, and l, the orbital quantum number. The principal quantum number also correlates to the size of the shell (atom): $n = 1$ is the smallest atom and higher numbers represent larger shells/atoms. The second quantum number l identifies subshells within each shell. For example, Na can be written as 1s:2; 2s:2; 2p:6; 3s:1; or, by convention: $1s^2\,2s^2\,2p^6\,3s^1$. Cl can be written as 1s:2; 2s:2; 2p:6; 3s:2; 3p:5; or, by convention: $1s^2\,2s^2\,2p^6\,3s^2\,3p^5$. Electrons that occupy the outermost filled shell—the so-called valence electrons—are responsible for bonding.

The first orbital closest to the nucleus is taken as the first orbital ($n = 1$) and is called the K shell, the next higher orbital is assigned $n = 2$ and is called the L shell, and so on. The electron also has an orbital (angular momentum) number l that takes values $l = 0, 1, \ldots, (n-1)$. As discussed in Appendix D, shapes of the electron orbital depend on the number l. When $l = 0$, the electron orbital is called the s orbital and is spherical. When $l = 1$, the electron orbital is called the p orbital and is dumb-bell shaped. This shape can be oriented in space in three different directions; the shapes are denoted three types p_x, p_y, p_z. When $l = 2$, the electron orbital is called d orbital and has a complex shape in the three-dimensional space. See Table 3.3. Electrons also have a spin assigned with them. There are only two ways an electron spin is oriented: either "up" or "down." Electrons have a tendency to pair with other electrons: a single electron orbiting with spin in state up, will have an affinity for another electron that has spin in state down. In the electronic configuration the first number shows the main shell (K, L, M, etc.), the second letter shows the shape of the orbital (s, p, d, etc.), and the next number shows how many electrons are there in the shell. In pictorial representations of atoms, as conceived by the chemist, spherical pictograms are given. Loosely, these can be interpreted as the s shells of the atom.

In Chapter 2 we introduced the periodic table. In the table, elements in the same column (elemental group) have similar properties. The group number indicates the number of electrons available for bonding. Electronegativity is a measure of how "willing"[1] atoms are to accept electrons. Electronegativity increases from left to right in the periodic table. Subshells with one electron have low electronegativity,

[1]Anthropomorphic terms such as "willing," "cooperative sharing," "gain," "give up," etc., are shown in quotes. These anthropomorphic words have no real scientific definition in the atomic context.

TABLE 3.2 Short List of Some Key Chemistry Concepts

Acid	Compound that releases hydrogen ions (H^+) in solution and/or that can accept a pair of electrons from a base.
Addition compound	Compound that contains two or more simpler compounds that can be packed in a definite ratio into a crystal. Hydrates are a common type of addition compound.
Alkaline earth	Oxide of an alkaline earth metal, which produces an alkaline solution in reaction with water.
Alkane	Series of organic compounds with general formula C_nH_{2n+2}. Alkane names end with –ane; examples include propane (with $n = 3$) and octane (with $n = 8$).
Anhydrous	Compound where all water has been removed.
Base	Compound that reacts with an acid to form a salt and/or produces hydroxide ions in aqueous solution. It can also be seen as a molecule or ion that captures hydrogen ions or that donates an electron pair to form a chemical bond.
Catalyst	Compound that accelerates the rate of a chemical reaction and is not itself consumed in the reaction.
Efflorescent	Substances that lose water of crystallization to the air. The loss of water changes the crystal structure, often producing a powdery crust.
Formula weight	Sum of the atomic weights of the atoms in an empirical formula. Formula weights are usually written in atomic mass units (u).
Halide	Compound or ion containing fluorine, chlorine, bromine, iodine, or astatine.
Hydrate	Addition compound that contains water in weak chemical combination with another compound.
Hydrocarbon	Organic compounds that contain only hydrogen and carbon.
Hygroscopic	Able to absorb moisture from air.
Molecular weight	Average mass of a molecule, calculated by summing the atomic weights of atoms in the molecular formula.
Polyatomic ion	Charged particle that contains more than two covalently or ionically bound atoms.
Polymer	Large molecule (molecular weight ~10,000 or greater) composed of many smaller molecules (monomer) covalently bonded together. A substance consisting of molecules characterized by the repetition (neglecting ends, branch junctions, and other minor irregularities) of one or more types of monomeric units.
Valence	Number of electrons needed to fill out the outermost shell of an atom. Example: a carbon atom has 6 electrons, with an electron shell configuration of $1s^22s^22p^2$. Hence, carbon has a valence of 4, since 4 electrons can be accepted to fill the 2p orbital. Some exceptions exist; therefore, the more general definition of valence is the number of electrons with which a given atom generally bonds or number of bonds an atom forms (e.g., iron may have a valence of 2 or a valence of 3).

TABLE 3.3 Quantum Numbers and Orbitals

n	l	Orbital Name	Orbital Shape	Total Number of Electrons Accommodated on Orbital	Nomenclature/ Electronic Configuration
1	0	s	Spherical	2	$1s^2$
2	0	s	Spherical	2	$2s^2$
	1	p	Dumb-bell	6	$2p^2_x, 2p^2_y, 2p^2_z$
3	0	s	Spherical	2	$3s^2$
	1	p	Dumb-bell	6	$3p^2_x, 3p^2_y, 3p^2_z$
	2	d	Complex	10	Topologically complex
etc.					

while subshells with one missing electron have high electronegativity. The following observations are pertinent:

- Inert gases with group number = 0, (He, Ne, Ar...), have filled subshells; these elements are chemically inactive.
- Alkali metals with group number = 1, (Li, Na, K...) have one electron in outermost occupied s subshell; these elements have a tendency to "give up" electrons, and so they are chemically active.
- Halogens with group number = 7 (F, Br, Cl...) are missing one electron in outermost occupied p shell (they have seven when the shell can accept eight electrons); these elements have a tendency to want to "gain" an electron, and so they are chemically active.

Metals (generally in the middle of the table with group number = 8) are electropositive, and so they can "give up" their few valence electrons to become positively charged ions.

3.2.2 Bonding

There is a definite way atoms of one element interact with other atoms (of the same element or other elements) in a chemical reaction. A chemical bond (also called bond, bonding, or chemical bonding) is a strong attraction between two or more atoms. Bonding is a critical concept: Bonds hold atoms in molecules and crystals together. A bond involves energetic couplings between molecules [31]. There are many types of chemical bonds, but all involve electrons that are either shared or transferred between the bonded atoms [83].

First, let us focus on the unit of measure. The electron volt (eV) is used in descriptions of atomic bonding. It is the energy lost and/or gained by an electron when it is taken through a potential difference of 1 V. Namely, $E = q \times V$. With $q = 1.6 \times 10^{-19}$ C and $V = 1$ V, we get 1 eV $= 1.6 \times 10^{-19}$ J. Next, we define (again) some basic terminology:

- *Ion*: a charged atom.
- *Anion*: a negatively charged atom. For example: Cl has 17 protons and 17 electrons: Cl: $1s^2\,2s^2\,2p^6\,3s^2\,3p^5$. By receiving an electron, 18 electrons are obtained, and a negative ion results: Cl^-. We now have: Cl^-: $1s^2\,2s^2\,2p^6\,3s^2\,3p^6$. Nonmetals typically form anions.
- *Cation*: a positively charged atom. For example: Na has 11 protons and 11 electrons $1s^2\,2s^2\,2p^6\,3s^1$. By "donating" an electron, 10 electrons are left, and a positive ion results: Na^+. We now have: Na^+: $1s^2\,2s^2\,2p^6$. Metals typically form cations.

In summary the ion that looses an electron is called a positive ion or a cation, and the ion that gets an electron is called a negative ion or an anion.

All bonding forces are due to electrostatic charge: Opposite charges attract. Like charges repel. Figure 3.1 shows the attraction and repulsion between atoms. Electronic configuration and valence of the atoms (valence is the number of electrons needed to fill out the outermost shell) play a determining role in the types of bonds. There is both a repulsive force and an attractive force. A point of equilibrium can be reached (until such time as an external force is applied that might change the equilibrium point, particularly if a new quantum state is reached). The repulsion between atoms in close proximity is related to the Pauli exclusion principle: When the electronic clouds surrounding the atoms begin to overlap, the energy of the system increases, giving rise to a repulsive force. The attractive force, more pronounced at

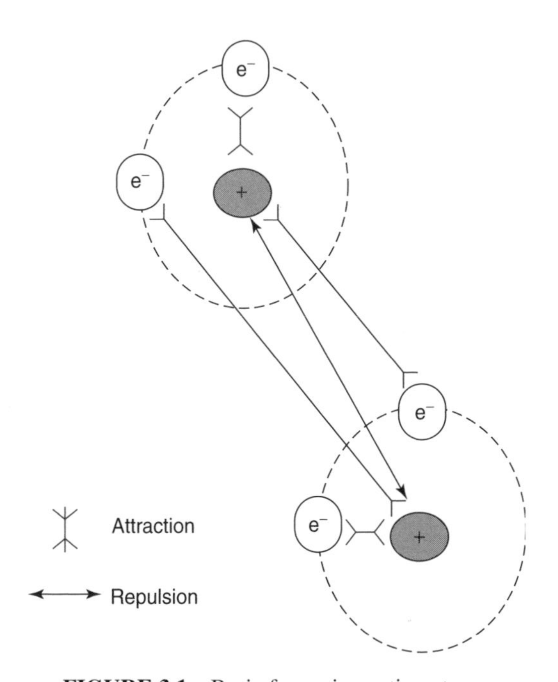

FIGURE 3.1 Basic forces impacting atoms.

(relatively) large distances, depends on the type of bonding. Some of the basic concepts associated with bonding are as follows [83]:

- *Covalent bond* A very strong attraction between two or more atoms that are sharing their electrons. In structural formulas, covalent bonds are represented by a line drawn between the symbols of the bonded atoms.
- *Covalent compound* A compound made of molecules (not ions.) The atoms in the compound are bound together by shared electrons. Also called a molecular compound.
- *Diatomic molecule* A molecule that contains only two atoms. All of the noninert gases occur as diatomic molecules, for example, hydrogen, oxygen, nitrogen, fluorine, and chlorine are H_2, O_2, N_2, F_2, and Cl_2, respectively.
- *Ionic bond* (also used if the forms of "ionically bound" and/or "ionic bonding") An attraction between ions of opposite charge. Potassium bromide consists of potassium ions (K^+) ionically bound to bromide ions (Br^-). Unlike covalent bonds, ionic bond formation involves transfer of electrons, and ionic bonding is not directional.
- *Ionic compound* A compound made of distinguishable cations and anions, held together by electrostatic forces.
- *Polyatomic molecule* An uncharged particle that contains more than two atoms.

Two types of bonding are of particular interest (see Table 3.4):

- *Primary bonding* Electrons are transferred or shared. This is a strong bond; it measures in the range 100–1000 kJ/mol or 1–10 eV/atom.
- *Secondary bonding* Van der Waals effects; physical bonding. No electrons are transferred or shared. This weaker bond results from interaction of atomic or molecular dipoles. It measures about 10 kJ/mol or 0.1 eV/atom.

A discussion of these bonds follows. As we go through the discussion, keep in mind that metals are impacted the most by metallic forces/bonding; ceramics depend on ionic and covalent forces/bonding; polymers[2] depend on covalent and secondary forces; and semiconductors rely on covalent and/or ionic bonds.

Ionic Bonding

Mutual ionization occurs by electron transfer (based on electronegativity values). Ions are attracted by strong coulombic interaction: Oppositely charged atoms attract.

[2]The first polymeric materials to attract recorded scientific interest were silk and cobwebs: In 1665 Robert Hooke suggested that the products of the silkworm and spider could be imitated by drawing a suitable gluelike substance out into a thread. This is basically the process used today in industry to manufacture synthetic polymer fibers such as rayon. It was not until the 1920s that it was understood that plastics consisted of linear molecular chains rather than disorderly conglomerates of small molecules [31].

TABLE 3.4 Generic Bonding Types

Features	Details/Subtypes	Example
	Primary Bonding	
Strong effect: electrons are transferred or shared	*Ionic:* strong Coulomb interaction among negative atoms (have an extra electron each) and positive atoms (lost an electron).	$Na^{+}Cl^{-}$
	Atomic bonds where electrons are *transferred* between the constituent atoms of a compound. Typically, ionic bonds are formed between metallic atoms having extra electron to share and nonmetallic atoms that are electron deficient.	
	Covalent: electrons are shared between the molecules, to saturate the valency.	H_2
	Atomic bond where electrons are shared. If the atoms in proximity share a pair of electrons, then the covalent bond is called a *single covalent bond.* If the atoms in proximity share two or three pairs of electrons, then the covalent bond is called *double* or *triple covalent bond*, respectively.	
	Metallic: the atoms are ionized, loosing some electrons from the valence band. Those electrons form a electron sea, which binds the charged nuclei in place.	
	Situation (occurring in metals only) where the valence electrons of metal atoms are shared by more than one neighboring atom. The metal atoms are held together by a "sea" of electrons floating around.	
	Secondary Bonding	
Weak effect; interaction of atomic/molecular dipoles; no electrons are transferred or shared	Fluctuating-induced dipole.	Inert gases, H_2, Cl_2, etc.
Very weak bonds usually found between gaseous compounds. In solids there is one example of a graphite crystal; each planar hexagonal structure in a graphite crystal is attracted to the other by van der Waals forces or bonds.	Permanent dipole bonds.	Polar molecules: H_2O, HCl . . .
	Polar molecule-induced dipole bonds (a polar molecule induces a dipole in a nearby nonpolar atom/molecule).	

An ionic bond (also known as electrovalent bond) is nondirectional: Ions may be attracted to one another in any direction. Ionic bonds are strong bonds; ionic compounds are crystalline in nature. Due to this crystalline nature, these compounds have high melting points. Because of availability of ions, such compounds are good conductors of heat, but they are nonconductors of electricity when solid, the atoms being locked together in a crystal [84].

Consider the example of NaCl, which was discussed in Chapter 2. Na has 11 electrons, 1 more than needed for a full outer shell, while Cl has 17 electrons, 1 less than needed for a full outer shell. During a chemical reaction, a bond is formed by considering the ions that are formed: by donating an electron, 10 electrons are left with Na, and a positive ion results: Na^+. One now has Na^+: $1s^2\ 2s^2\ 2p^6$. By receiving an electron, 18 electrons are obtained by Cl, and a negative ion results. One can say that "Na shrinks and Cl expands." See Figure 3.2, top. The electron transfer reduces the energy of the system of atoms, that is, electron transfer is energetically sustainable. Other examples of ionic compounds include magnesium oxide (MgO), potassium chloride (KCl), and iron oxide (FeO).

Table 3.5 shows the electron structure, along with an "electronic dot notation," of some atoms. For example, the Cl shown in boxed bracket represents the dot structure for a chlorine ion. Using the electron-dot structure notation the reaction $Na^+ + Cl^- \rightarrow NaCl$ can be written as:

$$Na^{\cdot} + \cdot\ddot{\underset{\cdot\cdot}{Cl}}\cdot \longrightarrow Na^+\left[\cdot\ddot{\underset{\cdot\cdot}{Cl}}\cdot\right]^-$$

Covalent Bonding

Covalent bonding is based on *cooperative sharing* of valence electrons. It can be described by orbital overlap. Covalent bonds are conspicuously directional: The bond is manifest in the direction of the greatest orbital overlap. The potential energy of a system of covalently interacting atoms depends not only on the type of atoms and the distances between atoms but also on the physical angles between bonds. The covalent bond model asserts that an atom can covalently bond with at most $8 - n'$, where n' is the number of valence electrons. Compounds that are formed due to covalent bonding of atoms are called *covalent compounds*. See Figure 3.2, middle. Covalent bonds are generally formed between nonmetals. Nitrogen gas (N_2), hydrogen gas (H_2), oxygen gas (O_2), and hydrochloric acid (HCl) represent some of the examples of covalent bonded compounds. Most of the carbon bonds in organic materials (cells, sugar, etc.) are examples of covalent bonds. Since covalent bonds are directional in space, the molecules that arise have definite shapes.

The physical length of covalent bonds (distance between the atoms forming the respective bonds) is shorter than the length of an ionic bond. It follows that covalent bonds are harder to break compared with ionic bonds. Due to the absence of free ions, covalent compounds are not good conductors of heat and electricity. The characteristics

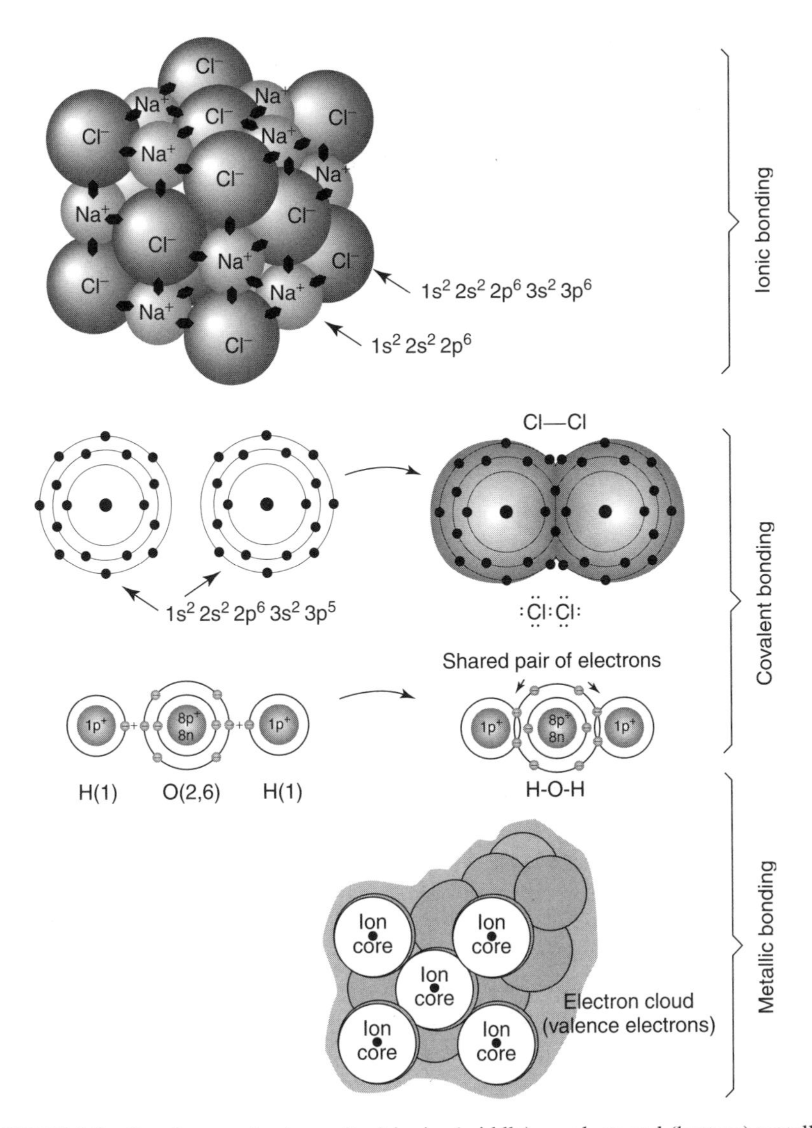

FIGURE 3.2 Bonding mechanisms: (top) ionic; (middle) covalent; and (bottom) metallic.

of covalent bonding are perhaps (better) appreciated when one considers the following: if one were to drop some ionically bonded compound (salt) into water one just ends up with salty water: The positive and negative charges on the sodium and chloride atoms are surrounded by water molecules that break the ionic bonding. On the other hand, if one were to drop a diamond into water, it remains a diamond because it has a much more resilient bond: a covalent bonding between its carbon atoms.

We now look at a few specific examples of covalent bonding.

TABLE 3.5 Electron Structure And Electronic Dot Notation For A Few Elements

Element (Symbol)	Atomic Number (Z)	Electronic Configuration K L M N	Number of Outermost Electrons	Valence	Ion	Electron Dot Notation (Atom)
Sodium (Na)	11	2 8 1 $(1s^2)(2s^22p^6)(3s^1)$	1	1+	Na^+ (cation)	Na·
Magnesium (Mg)	12	2 8 2 $(1s^2)(2s^22p^6)(3s^2)$	2	2+	Mg^{2+} (cation)	Mg:
Aluminum (Al)	13	2 8 3 $(1s^2)(2s^22p^6)(3s^23p^1)$	3	3+	Al^{3+} (cation)	·Al·
Fluorine (F)	9	2 7 $(1s^2)(2s^22p^5)$	7	1−	F^- (anion)	:F:
Chlorine (Cl)	17	2 8 7 $(1s^2)(2s^22p^6)(3s^23p^5)$	7 7	1− 1−	Cl^- (anion)	:Cl:
Oxygen (O)	8	2 6 $(1s^2)(2s^22p^4)$	6	2−	O^{2-} (anion)	:O:

- Consider the Cl_2 molecule. We have Cl: $1s^2\ 2s^2\ 2p^6\ 3s^2\ 3p^5$; $n' = 7$, $8 - n' = 1$, therefore, it can form only one covalent bond. See Figure 3.2, middle.
- As another example, consider the ethene molecule (C_2H_4) that we discuss in Appendix D. Since C has the electron distribution $1s^2\ 2s^2\ 2p^2$, $n' = 4$, hence, the number of covalent bonds is $8 - n' = 4$.
- As an example, the ethylene molecule has the form

```
H   H
|   |
C = C
|   |
H   H
```

while the polyethylene molecule has the form

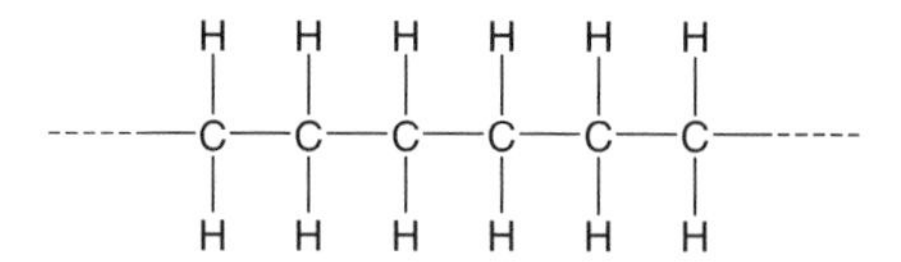

The diamond form of carbon has four covalent bonds with four other atoms.

- As a final example, consider an oxygen (O) atom. The electronic configuration is: two electrons in the K shell and six electrons in the L shell; that is, $(1s^2)(2s^22p^4)$. The oxygen atom would rather "borrow" two electrons from some other atom to complete its last shell than give up its six electrons. Hydrogen (H) atom has only one electron and needs to "borrow" only one additional electron to complete its first shell. Two atoms of H and one atom of O "fulfill each other's needs" by sharing electrons in their outermost orbitals and form H_2O or a water molecule. In an H_2O molecule, the electrons are not totally "given up" but are shared by each of the neighboring atoms.

Metallic Bonding

In these arrangements, the valence electrons are detached from atoms and spread in an "electron field (sea)" that "holds" the ions together. A metallic bond is nondirectional: Bonds form in any direction, leading to dense packing of atoms. See Figure 3.2, bottom. Metals consist of a lattice of positive ions through which a cloud of electrons moves. The positive ions tend to repel one another, but are held together by the negatively charged electron cloud. These mobile electrons make metals good conductors of heat and electricity. Because of the fact that the positive ions in a metal are not held by rigid bonds, they are capable of sliding past one another if the metal is deformed, making metals malleable and ductile.

Secondary Bonding

Secondary bonding is based on van der Waals forces (named after the Dutch physicist Johannes van der Waals), hence, it is a physical phenomenon. Bonding results from interaction of atomic or molecular dipoles, rather than on chemical bonding that involves the transfer of electrons. Molecules are able to attract each other at moderate distances and repel each other at close range. The attractive forces are collectively called *van der Waals forces*.

A van der Waals force is a weak physical force that holds together two molecules or two different parts of the same molecule. These forces arise from electric dipole interactions (See Table 3.6 partially based on [85]). They can lead to the formation of stable but weakly bound molecules or clusters. Van der Waals forces are much weaker than the chemical bonds discussed earlier: Random thermal motion (even at room temperature) can usually overcome or disrupt them. Intermolecular forces are responsible for many properties of molecular compounds, including crystal structures (e.g., the shapes of snowflakes), melting points, boiling points, heats of fusion and vaporization, surface tension, and densities. Intermolecular forces impact large molecules like enzymes, proteins, and DNA, by molding these molecules into the shapes required for biological activity. For example, water would not condense from vapor into solid or liquid forms if its molecules did not attract each other [85].

Considering molecules that have permanent dipoles and molecules that can have dipoles induced by the electric fields of other molecules, there are three possible mechanisms recognized in the formation of the van der Waals bonds [31, 86]: (1) the orientation effect in which molecules rearrange themselves in their mutual electrical fields, the rearrangements involving reorientations of whole molecules; (2) the static induction effect in which molecules that are static monopoles (ions) or dipoles may induce a static rearrangement of the electron distribution of other molecules; and (3) the dynamic induction effect, or "dispersion" effect in which any molecule, polar or nonpolar, may induce in other molecules transient electron distribution rearrangements that are time variant.

TABLE 3.6 Terms Related to Secondary Bonding

van Der Waals force	Force acting between nonbonded atoms or molecules. Includes the following forces: dipole–dipole, dipole–induced dipole, and London forces.
Dipole–dipole interaction	A dipole–dipole force based on electrostatic attraction between oppositely charged poles of two or more dipoles.
Electric dipole moment	Measure of the degree of polarity of a polar molecule. Dipole moment is a vector with magnitude equal to charge separation times the distance between the centers of positive and negative charges. Chemists point the vector from the positive to the negative pole; physicists point it the opposite way. Dipole moments are often expressed in units called debyes.
Hydrogen bond	Relatively strong dipole–dipole force between molecules X–H···Y, where X and Y are small electronegative atoms (usually F, N, or O) and "..." denotes the hydrogen bond. Hydrogen bonds are responsible for the unique properties of water, and they loosely pin biological polymers like proteins and DNA into their characteristic shapes.
Intermolecular force	Attraction or repulsion between molecules. Intermolecular forces are much weaker than chemical bonds. Hydrogen bonds, dipole–dipole interactions, and London forces are examples of intermolecular forces.
London force	Intermolecular attractive force that arises from a cooperative oscillation of electron clouds on a collection of molecules at close range. These forces (also known as transitory forces) arise when electron clouds oscillate in step on two molecules at close range. Bond vibrations in molecules may produce the oscillations or they may be triggered by random, instantaneous pile-ups of electrons in atoms.

Permanent dipole moments exist in some types of molecules (e.g., H_2O, HCl), called *polar molecules*, that arise due to the asymmetrical arrangement of positively and negatively charged regions. Bonds between adjacent polar molecules—known as *permanent dipole bonds*—are strongest among secondary bonds. Polar molecules can induce dipoles in adjacent nonpolar molecules and a bond is formed due to the attraction between the permanent and induced dipoles.

Even in electrically symmetric molecules/atoms an electric dipole can be created by fluctuations of electron density distribution. The fluctuating electric field in one atom is picked up by the electrons of an adjacent atom, and it can induce a dipole momentum in this atom. This bond due to fluctuating induced dipoles is the weakest of these kinds of bonds (inert gases, H_2, Cl_2).

As examples, consider the hydrogen bond in water, which we have just alluded to. This so-called hydrogen bond is a secondary bond formed between two permanent dipoles in adjacent water molecules: the H side of the molecule is positively charged and can bond to the negative side of another H_2O molecule (the O side of the H_2O dipole).

3.2.3 Basic Formulation/Machinery of Chemical Reactions

Here we provide at a very high level some principles that drive (explain) chemical reactions. Bonding, valance, and catalysis are fundamental factors.

The terms and concepts that follow define some of the key (empirical) mechanisms and concepts involved in chemical process and reactions [83]:

- *Empirical formula* Formula that shows which elements are present in a compound, with their mole ratios indicated as subscripts. For example, the empirical formula of glucose is CH_2O, which means that for every mole of carbon in the compound, there are 2 moles of hydrogen and one mole of oxygen.
- *Law of conservation of mass* This law states that there is no change in total mass during a chemical change. The demonstration of conservation of mass by Antoine Lavoisier in the late-18th century was a milestone in the development of modern chemistry.
- *Law of definite proportions* This law states that when two pure substances react to form a compound, they do so in a definite proportion by mass. For example, when water is formed from the reaction between hydrogen and oxygen, the "definite proportion" is 1 g of H for every 8 g of O.
- *Law of multiple proportions* This law states that when one element can combine with another to form more than one compound, the mass ratios of the elements in the compounds are simple whole-number ratios of each other. For example, in CO and in CO_2, the oxygen-to-carbon ratios are 16 : 12 and 32 : 12, respectively. Note that the second ratio is exactly twice the first, because there are exactly twice as many oxygen atoms in CO_2 per carbon as there are in CO.
- *Molecular formula* Notation that indicates the type and number of atoms in a molecule. The molecular formula of glucose is $C_6H_{12}O_6$, which indicates that a molecule of glucose contains 6 atoms of carbon, 12 atoms of hydrogen, and 6 atoms of oxygen.
- *Molecular model* (also called stick model, ball-and-stick model, space-filling model) Graphical representation of a molecule. The model can be purely computational or it can be an actual physical object. Stick models show bonds, ball-and-stick models show bonds and atoms, and space-filling models show relative atomic sizes.
- *Stoichiometry* Branch of chemistry that quantitatively relates amounts of elements and compounds involved in chemical reactions, based on the law of conservation of mass and the law of definite proportions. (Also it can refer to the ratios of atoms in a compound or to the ratios of moles of compounds in a reaction.)
- *Structural formula* Diagram that shows how the atoms in a molecule are bonded together. Atoms are represented by their element symbols and covalent bonds are represented by lines. The symbol for carbon is often not drawn. Most structural formulas do not show the actual shape of the molecule (they are like floor plans that show the layout but not the 3D shape of a house).

One important concept that relates to facilitating chemical reactions is catalysis (and the agent is the catalyst). The term *catalysis* was invented in 1835 by the chemist J. J. Berzelius, and the term was later amended by W. Ostwald, who proposed the more modern definition: "A catalyst is a substance that accelerates the rate of a chemical reaction without being part of its final products." Essentially, the catalyst acts by forming intermediate compounds with the molecules involved in the reaction, providing an alternate and more rapid path to the final products. Hence, catalysis is a process whereby the rate of a particular chemical reaction is hastened, sometimes enormously so, by the presence of small quantities of a substance that does not itself seem to take part in the reaction. For example, powdered platinum will catalyze the addition of hydrogen to oxygen and to a variety of organic compounds [87]. Catalysis is critical: In biological systems enzymes are essential catalysts for various biosynthetic pathways; in the chemical and petroleum industries key processes are based on catalysis; in environmental chemistry catalysts are essential to breaking down pollutants such as automobile and industrial exhausts. If the catalyst and the reacting species are in the same phase (e.g., in a liquid), then the process is known as *homogeneous catalysis*; more relevant in industrial processes is *heterogeneous catalysis*, where the catalyst is a solid and the reacting molecules interact with the surface of the solid from the gaseous or liquid phases [87]. Catalysis would deserve several chapters (books) to fully describe. As we saw in passing in Chapter 1, it is one of the top-five areas of research and publication in the context of nanotechnology.

3.2.4 Chemistry of Carbon

Chemistry of carbon compounds is known as organic chemistry. Organic chemistry encompasses the study of all carbon–hydrogen compounds (except as noted earlier). Carbon and its chemistry are very important: Living cells, food, petrochemicals, cooking gas, and so on are all comprised of carbon atoms, along with other atoms. The carbon atom is unique: It can form long-chain molecules (this ability called *catenation.*) Carbon compounds exist by forming covalent bonds. These compounds have low melting points, are generally insoluble in water, and are inflammable (inorganic compounds, by contrast, usually dissolve in water and have high melting points). Some nanotechnology materials are carbon based.

As we can see in Appendix D, carbon has six protons, six neutrons, and six electrons. The electronic configuration is two electrons in the K shell, and four electrons in the L shell. In theory, while forming compounds, carbon should either give up four electrons or borrow four electrons. However, carbon does not form ionic bonds; the mechanism, instead, is to share its four electrons with other atoms and form covalent bonds instead. When a carbon atom forms a compound, it always forms covalent bonds. Carbon covalent bonds are the strongest in nature. The six electrons of carbon are distributed as shown in the top row of Figure 3.3.

Carbon exhibits tetravalency: Since the 2s and the 2p orbitals are very close in energy, one electron from the 2s orbital jumps to the $2p_z$ orbital. The one 2s and three 2p orbitals mix together and give rise to four new altogether different types of orbitals (as seen in the lower part of Fig. 3.3). This arrangement, seen only in the carbon atom,

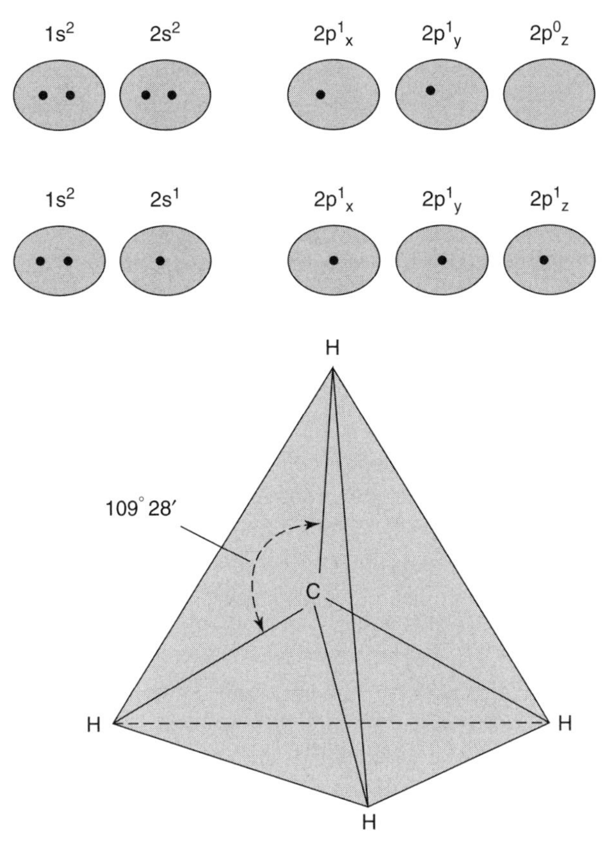

FIGURE 3.3 Carbon orbits.

is called *hybridization*. The four orbitals are at an angle of 109°28′, giving rise to what is labeled as tetrahedral-type hybridization and shown as sp^3. There also are sp^2 and sp types of hybridization. The sp^3 hybridization is depicted as in the top portion of Figure 3.4. The methane CH_4 molecule is an example of sp^3 hybridization. The methane molecule has a tetrahedral shape. The C atom is at the center of the tetragon and the four H atoms in the four corners of the tetragon. Each carbon bond in methane makes an angle of 109°28′ with the other bonds, as seen in the bottom portion of Figure 3.3. In a sp^2 hybridization, two of the four carbon bonds are parallel; the sp^2 hybridization is depicted as shown in the middle portion of Figure 3.4. The sp^2 hybridization leads to carbon double bonds. The angle between the three directions of the bonds is 120°. An example of sp^2 hybridization is the molecule of ethene ($CH_2{=}CH_2$). The molecule of ethyne is an example of sp hybridization; here three of the bonds lie parallel to each other ($CH{\equiv}CH$) and the angle between the direction of the bonds is 180°. The carbon chains we mentioned above are formed because carbon atoms form tetravalent bonds with other carbon atoms. This structure is repeatable endlessly without disturbing the stability of the bonds and the compounds formed.

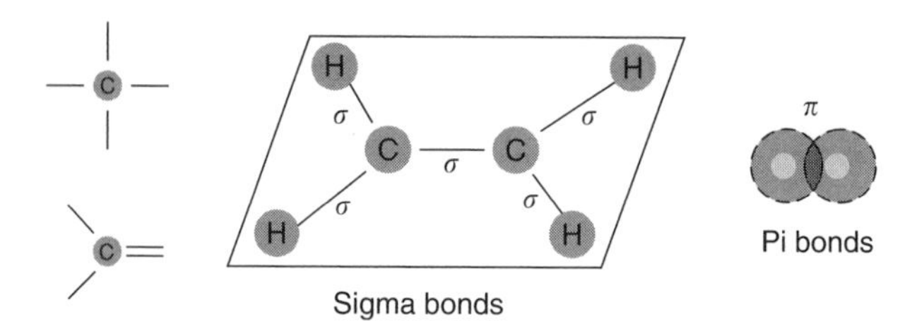

FIGURE 3.4 Representation of covalent bonds and depiction of sigma (σ) and pi (π) bonds.

There are two types of covalent bonds for carbon: *sigma* and *pi bonds* (see bottom two diagrams in Fig. 3.4):

- Sigma (σ) bond: covalent bonds are linear or aligned along the plane containing the atoms. Sigma bonds are strong and the electron sharing is maximum. Example: methane (CH_4) (it has four σ bonds).
- pi (π) bond: electron orbitals overlap laterally. The resulting overlap is not maximum, hence, it follows that the bonds are relatively weak. Example: molecule ethene (C_2H_4) (it has four σ bonds and two π bonds).

Carbon can form chains with branches and subbranches (open-chain compounds are known as *aliphatic compounds*), as well as rings with other rings attached to them (closed-chain compounds are known as *cyclic compounds*; when the cycle is six, the compound is known as an *aromatic compound*[3]—as the name suggests these give off an aroma.) The rings are formed with six carbon atoms only. No other element in the periodic table bonds to itself in an extended network with the strength of the carbon–carbon bond [88]. Figure 3.5 depicts some basic patterns of chains and rings; Figure 3.6 depicts classical carbon forms for pure carbon, while Figure 3.7 depicts some of the forms discovered in the late 1980s and early 1990s (Figs. 3.6 and 3.7 depict allotropes of carbon; see main Glossary). Aliphatic compounds (compounds forming carbon–carbon chains) are found in fats; the alkanes (sp^3 hybridization), alkenes (sp^2 hybridization), and alkynes (sp hybridization) fall under this category (one finds a large number of alkanes, alkenes, and alkynes in nature). As noted, compounds form closed rings along with the branches of the rings are called cyclic or aromatic compounds. The number of variation of ringed compounds is large (perfumes and fruits/flowers smells are due to such compounds).

Next, we focus briefly on electrical conductivity. Carbon in a planar graphene sheet is bonded in such a way that one electron per carbon atom is freed up to move freely, rather than stay near its "base" atom. This is the situation in metals, where some electrons are not bound to their donor atom but can easily be pulled in different directions under the influence of an electric field. The quantum mechanics of graphene

[3]To be considered aromatic the compound also has to contain alternating single and double bonds.

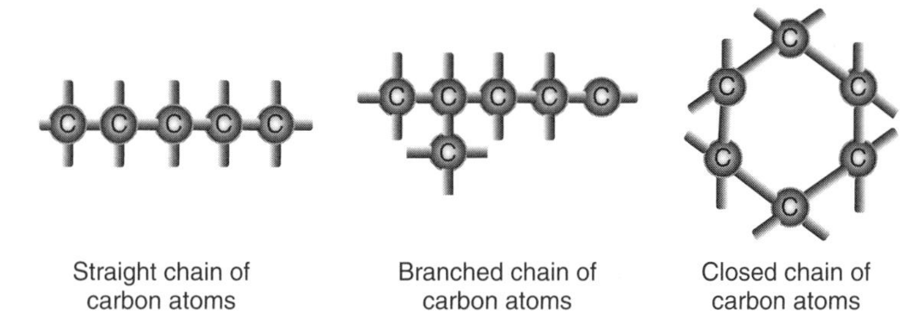

FIGURE 3.5 Carbon chains/rings.

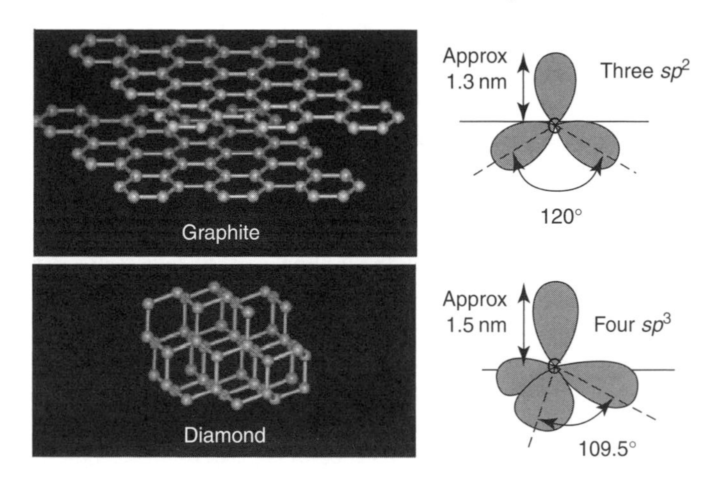

FIGURE 3.6 Carbon forms (classical).

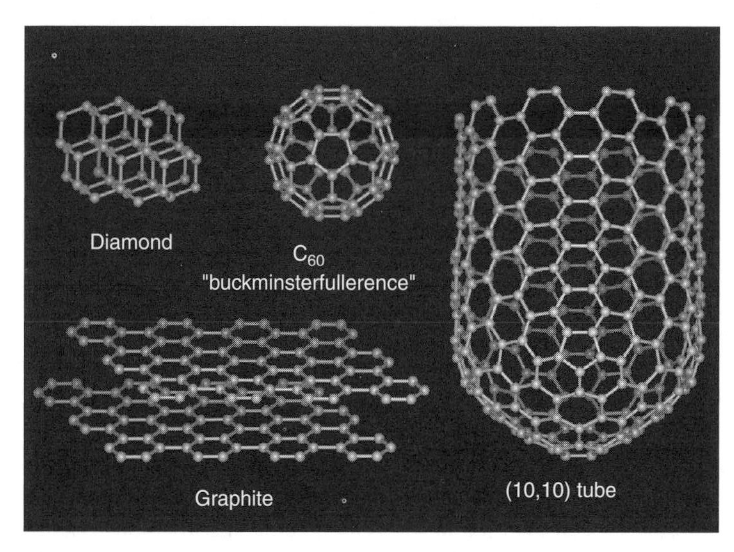

FIGURE 3.7 Carbon forms (classical and more recent discoveries). (Courtesy: Carbon Nanotechnologies Incorporated).

result in a semimetal: the in-plane conductivity is only moderate and similar to that of a poorly conducting metal, such as lead [89]. On the other hand, the carbon structures (nanotubes) discussed in the next chapter have excellent electrical conductivity characteristics.

3.2.5 Graphical View of the Atomic Structure of Materials

In this section we briefly look at some key structures that result from the bonding mechanisms discussed earlier. The characterization of the materials is performed using modern analytical methods. These include X-ray diffraction (XRD), high-resolution scanning electron microscopy (HRSEM), high-resolution transmission electron microscopy (HRTEM), electron microprobe analysis (EMPA), secondary ion mass spectroscopy (SIMS), atomic force and scanning tunneling microscopy (AFM/STM), Mössbauer spectroscopy (MS), depth-sensitive conversion electron Mössbauer spectroscopy (DCEMS), Auger electron spectroscopy (AES), X-ray photoelectron spectroscopy (XPS), ion scattering spectroscopy (ISS), nitrogen adsorption, small-angle neutron scattering (SANS), and extended X-ray adsorption fine structure (EXAFS) [16] (also see Appendix F). In addition, several accelerator facilities such as Rutherford backscattering spectroscopy (RBS) or nuclear reaction analysis (NRA), can be used for ion beam analysis of thin-film structures and nanocrystalline materials after ion implantation and ion beam modification.

We start with shapes that arise from bonding between electrically charged ions (ionic bonding); then we look at covalent bonding. Superconducting and magnetic materials are also briefly discussed. This section is partially based on [90].

Packing of Atoms

Three basic atomic packing structures are as follows (see Fig. 3.8):

- *Cubic close-packed (ccp) atomic structure* This structure is comprised of atoms that reside on the corners of a cube, with additional atoms residing at the centers of each cube face [also called *face-centered cubic (fcc)*]. The symmetry is described by nomenclature as Fm-3m where F means face-centered, m signifies a mirror-plane (there are two) and -3 indicates that there is a three-fold symmetry axis (along the body diagonal) as well as inversion symmetry. Many metals have this fcc structure, for example, gold.
- *Hexagonal close-packing (hcp) atomic structure* This structure is comprised of layers with stacking (e.g., the structure of sodium at low temperatures). The formation dynamic between ccp and hcp is determined by longer-range forces between the atoms.
- *Body-centered cubic (bcc) atomic structure* This structure is comprised of a unit cube with atoms at the corners and center of the cube. The bcc structure is less closely packed than fcc or hcp. Often bcc is the high-temperature form of metals (these being close-packed at lower temperatures). For example, the structure of iron (Fe) can be either ccp or bcc depending on the temperature, while

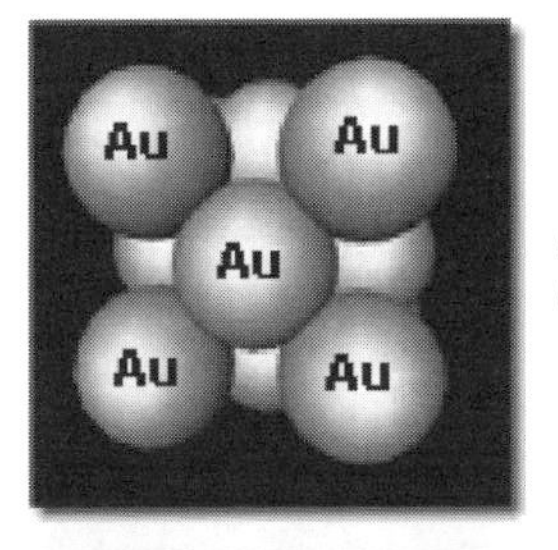

Gold
Cubic close-packed (*ccp*)

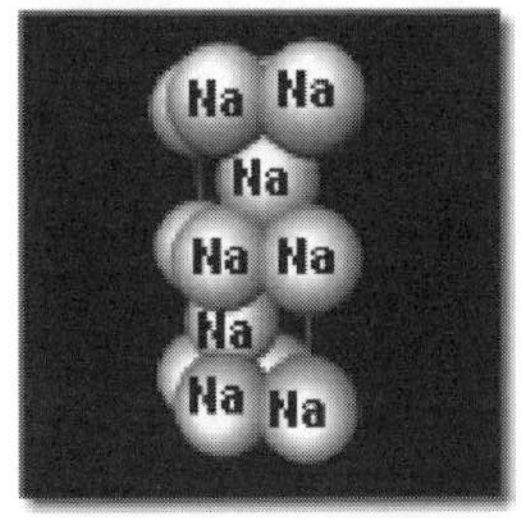

Sodium
Hexagonal close-packing (*hcp*)

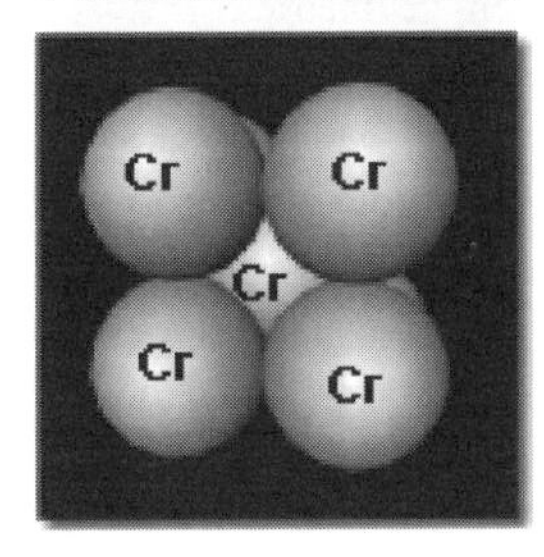

Chromium
Body-centered cubic (*bcc*)

FIGURE 3.8 Examples of close packing.

chromium is consistently bcc. Usually bcc-based metals are harder and less malleable than close-packed metals such as gold. When the metal is deformed, the planes of atoms must slip over each other, and this is more difficult in the bcc structure.

Different Sized Atoms Packing Together

Typically when two or more different atoms combine to form molecules, the packing is determined by the larger atoms, as illustrated in the case of lithium chloride (LiCl) (see Fig. 3.9). Lithium is a relatively small atom (in the sense that it only has a 2s shell) and the larger chlorine atoms (in the sense that it has a 3p shell), just pack together with the ccp structure, leaving the small lithium atoms to fit into the octahedral "holes." Each space (hole) occupied by a lithium atom is surrounded by six chlorine atoms at the vertices of an octahedron. Crystallographers represent the "big" atoms as small spheres to emphasize the "coordination polyhedrae" to help one understand the coordination of atoms (their nearest neighbors). Common kitchen salt (NaCl) is also shown in Figure 3.9. The structure of sodium chloride (NaCl) can be

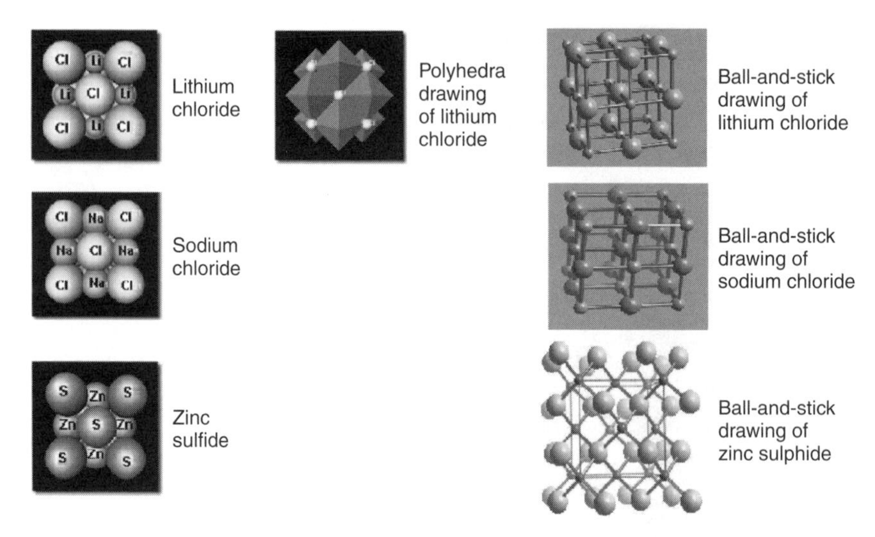

FIGURE 3.9 Two-atom geometries.

regarded as a cubic packing of almost equal spheres. Notice the actual radii at the physical level are:

	1s	2s	2p	3s	3p	Actual Radius
Li	2	1				1.55 Å
Na	2	2	6	1		1.9 Å
Cl	2	2	6	2	5	0.97 Å

In the ccp structure, in addition to the octahedral holes described above, there are also tetrahedral holes. For example, Figure 3.9 depicts the structure of zinc sulfide (ZnS). The Zn atom prefers to occupy these tetrahedral holes, where it is surrounded by only four S atoms; while one can depict the coordination polyhedrae around zinc, in this case it is better to emphasize the actual bonds between the Zn and S atoms, using a so-called ball-and-stick model, which is what is shown in Figure 3.9 [90].

Other geometries are possible, as shown in Figure 3.10 (this figure is not exhaustive in terms of possible geometries). For example, zinc oxide (ZnO) has a hcp packing of oxygen anions (the coordination of Zn is still tetrahedral.) When the cations are large, such as those of calcium, a structure like that of CaF_2 fluorite is the topological outcome. The TiO_2 rutile structure is typical of quadrivalent metals or divalent metal fluorides; here, the Ti cations are in octahedral holes between the oxygen anions, which is easily seen in the coordination octahedrae.

Ferroelectrics and Antiferroelectrics

Perovskites include materials that fall in the ferroelectric and superconductor categories. These substances have generic formulas ABX_3 (e.g., $BaTiO_3$). Figure 3.11

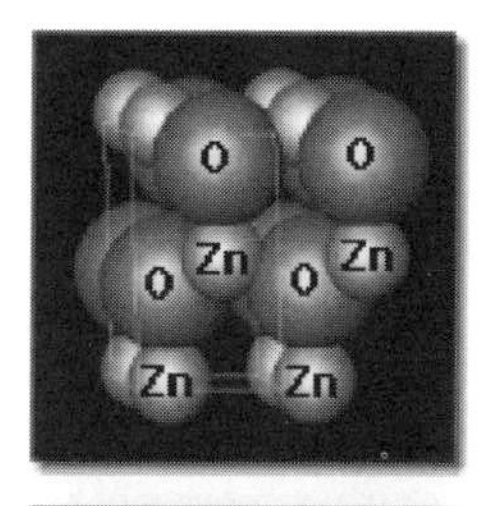

Zinc oxide (ZnO)

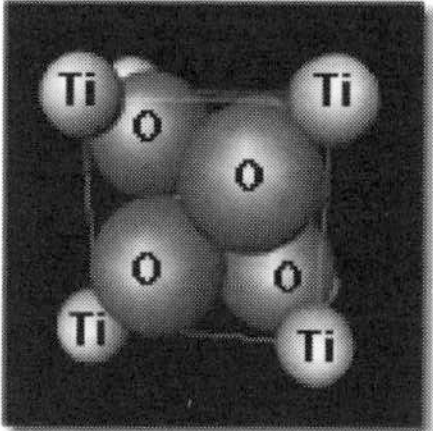

Rutile (TiO_2)

Polyhedra drawing
Rutile (TiO_2)

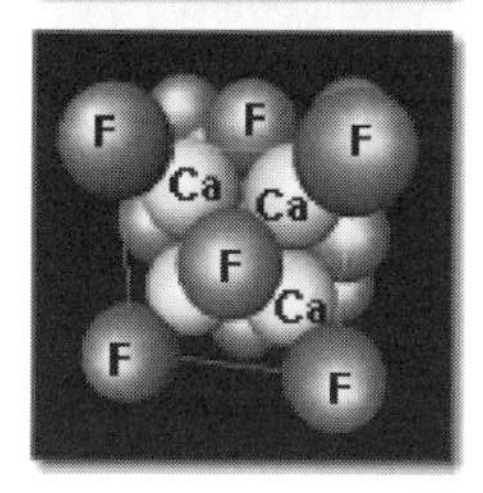

Fluorite (CaF_2)

FIGURE 3.10 More two-atom geometries.

Perovskites
($BaTiO_3$)

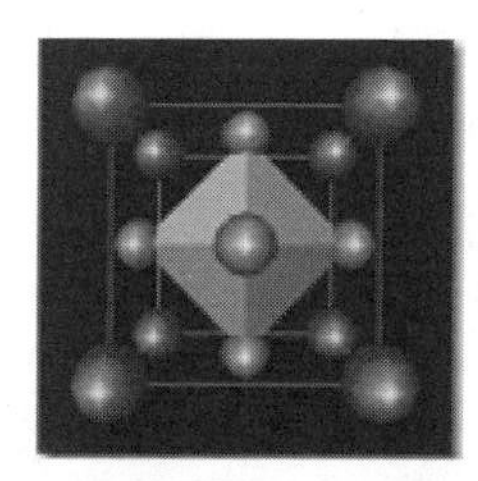

Polyhedra drawing
Perovskite
($BaTiO_3$)

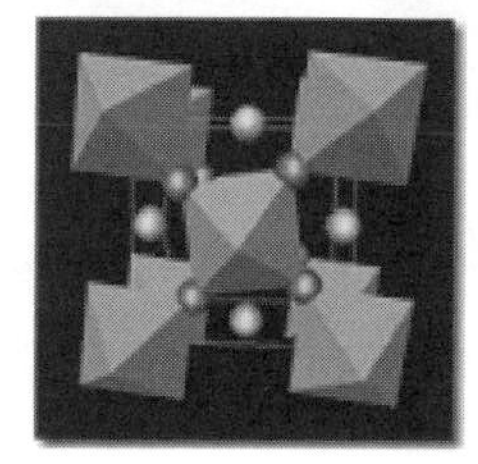

$NaNbO_3$

FIGURE 3.11 Ferroelectric types.

depicts the structure (along with the BX_3 octahedra enclosing the B cation). The large A cations and the X anions, often oxygen, are ccp structures, with the smaller B cations occupying the octahedral holes between the X anions. The stability of the structure depends on the relative ionic radii: if the cations are too small for close packing with the oxygen atoms, they can be displaced slightly. Since these ions carry electrical charges, such displacements can result in a net electric dipole moment (opposite charges separated by a small distance); the material is said to be *ferroelectric*. An alternative type of structural transition, called *antiferroelectric*, is also common in perovskites (also see Fig. 3.11 for this structure.) This occurs when/if the A cation is too large for close packing; the net result is that the X cations are displaced, however, the BX_6 octahedrae are relatively rigid units connected at their apexes, and they twist together as seen in the case of $NaNbO_3$ [90].

Chemical/Covalent Bonding (and Carbon)

Building on the Section 3.2.2, we now look at examples of covalent bonding between atoms (the reader may want to refer to Fig. 3.12). (Strong) covalent bonds are usually represented by drawing them as sticks between the atoms. The covalent bonding in diamond consists of electrons that are "intimately" shared between the carbon atoms. Other important materials, such as silicon and germanium used for computer chips

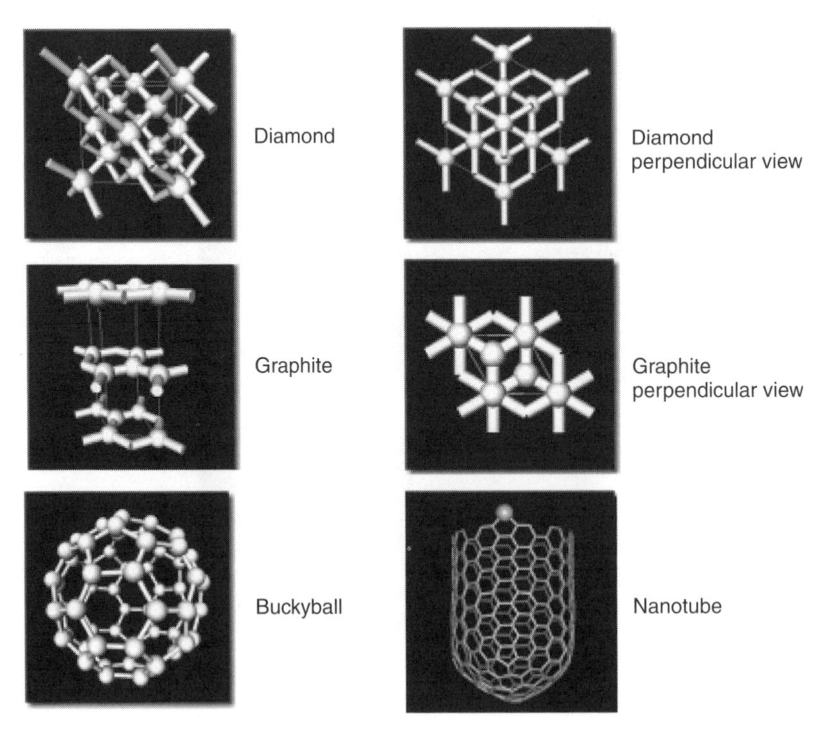

FIGURE 3.12 Covalent bonding in carbon.

also have the diamond structure. (Unfortunately?) there is another alternative arrangement of carbon: plain old graphite. The carbon atoms in graphite are strongly joined by covalent bonds, but only within a plane, unlike the 3D network of bonds in diamond; these planes of carbon atoms simply stack together one on top of the other, with relatively weak forces between them.

In the past 25 years a large number of new carbon structures with interesting properties have been discovered. The buckyball, already introduced in Chapter 1, consists of 60 carbon atoms bonded together to form a hollow sphere. Larger spheres and ellipsoids can also be synthesized; hollow nanotubes of carbon expressed as graphite layers rolled up to form microscopic pipes can be manufactured. These new *fullerenes* materials have interesting physical and chemical properties that are part of nanotechnology science. These are discussed in Chapter 4.

Catalysts and Sieves

Crystal structures can form networks of atoms. These materials, called zeolites (see Fig. 3.13), can be used as microscopic filters and also to break up molecules (molecular sieves—e.g., purify water or to separate out molecules of different sizes), or to join them together (catalysts). Zeolites swell and lose water from their porous structure when heated. Over 600 zeolites exist, and new synthetic zeolites are developed and patented on a routine basis. They have many industrial applications due to their unique architecture. As an example, ZSM-5 (Al_2O_3-SiO_2 artificial zeolite) is a catalyst for converting methanol into gasoline; the catalytic activity of ZSM-5 is due

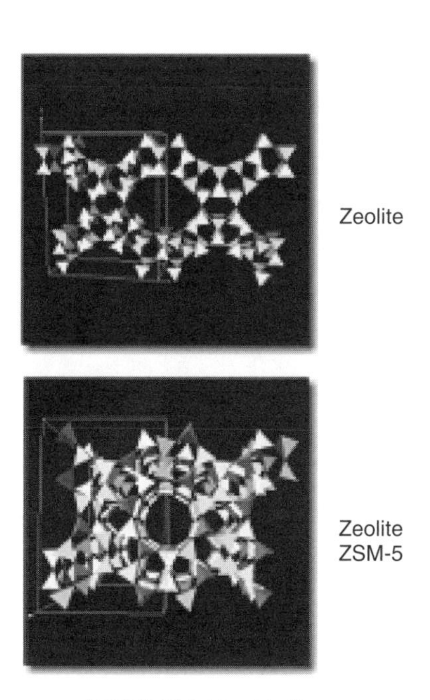

FIGURE 3.13 Zeolites.

both to its acidity and to size/shape of the channels that hold the intercalated molecules. Linde Zeolite-A (LZA) is used in washing powders to remove calcium and magnesium ions from hard water [90].

Superconductors

Metals can conduct electrical currents because, as hinted at in Chapter 2 and above, the electrons are relatively free. Oxides (e.g., silica) are normally electrical insulators because the electrons are intimately associated with the individual bonds or ions. Some oxides exist that can become metallic conductors or even superconductors. Of particular interest are those that contain "mixed-valence" atoms (e.g., copper) that can give up a variable number of electrons when bonding. These kinds of materials have zero electrical resistance even above the temperature of liquid air; this was thought to be impossible just a few years ago.

One example follows. Consider the following ceramic oxide superconductor $YBa_2Cu_3O_7$ (YBCO) and examine the coordination of copper (Cu) in Figure 3.14. There are two kinds of copper atoms: (i) those that are coordinated by four oxygen atoms (squares), typical of divalent Cu^{2+}, and (ii) those that have a fifth oxygen atom (pyramids). When one heats this superconductor in the absence of oxygen, it loses one of its oxygen atoms and becomes the insulator $YBa_2Cu_3O_6$. Similar materials that conduct at even higher temperatures can be made by replacing the CuO_4 chains by layers of other materials, such as heavy-metal oxides. The

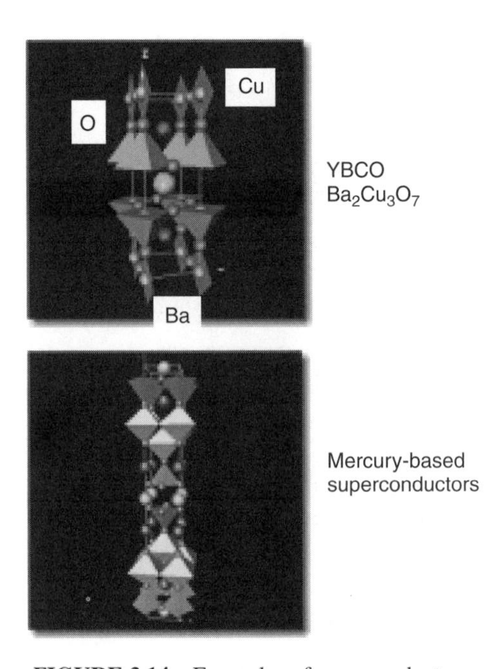

FIGURE 3.14 Examples of superconductors.

TABLE 3.7 Key Superconducting Concepts

Josephson effect	Effect observed in two superconductors that are separated by a thin dielectric when a steady potential difference V is applied. An oscillatory current is set up with a frequency proportional to V.
Meissner effect	When a superconducting loop (see superconductivity) or hollow tube, in a weak magnetic field, is cooled through its transition temperature (T_c) the magnetic flux is trapped in the loop, this is the Meissner effect. The flux is constant, being unchanged by variations in the external field. It is sustained by supercurrents circulating around the loop, any field variation is countered by the induction of an appropriate supercurrent.
Superconducting quantum interference device (SQUID)	One of a family of devices capable of measuring extremely small currents, voltages, and magnetic fields. Based on two quantum effects in superconductors: (1) flux quantization and (2) the Josephson effect.
Superconductivity	Phenomenon occurring in many metals and alloys. Superconductivity is the ability of certain materials to conduct electric current with near-zero resistance: If these substances are cooled below a transition temperature, T_c, close to absolute zero, the electrical resistance becomes vanishingly small.

highest superconducting temperature (T_c) so far obtained is for a material based on mercury oxide; here T_c is 50% higher than in YBCO. Table 3.7 lists some key superconducting concepts [51].

Magnets

Unpaired electrons orbiting an atom act as small electromagnets pointing in various directions. When magnetic moments or "spins" point in the same direction in all the atoms, the material itself behaves like a magnet and it is called a *ferromagnet*. Because a material like iron consists of many magnetic crystallites whose magnetic moments cancel each other until they are aligned, iron is not normally a magnet, however, it can be "magnetized" by other magnets. If the atomic-scale "spins" are in opposite directions, they cancel out, and the material is called an *antiferromagnet*. Magnetite (Fe_3O_4) is a well-known magnet: It is one of the common oxides of iron and is also cubic, with iron in two valence states. Many magnetic structures are more complex. Neutron diffraction has been used to identify the topology of these more complex magnetic structures. Magnets are used in motors of all types and sizes and in communication equipment (e.g., relays). In recent years, giant magnetoresistance (GMR) oxides have being used by manufacturers to make computer hard drives of much higher capacity. The phenomenon of GMR relates to the decrease of electrical resistance of materials when exposed to a magnetic field. This phenomenon was first observed (in the late 1980s) in multilayer ferromagnetic/nonferromagnetic thin-film systems; GMR has now also been observed in granular nanocrystalline materials [91].

3.3 CONCLUSION

This chapter introduced very basic concepts in the field of chemistry. While very superficial compared with a standard college chemistry book (e.g., [82]), it provides enough background to discuss basic nanotechnology concepts as we do in the rest of the book.

CHAPTER 4

Nanotubes, Nanomaterials, and Nanomaterial Processing

In the past few years, nanotechnology research has been carried out in the context of understanding and developing of nanoscale molecular assemblies and related new products. Nanostructures have attracted considerable attention of late because of their prominent properties, and their applications to novel devices that have advanced features have been studied and planned. While the scale may be small, the number of atoms in these assemblies can be large. Nanomaterials' features depend on quantum effects and typically involve movement of a small number of electrons to support specific usable actions and phenomena. The previous chapters (and Appendices A through F) provided *some* (but in no way, all) of the basic science for nanotechnology at the physics and chemistry level. From this chapter forward we concentrate on nanotechnology per se from an *engineering and applications* point of view. The theory provided in Chapters 2 and 3 and Appendices D and E is routinely employed by nanotechnology researchers to study, develop, and explain the behavior of nanostructures; however, we will not utilize the mathematics of quantum theory in the chapters that follow but sensitize the reader at this juncture that such methods are fundamental to any understanding and/or advancements in this field. The reader should generally be able to follow this material, and the material that follows, even if he or she skipped some portions of the previous chapters and/or appendices.

At a basic stage, nanotechnology requires an understanding of elemental carbon materials on an atomic level. The following list identifies some areas of research and development (R&D) interest: carbon nanostructures (e.g., carbon molecules, carbon clusters, and carbon nanotubes), bulk nanostructured materials, nanofabrication techniques (e.g., nanoimprint, soft lithography, scanning probe microscopy, and traditional lithography), self-assembly methods, nanostructured ferromagnetism, and organic compounds and polymers. Nanomaterials come in all shapes and sizes: They can range from small molecules to complex composites and mixtures [92]. The mechanical, thermal, and electrical properties of carbon nanostructures allow a wide gamut of applications. Besides carbon there are other elements that are beginning to be important as

Nanotechnology Applications to Telecommunications and Networking, By Daniel Minoli

nanostructures; this chapter, however, generally only provides an overview of carbon structures, such as fullerenes and nanotubes.

It is worth noting that there are R&D efforts underway to integrate nanoscience and electronics. From an electronic switching perspective, nanotechnology devices operate faster than larger mesoscopic components (the mesoscopic scale ranges from 10 to 10,000 nm, that is, 0.01–10 μm). Today's computers use conventional electronics that require costly fabrication techniques; molecular electronics has the potential to shrink devices to the nanoscale with improvements in the power consumption and speed profiles. Techniques have already been developed for the production of structures on a molecular level by suitable sequences of chemical reactions or lithographic techniques (e.g., the latter allows one to manipulate individual atoms on surfaces using a variant of the atomic force microscope*). These and other techniques have been used to manufacture, for example, high-density data storage devices. One is also interested in developing nanocomputers: computers whose fundamental components measure only a few nanometers in size; at press time state-of-the-art computer components are no smaller than about 40–100 nm. Researchers believe that nanoscale devices may lead to computer chips with billions of transistors, instead of millions—which is the range in today's semiconductor technology [93, 94, 95]. This may eventually allow one to develop quantum computers. Quantum computers and memory devices are expected to have applications in cryptography and information technology within a few years [45, 96]. These applications will be explored in Chapter 5 (nanophotonics) and Chapter 6 (nanoelectronics).

4.1 INTRODUCTION

There are three distinct but interdependent nanotechnology areas [97]:

- *"Wet" nanotechnology*: This is the study of biological systems that exist primarily in a water environment. The functional nanometer-scale structures of interest here are genetic material, membranes, enzymes, and other cellular components; the success of this nanotechnology is demonstrated by the existence of living organisms whose form and function are governed by the interactions of nanometer-scale structures.
- *"Dry" nanotechnology*: This is the study of fabrication of structures in carbon (e.g., fullerenes and nanotubes), silicon, and other inorganic materials. This area is based on and derives from surface science and physical chemistry. Unlike the wet technology, dry techniques make use of metals and semiconductors. The active-conduction electrons of these materials make them too reactive to operate in a wet environment, but these same electrons provide the physical properties that make dry nanostructures promising as electronic, magnetic, and optical devices. An R&D objective of many in the industry is to develop dry structures that possess some of the same attributes of the self-assembly that the wet ones exhibit.

* The atomic force microscope is discussed in Appendix F.

- *Computational nanotechnology*: This is the modeling and simulation of complex nanometer-scale structures. The predictive and analytical power of computation is critical to success in nanotechnology.

As implied, this chapter focuses on dry nanotechnology and on fabrication of structures based on carbon. Nanoparticles have distinctly different properties compared with bulk materials because the number of atoms or molecules on their surface is comparable to that inside the particles; therefore, nanoparticles can be used to develop materials with unique properties [98]. Figure 4.1 depicts some key commercial areas of current commercial interest in nanotechnology.

Elemental carbon is the simplest example of nanotechnology-usable materials; these carbon nanomaterials are based on covalent bonding. Fullerene research lead to the discovery of the C_{60} buckyball (a molecule comprised of 60 carbon atoms arranged in a soccer ball shape) by Robert F. Curl, Harold W. Kroto, and Richard E. Smalley. Since the identification of the buckyball in 1985, the field of fullerenes has experienced significant expansion: the discovery lead, soon thereafter, to the synthesis of a large class of new molecules. Volume availability of fullerenes was achieved in 1990.

Richard Smalley postulated in 1990 that, in principle, a tubular fullerene should be possible: These are capped at each end by the two hemispheres of C_{60} and are connected by a straight segment of tube, with only hexagonal units in their structure [89]. Carbon nanotubes were indeed discovered in 1991 by the Japanese researcher Sumio Iijima of NEC [99, 100, 101]: In 1991 he observed multiwall nanotubes (concentric cylinders of carbon, typically 10–100 nm in diameter); and, in 1993 he and Donald Bethune at IBM independently observed experimentally single-wall nanotubes.

Each fullerene (e.g., C_{60}, C_{70}, and C_{84}) possesses the essential characteristic of being a pure carbon cage, with each atom being bonded to three others as in graphite; however, unlike graphite, every fullerene has exactly 12 pentagonal faces with a varying number of hexagonal faces (e.g., buckyball—C_{60}—has 20) [89]. Single-wall nanotube molecules are fullerenes with relatively high count (e.g., C_{100}, C_{540}, etc.); in fact, in these cases they are macromolecules. Nanotube molecule of pure carbon are linked together in a hexagonally bonded chain to form a hollow cylinder;

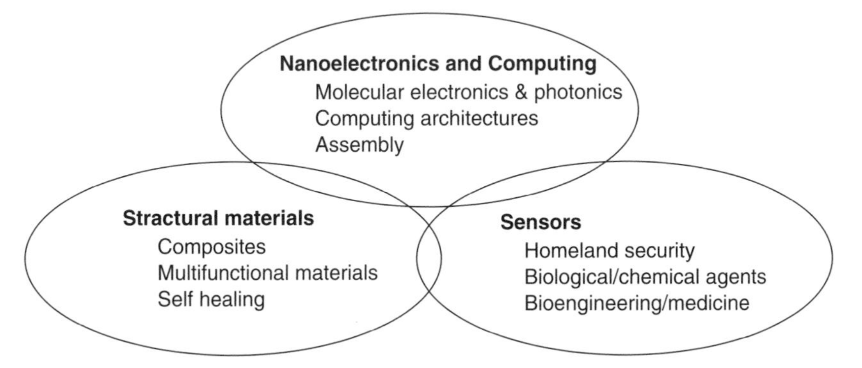

FIGURE 4.1 Key commercial areas of focus.

these materials constitute a new type of pure carbon polymers. See Figure 4.2. Some view nanotubes based on carbon (or other elements) as the most significant spin-off product of fullerene research. As mentioned, other nanotechnology materials (e.g., containing boron and nitrogen) are also becoming important from a research point of view and provide alternative components with unique mechanical and electronic properties [102, 103] (these other materials, however, are not described here).

Table 4.1 identifies some key nanotechnology terms and concepts used in this chapter and the ones that follow; refer to the Glossary at the end of the book for other

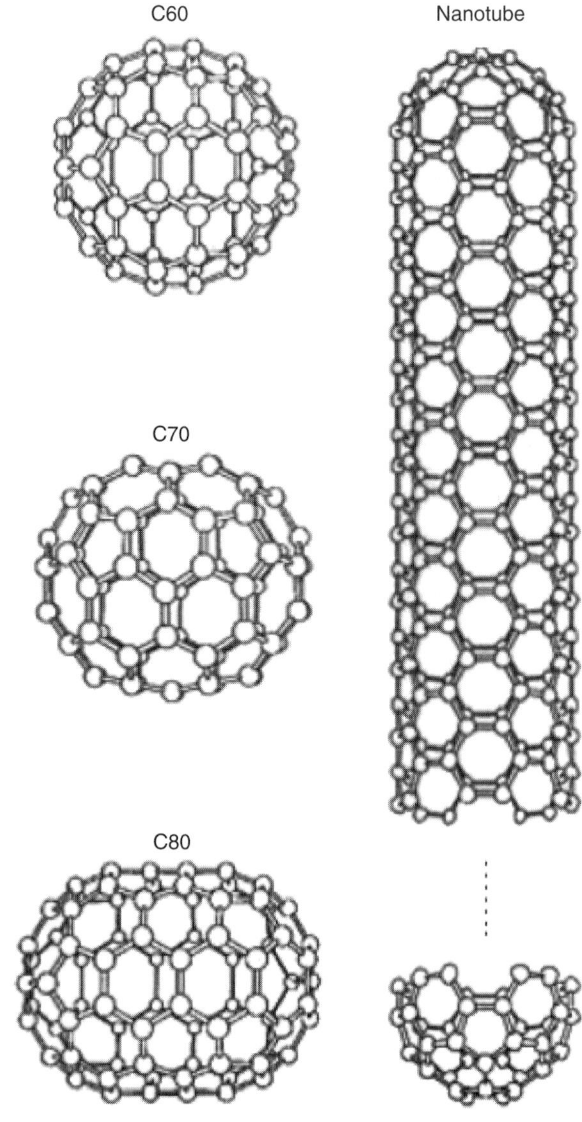

FIGURE 4.2 Fullerines.

TABLE 4.1 Some Key Nanotechnology Terms and Concepts

Term	Description
Fullerene	Molecular form of pure carbon that has a cagelike structure. The most common is buckminsterfullerene (buckyball) with 60 carbon atoms arranged in a spherical structure. Much larger fullerenes also exist. Fullerenes were discovered in 1985.
Molecular nanotechnology (MNT)	Techniques for the control of the structure of matter based on molecule-by-molecule engineering; the products and processes of molecular manufacturing.
Nanoarray	Nanoscale array (approximately 1/10,000th of the surface area occupied by a conventional microarray) for biomolecular analysis.
Nanobarcode	Technology that utilizes cylindrically shaped colloidal metal nanoparticles, where the metal composition is alternated along the length, and the size of the individual metal segment is controllable.
Nanobiotechnology	Molecular nanotechnology tools and processes to develop devices for studying biosystems.
Nanobubbles	Tiny air bubbles on colloid surfaces intended to reduce drag.
Nanochips	"Next-gen" computer chips making use of nanotechnology principles and structures. They enjoy higher density, greater speed, and lower cost. A nanocomputer is a computer made from nanochip components.
Nanochondria	Nanoscale machines symbiotically existing inside living cells and participating in their biochemistry.
Nanocones	Carbon-based nanostructures with fivefold symmetry; nanotube caps (but also available as freestanding structures). Nanocones form due to defects in graphene sheets.
Nanocrystals	Aggregates ~100 to ~10,000 atoms that combine into a crystalline form of matter measuring ~10 nm in diameter. Nanocrystals (aka nanoscale semiconductor crystals) are larger than molecules but smaller than solids; they exhibit chemical and physical properties that are a hybrid between those of molecules and solids. Nanocrystals are also known as quantum dots nanocrystals (QDNs). At the commercial level, these nanoparticles are high-precision nanoscale semiconductors that can be engineered to meet the needs for new fluorescent or photonic materials in biotechnology, optical transistors and switches, optical computing, photovoltaics, light-emitting diodes (LED), and lasers [104].
Nanocubic coating	Very thin layer coating to support improved digital memory storage (e.g., TB range for tape and GB range for floppy diskettes).
Nanoelectronics	Nanoscale electronic systems; includes molecular electronics and nanoscale devices and semiconductor devices.
Nanoelectromechanical systems (NEMS)	Nanoscale electromechanical devices; comparable to MEMSs but at the reduced physical dimensions.

TABLE 4.1 *(Continued)*

Nanofabrication (nanomanufacturing)	Construction of items using nanoscale engineering.
Nanofluidics	Approaches for controlling nanoscale amounts of fluids. Uses nanogates, which are devices that precisely outputs the flow of tiny amounts of fluids.
Nanoimprinting	Set of techniques comparable to traditional molds (masters) or form-based printing technology, but that uses masters with nanoscale dimensions. There are two techniques: (i) employs pressure to make indentations in the form of the mold on a surface, and (ii) employs the application of "transfer materials" applied to the mold to stamp a pattern on a surface.
Nanoindentation	Nanoscale indentation processes (either for hardness testing or for some atomic-level modification of a material).
Nanolithography	Imprinting at the nanoscale.
Nanomachining	Process similar to traditional machining where the goal is to remove or modify portions of the structure, but done on a nonoscale; the goal is changing the structure of nanoscale materials or molecules.
Nanomanipulation	Nanoscale process of manipulating items at an atomic or molecular level in order to produce precise structures.
Nanomanipulator	Virtual reality (VR) methods to provide a way to study/interact with the atomic world.
Nanomaterials	Bottom-up (quantum theory) designed materials where one engineers structures and functional capabilities from the ground up; materials are designed and assembled in controlled molecular fashion. Materials include nanoparticles, nanofilms, and nanocomposites.
Nanomedicine	Medical applications of nanotechnology. Specifically, the monitoring, repairing, construction, and control of (human) biological systems at the molecular level, utilizing engineered nanodevices, nanostructures, and nanopharmaceuticals.
Nanooptics	Nanoscale-level phenomena originated by the interaction of light and matter.
Nanopharmaceuticals	Nanoscale particles used for drug delivery applications.
Nanophase carbon materials	Form of matter where small clusters of atoms form the building blocks of a larger structure. Examples include carbon nanotubes, nanodiamond, and nanocomposites.
Nanoprobe	Nanoscale machines used to image, manipulate, and treat biological functions (typically in a living body).
Nanoreplicators	Set of nanomachines capable of self-replication.
Nanorods	Nanoscale rods (e.g., multiwall carbon nanotubes) of conducting/semiconducting materials.
Nanoropes	Nanotubes connected and strung together.
Nanosensors	Nanoscale-size sensors.

Nanosources	Sources that emit light from nanoscale components.
Nanoterrorism	Use of nanotechnology products to carry out terrorist acts.
Nanotransistor	Transistor measuring only several dozen nanometers (e.g, 60 nm) where the layer of insulation between a gate and a source and drain is around one nanometer in thickness (the equivalent of three atoms).
Nanotube (carbon)	1D fullerene (a convex cage of atoms with only hexagonal and/or pentagonal faces) with a cylindrical shape. Sheets of graphite rolled up to make a tube. Graphitic layers seamlessly wrapped to cylinders. A new class of carbon materials consists of closed (sp^2 hybridized) carbon chains, organized on the basis of 12 pentagons and any number of hexagons. More generally, any tube with nanoscale dimensions, e.g., a boron-nitride-based tube.
Nanowires	Nanoscale rods of some length made of semiconducting materials. Long-chain molecule capable of carrying a current. Microscopic wires from layers of different materials. Wires that are structured like "regular wires" but are at the nanoscale. Electrical conductors that function like wires but are at the nanoscale; can be used to manufacture faster computer chips, higher-density memory, and smaller lasers. Wires have been manufactured in the 40- to 80-nm diameter range.
Quantum dot (QD)	Nanometer-scale "boxes" for selectively holding or releasing electrons; the size of the box can be from 30 to 1000 nm, but more advanced QDs measure only 1–100 nm across [40, 41]. QDs are semiconductor structures where the electron wave function is confined in all three dimensions by the potential energy barriers that form the QD's boundaries [52]. Something (usually a semiconductor island) capable of confining a single electron, or a few, and in which the electrons occupy discrete energy states just as they would in an atom [42]. Grouping of atoms so small that the addition or removal of an electron will change its properties in a significant way [36]. They are small metal or semiconductor boxes that electrostatically confine/hold a specified number of electrons (the number can be adjusted from zero to several hundred by changing the dot's electrostatic status).

terms. Nanoscale microstructures are investigated experimentally using electron microscopies (SEM—scanning electron microscopy—and STM—scanning tunneling microscopy) and atomic force microscopy (AFM). (These tools and techniques are discussed in greater detail in Appendix F.)

Before proceeding further, we identify two key approaches to the manufacturing of nanostructures. Nanostructured materials can be synthesized by either bottom-up or top-down processes. The bottom-up approach starts with atoms, ions, or molecules as "building blocks" and assembles nanoscale clusters or bulk material from these building blocks [105].

Top-down assembly is the building of nanostructures and materials by mechanical methods (e.g., molding, machining, and lasers-based tools) and by bulk technology. Bulk technology is the (chemical/physical) technology where atoms and molecules are manipulated in bulk, rather than individually, as would be the case in nanotechnology. Table 4.2 depicts some the traditional fabrication techniques. The top-down methods for processing of nanostructured materials involve starting with a bulk solid and then obtaining a nanostructure by structural decomposition. One such top-down approach involves the lithography/etching of bulk material analogous to the processes used in the semiconductor industry wherein devices are formed out of an electronic substrate by pattern formation (such as electron beam lithography) and pattern transfer processes (such as reactive ion etching) to make structures at the nanoscale. Notice, however, that the wave nature of light imposes a fundamental resolution limit on optical lithography: objects and features much smaller than the wavelength of light are difficult to image and produce. There are also methods for the preparation of nanostructured materials from the bulk by the use of mechanical/thermal methods; for example, severe plastic deformation to introduce defects and/or dislocations into the material, which can then aggregate into nanoscale grains (with or without the aid of external thermal sources) [105].

TABLE 4.2 Traditional Fabrication Techniques (Partial List)

Bulk micromachining	Fabrication process of creating structures by etching into (and through) silicon wafers.
Etching	Removal of material from a surface. Wet etching of silicon uses a chemical bath (usually potassium hydroxide). Dry etching uses gas, plasma, or the blasting of particles.
LIGA	(German-language acronym from the words for lithography, electroplating, and molding). Micromachining technique used to create tall, straight-walled structures for microsystems.
Lithography	Process of copying a pattern onto a surface using light, electron beams, or X-rays (etymology: "writing on small rocks").
Mask	Pattern used in lithography that determines which areas are exposed and which are not.
RIE (reactive ion etching)	Form of dry etching where ions are blasted at a wafer's surface.
Surface micromachining	Fabrication process for MEMS based on standard CMOS microelectronic processes. MEMS structures are photolithographically patterned in alternating layers of deposited polysilicon and silicon dioxide, and then are "released" by dissolving away the silicon dioxide layers.

Self-assembly is a bottom-up assembly method where individual components of a structure come in proximity, usually by way of a solution or gas; individual components connect to each other based on their structural (or chemical) properties. This technique is used by chemists in attempting to create structures by connecting molecules. If molecules have suitable complementary surfaces, they can bind, and, thus, assemble to form a specific structure. Bottom-up, as noted, refers to the construction of larger objects from smaller building blocks, namely, the construction of nanostructures using atoms and molecules, rather than bulk materials. The assembly of molecular-materials in nanoscale architectures is a crucial step for the future molecular-scale electronics [106]. Hence, in summary, top-down methods create nanostructures out of macrostructures, while bottom-up methods entail self-assembly of atoms or molecules into nanostructures. Nanostructures encompass not only with engineered materials such as semiconductor quantum wells, quantum wires, and quantum dots, but also with domain formation and self-assembly in polymers, molecular crystals, Langmuir–Blodgett films, and organic self-assembled monolayers and multilayers [107].

4.2 BASIC NANOSTRUCTURES

4.2.1 Carbon Nanotubes

Basic nanostructures of interest include carbon nanotubes; these structures are also known as buckytubes, single-wall carbon nanotubes (SWCNTs), or single-wall nanotubes (SWNTs). SWNT is the term used here. As discussed earlier, extensive research has taken place since the early 1990s in the field of SWNT: The past decade has seen great interest in fullerenes and related carbon nanomaterials. See Table 4.3 for a partial timeline of some key advancements in the field. There have been rapid developments in the past decade in understanding the chemistry and physics of carbon nanotubes, and there is interest in both the materials science and electronics communities concerning possible applications of these unique structures [31].

Carbon nanotubes consist of graphitic layers seamlessly wrapped to cylinders: the SWNT has a cylindrical structure made of a single graphene sheet with a diameter of about 1 nm. A carbon nanotube is a single molecule of graphite, a large macromolecule, shaped in a cylindrical sheet (a hexagonal lattice of carbon); the ends of the cylinder are terminated by hemispherical caps. See Figure 4.3; Figures 1.6, 3.5, 3.6, 3.7, and 3.12 also provided additional pictorial views. The molecular formulation of single-wall carbon nanotubes requires that every atom be in the right place; this structure provides single-wall carbon nanotubes with their unique properties. Carbon-based SWNT can be metallic or semiconducting; they offer interesting possibilities to create future nanoelectronics devices, circuits, and computers.

Carbon nanotubes are a macromolecule of carbon (one can visualize this as a sheet of graphite rolled into a cylinder; graphite looks like a sheet of "chicken wire," a tessellation of hexagonal rings of carbon). Different kinds of nanotube are defined by the diameter, length, and chirality (twist). In addition to single cylindrical SWNTs nanotubes, one also has multiple-wall nanotubes (MWNTs) with cylinders inside the other cylinders. The length of the nanotube can be millions of times greater than its

TABLE 4.3 Partial Time Line of Carbon Nanotubes Advancements

Year	Advancement
1991	Discovery of multiwall carbon nanotubes
1992	Conductivity assessment of carbon nanotubes
1993	Structural rigidity assessment of carbon nanotubes
1993	Synthesis of single-wall nanotubes
1995	Nanotubes applications as field emitters
1996	Ropes of single-wall nanotubes
1997	Quantum conductance assessment of nanotubes
1997	Hydrogen storage in nanotubes
1998	Chemical vapor deposition synthesis of aligned nanotube films
2000	Thermal conductivity assessment of nanotubes
2000	Macroscopically aligned nanotubes
2001	Superconductivity of carbon nanotubes
2001	Single-wall carbon nanotube directly grown onto silicon tip
2001	Flat-panel display prototype using gated carbon nanotube field emitters
2002	Magnetic systems with carbon nanotubes
2002	Integration of suspended carbon nanotube arrays into electronic devices and electromechanical systems
2002	Carbon nanotube memory devices of high charge storage stability
2003	Carbon nanotube tips for thermomechanical data storage
2003	High-mobility semiconducting nanotubes
2003	Room temperature fabrication of high-resolution carbon nanotube field emission cathodes by self-assembly
2004	Growth of high-quality single-wall carbon nanotubes

diameter (tubes can be 1–2 mm long, and recently researchers have developed technique for growing nanotubes in straight structures as long as half an inch [75].)

The fortuitous nature of the carbon bonding apparatus (Chapter 3) supports the "molecular perfection" (also called "ideal configuration") of nanotubes, giving rise to allotropes with valuable properties such as electrical conductivity [108], thermal conductivity, strength, stiffness, and toughness. This new class of carbon materials consists of closed (sp^2 hybridized) carbon chains, organized on the basis of 12 pentagons and any number of hexagons except for the number one [109]. In SWNT the delocalized π electron donated by each C atom is free to move about the entire structure, rather than stay "home" with its donor atom, giving rise to the first known molecule with metallic-type electrical conductivity [88]. Furthermore, the high-frequency carbon–carbon bond vibrations provide an intrinsic thermal conductivity that is higher than the conductivity of diamond. In diamond the thermal conductivity is the same in all directions; SWNTs, on the other hand, conduct heat down the tube axis. As noted in Chapter 3,

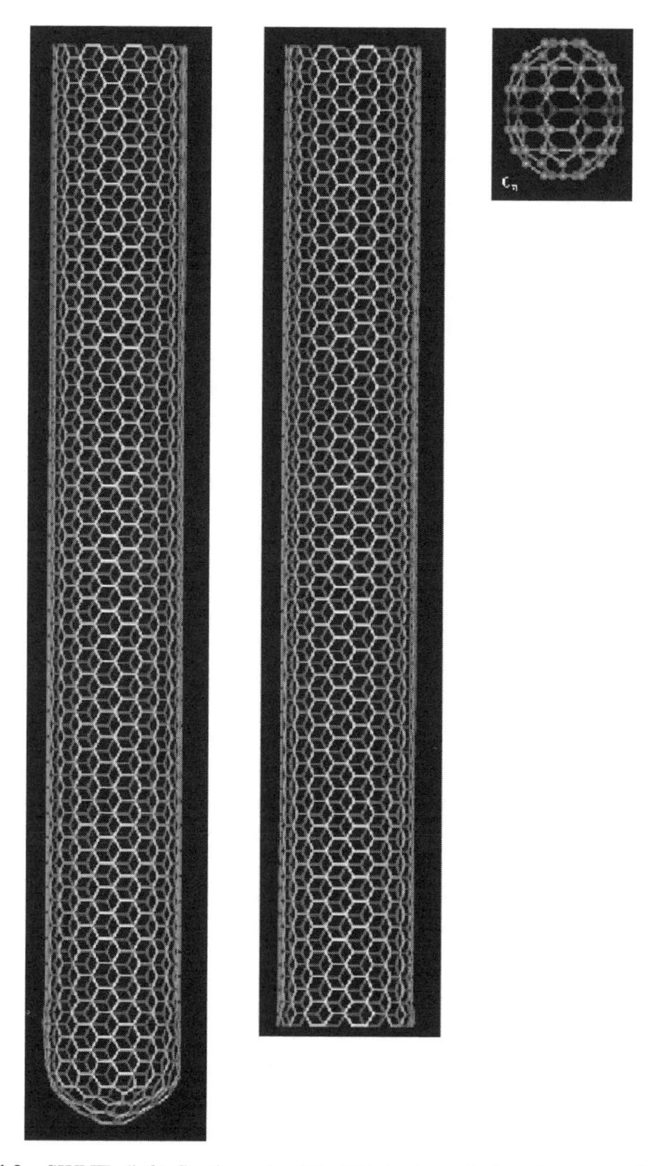

FIGURE 4.3 SWNT: (left) Section of a (10, 10) single-walled carbon nanotube with cap, (center) section of a (10, 10) single-walled carbon nanotube without cap, and (right) shortest nanotube. (Courtesy: Carbon Nanotechnologies Incorporated).

no other element bonds to itself in an extended chain with the strength of the C–C bond.

Regular graphite has atoms stacked on top of one another, but they slide past each other and can be separated easily; when coiled, the carbon arrangement becomes very strong. Carbon nanotubes have interesting and useful physical properties; for example, they are 100 times stronger than steel while being 6 times lighter. Nanotubes high stiffness toward bending exceeds that of known materials. Individual nanotubes may be the thinnest manmade structures that are stiff enough to be self-supporting [110]. Nanotubes are also very good electrical conductors. Carbon nanotubes have the same electrical conductivity as copper (they have the ability to be either semiconducting or metallic, depending on the "twist" of the tube). Nanotubes are light, thermally stabile, and are chemically inert. Furthermore, nanotubes can resist to very high temperatures (up to 1500°C under vacuum.) SWNTs also have a high surface area (~1000 m^2/g). Nanotubes are the best-known electron field emitter. SWNT can be modeled as a 1D electron transport (see Glossary). According to theoretical studies, such systems should be sensitive to Coulomb repulsion and become insulators at low temperatures; however, researchers have demonstrated that it is possible to create superconducting junctions when carbon nanotubes are connected to superconducting contacts, a phenomenon called *proximity-induced superconductivity* [31]. Some of the physical properties of carbon nanotubes are still being discovered.

Nanotubes are "molecularly ideal" (some say "molecularly perfect"), which implies that they are free of property-degrading flaws in the structure; their material properties can, therefore, approach closely the high theoretical levels intrinsic to them. Nanotube molecules can be manipulated chemically and physically. As polymers of pure carbon, nanotubes can be reacted and manipulated using the well-developed chemistry of carbon (this provides opportunity to modify the structure and to optimize solubility and dispersion) [88]. Carbon nanotubes are ideal for investigation at the interface of atomic and nanoscopic physics [111].

The following list provides some of the physical parameters of nanotubes. These kinds of molecules are remarkable for macroscopic devices of this size: Nanotubes are only a few nanometers in diameter (1.2 nm for SWNTs) and are up to about 1 mm in length (hence, single-wall nanotube lengths are typically thousands of times their diameters). The carbon bond length is 1.42 Å; the C–C tight-bonding overlap energy is on the order of 2.5 eV. Nanotubes are not always perfectly cylindrical because they can bend on themselves or even support Y-junctions.

Diameter of SWNTs:	12–14 Å (1.2–1.4 nm)
Distance from opposite carbon atoms (segment 1):	2.83 Å
Analogous carbon atom separation (segment 2):	2.456 Å
Parallel carbon bond separation (segment 3):	2.45 Å
Carbon bond length (segment 4):	1.42 Å

Nanotubes can be categorized in different *types*; these types are characterized by the vector $R = (n,m)$ where n and m are integers of the vector equation $R = na_1 + ma_2$. The values of n and m determine the chirality, or "twist," of the nanotube. The chirality affects the conductance of the nanotube, its density, its lattice structure, and other physical properties. For example, a SWNT is considered metallic if the value $n - m$ is divisible by 3; otherwise, the nanotube is semiconducting (note that when tubes are formed with random values of n and m, one can expect that two-thirds of nanotubes to be semiconducting, while the other third to be metallic, a fact that is

TABLE 4.4 Physical (Electrical, Thermal, Elastic) Properties

Metallic behavior	(n, m); $n - m$ is divisible by 3
Semiconducting behavior	(n, m); $n - m$ is not divisible by 3
Conductance quantization	$n \times (12.9\ k\Omega)^{-1}$
Resistivity	$10^{-4}\ \Omega$-cm
Maximum current density	10^{13} A/m^2
Thermal conductivity	~2000 W/m/K
Phonon mean free path	~100 nm
Relaxation time	~10^{-11} s
Young's modulus (elastic) (SWNT)	~1 TPa
Young's modulus (elastic) (MWNT)	1.28 TPa
Maximum tensile strength	~30 GPa

confirmed empirically) [112, 113, 114, 115]. If one (imaginarily) slits open a nanotube by cutting all the bonds along any straight line parallel to its axis, it uncurls to form a strip of graphite. The lattice vector connecting the sides of the strip is (n,m); this vector completely characterizes the tube. There are an infinite number of possible tubes characterized by different lattice vectors [116]. If $n = m$ ("armchair tubes") or $m = 0$ ("zigzag" tubes), there is n-fold rotational symmetry; otherwise, the tube has helical symmetry (it is chiral). The theory implies that armchair tubes, having $n = m$, should be metals; those with $n - m = 3i$ (i an integer) should be narrow-gap semiconductors, and the rest should be insulators[1] [116, 117, 118].

Table 4.4 provides a tabulation of some electrical and thermal properties of nanotubes. Nanotubes can vary in size; larger nanotubes [e.g., a (20, 20) tube], tend to bend to some degree under their own weight [119]. See Figure 4.4 for an example of a nanotube that bends on itself [120]. Many materials are stiff but they are also brittle; SWNTs, on the other hand, are stiff but also tough: they can stretch 20% of their normal length and can be bent over double and/or tied into a knot and then released with no resulting defect. The combination of stiffness and toughness makes SWNT the strongest known fibers (as noted, about 100 times stronger than high-strength steel at one-sixth the weight of steel—normalized strength-to-weight ratio with steel is over 460, as seen in Table 4.5 [89]). The current price for SWNT material is \$100/g for small quantities (50 g or less), \$50 for larger quantities.

A key capability of single-wall nanotubes relates to the electrical conductivity. As we noted in the previous chapter, carbon in a planar graphene sheet has poor electrical conductivity. Interestingly, when rolled into a perfect tube and the graphene carbon–carbon bonds are perpendicular to the tube axis, the resulting electronic structure becomes that of a true metal; this is the first and only known instance of a molecule being a true metallic conductor. Other ways of rolling up the grapheme sheet produce semiconducting tubes with such a small gap that at a few degrees above

[1]Ab initio simulations have been conducted on the deformation of carbon nanotubes; one finds that the electric conductivity of the carbon nanotubes shows transitions from metallic to semiconducting and vice versa under axial tension; one also finds that the semiconducting nanotubes become metallic under radial compression, while the metallic nanotubes do not show the transition [121].

FIGURE 4.4 Nanotube with bends.

absolute zero they have a high conductivity; yet others are similar to silicon in their conductivity [89, 122]. Recently, scientists were also able to cause an individual carbon nanotube to emit light; this may help materialize many of the proposed applications for carbon nanotubes, such as in electronics and photonics development.

As we discussed above *single-walled* carbon nanotubes exhibit either a metallic or a semiconducting character depending on their helicity. Metallic tubes have micron-long paths and behave as long ballistic conductors (see Glossary). The intrinsic properties of conducting *multiwalled* nanotubes, on the other hand, were considered "elusive" at press time. Indeed, reported ballistic, diffusive, or insulating experimental behaviors remain difficult to relate to the number and helicities of constitutive shells and interlayer coupling. It is usually assumed that the outermost shell, in contact with metallic electrodes, determines the metallic or semiconducting character of multiwalled carbon nanotubes [123]. Depending on how they are fabricated,

TABLE 4.5 SWNT Comparison

Material	Density (g/cm^3)	Normalized Strength-to-Weight Ratio (to steel)
Single-wall nanotube	1.4	462
Multiwall nanotube	1.8	15
Graphite fiber	1.6	13
Titanium	4.5	2
Aluminum	2.7	2
Steel	7.8	1

carbon nanotubes can act as metallic or semiconducting substances. However, researchers have recently shown that a coaxial magnetic field can be used to convert nanotubes from metallic to semiconducting and vice versa. As noted, the manner in which the sheets are rolled and seamed determines whether the tubes are metallic or semiconducting; while one cannot undo the seam and rejoin it when one wants to change the electronic properties of the nanotube, one can tune these materials not by restructuring the molecules themselves but by moving their energy levels with a strong magnetic field [124]. MWNT with diameter of about 30 nm allows one to apply a magnetic field strong enough to significantly modify the energy spectrum and convert the nanotube's electronic properties. As an electron moves, the wave actually takes multiple paths, including ones that encircle the nanotube and the magnetic flux threading it. Depending upon the strength of the magnetic field, the properties of the molecule will change from metallic to semiconducting, and back again [124].

The electronic properties of MWNTs are rather similar to those of SWNTs because the coupling between the cylinders is weak in MWNTs. Because of the nearly one-dimensional electronic structure, electronic transport in metallic SWNTs and MWNTs occurs ballistically (i.e., without scattering) over long nanotube lengths, enabling them to carry high currents with essentially no heating. Phonons also propagate easily along the nanotube [125].

The SWNTs also self-assemble into "ropes" of up to 100 aligned tubes, running side by side, branching and recombining; these ropes form long conductive pathways that can be put to use in making electrically conductive compounds. Ropes of carbon nanotubes are bundles of tubes packed together in an orderly manner. SWNTs pack into a close-packed triangular lattice with a lattice constant of about 1.7 nm; the density, lattice parameter, and interlayer spacing of the ropes is dependent on the chirality of the tubes [126, 127]. Typical rope diameter are ~20 nm. Figures 4.5 and 4.6 depict nanoropes. Also, SWNTs have the ability to be precisely derivatized: Nanoscale Y-type tubes have been designed. Figure 4.7 shows a MWNT [128]; quite a variety of MWNT arrangements exist [129].

The SWNTs are grown by several techniques in the laboratory. The formation of a SWNT is relatively straightforward and a reasonable quantity of SWNTs of high purity can be obtained. One approach uses the degradation of high-pressure CO with Fe(CO) at about 1000°C; additional information on manufacturing methods is provided in Section 4.3. There have been demands for structural control of SWNTs along with better methods of controlling the diameter and chirality. Although chirality control is difficult, partial success in diameter control has been achieved of late by controlling the furnace temperature or choosing the metal catalysts in the laser ablation process [130].

4.2.2 Nanowires

Nanowires [also called quantum wires (QWR)] are 1D nanoscale molecular structures with electrical and/or optical properties. Nanowires, which we introduced in Chapter 1 as an example of a nanostructure, have attracted extensive interest in recent years because of their unusual quantum properties and potential use as nanoconnectors and

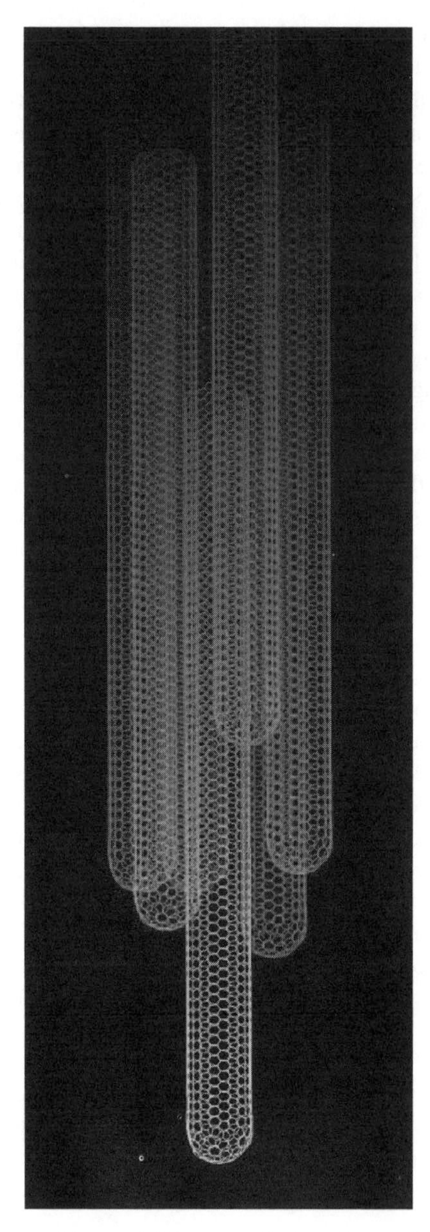

FIGURE 4.5 Nanoropes (diagram). (Courtesy: Carbon Nanotechnologies Incorporated).

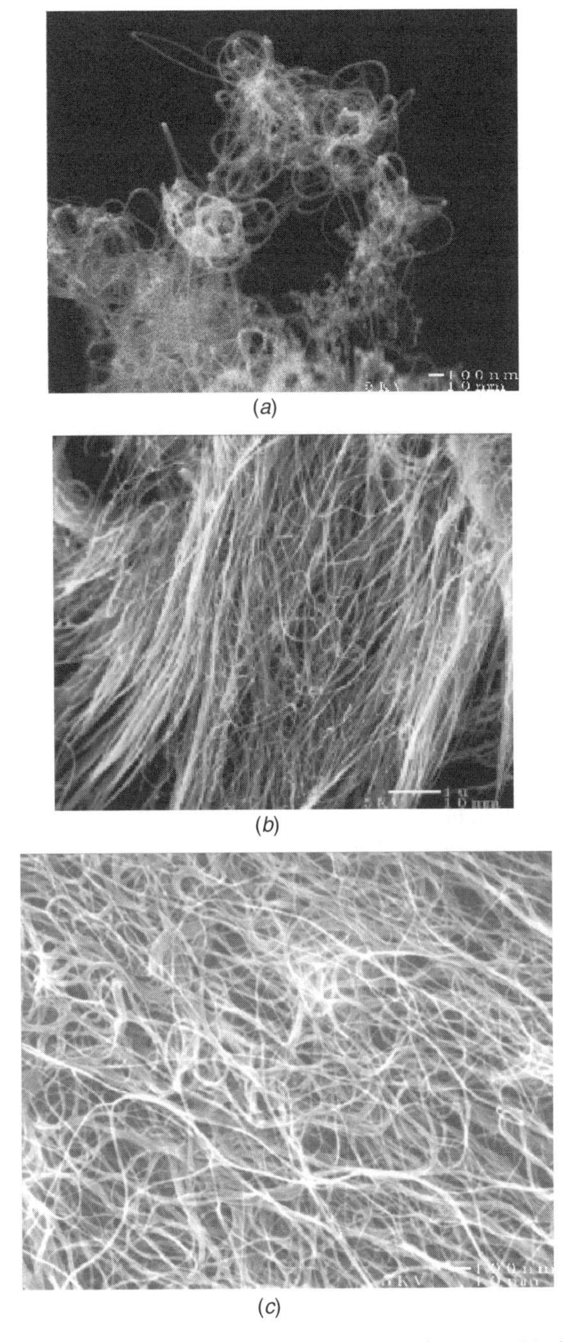

FIGURE 4.6 Nanoropes (scanning electron microscope images): (*a*) General example; (*b*) Torn edge of a bucky paper showing alignment of ropes; (*c*) SEM image of a bucky paper, prepared from filtering buckytubes. (Courtesy: Carbon Nanotechnologies Incorporated).

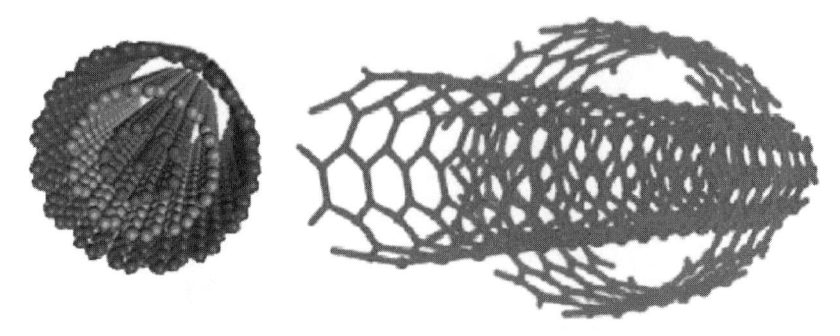

FIGURE 4.7 Strand of carbon atoms (center) tunneled in a multiwalled carbon nanotube.

nanoscale devices. In order to have enhanced physical properties, the wires must be of small diameter, must have high aspect ratio (i.e., ratio of length to thickness), and must be uniformly oriented [131]. Nanowires are *not* (as of yet) based on nanotube materials, but a nanorope could theoretically be used, at some point in the future, to construct a nanowire [132, 133]. Nanowires have potential applications in areas ranging from probe microscopy to nanoelectronics. Nanowires are considered one of the key components for the envisioned creation of *molecular electronics chips*: They can be assembled in grids and become the basis of nanoscale logic circuits [134, 135]. Clearly, nanowires need to have physical connection capabilities at each end of the conductor/wire. The stability of the nanowires, however, is a concern since metal nanowires of approximately 1 nm diameter may exist only transiently in some manufacturing instances (less than 10 s in ultrahigh vacuum); work is underway to ameliorate this situation [131].

We have already mentioned that carbon nanotubes are the ultimate candidate to replace (some) wires inside chips; this is expected to be feasible in the early part of the 2010 decade. As noted, carbon nanotubes conduct electricity well and are very small, allowing manufacturers to deploy billions of transistors onto a single chip. At this time nanotubes can only be manufactured in small numbers in labs, and cost-effective mass-marketing techniques remain a future goal. The current generation of nanowires is based on other mechanisms and compositions.

Engineered nanowires that are 50–100 nm in diameter and 1 μm to 10 cm in length are now being developed that perform ferroelectric, dielectric, or sensor functions in nanoelectronics. Semiconductor nanowires can transport electrons and holes, hence, they can function as building blocks for nanoscale electronics [136] (recently some researchers have reported the synthesis of single-crystalline silver nanowires of atomic dimensions, namely, having 0.4 nm width [131]). There are a number of methods of producing 1D nanoscale structures, but so far none of them has allowed sufficient control of structure parameters to be satisfactory; hence, researchers are looking into finding effective ways to synthesize nanomaterials such as ceramics, metals, and composites, in nanowires (1D) and thin-film (2D) forms (thin films are discussed in a section that follows).

As an illustrative example, 50-nm-diameter nanowires have been manufactured using 19-nm core of pure silicon and a shell of silicon doped with boron [137]. Other nanowires reported include a 26-nm germanium core with a 15-nm doped silicon shell; a 21-nm doped silicon core with a 10-nm germanium shell; and a 20-nm silicon core with a 30-nm middle layer of germanium and a 4-nm outer shell of doped silicon [137]. Nanowires are relatively easy to produce and can have very different shapes. They are often thin and short "threads" but can also have other manifestations. Gallium-arsenide quantum wires have been produced (these are referred to as GaAs/AlGaAs quantum wire). Nanowires are being considered for use in next-generation computer chips, as illustrated in Figure 4.8.

The propagation of electromagnetic energy has been demonstrated along noble metal stripes with widths of a few microns [138]; propagation has also been demonstrated along nanowires with subwavelength cross sections and propagation lengths

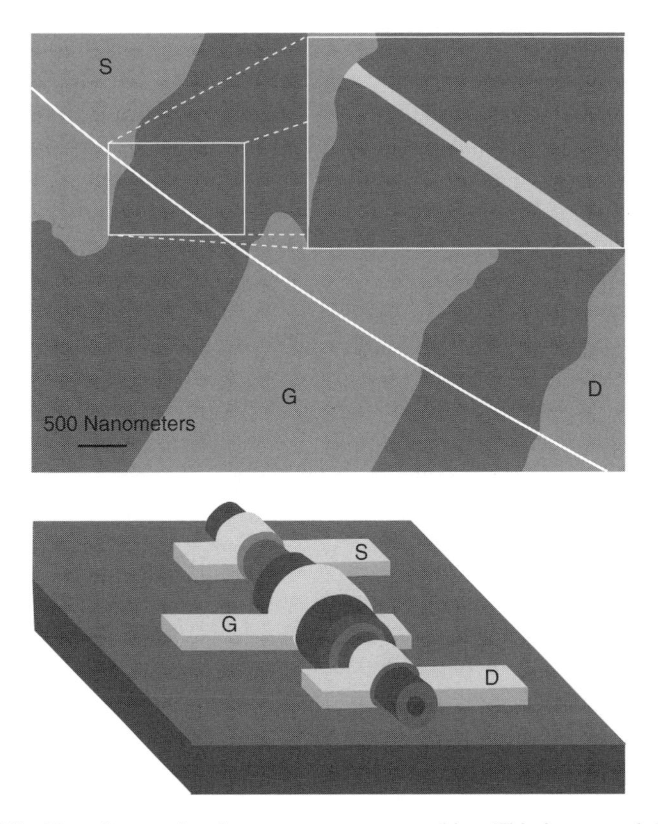

FIGURE 4.8 Use of nanowires in next-gen computer chips. This image and the diagram below it show a transistor made from a multilayer nanowire. The core of the nanowire is doped silicon and the first layer is germanium. The second layer is the insulator silicon oxide. The outer layer is doped germanium. S is the source electrode, G is the gate electrode, and D is the drain electrode. (Source: Lieber Group, Harvard University).

of a few microns [139, 140, 141]. Metal nanowires can also be used to "transmit" photons [142] (this topic is discussed in Chapter 5). The optical properties of metal nanowires can be optimized for particular wavelengths of interest, and nonregular cross sections and coupling between closely spaced nanowires allows a tuning of the optical response [142, 143, 144].

As noted, semiconductor nanowires are of interest. Along these lines, researchers report that boron- and phosphorous-doped *silicon nanowires* can be used as building blocks to assemble three types of semiconductor nanodevices: (i) passive diode structures consisting of crossed p- and n-type nanowires exhibit rectifying transport similar to planar p-n junctions. (ii) Active bipolar transistors, consisting of heavily and lightly n-doped nanowires crossing a common p-type wire base, exhibit common base and emitter current gains. In addition, (iii) p- and n-type nanowires have been used to assemble complementary inverterlike structures [135].

Nanowires also will find applications to optics. While optical studies of 1D nanostructures have in the past focused primarily on lithographically and epitaxially defined quantum wires embedded in a semiconductor medium, free-standing nanowires (e.g., indium-phosphide-based nanowires) have several attractive differences from these systems, including a large variation in the dielectric constant between the nanowire and the surrounding medium, and a cylindrical and strongly confined potential for both electrons and holes [145]. Free-standing nanowires can be used to create polarization-sensitive nanoscale photodetectors that may prove useful in integrated photonic circuits, optical switches and interconnects, near-field imaging, and high-resolution detectors.

With ongoing research, a new practical science of bottom-up nanoelectronics is beginning to emerge, where 1D nanowire building blocks of inorganic conductors, semiconductors, and dielectrics are used to create nanocomponents such as transistors, diodes, capacitors, and resistors, which are then assembled into dense circuits. With this arrangement the possibility exists of achieving 1 terabits/cm^2 density [146]. Nonowires will be discussed again in Chapter 6.

4.2.3 Nanocones

Single-walled carbon cones, as depicted in Figure 4.9, were first reported by Harris and co-workers in 1994 [147]. They were produced by high-temperature heat treatments of fullerene soot. Sumio Iijima's group later showed that this structure can also be produced by laser ablation of graphite (Iijima's group gave them the name "nanohorns"). It has been shown that nanohorns (nanocones) have good adsorptive and catalytic properties and that they can be used as components for high-quality fuel cells.

4.2.4 Applications of Nanotubes, Nanowires, and Nanocones

Fullerene science in general, and nanotubes in particular, open ways for constructing new materials with predetermined properties on the atomic scale and actually create the fundamentals of material science in the 21st century [148]. The number of

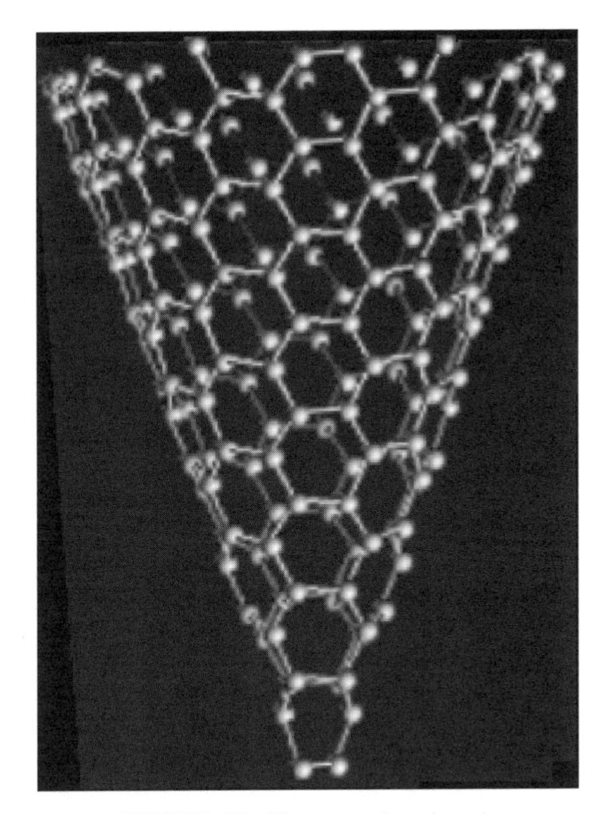

FIGURE 4.9 Nanocones (nanohorns).

applications of nanotubes is growing on a regular basis. The "extraordinary" technical characteristics of nanotubes imply that there are numerous potential applications. Applications include the ones shown in Table 4.6 (based loosely on [88]). Applications of carbon nanostructures include field emission displays (field emission), electron sources (nanoelectronics, superconductivity), transistors, chemical and biological sensors [149], hydrogen storage [150], fuel cells, supercapacitors, secondary batteries, nanocomposites (composite materials), lubricants, and biotechnology. For example, Samsung has already demonstrated a flat-panel display prototype using nonotubes as the field emission source. Fuel cells can be developed having an electrode and/or catalyst formed from a nanocarbon material, such as fullerene, carbon nanotube, carbon nanohorn, carbon nanofiber, or metal-encapsulated fullerene.

Manufacturing electronic circuitry at the nanoscale level by building integrated components at the molecular level is a topic of fairly sustained research. For example, in 2001 IBM announced the fabrication of the world's first array of transistors made from carbon nanotubes [151], and a lot of work has taken place since. This topic is revisited in Chapters 6.

In addition to making use of the electrical conductivity of SWNT, there are also applications that exploit the thermal and mechanical properties. For example, SWNT

TABLE 4.6 SWNT Applications (Partial List)

Field Emission	Because of their high electrical conductivity, SWNTs are excellent field emitters; in fact, they are the best known field emitters of any material. Note that the sharpness of the tip of a SWNT is at the nanoscale (as seen in the figures in this chapter). Clearly, the sharper the tip, the more concentrated will be the electric field, leading to field emission. The sharpness of the tip also implies that the SWNTs emit at low-voltage levels, while at the same time there will be high current density (some claim high as 1000 A/cm^2). An application of these features now receiving interest is in field emission flat-panel displays where high current density and low turn-on and operating voltage are very desirable. Other applications include electron micro scope sources, general cold-cathode lighting sources, and lightning arrestors [152, 153, 154, 155, 156, 157, 158, 159].
Conductive plastics	In structural applications (as a replacement for metal), plastics have seen major penetrations in the past several decades. However, plastics are electrical insulators; there are situations where electrical conductivity is required. This can now be addressed by loading plastics with conductive fillers of the nanotube kind (the loading required to provide the necessary conductivity using con ventional products is typically high, resulting in heavy parts, which is often undesirable) [160].
Energy storage	SWNTs have functional characteristics that are desirable for materials used as electrodes in capacitors and batteries. Wireless and mobility electronics are driving the need for improved energy storage; SWNTs' high strength and toughness (in relation to weight) are valuable for fuel cells that are deployed in transportation applications where durability is important. SWNTs have a high surface area, good electrical conductivity, and linear geometry that makes the surface well-accessible to the electrolyte. SWNTs have the highest reversible capacity of any carbon material for use in lithium-ion batteries. SWNTs also have applications in a variety of fuel cell components, again because of the high surface area and thermal (they may be used in gas diffusion layers as well as current collectors)
Conductive adhesives and connectors	Same characteristics that make SWNTs attractive as conductive cores for use in shielding make them attractive for electronics connectors.
Molecular electronics/ microelectronics	Next-gen components. Manufacturing electronic circuits at the nanoscale by building at the molecular level is receiving a lot of attention of late. Two approaches are contemplated: (i) as limiting process, as traditional methods lead to sub-0.1-μm components and (ii) as new technology, e.g., quantum computers. As dimensions shrink to the nanoscale, the interconnects between switches and other active devices become increasingly important. Y-type SWNT make them the ideal candidates for the interconnects in molecular electronics. In addition, they can be used to buildout as logical switching elements directly at the nanoscale [95, 161, 162, 163, 164, 165, 166, 167, 168, 169, 170, 171, 172, 173, 174, 175, 176, 177].

TABLE 4.6 (*Continued*)

Thermal materials	Best-in-class anisotropic thermal conductivity of SWNTs gives rise to heat sink/conduit applications, where heat needs to be removed. Electronic systems, such as blade computers, supercomputers, and advanced computing, can generate surface heat as high as 100°C. Aligned structures and ribbons of SWNTs can be utilized to develop efficient heat conduits [178].
Structural composites	Mechanical properties of SWNTs (stiffness, toughness, and strength) support a plethora of applications, including advanced composites.
Fibers and fabrics	Superstrong fibers spun of SWNTs (including SWNT ropes) will have applications ranging from body armor and vehicle armor to transmission line cables to textiles.
Catalyst supports	SWNTs have a high surface area, and their ability to attach chemical compounds to their sidewalls make them useful for catalytic applications.
Biomedical applications	Significant opportunities exist in the medical/biomedical fields (this topic is not emphasized here). Cells have been shown to grow on SWNTs but do not adhere to the SWNTs. Hence, SWNTs can be used as coatings for prosthetics, vascular stents, and neuron growth and regeneration, to list just a few [179, 180].

have been used to enhance the primary components of composites, as well as polymer resins and fibers. SWNT nanotubes can be blended with yet other materials to improve existing properties or provide new ones. The resulting materials are called *composites nanostructured materials*. Carbon nanotubes are considered ideal reinforcing fibers since, as we have noted, they have high aspect ratios, are very strong individually, and are very light. The use of nanotubes as additives in thermoplastics and thermosets, for example, is now undergoing rapid development as sufficient quantities of high-quality single-wall nanotube material are becoming available to permit such investigations [89]. Recent work has focused on developing ceramic and polymer composites with multiwalled and single-walled carbon nanotubes as fillers for fabricating high-strength, high-toughness, and lightweight composites [181]. Nanofibers can also be for clean-room products, filtration, surgical gowns, biomedical devices, and specialty fabrics.

The SWNTs can also be used as additives for *conductive* plastics. Conductive plastics can be utilized as antistatic, electrostatic, dissipative, and electromagnetic shielding and/or absorbing materials. For example, electromagnetic interference shielding is important in laptop computers and other electronic devices to prevent interference with and from other electronic equipment. Currently, there is no suitable pure-plastic material for this application (e.g., for electronic equipment cases), and metal is typically added to the plastic to provide this function; unfortunately, this adds significant weight and manufacturing cost. SWNT-filled plastics can also be employed in electromagnetic interference shielding for military applications and radar absorption. SWNT-based materials represent the future aerospace vehicle construction material of choice based on predicted strength-to-weight advantages and inherent multifunctionality [182]. SWNT-based composites are still in their infancy.

Nanoparticles can be incorporated into polymeric coatings to facilitate measurable improvements in targeted properties, for example, scratch resistance, UV resistance, conductivity, and the like. Commercial-grade processing methods have been developed in the recent past to control the average particle size and particle size distribution of the dispersed nanoparticles; these integrated technologies allow transparent coatings containing nanoparticles to be formed in a plethora of resin formulations [183].

As pure carbon molecules, SWNTs offer a wide variety of derivatives for compatibilizing, dispersing, and coupling with a host polymer; furthermore, SWNTs can be considered a new type of polymer because the molecules can be modified chemically and regarded as a backbone polymer—the basis of new classes of block-and-graft copolymers [89].

Nanotubes are especially promising for biomedical applications because one is able to tailor them for specific parts of the body. For example, nanotubes that assemble themselves using the same chemistry as DNA may be used for creating better artificial joints and other body implants. Researchers have discovered that the self-assembling nanotubes represent an entirely new and potentially superior material to use for artificial body parts [184, 185]. Bone cells (osteoblasts) attach better to nanotube-coated titanium (Ti) than they do to conventional titanium used to make artificial joints. Bone cells and cells from other parts of the body attach better to various materials that possess surface bumps about as wide as 100 nm; conventional titanium used in artificial joints has surface features on the scale of microns, causing the body to recognize them as foreign and prompting a rejection response. Helical rosette nanotubes (HRN) are a new class of self-assembled organic nanotubes possessing biologically inspired nanoscale dimensions. Because of their chemical and structural similarity with naturally occurring nanostructured constituent components in bone such as collagen and hydroxyapatite, researchers believe that an HRN-coated surface may simulate an environment that bone cells are accustomed to interacting with [184, 185].

Other applications envisioned include the following [2]: solar cells in roofing tiles and siding that provide electricity for homes and facilities; this can result in a much cleaner environment due to greater use of solar energy. Prototype tires exist today that provide improved skid resistance, reduced abrasion, and resulting longer wear, although a date for market introduction had not been announced as of press time. The nanocomposites being used in tires can be used in other consumer products such as high-performance footwear, exercise equipment, and car parts (belts, wiper blades, and seals). (Nanocomposites, not just SWNT-based nanocomposites are discussed further in a subsection that follows.) New commercial applications of nanotechnology that are expected within 5 years in the pharmaceutical and chemical industries include advanced drug delivery systems (e.g., implantable devices that automatically administer drugs and sense drug levels) and medical diagnostic tools, such as cancer tagging mechanisms. Finally, some researchers also contemplate nanoscale machines comprised of gears built with nanotubes and close variants, as illustrated in Figure 4.10.

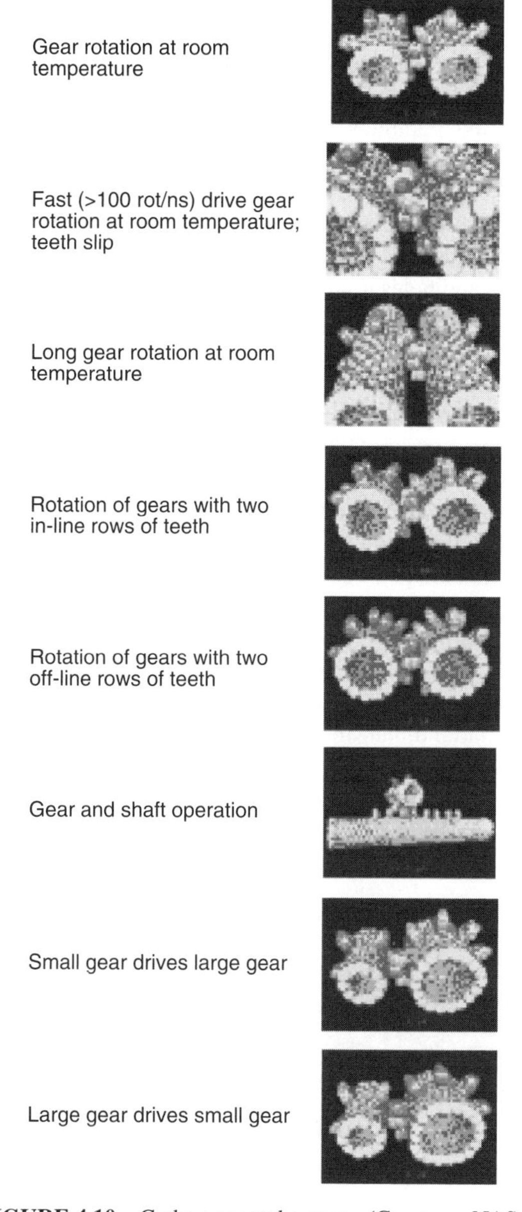

FIGURE 4.10 Carbon nanotube gears. (Courtesy: NASA).

4.2.5 Quantum Dots

Technological advances of the last two decades have opened windows of opportunity for designing and controlling electronic states in novel systems. Quantum dots (QDs) are nanometer-scale "boxes" for selectively holding or releasing electrons

[40, 41]. They are small physical devices that contain a "tiny droplet" of free electrons [186, 187, 188, 189, 190]: small metal or semiconductor boxes that hold a specified number of electrons (the number can be adjusted from zero the several hundred by changing the dot's electrostatic status). See Figure 4.11 [191, 192] (QDs generally look more like pyramids than actual dots). QDs are a grouping of atoms so small that the addition or removal of an electron will change its properties in a significant way [36]. QDs are a semiconductor structures where the electron wave function is confined in all three dimensions by the potential energy barriers that form the QD's boundaries [52]. Specifically, QDs are semiconductor structures that confine the electrons and holes to a volume of the order of 20 nm^3 (or slightly larger in some cases). These structures are similar to atoms, but they are more than an order of magnitude larger. Hence, using nanoscale techniques it is feasible to manipulate their quantum wave functions; with this ability, promising applications can be developed, such as quantum logic gates [193]. There is a plethora of interesting phenomena that have been measured in QD structures over the past decade [43, 44].

Modern semiconductor processing techniques permit the artificial creation of quantum confinement (quantum confinement in all three spatial dimensions) of only a few electrons; such finite fermion QD systems have much in common with atoms, yet they are manmade structures, designed and fabricated in the laboratory [194]. The fabrication of QDs in semiconductor devices has led to the invention of single-electron transistors and controllable single-photon emitters; such "designer atoms" are ideal settings for the control and manipulation of electronic states [111]. As noted, QDs are small semiconductor or metal structures in which electrons are confined in all spatial dimensions; as a consequence, discreteness of the energy and charge arises—for this reason QDs are often referred as *artificial atoms*. In contrast to real atoms, different regimes can be studied by continuously changing

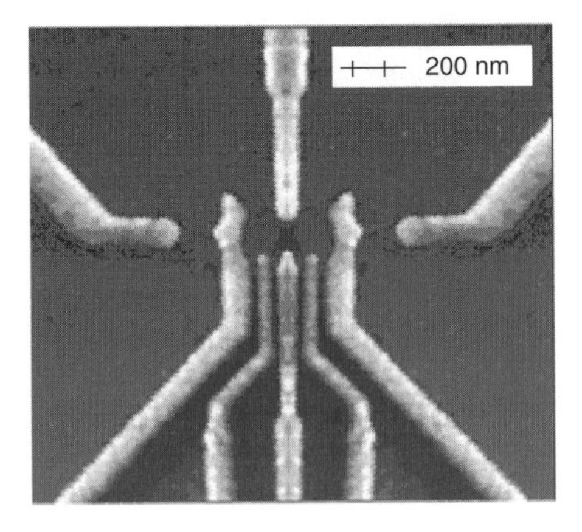

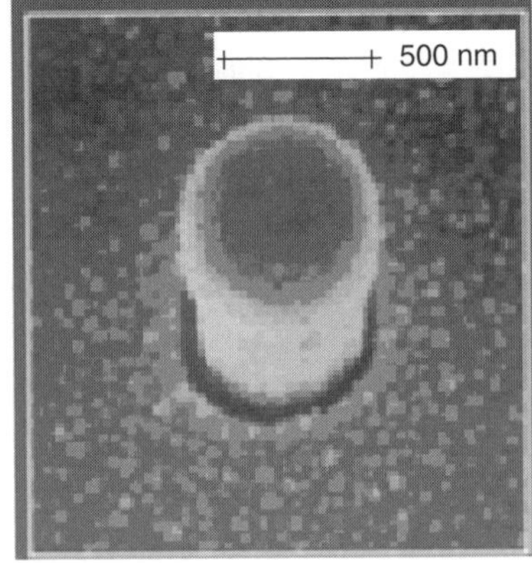

FIGURE 4.11 Quantum dot. (Courtesy: Kavli Institute of Nanoscience Delft, The Faculty of Applied Sciences at Delft University of Technology).

the applied external potential. A QD array can be considered as an *artificial molecule* or *artificial crystal* [195]. Specifically, the physics of QDs shows many parallels with the behavior of naturally occurring quantum systems in atomic and nuclear physics. As in an atom, the energy levels in a QD become quantized due to the confinement of electrons; unlike atoms, however, QDs can be easily connected to electrodes and are therefore excellent tools to study atomiclike properties [43, 44].

Quantum dots are promising systems due to their physical properties, as well as their potential applications in electronic devices. A QD's electronic response, like that of a single atom, is manifest in its discrete energy spectrum, which appears when electron–hole pairs are excited; although the wave function of a QD electron and its corresponding hole extends over many thousands of lattice atoms, the pair (termed an *exciton*) behaves in a quantized and coherent fashion [52]. Over the past 10 years QDs have evolved from laboratory constructs to the building blocks for a future computer applications. QD elements can also be used for next-generation telecom devices and can be incorporated into optical circuits for high-speed signal processing applications in optically routed networks [195]. Resesarchers believe that if QDs can be integrated onto a chip, their unique electrical properties can be put to work to perform functions similar to those of conventional transistors, while requiring only a small fraction of the space; hence, QDs may allow a computer processor many times more powerful than current supercomputers to be constructed on single chips. Self-assembled QDs can also act as lasers; recent advances in QD lasers validate that self-assembled QDs provide an opportunity for the development of new electronic and optoelectronic devices [196].

The size and shape of QD structures, and, therefore, the number of electrons they contain, can be precisely controlled: A QD can contain a single electron or a collection of several thousand electrons. The size, shape, and composition of QDs can be tailored to create a variety of desired properties; these "artificial atoms" can, in turn, be positioned and assembled into complexes that serve as new materials. Below we discuss briefly the parallels between atoms and QDs (artificial atoms). Researchers that work on QDs anticipate that a host of complex, customized QD-based materials will become available [52]. QDs' properties can be changed in a controlled way by electrostatic gates, changes in the dot geometry, or applied magnetic fields [186, 187, 188, 189, 190, 194]. For example, work by Loss and DiVincenzo [186] suggests that the spin of a single electron confined in a QD could be used as a quantum bit (qubit), the building block of a quantum computer. Lateral QDs (e.g., like the one of Fig. 4.9) are best suited as all the tunnel barriers can be freely controlled using electrostatic gates [187, 188, 189, 190].

Quantum dots are considered 2D analogies for real atoms by researchers; but because they have much larger dimensions, they are suitable for experiments that can not be carried out in atomic physics. The 3D spherically symmetric potential around atoms yields degeneracies known as shells, 1s, 2s, 3s, 3p, and so forth; each shell can hold a specific number of electrons. The electronic configuration is particularly stable when these shells are completely filled wih electrons, occurring at "magic" atomic numbers 2, 10, 18, 36, and so on. In a similar way, the symmetry of a 2D, disk-shaped QD leads to a shell structure with magic numbers 2, 6, 12, 20, and so on. The lower degree of symmetry in 2D results in a different sequence of magic numbers than in 3D. By measuring electron transport through QDs, a periodic table of artificial 2D elements can be obtained, as shown in Figure 4.12 [43, 44].

Researchers at The Delft Spin Qubit Project at the Kavli Institute of Nanoscience [43, 44] describe the process as follows: For this purpose, dots are connected via potential barriers to source and drain contacts. If the barriers are thick enough, the number of electrons on the dot, N, is a well-defined integer. This number changes when electrons tunnel to and from the dot. However, due to Coulomb repulsion between electrons, the energy of a dot containing $N + 1$ electrons is larger than when it contains N electrons. Extra energy is therefore needed to add an electron to the dot. Consequently, no current can flow, which is known as the Coulomb blockade. The blockade can be lifted by means of a third electrode closeby, known as the gate contact. A negative voltage applied to this gate is used to supply the extra energy and thereby change the number of free electrons on the dot. This makes it possible to record the current flow between source and drain as the number of electrons on the dot, and hence its energy is varied. The Coulomb blockade leads to a series of sharp peaks in the measured current. At any given peak, the number of electrons on the dot alternates between N and $N + 1$. Between the peaks, the current is zero and N remains constant. The distance between consecutive peaks is proportional to the so-called addition energy, which is the difference in energy between dots with $N + 1$and N electrons. The magic numbers can be identified because significantly higher voltages are needed to add the 2nd, 6th, and 12th electron.

Quantum dot composites can be used for next-generation telecom devices and can be incorporated into planar lightwave circuits for high-speed signal-processing applications in optically routed networks. QDs can be utilized in the manufacturing of telecom lasers that operate well at relatively high temperatures (thereby eliminating the need for cooling of the laser, which adds cost to the system): QD lasers can

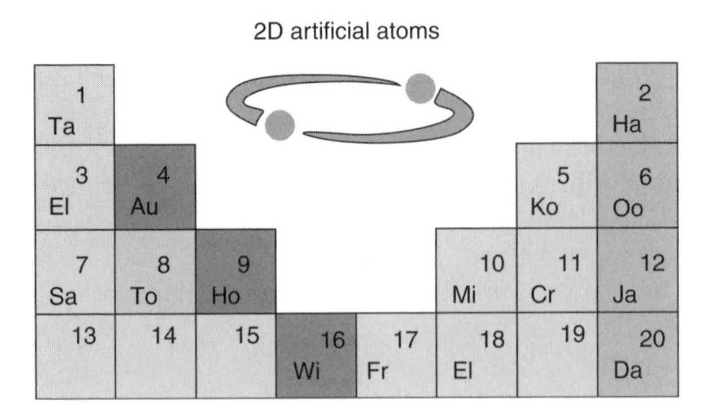

FIGURE 4.12 Periodic table of artificial 2D elements. (Courtesy: The Delft Spin Qubit Project, Kavli Institute of Nanoscience at Delft).

operate at 100°C. However, developers see a 2007–2009 commercial realization rather than a short-term rollout: The main challenge here relates to manufacturability and reliability. Material growth (creating the QDs on a wafer) is the issue: One needs to be able to grow them in a uniform and predictable manner, in high-yield environments. Given that there has been a slowdown in the consumption (and development) of (commercial-level) optical telecom technology in the 2001–2005 time frame, the overall engineering production cycle has eased a bit of late; however, the production cycle is expected to resume its historical trend in the near term. It should be noted, though, that other (non-QD-based) uncooled-laser technology is emerging of late; a number of manufacturers have announced uncooled distributed feedback (DFB) lasers that operate in the 100–120°C range and can transmit information at 10 Gbps.

As noted above, QDs can also be used for information technology (IT) applications. Traditional IT is based on incoherent processes in conventional semiconductor devices. To facilitate future quantum-based IT, which is based on coherent phenomena, a new type of "hardware" is required. Semiconductor QDs are promising candidates for the basic device units for quantum information processing; one approach is to exploit optical excitations (also known as excitons) in QDs (as previously indicated, an exciton is a quasi-particle comprised of a negatively charged electron bound together with a positively charged "hole") [197]. This topic is revisited in Chapter 6.

4.2.6 Quantum Dots Nanocrystals

Somewhat related but not exactly identical to QDs are quantum dots nanocrystals (QDNs). QDNs are "tiny" (5–10 nm) semiconductor nanocrystals that have the ability to glow in various colors when excited by laser light. QDNs are used to tag biological molecules [198]. QDNs' cores contain paired clusters of atoms that combine to create a semiconductor; the clusters are surrounded by a shell made of an inorganic substance to protect the clusters. The cluster releases light of a specific color when stimulated by ultraviolet light. QDNs in solid matrix (composite) materials allow product developers to control the form factor of nanocrystals. QDN matrix material allows creation of films, beads, fibers, and micron-sized particles for numerous applications. QDN composites accelerate engineering and development of nanocrystal applications including photonics, LEDs, ink, and paints [199]. See Figure 4.13 for one example of a QDN [198].

Because QDNs are nanosized semiconductor crystals that fluoresce in several different colors upon excitation with a laser source, they are finding applications in biotechnology (biological reagents and cellular imaging), and in engineering (LEDs, lasers, and telecommunication devices such as optical amplifiers and waveguides) [199]. Quantum dot composites accelerate engineering and development of nanocrystal applications including photonics, LEDs, ink, and paints; the spectral features, that is, wavelength and intensity, of fluorescence generated from semi-conductor nanocrystals can be used for coding information; unlike the 1D and 2D barcodes, the information carrier is applied to a very small area and is inconspicuous; the information is retrieved by a fluorospectrometer. This technology can be applied to small products labeling, document security, and object identification [198, 200, 201, 202].

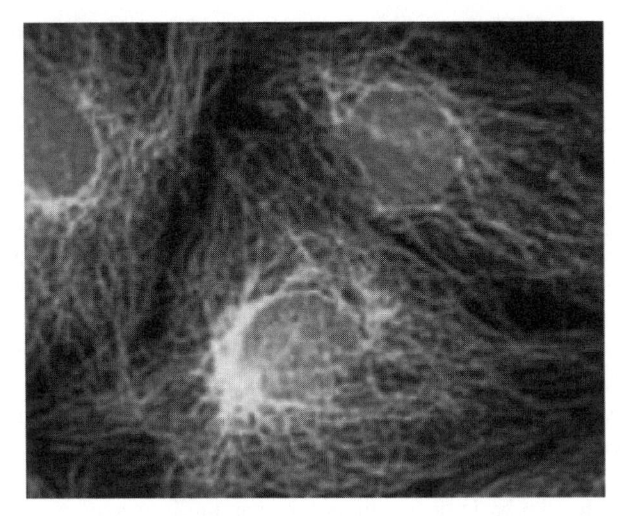

FIGURE 4.13 QDN. (Courtesy: Quantum Dot Corporation).

Quantum dot nanocrystals can be utilized as nanosized markers to visualize DNA sequences, proteins, or other molecules and track them in the cell; the QDNs glow in a variety of colors and are up to 1000 times brighter than conventional fluorescent dyes [203]. These crystals also enable the researchers to generate real-time video-clips of signal transmission in receptors, which are important targets for many antitumor drugs.

The recent research has been directed to achieving better control of QDN self-assembly with the goal of eventually using these unique materials for quantum computing. Semiconductor quantum dots combine many of the properties of atoms, such as discrete energy spectra, with the capability of being easily embedded in solid-state systems [199].

The semiconducting materials include cadmium selenide (CdSe), cadmium sulfide (CdS), zinc selenide (ZnSe), and zinc sulfide (ZnS). QDNs can be made from a single compound, such as CdSe or ZnS, or from multiple compounds in a specific manner such as CdSe–ZnS core–shell configuration. For the same materials system, the smaller the QDN particles, the shorter the emitted fluorescent wavelength. For example, CdSe nanoparticles with a nominal diameter of 2.8 nm show the fluorescence at 535 nm,* while CdSe QDNs of 5.6 nm diameter have an emission centered at 640 nm. QDNs of lead selenide (PbSe) of various diameters can emit fluorescence in the near-infrared range. A properly designed mixture of QDNs with different emission wavelengths can emit light with spectral features that represent a set of data. The idea of using QDNs for spectral coding has been demonstrated in the bioanalytical area. Prototype systems capable of retrieving the information coded with QDNs on the surface of an object, including the passport page, ID card, and even a nail of a human finger, have been advanced [200, 201, 202, 202a, 202b, 202c, 202d].

*The visible light region spans the wavelength from around 400 nm (violet) to around 700 nm (red).

4.2.7 Ultrananocrystalline Diamond

Ultrananocrystalline diamond (UNCD), a form of industrial diamond, captures many of the best properties of natural diamond in thin-film form. UNCD has unique properties not found in any other carbon-based material and can be considered a new allotrope of carbon along with diamond, graphite, and fullerenes. UNCD is currently being evaluated for a variety of applications including MEMS (RF and optical-MEMS, BioMEMS), cold-cathode electron sources, chemical process pump seals, bioelectrochemical electrodes, among others. The extreme hardness of UNCD makes it ideal for MEMS: UNCD is harder than any other material currently used in macro- or microdevices. It is highly wear resistant, lasting 10,000 times longer than silicon in MEMS devices [204].

4.2.8 Diamondoids

Diamondoids are diamond molecules that possess the same rigid carbon framework as diamond, making them attractive materials for nanometer-scale construction and other applications. Adamantane ($C_{10}H_{16}$) is the smallest member of the diamondoids family, consisting of one diamond crystal subunit. Diamantane contains two diamond crystal subunits and triamantane contains three. Recently, researchers discovered larger members of this class of compounds. These are known as higher diamondoids [205]. These diamondoids include tetramantanes through undecamantane, 1- to 2-nm hydrogen-terminated diamonds containing 4–11 face-fused diamond crystal subunits. These molecules encompass a variety of 3D shapes, including rods, helices, and discs, arising from different ways of face-fusing diamond crystal subunits. The expectation is that higher diamondoids make possible new applications in a wide range of fields, including pharmaceuticals, microelectronics, and optics, employing polymers, films, and engineered crystals [205].

4.2.9 Nanocomposites

Within the class of nanoscale materials (e.g., see Table 4.7 inspired by [206]), nanocomposites are of major interest. In nanocomposites the components are mixed at the nanoscale, resulting in materials that often have properties that are superior to conventional microscale composites and, at the same time, can be synthesized using simple and inexpensive techniques (see Table 4.8 [206]). A nanoscale dispersion of sheetlike inorganic silicate particles in a polymer matrix, for example, is superior to either constituent in such properties as optical clarity, strength, stiffness, thermal stability, reduced permeability, and flame retardancy (see Table 4.9 for a list of properties of interest [17, 206]). Recent examples of composite materials with unique characteristics include magnetic and semiconductor multilayers made from magnetite nanoparticles, which can also be combined with carbon nanotubes and clay platelets [207]. Nanostructured materials include metals, composites, polymers, liquid crystals, clusters, quantum dots, colloids, and nanotubes; nanostructures include fullerenes, nanotubes, clusters, layers, quantum dots, thin films, surfaces, and interfaces [208].

TABLE 4.7 Taxonomy of Nanocomposites

1. *Bulk Metal and Ceramics Nanocomposites*
- Ceramic/metal nanocomposites
  - Nanocomposites by mechanical alloying
  - Nanocomposites from solgel synthesis
  - Nanocomposites by thermal spray synthesis
- Metal matrix nanocomposites
- Bulk ceramic nanocomposites for desired mechanical properties
- Thin-film nanocomposites: multilayer and granular films
- Nanocomposites for hard coatings
- Carbon nanotube-based nanocomposites
- Functional low-dimensional nanocomposites
- Inorganic nanocomposites for optical applications
- Inorganic nanocomposites for electrical applications
- Nanoporous structures and membranes: other nanocomposites
- Nanocomposites for magnetic applications
  - Particle-dispersed magnetic nanocomposites
  - Magnetic multilayer nanocomposites
- Nanocomposite structures having miscellaneous properties

2. *Polymer-Based and Polymer-Filled Nanocomposites*
- Nanoscale fillers
- Nanofiber or nanotube fillers
  - Carbon nanotubes
  - Other nanotubes
- Platelike nanofillers
- Equi-axed nanoparticle fillers
- Inorganic filler polymer interfaces

3. *Biomimetic-Inspired Nanocomposites*
- Natural nanocomposite materials
  - Biologically synthesized nanoparticles
  - Biologically synthesized nanostructures
- Biologically derived synthetic nanocomposites
  - Protein-based nanostructure formation
  - DNA-templated nanostructure formation
  - Protein assembly
- Biologically inspired nanocomposites
  - Lyotropic liquid-crystal templating
  - Liquid-crystal templating of thin films
  - Block-copolymer templating
  - Colloidal templating

The key to the synthesis of nanocomposites is how the silicates are made to disperse in the polymer; since the organic and inorganic components are typically immiscible (like water and oil), the silicate surface is modified by attaching surfactant molecules [17]. Polymer-based nanocomposites are also being developed for electronics applications such as thin-film capacitors in integrated circuits and solid polymer electrolyes for batteries. There are nature-inspired nanocomposites (e.g., nanobiocomposites, biomimetic nanocomposites, and biologically inspired

TABLE 4.8 Processing of Polymer Nanocomposites

Layered filler polymer composite processing
Polyamide matrices
Polyimide matrices
Polypropylene and polyethylene matrices
Liquid-crystal matrices
Polymethylmethacrylate/polystyrene matrices
Epoxy and polyurethane matrices
Polyelectrolyte matrices
Rubber matrices
Nanoparticle/polymer composite processing
Direct mixing
Solution mixing
In situ polymerization
In situ particle processing ceramic/polymer composites
In situ particle processing metal/polymer nanocomposites
Modification of interfaces
Modification of nanotubes
Modification of equi-axed nanoparticles
Small-molecule attachment
Polymer coatings
Inorganic coatings

TABLE 4.9 Properties of Polymer Composites

Mechanical properties
Modulus and the load-carrying capability of nanofillers
Failure stress and strain toughness
Glass transition and relaxation behavior
Abrasion and wear resistance
Permeability
Dimensional stability
Thermal stability and flammability
Electrical and optical properties
Resistivity, permittivity, and breakdown strength
Optical clarity
Refractive index control
Light-emitting devices
Other optical activity

nanocomposites) that scientists are trying to emulate for industrial applications (as an example, the shell of some mollusks like abalone have alternating layers of calcium carbonate and a rubbery biopolymer that make it twice as hard and a thousand times tougher than its components).

To attain higher-speed communication, one needs to use photons, rather than electrons, to transmit signals. The nanocomposite materials used for optical components are called photonic crystals. These materials are two-phase periodic structures with a large difference in refractive index between phases. Block copolymers offer a convenient route to create self-assembled photonic materials, but native refractive index contrast is low. To address this, researchers have proposed methods for embedding nanoparticles within one microphase to create dielectric contrast. In fact, these nanoparticles do provide the necessary contrast, but they also alter the onset and growth of defects, such as crazes; these defects or crazes naturally will affect the dielectric properties. Research is currently underway to understand the design constraints [209] (the field of nanophotonics is discussed in greater depth in Chapter 5.)

Layered nanocomposites with a high degree of organization can be prepared from polymers and a variety of nanocolloids such as nanoparticles, nanowires, nanotubes, clay platelets, and proteins by thin-film deposition technique known as layer-by-layer assembly. Here, (sub)monolayers of the organic and inorganic materials are deposited in regular stacks determined by the deposition protocol [207]. Control of distance and orientation of nanocolloids in the multilayers affords fine tuning of the composite properties—optical, electronic, and magnetic. The technique also makes possible the preparation of the ordered composites from core–shell nanoparticles. The successful deposition of the multilayers with complex nanostructures can also be extended to include biological applications, which include implantable sensors and artificial tissues [207].

4.2.10 Thin-Films

At the nanoscale level, thin films are 2D nanostructures. Thin-film nanocomposites are typically composed of nanocrystalline grains with sizes in the range 3–10 nm of a hard, ceramic material in a matrix of an amorphous or crystalline material. Compared with existing technologies, thin nanocomposite films offer a much wider variety of opportunities for structural control: The structure may be tailored not only at the molecular and microlevel but also at intermediate scales. This offers the potential of hierarchical control similar to the more sophisticated morphologies present in materials synthesized in nature [210]. For example, the architecture of mother-of-pearl, consisting of alternating tablets of aragonite (a few hundreds of nanometers thick) and thin organic films (a few tens of nanometers thick) gives the mollusk shell exceptional strength without the brittleness associated with pure inorganic phases. While this construction model has been of interest to material scientists, research involving organic/inorganic interfaces, thin layers, and lamellar heterostructures has expanded beyond their mechanical properties to include structural, electronic, and optical properties of mesoscale composites [211].

There are major potential application opportunities for thin nanocomposite films in the areas of new paints, coatings, diffusion barriers, and functional polymer films; also, a large variety of processing strategies are possible, including solvent casting, water suspensions, and UV and thermal curing [210]. For example, there is a lot of interest in studying ferroelectric/oxide thin-film structures that are nanopatterned and/or compositionally graded. The properties of these ferroelectric films are expected to have a broad technological impact on applications such as dynamic random-access memories (DRAMs), nonvolatile RAMs (NVRAMs), thermal imaging, piezoelectric displays, electronic data storage, sensors, transcapacitors, and MEMSs [146]. GMR are also based on layers of small ferromagnetic single-domain particles with randomly oriented magnetic axes in a nonmagnetic matrix; as noted briefly in Chapter 3, GMR relates to the decrease of electrical resistance of materials when exposed to a magnetic field [91].

The unique microstructures of nanocrystalline materials suggest that these materials have the potential of exhibiting exceptional mechanical properties. As a consequence, they have been attracting wide attention in materials research [212]. For example, capacitors in which the main dielectric layers are made from sintered nanocrystalline $BaTiO_3$ (grain sizes less than about 100 nm) have been fabricated and tested in a continuing effort to increase energy densities, breakdown potentials, and insulation resistances beyond those of prior commercial capacitors that contain coarser-grained sintered $BaTiO_3$ [213]. Inorganic/organic thin films include:

- Organic/inorganic interfaces and biomimetic thin films
- Polymer films on inorganic substrates
- Patterned polymer and inorganic films on inorganic substrates
- Organic monolayers and organic/organic nanolaminate films
- Inorganic-on-inorganic thin films and microlaminates

The synthesis of inorganic-on-organic thin films and nanolaminates can be taxonomized into two general approaches [211]: (i) nanoparticle-based synthesis, where the inorganic phase is preformed and (ii) molecular precursor-based synthesis, where the inorganic phase forms in situ, either by precipitation or hydrolysis/condensation reactions. Synthesis methods using preformed inorganic particles generally fall into four categories: (a) Langmuir–Blodgett (LB) deposition (already used successfully with a variety of near-symmetric nanoparticles; recent examples include semiconductors, ferroelectrics, and metals transferred onto oxide substrates), (b) covalent self-assembly, (c) alternating sequential adsorption, and (d) intercalation of organics into layered inorganic structures.

Alkanethiol self-assembled monolayers (SAMs) are one type of nanometer-scale chemical systems that are being studied for applications such as chemical sensing, biological assays, and molecular electronics [214]. For over 15 years, scientists have studied alkanethiol monolayers because they allow researchers to pattern molecules on surfaces in reproducible ways. Alkanethiol SAMs form spontaneously on gold surfaces; the molecules are anchored strongly to gold by their sulfur atoms resulting in densely packed, single-molecule-thick thin films. The functional group on the other end of the alkanethiol molecule can be tailored to obtain desirable surface properties of the monolayers; for example, DNA chips used to identify unknown DNA sequences are created using alkanethiol chemistry, with DNA serving as the functional group that can capture complementary DNA molecules for analysis [214].

4.2.11 Nanofoam

Recently, researchers announced that they had created a new form of carbon: a spongy solid that is extremely lightweight and is attracted to magnets. The new structure was created when physicists bombarded a carbon target with a laser capable of firing 10,000 pulses a second. As the carbon reached temperatures of around 10,000°C, it formed an intersecting web of carbon tubes, each just 1 nm in diameter. The researchers have called the solid a *nanofoam* [215]. Applications are still evolving.

4.2.12 Nanoclusters

Nanoscale metal and semiconductor particles are of interest because they mark a material transition range between quantum and bulk properties. With decreasing particle size, bulk properties are lost as the continuum of electronic states becomes discrete (the quantum size effect) and as the fraction of surface atoms becomes large. The electronic and magnetic properties of metallic nanoparticles and nanoclusters have new characteristics that can be utilized in novel applications. Applications range from nonlinear optical switching and catalysis to high-density information storage [216]. A number of methods have been developed to synthesize metal nanoparticles, but there currently continues to be major challenges with scaling-up of these methods.

4.2.13 Smart Nanostructures

Researchers from Duke University have described building so-called *smart nanostructures*, including nanoscale *nanobrushes* that can selectively and reversibly sprout from surfaces in response to changes in temperature or solvent chemistry. Using an atomic force microscope (AFM) (see Appendix F) researchers reportedly created reprogrammable "nanopatterns" of large biologically based molecules that can potentially be used to analyze the protein contents of individual cells, among other uses [217].

4.2.14 Environmental Issues for Nanomaterials

Some nanomaterials may have environmental effects that require proper handling and disposing of the material. In effect, this is no different than chemicals currently being used in the production of standard highly integrated chips. Recent toxicology studies appear to show that some adverse effects of nanoparticles (buckyballs in particular) on aquatic animals may exist (could apparently cause brain damage in fish), but the issue needs further study according to most observers. The toxicity of the nanoparticles was not conclusive and, in fact, it was suggested that the clumping was a good sign, since it prevented the nanotubes from reaching deeper into the lungs [218]. Eva Oberdorster, the study's leader and an environmental toxicologist with Southern Methodist University, was quoted as saying "I want to emphasize that the benefits of nanotech are great, and we definitely should not put the brakes on positive nanotechnology research"; buckyballs show promise for electronic, industrial, and pharmaceutical applications [219, 220].

4.3 MANUFACTURING TECHNIQUES

Earlier we mentioned the bottom-up (self) assembly approach, as well as the top-down approach that involves the lithography/etching of bulk material similar to the processes currently used in the semiconductor industry. Example of top-down tools include (but are not limited to) the atomic force microscope and scanning tunneling microscope (STM); these tools have allowed researchers to both detect and manipulate individual atoms [9]. Top-down methods create nanostructures out of macro-structures, while bottom-up methods entail self-assembly of atoms or molecules into nanostructures. Self-assembly has the advantage of mimicking biological systems. This section briefly looks at these two classes of manufacturing methods.

Microfabrication techniques are being increasingly used to manufacture nanoscale devices as well as conventional optoelectronic structures. However, one must find an efficient method of manipulating molecules on an industrial scale, if nanotechnology is to be economically feasible: The major limitation of the positional assembly method is its low throughput and cost because even though one can bind molecules with an STM, it is not (yet) economically feasible.

Traditional manufacturing techniques for nanomaterials are based on vapor-phase processes such as molecular beam epitaxy (MBE), chemical vapor deposition (CVD), and DC and RF magnetron sputtering for thin films and multilayers; and on chemical vapor synthesis (CVS) and inert gas condensation (IGC) for clusters and nanocrystalline materials [713]. Self-assembly relies on weak atomic and molecular interactions to hold the macromolecule together and is efficient at building complex molecules [9, 221]. Table 4.10 identifies some of the major approaches and techniques for nanomaterial synthesis and assembly, while Table 4.11 summarizes some of the fabrication processes for nanodevices/nanostructures.

A variety of manufacturing techniques are surveyed below. Some of these apply to the manufacturing of SWNT; other techniques apply to a variety of nanomaterials.

4.3.1 General Approaches

Fullerenes and carbon nanomaterials can be prepared from a number of starting materials by diverse techniques. Carbon nanotubes manufacturing techniques include the following three main techniques:

- Carbon arc discharge (CA)
- Pulsed-lased vaporization (PLV)
- Chemical vapor deposition (CVD) (see Table 4.12 for a description of traditional methods, including CVD, as used in semiconductor/microchip fabrication)
- High-pressure CO conversion (HiPco)

TABLE 4.10 Fabrication Techniques for Nanodevices/Nanostructures

Top-Down Approaches
Scanning probe microscope (SPM)
Dip-pen nanolithography (DPN)
Extreme ultraviolet (EUV)
Electron beam nanolithography
Molecular beam epitaxy (MBE)
X-ray nanolithography
Focused ion beam
Light coupling nanolithography:
Imprint nanolithography
Bottom-up Approaches
Selective growth
Self-assembly
Scanning tip manipulation

TABLE 4.11 Nanomaterials Synthesis and Assembly

	Approaches to Manufacturing Nanodevices
Biomimetic approach	"Copying," "emulating," "imitating," or "learning" from nature. The biomimetic approach for manufacturing nanosystems was proposed by Feynman and Drexler. Nanotechnology has been linked with biological systems since its inception by Richard Feynman in his 1959 speech. In 1981, Eric Drexler published on molecular engineering and manipulation on the atomic scale and focused on protein synthesis as a mechanism for creating nanoscale devices [222]. The advantage of the biomimetic approach is that nature has already proven that it is possible to make complex machines on the nanoscale [9].
Inorganic methods	Richard Feynman also mentioned the possibility directly manipulating atoms and molecules. Major progress has been made in the last half-century, and, at this time, a number of approaches, including among others, conventional photolithography-based approaches, AFM-based approaches, and STM-based approaches can be utilized in conjunction with *inorganic* materials [9].
	Assembly of Nanodevices
Positional assembly	Historically, machinery in the macroworld is assembled by physically bringing components together and then fastening them in a process referred to as positional assembly. In the nanoscale, the idea of bringing atoms together and "fastening" them becomes somewhat challenging. However, there appears to be no fundamental barrier for using positional assembly to create

TABLE 4.11 (*Continued*)

	nanodevices; however, we have yet to match the efficiency and precision of biological positional assemblers [9].
Self-assembly	Molecular self-assembly is a process in which molecules spontaneously form ordered aggregates and involves no human intervention; the interactions involved usually are noncovalent [223]. Molecular self-assembly is ubiquitous in biology; examples include protein folding, formation of nucleic acid structures and macromolecules such as the ribosome [9, 223, 224].

TABLE 4.12 Subset of Basic Traditional Growth Techniques (Semiconductor/Microchip Methods)

Chemical vapor deposition (CVD)	Chemical growth process where a reaction between gaseous reactants creates products that are deposited as solid. It is a fabrication technique used to deposit thin layers of material on a substrate. Often, a crystalline substrate is employed leading to epitaxial growth of the deposited material. Deposition occurs when a heated substrate is placed in a stream of vapor containing materials that react on the hot substrate surface, leading to growth. CVD is the general term used to describe MCVD, OVD, VAD and other preform manufacturing processes (for optical materials) where chemical vapors (presence of heat) are deposited on the surface of a substrate. Metalorganic chemical vapor deposition (MOCVD) is a type of chemical vapor deposition process where metalorganic materials (group III elements such as Ga and Al) are employed as source gases for the deposition process. One of the two main techniques for fabricating epitaxial layers of compound semiconductors [the second technique being molecular beam epitaxy (MBE)].
Molecular beam epitaxy (MBE)	Process used to make compound (multilayer) semiconductors. The process consists of depositing alternating layers of materials, layer by layer, one type after another (such as the semiconductors gallium arsenide and aluminum gallium arsenide). A form of vacuum evaporation where the vacuum levels (referred to as *ultra high vacuum* conditions) are on the order of 10^{-11} torr, which permit molecular flow (i.e, molecules from the source arrive at the substrate without suffering collisions with other molecules). The sources provide beams of material used for deposition. MBE is one of the most sophisticated epitaxial techniques available, offering high flexibility during growth and highest quality material. Considering the slow growth rates, this technique permits atomically engineered device structures (specifically, *nanostructures*) to be fabricated.

Currently, processes for carbon nanotube synthesis use either of the following methods:

- Group 1: arc, laser, solar, plasma (vapor) methods
- Group 2: catalytic methods

Approaches in group 1 employ high-temperature methodologies. Here, the carbon nanotube formation process is based on the sublimation and recondensation of the carbon precursor. Within group 1, the first three processes have reached their limits in production rate. A fundamentally new approach is employed with the plasma process. Here, the carbon mass flow is no longer limited by a physical ablation rate but is freely adjustable. Moreover, the process is operated at atmospheric pressure and the carbon nanotube soot is extracted continuously. Plasma technology shows a significant potential for the continuous production of bulk quantities of carbon-based nanotubes having controlled properties and novel structures [225].

The current bottom line is that manufacturing techniques have limited the development of production of methods to produce large quantities of SWNTs, including the laser ablation/oven technique, the arc vaporization process, and the chemical reaction methods such as the high-pressure carbon monoxide (HiPco) disproportionation and combustion techniques. Currently, there is insufficient understanding of the growth mechanisms to be able to translate the mechanisms into better production methods [226].

Researchers (including Richard E. Smalley's group at Rice University) have developed HiPco processes to produce SWNT from gas-phase reactions of iron carbonyl in carbon monoxide at high pressures (10–100 atm). The HiPco process is a relatively clean process, that is, SWNT is obtained without the graphitic deposits and amorphous carbon; and, the process has the potential for producing SWNT in reasonable quantities (currently 10 g a day) since it is from a gas-phase reaction [227, 227a].

The synthesis of nanotubes by laser evaporation of graphite enriched by a nickel–cobalt alloy is relatively uncomplicated when measured by today's technology standards. The 80–90% efficiency of nanorope production from the raw material is advantageous, given that it implies a catalytically assisted self-assembly mechanism on the atomic scale. Still, significantly more research is needed before bulk production can be expected [110].

Among the starting materials tested for SWNT manufacturing, different types of coals have been examined by the arc discharge method. The advantage of preparing fullerenes and carbon nanomaterials with this method lies both in cost and in the fact that coal contains weak covalent bonds linking macromolecular units together (whereas the covalent bonds in graphite are all the same strength). The weak bonds present in coal and coal-based cokes may break preferentially, resulting in the formation of larger fragments other than mono or dimeric carbon species [109]. Currently, multiwall nanotubes can be produced by these methods, but they have a high percentage of structural defects; these defects affect the material properties of the substance (e.g., strength or electrical conductivity.) Improvements, clearly, are being sought.

There is interest in characterizing the purity and dispersion of nanotube material. No single experimental technique exists that can reliably determine the purity of SWNT material, nor can one assesses the composition and properties of the impurities that are generated during SWNT production. The same is true for assessing the degree of dispersion in liquids and matrices that can influence the SWNT interactions with other materials. Understanding the properties that facilitate purification and dispersion are critical to the development of all SWNT applications and for further improving the current purification protocols. Researchers worldwide resort to a combination of available techniques to characterize SWNTs; however, such a variety of techniques combined with differences in methodology and interpretation complicates the data comparison of SWNT materials. Furthermore, thorough criteria for assessing dispersion are lacking [228]. In general terms, quality and reproducibility of the materials are achieved by establishing clean conditions such as ultra-high-vacuum environments, computer control of the synthesis parameter, and in situ analysis techniques such as reflection high-energy electron diffraction (RHEED) [16].

In conclusion, the commercial potential of SWNTs is limited by the (current) inability to produce large quantities of the nanomaterial. The development of a reliable source of large quantities of SWNTs depends on better production methods; however, the growth mechanisms of single-wall carbon nanotubes are not well understood as of yet. The following list identifies some key goals related to nanotube manufacturing [226]: (i) Ensure a reliable source of nanotubes with controlled properties (length, purity, diameter, chirality) using diagnostics, parametric studies, and modeling to understand and improve processes; (ii) develop and employ characterization techniques to examine nanotubes and nanoscale materials; and (iii) develop processing methods for nanotubes from various sources to enhance structural, thermal, electrical, and chemical properties.

4.3.2 Self-Assembly Methods

Self-assembly refers to the ability of (specific) organic or inorganic molecular structures to spontaneously organize into ordered 1D, 2D, or 3D arrays under appropriate processing conditions. It needs to be noted, however, that the majority of current self-assembly mechanisms lack the reliability and reproducibility necessary for the deposition of individual nanotubes and wires controllably and consistently on microprocessor chips; the current techniques result, in many cases, in bundling, overlapping, and/or crossing of multiple tubes and wires. In addition, self-assembly fabrication techniques are, in general, incompatible with prevailing semiconductor processing procedures. These incompatibilities include vapor deposition approaches for carbon nanotubes that require high processing temperatures and liquid-phase deposition approaches that are not amenable to high-volume processing on wafer platforms [229].

As noted, in the context of self-assembly, the challenge is for scientists and researchers to uncover, understand, and develop repeatable processes for self-assembly; specifically how self-assembly is achieved, how it can be controlled, and how it can be effectively applied to strategies for nanofabrication. Nature uses self-assembling materials for nanostructures as the components for living cells; however, human attempts at similar structures are limited to building self-assembling nanoscale materials a few atoms or molecules at a time. A variety of examples of self-assembly can be observed in biology, in the fields of embryology and morphogenesis; in chemistry, where groups of molecules form more loosely bound supramolecular structures; and in robotics, where some efforts are aimed at producing and programming robots capable of self-assembly. In constructing nanostructures and nanoelectronic devices, chemical self-assembly has become an important factor in building supramolecular nanostructures with useful electrical properties. One example is in the field of X-ray nanolithography, which is well-suited for generating patterns on the submicron scale. However, it is not the best method for accurately manipulating structures that are less than approximately 30 nm wide; therefore, new self-assembly techniques are now being investigated for going beyond the limits of this technology [22]. Several self-assembly approaches have been studied and employed over the years. Some of these approaches are discussed in this section.

Biomimetics and molecular self-assembly (MSA) are now attracting interest from many researchers and industries. These self-organized systems are important in nanotechnology research because of their potential use in the bottom-up development of functional supramolecular structures. The construction of complex assemblies using molecular entities is a key point for the development of novel functional materials [230].

Biomimetics refers techniques that aim at copying, emulating, imitating, or learning from nature; it is the study and process of optimizing functions in nature to look for design solutions in biology. Nanoscale devices already exists in nature; hence, researchers have a plethora of components and techniques already available. Biomimetics is the study of the structure and function of biological substances with the goal of making artificial products that mimic nature. It also refers to the process of designing molecules, molecular assemblies, and macromolecules having biomimetic functions; hence, biomimetic materials are materials that copy, emulate, or imitate nature. Biomimetics provides a set of techniques used to develop novel synthetic materials, processes, and sensors through advanced understanding and exploitation of design principles found in nature.

The MSA method is an efficient mechanism for the self-regulated creation and/or fabrication of nanoscale elements and machinery; it is a chemistry-based method for assembling atomically precise materials, similar, in some ways, to biological-based mechanism. Applications include, among others, electronics, energy, material sciences, and medicine. Learning from nature, one can improve biocompatability, produce more complex electronic circuits, engineer material surfaces to have many different properties, efficiently harvest energy from natural light, and provide locomotion and motility to nanoscale objects and mechanisms [231]. MSA research for nanotechnology applications, however, is just at its infancy. Some see chemistry and polymer science of the future as being directed at making materials that are able to orchestrate their own growth in an autonomous fashion: with self-assembly one lets nature do most of the work.

The MSA technique is an important and exploitable method for assembling atomically precise materials and (in the future) atomically precise devices. Biological organisms (including humans) are composed of molecular building components (e.g., nucleic acids, proteins) that they are able to assemble into well-organized cellular or other structures. Leveraging MSA will provide, over time, a cost-effective mechanism for the creation and/or fabrication of nanoscale machinery and microelectronics. Macroscale-level self-assembly exists throughout the natural world, and self-assembled monolayers are already found in a number of industries (e.g., 3M manufactures a polish that uses a self-assembling monolayer system to prevent silver from tarnishing). Now, the idea is to look at these approaches as a discrete set of processes that can enable nanotechnology to engineer and fabricate devices at a molecular level [232]. Although the microfabrication of electronic circuits currently uses solid-state or inorganic materials, one would want to be able to use organic and biological materials for electronic purposes. Some areas of interest in MSA include:

- Fluid self-assembly (FSA)
- Labs-on-a-chip
- Material self-assembly
- Smart plastics

Fluid self-assembly has already been used to deposit integrated circuits across plastic substrates. Here, the transistors are made on standard silicon wafers and are etched to separate them; then the transistors are floated into place across a large surface area covered with holes shaped like the transistors. Once a circuit lands in a hole, it lines itself up perfectly because it only fits one way. The process uses a thin, light, flexible, low-cost plastic film and allows continuous flow, roll-to-roll (web) processing [232]. Monochrome display for smart cards have already been manufactured using this technique.

According to some, tailor-made, submicrometer particles in colloidal suspensions will be the building blocks of a new generation of nanostructured materials with unique physical properties; the basis of this prediction is that the macroscopic physical properties of colloidal suspensions can be influenced by tuning the interactions between their building blocks [233]. If the forces between macromolecules in solution were similar in shape to those between gas atoms, then all colloidal suspensions would exhibit the same phase diagram as an assembly of argon atoms; however, the forces between macromolecules in solution manifest themselves in a variety of ways. One is able to "tune" these forces by choosing an appropriate combination of solvent, solute, and additives. It follows, according to some, that far from being simply a scale model for atomic fluids, colloidal suspensions can form new states of matter, the building blocks of which are large "designer atoms" [233, 234].

Self-organized supramolecular organic nanostructures have potential applications that include molecular electronics, photonics, and precursors for nanoporous catalysts. It follows that understanding how self-assembly is controlled by molecular architecture will enable the design of increasingly complex structures [235]. Molecular self-assembly into a variety of bulk phases with 2D and 3D nanoscale periodicity, such as cubic, cylindrical, or mesh phases, has been researched in lyotropic

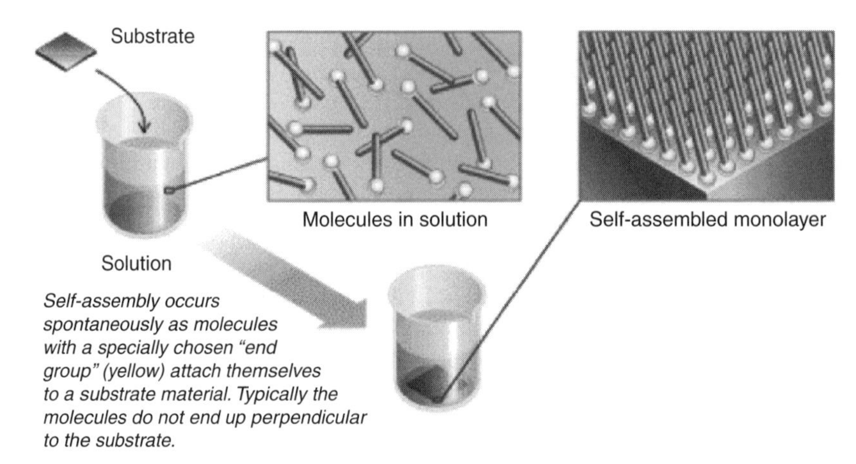

FIGURE 4.14 Bottom-up synthesis of a self-assembled monolayer (SAM). Self-assembly occurs spontaneously as molecules with a specially chosen "end group" (yellow) attach themselves to a substrate material. Typically the molecules do not end up perpendicular to the substrate. (Courtesy: Esko Forsén, Technical University of Denmark, Department of Micro and Nanotechnology).

(e.g., surfactant–water) liquid crystals (LCs) [236], block copolymers [236a, 236b, 236c], and thermotropic (solvent-free) LCs [236d]. Figure 4.14 [237] depicts for illustrative purposes one example of a self-assembly technique.

Self-assembly is also usable for lab-on-a-chip, which is a commercial sensor array for high-throughput screening in genomics and drug development. Some vendors already market such systems; plans exist to extend the technology further by combining electronic concentration, focusing, transport, and analysis of a biological sample on an integrated device [232]. Researchers have used the concept to allow molecules to self-assemble in order to identify molecules that bind to receptors; these methods may find application in the production of next-generation chemicals (other than drugs) by developing a group of molecules that can self-assemble and by developing a selection system to pull out the molecules that are desirable. Currently, there is information in chemical signatures that is not being used because it is costly to extract such information; hence, efforts have been underway to develop sensors about the size of a palm-top computer (based on the lab-on-a-chip concept) that is able to detect explosives and chemical warfare agents. The midterm goal is building labs-on-a-chip that simultaneously identify hundreds of liquids and gases.

The feasibility and commercial promise of labs-on-a-chip and fluidic self-assembly techniques have led to more advanced plans that seek to exploit MSA to construct new materials and control their properties. The ability to directly monitor self-assembly and self-organizing processes, to probe single particles and nanoscale architectures, and to perform sophisticated physical measurements with high spatial resolution is central to the development of such advanced materials. These materials are expected to have applications in light-emitting diodes, optical memories, switching devices, and improved chemosensors (e.g., intelligent spray-on materials that assemble themselves into circuits, as a new, quicker way to build sensor chips for handheld chemical analyzers) [232]. For example, one can spray sensor arrays onto a silicon substrate using an ink-jet printer; sprayed in a line on the substrate, the pores spontaneously form a waveguide measuring about 2.5 nm in diameter and become solid when heated. The technique may also prove useful for molecular filters, nanocomposite materials, and passive circuits for devices with dielectric constants lower than those made using conventional lithography, all of which are critical to realizing molecular-scale electronic devices. There is interest in applying these techniques to patterned nanostructures for use as optical waveguides, laser/waveguide microarrays, and optical fiber microring lasers [232]. At Northwestern University, for example, researchers have developed a dip-pen nanolithography (DPN) technique based on an atomic force microscope that enables the researchers to draw fine lines one molecule high and a few dozen molecules wide; this nanoplotter lays down a series of molecule only 15 nm wide, with 5 nm separating each line.

Smart plastics are plastic materials that assemble themselves into photonic crystals using hierarchical self-assembly. Many research groups have built photonic crystals, but the standard manufacturing approach usually involves laborious and expensive fabrication processes. "Smart plastics" are the first photonic crystals to grow themselves. The control of electrons led to the microelectronics evolution; one would like to be able replicate these advancements using photons. To achieve these advancements one needs materials such as photonic crystals that allow one to trap light and control the way it propagates. Light can carry much more information than electrons. Applications include optical data storage (including holographic storage), telecommunications, high-resolution video imaging, ultrafast switching, logic devices, and solar power generation. (The Oak Ridge National Laboratory has reportedly used self-assembly to create spinach-based optoelectronic circuits; the goal of the research is to preformat, predesign, and preink substrates onto which molecular components can be assembled into functional devices—this being done in a less expensive and less energy-intensive manner than conventional lithography) [232].

Another promising building block is the organic semiconductors (poly- and oligothiophenes). These structures have been studied extensively in the last few decades. π-Conjugated macrocycles are of interest as modular building blocks for the assembly of new materials and supramolecular chemistry. There is an expectation that these macrocycles will play an important role as key components (e.g., "molecular circuits") for emerging nanoelectronic devices [230, 238, 239].

As noted, self-assembling materials have been around for a number of years. But until recently scientists only had figured out how to get layers of material to stack one on top of another, building a structure that is essentially flat (2D). In 2003 IBM and Columbia University constructed a 3D material that builds itself: IBM T.J. Watson Research Center and Columbia University announced the development of self-assembled bimodal superlattice of nanocrystals. The iron oxide molecules were 11 nm, with each molecule being 60,000 atoms across; and the lead selenium quantum dots were 5 nm across, each of which contains 3000 atoms. By varying the experimental conditions, the researchers could "coax" them to assemble into a uniform lattice [240].

Some researchers hold a different view than that offered above (e.g., [32]). Self-replication and nanotechnology are often discussed together, since it has been suggested that no large-scale nanotechnology industry can develop without self-replication. The self-replicating entities of the biological arena are often used as an existence proof for the possibility and inevitability of self-replicating machinery. Manmade self-replicating machinery, however, has not seen much progress since the demonstrations of the first rudimentary self-reproducing machine demonstrations [241, 242]. The concept of molecular assemblers (as envisioned by Eric Drexler) is rejected by some as "unworkable and unnecessary" [15]. Fortunately, completely self-replicating systems may not be required for near-term practical nanotechnology: More basic systems containing some (but not a totality of) replicating aspects may (and have) become available. The future outcome of these manufacturing methods will determine who is right in this context.

In fact, Eric Drexler and Chris Phoenix (Directors of Research at the Center for Responsible Nanotechnology) recently published the article "Safe Exponential Manufacturing" in the Institute of Physics' *Nanotechnology* journal, that analyzes risks, concerns, progress, misperceptions, and safety guidelines for future MNT development [243]. As we saw in Chapter 1, Drexler introduced the concepts of nanotechnology through his 1981 article in the *Proceeding of the National Academy of Sciences (PNAS)* and his 1986 book *Engines of Creation*. The *PNAS* article was based on a biological model of molecular machine systems—hence the early focus on self-replication—but the logic of the technology led to the very different, nonbiological approach described by *Nanosystems* in 1992 and in the more recent literature. Drexler now notes: "Research and thinking in this area has come a long way since the earlier works. Molecular machine systems can be thoroughly nonbiological, and self replication is not necessary." In particular, it turns out that developing manufacturing systems that use tiny, self-replicating machines would be needlessly inefficient and complicated. The simpler, more efficient, and more obviously safe approach is to make nanoscale tools and put them together in factories big enough to make what a manufacturer has in mind. Throughout history, people have used tools to make more and better tools. That is how one progressed from blacksmith's tools to automated industries; the natural path for nanotechnology is similar. Since the publication of *Nanosystems*, the focus for Drexler and his colleagues has been on desktop-scale manufacturing devices. This nanofactory is based on the convergent assembly architecture where small parts are put together to form larger parts, starting with nanoscale blocks. The machines in this would work like the conveyor belts and assembly robots in a factory, doing similar jobs. With the fear of runaway replicators now in better perspective, attention on molecular nanotechnology can be directed to more important issues, including how the technology will be used, and by whom. Molecular nanotechnology will, according to proponents, introduce a clean, large-scale manufacturing capacity; these systems will affect all areas of society including medicine, the environment, national security, space development, economics, intellectual property, and privacy [243].

4.4 SYSTEM DESIGN

The ultimate challenge is to take the materials discussed in this chapter and incorporate them for use in actual microelectronic systems. Fortunately, a lot of research and related work is underway, for example, see [95, 161, 162, 163, 164, 165, 166, 167, 168, 169, 170, 171, 172, 173, 174, 175, 176, 177]. In this context, the invention and development of scanning probe microscopy has given researchers the ability to image matter to the atomic scale and opened new perspectives from semiconductors to biomolecules, and new methods are being developed to modify and measure the microscopic landscape in order to explore its physical, chemical, and biological features [244].

Observers expect a profound change to microelectronics in the next few years. In the 1980s and 1990s digital design was almost entirely separate from manufacturing, but going forward, as manufacturing issues begin to limit chip performance, digital design engineers must reconnect and work together with process engineers [245]. The issue is that while individual transistors may be manufacturable at 45-nm design rules, but system-level chips composed of millions of 45-nm transistors may not be. Hence, the challenge is: "What tools will be needed to automate the arrangement of 100 million (or even a billion) transistors, while also taking into account the physics implications of layouts resulting from an increasingly complex manufacturing process? Design below 100 nm 'is not business as usual' " [245]. Developers are now contemplating efforts whereby bottom-up manufacturing, device test, device yield, and mask and lithography considerations become embedded in the design process from the very beginning; these more symbiotic approaches are in preparation for the eventual use of nanotechnology-based building elements expected 5–10 years out.

Two current trends in microelectronics are (i) the reduction of the dimensions and (ii) the search for new devices based on new phenomena, as in the case of a single-electron transistor (SET) that is based on the Coulomb blockade effect. The near-term limitation related to the development of devices of smaller dimensions, is the fact that we are reaching the resolution limits of conventional lithography techniques. Among new manufacturing approaches, atomic force microscope lithography is one of the more promising because of the fact that good resolution that can be obtained (less than 10 nm, with improvement capabilities when using carbon nanotubes). Furthermore, AFM lithography is fully compatible with CMOS technology. In addition, the use of SOI as a substrate ensures reproducible devices with very thin monocrystalline films (15 nm) and with good interface quality [246]. The topic of microelectronics is revisited in Chapter 6, while the topic of high-resolution tools is covered in Addendix F.

4.5 CONCLUSION

This chapter looked at the basic building blocks now being developed, via a number of manufacturing techniques, to support developments in next-generation optics (nanophotonics) and electronics (nanoelectronics). Basic building blocks include nanotubes, nanowires, quantum dots, and other nanocomposites. Specific applications of these technologies to telecommunications and computing are discussed in the chapters that follow.

CHAPTER 5

Nanophotonics

Photonics is the science of shaping (molding) the flow of "light"[1] for the purpose, among others, of transporting a modulated signal, amplifying a signal, generating an electric signal from a group of photons, or supporting a logical Boolean function. Nanophotonics[2] is the manipulation of light at a spatial scale considerably smaller than its wavelength, which is typically[3] 1300–1500 nm for telecom applications. From another perspective, nanophotonics is the study of the interaction of light and matter at the nanometer scale; it is the use of photons instead of electrons to support transmission and computing. Nanophotonics is also defined as a technology to fabricate and operate nanoscale photonic devices.

This chapter focuses on an overview of the basic concepts and application of nanoscale photonic technologies and structures. A number of important recent developments in nanophotonic materials and nanophotonic devices are covered. Only the basic concepts are discussed and not the physics itself. We have made an effort to define and describe the many terms used in this chapter in the chapter glossary as well as in the more inclusive Glossary at the end of the book. The reader may find it useful to refer to the glossaries while reading the text.

5.1 INTRODUCTION AND BACKGROUND: A PLETHORA OF OPPORTUNITIES

A photonic device is a device that uses photons in a function-specific (e.g., switching element, logic element) and function-rich manner. In turn, there is a desire to develop complex multicapability components that can be used in engineering applications. Photonic microchips of various degrees of complexity and functionality are now beginning to appear at the commercial level. While nanophotonics is not strictly

[1]For telecom applications the signal is actually below the visible range, more specifically in the infrared range; hence, the term "light" is (often) used euphemistically in this chapter.
[2]The term nanooptics is also used by some.
[3]Operation at 850 nm is also of interest for other applications.

Nanotechnology Applications to Telecommunications and Networking, By Daniel Minoli

necessary to develop all-optical components, the major thrust of the forward-looking research is now in the realm of nanoscience; ongoing work is directed at advancing the fundamental basic science as well as advancing the technology and its commercial deployment.

At the basic science end one finds, among others, research on quantum optics (including atom–photon interactions in the optical near field and the potential applications for atom trapping and manipulation), research on photonic crystals and photonic bandgaps, and research on plasmonics. At the technological end one finds engineering initiatives in, among others, new optical materials (e.g., photonic crystal fibers), near-field microscopy, nanolithography, integrated optoelectronic devices, and high-density optical transmission, computing, and data storages [193].

The field of photonics has seen a number of efforts, yet unsuccessful to the present, to develop a universal, fully integrated platform for optical functions that from a capabilities perspective is at least comparable to the functionality provided by traditional electronics-based integrated circuits, but preferably with higher performance in the speed and throughput dimensions [247]. Nanophotonics-based approaches are expected to lead, in the relative short term, to the development of high-bandwidth, high-speed, and ultraminiturized optoelectronic components. Industry stakeholders believe that nanophotonics technology has the potential to revolutionize telecommunications, computation, and sensing. Developers are beginning to use nanoscale fabrication techniques to make the silicon components for optical repeaters, switches, and routers; the next step is to move to all-optical devices [248]. In particular, nanophotonic integrated circuits are expected to perform optical-switching functions that currently require large assemblages of components; in fact, up to now optical switches have been electrooptical and/or electro-mechanical-optical in nature (e.g., MicroOptoElectroMechanical Systems). Off-the-shelf photonic chips that support all-optical switching are expected be available commercially by 2010 or thereabouts, and the optical computer may become a reality by 2015.

Recent advances in top-down and bottom-up nanofabrication techniques (as was discussed in the previous chapter) as well as electromagnetic computational methods have created conditions for scientific progress with the potential to yield a new class of nanophotonic materials for subwavelength optical components [249]. Nanolithography (or nanoimprinting) is already being utilized at this time in the production of subwavelength optical components [250]. The goal of nanotechnology is to exploit the new properties of nanomaterials by acquiring control of structures and devices at atomic, molecular, and supramolecular levels and to establish how to efficiently manufacture and use these structures and devices. Newly developed nanophotonic structures provide the basic building blocks for an all-optical circuit, where passive and active components are integrated on a single chip. In photonic microchips, flows of electrons are replaced by beams of photons (light). There is interest both in the transmission and in the processing (e.g., switching) of photons and/or coded information.

The information-carrying capacity of a strand of standard single-mode fiber stood at around 7 Tbps at press time for *laboratory* systems (about 1 Tbps for field-deployable *systems*). Over the last decade, transmission capacity has been doubling every 9–12 months. If such growth is sustained, one will reach the theoretical limit of communications capacity on conventional glass fiber (generally thought to be around 100 Tbps) by 2010. If (when) there is renewed end-user/commercial demand[4] for higher bandwidth compared with current pragmatic levels, typically in the T1-(1.544-Mbps)-to-OC3 (155-Mbps) ranges for most enterprises, new technologies will be required; therefore, nanophotonics in general and photonic crystal fibers in particular may be a possible solution. In addition to advances in transmission mechanisms, one also needs complementary advances in switching systems.

Emerging photonic technologies that are expected to be (or become) important in this context include, among others, (1) nanometer-scale optical devices that are able to operate below the wavelength of light; (2) (an enabler of such devices), the creation of planar-silicon optical-bandgap materials, including photonic crystals (PCs) and photonic crystal fibers (PCFs); and (3) plasmonics (plasmonic devices are devices that make use of optical properties of metallic nanostructures and near-field phenomena). Maintaining the stability of interfaces and the integration of these nanostructures with other elements at micrometer-length dimensions (and/or mesoscopic level) is intrinsic to achieving commercial success for the emerging metamaterials and nanostructures [2, 33]. Ultimately one is interested in "laser-on-a-chip" applications.

Nanophotonics encompasses the areas of optical response theory, nonlinear optical properties, luminescence, photonic crystals, and quantum optics [208]. Nanophotonics (including completely optical chips) can open the door to new communications and computer technologies that are free from traditional microelectronics-imposed limitations [251]. As noted above, optical data links with field-deployable bandwidth in the 10–100-Tbps range and that reach in the tens of kilometers will be needed in the near term to deploy telecom and next-generation computer networks serving metro and long-haul geographies; nanophotonics will facilitate these deployments. On the computing front, researchers are looking to fabricate all the components for silicon-optical computers that use light instead of electrons; wires have already been replaced in laboratory experiments by beams of light routed on-chip through air by silicon waveguides controlled by electrooptical switches and routed off-chip by a "pin hole" lens connecting to standard optical fibers. For example, Cornell University is developing photonic chips in nanoscale silicon by routing light through "slot waveguides" filled with air, vacuum, or new organic polymers [248, 248a, 248b].

The scientific and commercial evolution towards nanoscience and nanotechnology (discussed throughout this book) makes it desirable to address optics issues at the nanoscale. Because the diffraction limit does not permit one to focus light to dimensions smaller than roughly half a wavelength ($\lambda/2$), traditionally it has not been possible to interact selectively with nanoscale features. In recent years, however, several new approaches have emerged to "shrink" the diffraction limit (e.g.,

[4]Demand for bandwidth by commercial enterprises reached a "soft spot" in the early part of the decade, and at press time it was not yet clear what the exact future trend would be, at leat in the short and medium term.

confocal microscopy) or even to overcome it (through near-field microscopy) For example, with tip enhancement techniques one is able to undertake Raman spectroscopy and multiphoton fluorescence imaging with a spatial resolution of less than 20 nm [193] (see Appendix F for a discussion of imaging tools and techniques applicable to various environments).

The advances cited thus far are just some of the areas of interest within the space of nanotechnology in general and nanophotonics in particular. The sections that follow survey some of the major and promising areas of research. As already suggested, the field of nanophotonics includes photonic crystals devices (where a high-index contrast lattice creates "photonic bandgaps" that forbid light propagation—photonic bandgaps and other resonant nanoscale structures can be used to control the propagation of light) and plasmonic devices (where surface plasmons in metals convey and/or concentrate optical energy), among several other devices and phenomena [33]. Hence, the concepts of photonic crystals, integrated photonic circuits, photonic crystal fibers, superprism effects, and optical properties of metallic nanostructures are addressed in this chapter.

5.2 GENERAL PHOTONICS TRENDS

We open this discussion with a short review of the middecade status of communications-related photonics in general; the sections that follow then look at nanophotonics. Because of the telecom perspective taken by this book, we approach optics from a transmission and computing angle. Table 5.1 depicts some basic concepts and key terms in the field of optics; these terms are grouped here to convey to the reader a sense of the range of applicable approaches and constituent technologies (the Glossary at the end of the book contains and describes additional concepts). A major portion of this research is currently at the mesoscopic scale (10–10,000 nm, that is, 0.01–10 μm), but nanoscale work is also taking place in earnest, as the rest of the chapter demonstrates. In microphotonic integrated circuits (at the mesoscopic scale) light is guided in optical waveguides that are made on a planar substrate. By proper materials design and engineering, passive devices such as optical splitters and multiplexers can be constructed, and by adding optically active impurities (such as rare-earth ions), optical amplifiers (e.g., erbium-doped fiber amplifier) and lasers can be realized. Microresonater designs have also been introduced where light can be stored and confined in small volumes at high intensities [252, 253] (microresonators are discussed later).

Current optical research is directed towards (1) the advanced *design and fabrication of optical fibers*, (2) *integrated optics*, (3) *optical amplifiers*, (4) *optoelectronic devices* and, as noted, (5) *nanostructures* [254]. In the paragraphs that follow each of these areas are briefly surveyed.

1. *Optical fibers* are used for information transport and also for optical amplification. Specific areas of recent interest include but are not limited to fiber fabrication methods, doped fibers, specialty fibers (e.g., polarization-preserving fibers,

TABLE 5.1 Basic Optical and Photonic Terms

Term	Definition
3R	Used in optical transmission systems: reshape, reamplify, retime.
Anisotropy (optical anisotropy)	Condition that refers to the fact that the refractive index of a crystal depends on the direction of the electric field in the propagating light beam. Hence, the velocity of light in a crystal depends on the direction of propagation and on the state of its polarization (i.e., the direction of the electric field). The outcome is that (except along certain special directions) any unpolarized light ray entering an optically anisotropic crystal breaks into two different rays with different phase velocities and polarizations. Contrast with *optically isotropic*, where the refractive index is the same in all directions. Examples of optically isotropic materials are most noncrystalline materials (e.g., glasses, liquids) and all cubic crystals. For all other classes of crystals (excluding cubic structures), the refractive index depends on the propagation direction and the state of polarization. When one views an image through an optically anisotropic crystal, one sees two images, each constituted by light of different polarization passing through the crystal.
Arrayed waveguide grating (AWG)	Optical multiplexing device that uses interference effects between different waveguides of progressively longer optical path length on a planar substrate (typically silicon). The interference effect directs each wavelength (typically arriving on different ports on the AWG) onto an output port that is coupled to a fiber output.
Bandgap	Range of frequencies where propagating modes of a signal or wave are absent. This also is applicable to optics.
Bandgap (electronics)	In a semiconductor material, the minimum energy necessary for an electron to transfer from the valence band into the conduction band, where it moves more freely [255].
Bragg angle/law/ condition	Bragg angle (θ) is defined by the expression $2d \sin \theta = n\lambda$, where θ is the angle between a crystal plane and the diffracted X-ray beam; λ *is* the wavelength of the X-rays, d is the crystal plane spacing, and n is the diffraction order (any integer). The Bragg law is the cornerstone of X-ray diffraction analysis because it allows one to make accurate quantification of the results of experiments carried out to determine crystal structure. It was formulated in 1912 by W. L. Bragg in order to explain the observed phenomenon that crystals only reflect X-rays at certain angles of incidence [256]. An X-ray diffraction occurs from a crystal structure only when the Bragg condition is satisfied; this condition depends on the angle of the incident X-ray beam as it enters the crystal structure and the direction at which the diffracted beam exits the structure.
Bragg grating	Filter that separates light into many colors under the principles of Bragg's law. Specifically, a fiber-based Bragg grating used in optical communications to separate wavelengths.
Cladding	Material that surrounds the core of an optical fiber. The core has a higher index of refraction than the cladding. The lower index of refraction in the cladding causes the transmitted light to travel downstream through the core.
Continuous wave (CW)	The constant optical output from an optical source when it is turned on but not (yet) modulated with a signal.
Continuous-wave (CW) laser	Continuously on laser. The laser is used as a light source for external test equipment or modulators; the laser is not modulated by drive voltage or current.

Course wavelength division multiplexing (CWDM)	A WDM system where only a few channels are needed or supported. Here wider wavelength spacing is possible compared to a dense WDM system. Do not need to be optically amplified; this typically reduces cost by allowing uncooled lasers and simpler termination equipment.
Diffraction grating	Optical device that behaves similarly to a prism. An array of fine, parallel, equally spaced reflecting or transmitting channels that mutually enhance the effects of diffraction to concentrate the diffracted light in a few directions determined by the spacing of the channels and by the wavelength of the light.
Diode laser	Also called *semiconductor laser*, a laser where the active element is a p–n semiconductor junction. When current flows across the junction, light is emitted from the edge of the chip in the plane of the junction.
Dispersion-shifted fiber (DSF)	Fibers optimized for operation at 1550 nm. Regular single-mode fibers exhibit lowest attenuation performance at 1550 nm and optimum bandwidth at 1310 nm. Are made so that both attenuation and bandwidth are optimal at 1550 nm.
Distributed feedback laser diode (DFB-LD)	Injection laser diode that has a Bragg reflection grating in the active region; the grating is used to suppress multiple longitudinal modes and enhance the properties a single longitudinal mode.
Dual-window fiber	For single-mode fibers, implies that the fiber supports operation at 1310 and at 1500 nm operation. For multimode fibers, the term means that the fiber is optimized for 850- and 1310-nm transmission.
Electroabsorption (EA) modulator	Optical device (e.g., $LiNbO_3$) used to electrically attenuate the laser light at microwave rates. Applicable at digital data rates to over 10 GHz [257].
Electroabsorption-modulated laser (EML)	Has integrated a CW laser and an EA modulator on the same semiconductor chip.
Erbium-doped fiber amplifier (EDFA)	Amplifier based on optical fibers doped with erbium. A device that uses doped fiber and a secondary pump laser to optically amplify a signal. Operate as basic transmission network elements that eliminate the need for intermediate regeneration and retransmission functions. The doped fiber can amplify light in the 1550-nm region when pumped by an external light source.
Extinction ratio	Ratio of the optical output in the "on" state (rated output power) to the optical power in the "off" state (threshold power).
Fabry–Perot laser diode (FP-LD)	Semiconductor laser diode that uses a "Fabry–Perot" filter. The filter selects wavelengths utilizing a light interference pattern produced by precisely spaced parallel surfaces.
Fiber Bragg grating (FBG)	Grating that consists of fiber segment whose index of refraction varies along its length; the variations of the refractive index constitute discontinuities that emulate a Bragg structure.
Fiber laser	Laser where the lasing medium is an optical fiber doped with rare-earth atoms to make it capable of amplifying light. Because of the fiber laser's low threshold power, laser diodes can be used for pumping.

TABLE 5.1 (*Continued*)

Holey fiber	New kind of optical fiber that has microscopic holes running along the length of the fiber and/or a hollow core in the center of the light guide. The holes give the fiber advantages for transmitting information: They make it is possible to control the physical properties of the light as it propagates through the fiber and/or the hollow core.
Injection laser	Another term for a semiconductor or laser diode.
Internal reflection	Reflection of an electromagnetic wave traveling in a medium 1 with high refractive index n_1 when it hits the boundary with a medium 2 of lower refractive index n_2 ($n_2 < n_1$); however, some of the energy may be transmitted (escape) into medium 2.
Laser	Originally an acronym for light amplification by stimulated emission of radiation. Laser-generated light is directional, spans a narrow range of wavelengths, and is more coherent than ordinary light.
Laser diode	Semiconductor diode lasers are the standard light sources in fiber-optic systems.
Laser diode wavelengths	Typical wavelengths for laser diodes are 1550 and 1310 nm. For WDM applications, laser diodes may be specified at different subwavelengths.
Light guide	Optical fiber or light-conducting material.
Light localization	State where light of a given frequency is totally confined to a small and finite region of space and cannot propagate except through a nonlinear interaction.
Light-emitting diode	Semiconductor diode that emits chromatically pure but incoherent light (spontaneous emission.) Light is emitted at the junction between p- and n-doped materials.
Linewidth	Width of laser beam frequency.
Mach–Zehnder (MZ) modulator	Intensity modulation (IM) approach that relies on an interference effect between two waveguides. By modulating the refractive index in one portion of the device, IM is achieved at the output.
Modal dispersion	Temporal dispersion arising from differences in the travel times that different modes (rays) take to travel through multimode fibers.
Modified chemical vapor deposition (MCVD)	A chemical vapor deposition (CVD) process for manufacturing preforms where glass layers are deposited on the inside surface of a starting tube.
MOEMSs (MicroOptoElectro Mechanical Systems) (aka MEMSs)	Refers to machines with micrometer-level moving parts that contain both electrical and mechanical components on silicon. Also referred to as microsystems, microstructures, microstructure technology, and mechatronics. Optical switching is possible with the aid of MEMS-based micromirrors which mechanically deflect the input optical signal into the desired output port directly with micromirrors mounted on tiltable cantilevers. Are already being used in components for telecom switches; also being used in projection display systems, optical displays, scanners, maskless lithography, and optical spectroscopy [250].

Term	Definition
Nanophotonics	Photonics is the science of shaping the flow of "light" for the purpose, among others, of transporting a modulated signal, amplifying a signal, generating an electric signal from photons, or supporting a logical Boolean function. Nanophotonics is the manipulation of light at a spatial scale smaller than its wavelength, which is typically 1300–1500 nm for telecom applications. Nanophotonics is the study of the interaction of light and matter at the nanometer scale; it is also defined as a technology to fabricate and operate nanoscale photonic devices. It also refers to a technology to fabricate and operate nanoscale photonic devices which utilize local electromagnetic interactions between a small nanoscale element and an optical near field. Since an optical near field is free from the diffraction of light due to its size-dependent localization and size-dependent resonance features, nanophotonics enables the fabrication, operation, and integration of nanoscale devices. Atom-photonics manipulates atoms by using an optical near field, which enables the fabrication of novel matter on the atomic scale [250].
Nonreturn to zero (NRZ)	Optical line coding where a 1 or 0 is designated by a constant level of opposite polarity. Used by used by SONET (Synchronous Optical Network—the transmission hierarchy utilized in the United States).
Non-zero-dispersion-shifted fiber (NZDSF)	Fiber to introduce a small amount of dispersion without the zero-point crossing being in the C-band (1528–1565 nm).
OEO	Electrical–optical–electrical conversions, for example in a transmission system.
Optical amplifier	Device that amplifies the input optical signal without converting it to electrical form.
Optical mode	"Ray" of light. Light entering a waveguide can be regarded as confined and is referred to as an optical mode. The properties of the optical mode are determined from the characteristics of the propagating light and the refractive indices of the absorbing cladding and/or substrate regions. Propagation of the confined mode can be defined unambiguously by a property of the mode called its *effective index*. Propagation is a function of the wavelength. Single-mode fibers transmit a single (one) mode of light. Multimode fibers transmit multiple modes. Telecom systems are based on single-mode fibers (multimode fibers find some applications in short-run applications in data centers, central offices, and interrack and/or collocated interrack cabling).
Optical multiplexer	Device that combines two or more optical wavelengths into a single output or fiber.
Optical pumping	Exciting the lasing medium by the application of light.
Optical waveguide	Any structure that can guide light, e.g., optical fiber, planar light waveguides, etc.
Photodiode	Diode that can produce an electrical signal proportional to the light falling upon it.
Photonic bandgap material	Non-light-absorbing material that contains a bandgap for electromagnetic waves propagating in any and all directions.

TABLE 5.1 *(Continued)*

Term	Definition
Photonic crystal	Non-light-absorbing material with a refractive index that exhibits periodic modulation in two or three orthogonal (vector) spatial directions.
Photonic switching	Use of photonic devices to make or break connections within integrated circuits, rather than electronic devices.
Photonics	Technology of generating and harnessing light and other forms of radiant energy whose quantum unit is the photon. The science includes light emission, transmission, deflection, amplification, and detection by optical components and instruments, lasers and other light sources, fiber optics, electrooptical instrumentation, related hardware and electronics, and sophisticated systems. The range of applications of photonics extends from energy generation to detection to communications and information processing [257].
Planar lightguide circuit (PLC)	Device (typically manufactured in wafer form, say over a silicon substrate) that is used to guide light, such as planar light waveguide. A circuit (waveguide) that is fabricated on flat material, such as a thin film.
Polarization-maintaining fiber (PMF)	Fiber where light is able to propagate in one mode and maintain a fixed polarization.
Polarization mode dispersion (PMD)	Light transmitted on a single-mode fiber is decomposable into two perpendicular polarization components. Distortion that results due to each polarization propagating at different velocity.
Pulsed laser	Laser that emits light in a series of pulses rather than continuously. Used for testing fiber systems.
Pump laser	Power laser used to drive optical amplifiers by exciting the rare-earth doped fiber.
Pumping	Addition of energy (thermal, electrical, or optical) into the atomic population of the laser medium, necessary to produce a state of population inversion.
Quantum well	In a diode laser, a region between layers of gallium arsenide and aluminum gallium arsenide, where the density of electrons is very high, resulting in increased lasing efficiency and reduced generation of heat [257].
Raman effect	Effect where part of the energy in a photon is transferred to (or from) the vibration/rotational energy of a molecule.
Rayleigh scattering	Scattering of radiation as it passes through a medium containing particles the size of which is small compared with the wavelength of the radiation.
Reflectance	Ratio of reflected light to light falling on the object.
Reflection	Return of radiant energy (incident light) by a surface, with no change in wavelength.

Refraction	Change of direction of propagation of any wave when it passes from one medium to another in which the wave velocity is different. Also, the bending of incident rays as they pass from one medium to another.
Refractive index	Ratio of the speed of light in a vacuum to the speed of light in a material.
Refractive index gradient	Change in refractive index with respect to the distance from the axis of an optical fiber.
Regenerator	Receiver–transmitter pair that receives a weak signal, reshapes it, then retransmits it.
Semiconductor laser	Laser where the injection of current into a semiconductor diode produces light by recombination of holes and electrons at the junction between p- and n-doped regions.
Single mode	Containing only one mode.
Snell's law	Principle that relates the angles of incidence and refraction to the refractive indices of the media. For example, if light is traveling in a medium with index n_1 incident on a medium of index n_2 and the angles of incidence and refraction (transmission) are θ_i and θ_t, then $n_1 \sin \theta_i = n_2 \sin \theta_t$.
Soliton	Optical pulse that regenerates to its original shape at certain points as it travels along an optical fiber. Solitons can be combined with optical amplifiers to carry signals very long distances [257].
Vertical-cavity surface-emitting laser (VCSEL)	A type of laser that emits light vertically out of the element, not out of the edge.
Waveguide	Structure that guides electromagnetic waves along its length. An optical fiber is an optical waveguide.
Waveguide dispersion	Portion of chromatic dispersion that arises from the different speeds light travels in the core and cladding of a single-mode fiber.
Wavelength division multiplexing	Multiplexing of optical signals by transmitting them at different wavelengths through the same fiber.
Window	Inspired by the optical loss of the fiber. Some of the bands are the S-band, defined in the range 1280–1350 nm; the C-band, defined in the range 1528–1565 nm; the L–band, defined in the range 1561–1620 nm.
Zero-dispersion wavelength	Wavelength at which the net chromatic dispersion of an optical fiber is nominally zero. This arises when waveguide dispersion cancels out material dispersion.

dispersion-shifted fiber, dispersion-compensating fibers, active fiber devices, holey fibers, and fiber Bragg gratings), and fiber fabrication processes (e.g., MCVD process and sol–gel process). Optical sources (laser diodes and LEDs), photodetectors, transmission media (principally for guided applications but also for unguided free-space optics), components (passive and active), and full-blown systems are all areas of current development. Some related emerging trends in photonics include the following. (i) New modulation formats: In high-end products the signaling format is starting a transition from the traditional return to zero/nonreturn to zero to a differential-phase shift keying (DPSK) format; quadrature phase shift keying (QPSK) is also currently under study, with the goal of achieving increased transport throughput. (ii) Transparent systems, that is, optical-signal-based systems without OEO conversions, MOEMSs (aka MEMSs), and PLC devices are still being investigated (and deployed), but the focus is increasingly on all-optical networks, especially through the use of simple single SOA (semiconductor optical amplifier) devices. And, (iii) polymer waveguides: New, highly nonlinear polymers are being investigated, along with new techniques to create waveguides (e.g., printing/burning waveguides into a 3D polymer with a laser).

2. *Integrated optics* deals with the modeling and fabrication of integrated and microoptical components and, in turn, systems. This area includes materials, processing, and fabrication. The integration of bulk and waveguide systems and the efficient coupling between source, waveguide, and detector are important to integrated and microoptics.

3. *Optical amplifiers* operating in a wide range of wavelengths are key to high-speed communications, particularly in ultralong-distance environments. The goal is to extend the functionality of the now-ubiquitous EDFAs. Issues here relate to design and fabrication of doped fiber amplifiers with dopants such as Nd, Er, Yb, among others; Raman fiber amplifiers; and double-clad fiber amplifiers.

4. The focus of work in reference to *optoelectronic devices* and nanostructures is on advances in optoelectronic devices, systems, and applications. Constituent subareas include electrooptic effects, material research, device design and fabrication, and packaging. A list of disciplines of active research include photonic switching, nonlinear optics, electrooptic materials, quantum well devices, modulators, multiplexers, sensors, integrated optics, instrumentation probes, optoelectronic integrated circuits, optoelectronic switching, and sources such as semiconductor lasers [254]. Distributed feedback (DFB) lasers have been used over the years for optical sources for fiber communication systems. In Chapter 4 we covered quantum dots (QDs) as another possible (uncooled) laser technology under the rubric of nanostructures.

For optoelectronic devices, there are efforts in the development of multifunctional photonic chips to support evolving high-capacity optical networks. Sought-after multifunctional capabilities include reconfigurable optical add/drop multiplexer (OADM) with switching, routing, amplification, and/or multiplexing capabilities.

Emerging high-capacity transparent optical networks plan to use flexible and transparent nodes based on multifunction photonics (e.g., optical switching fabrics, reconfigurable OADMs) to route optical signals while minimizing OEO conversions. The integration of passive and active photonic components to support some or all of the multiple functionalities just listed on a PLC chip, along with embedded electronic control functions, enables developers to bring to market compact, low-cost optical subsystems on an ultradense footprint format.

MicroOptoElectroMechanical systems are another example of optoelectronic device technology that has experienced considerable progress during the past decade. The MOEMSs are small integrated devices or systems that combine electrical and mechanical components (MOEMS are MEMS devices with integral optical components such as mirrors, lenses, filters, laser diodes, emitters, or other optics.) The basic MEMS concepts were developed in the 1980s and the technology has continued to make progress over time. They range in size from the submicrometer level to the millimeter level, and there can be any number, from a few to millions, in a particular system. The next level of development is expected to be at the nanoscale. The MEMSs in general, and MOEMSs in particular, extend the fabrication techniques developed for the integrated circuit industry to add mechanical elements such as beams, gears, diaphragms, and springs to devices [258]. Examples of MOEMS device applications include inkjet printer cartridges, accelerometers, miniature robots, microengines, locks, inertial sensors, microtransmissions, micromirrors, microactuators, optical scanners, fluid pumps, transducers, and chemical/pressure/flow sensors. New applications are emerging as the existing technology is applied to the miniaturization and integration of conventional devices. The MOEMSs are already being used in components for telecom switches: Optical switching is possible with the aid of micromirrors that mechanically deflect the input optical signal into desired output ports (mirrors are mounted on tiltable cantilevers). The next evolutionary step would be to develop and make use of *nanoscale NEMSs* and/or make use of nonmechanical tilting. The technology at press time was seeing the introduction of MOEMS-based switches that incorporate planar light circuits on a single silicon chip. The MOEMSs/NEMSs are electromechanical micro/nanorange devices that can be used as sensors, actuators, and light-switching systems. In addition to traditional telecom applications, MOEMS/NEMSs can be used in video-routing/video-on-demand applications. For example, companies have developed MOEMS-based switches that control the flow of light on a single silicon chip; the technology allows one to shrink what normally would require racks of equipment into a 1-in.^{2} package. The MOEMS-based optical components can act as variable optical attenuators and/ or photonic switchers.

A recent trend in optoelectronics relates to chip-scale integration of optical and electronic circuitry with IC on PLCs and/or PLC on ICs. Increasingly, electronics will be integrated with PLCs to reduce component size and to provide embedded intelligence. This integration may be viewed either as "electronics-enhancing photonic functionality" or, alternatively, as "a photonic overlay to enhance electronic functionality." Silicon microphotonics is the high-density integration of individual

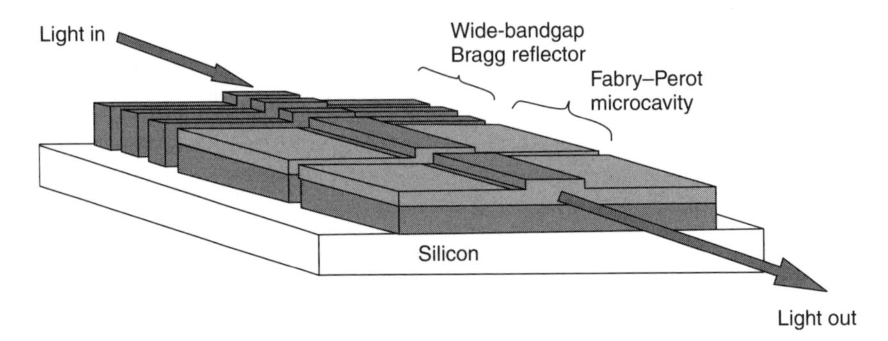

FIGURE 5.1 Example of integrated optics-on-silicon technology.

optical components on a single chip [259]. In planar optical waveguides, the light is confined to substrate-surface channels and routed onto the chip [260]; these channels are typically less than 10 μm across and are patterned using microlithography techniques. With appropriate optical circuits based on these channel guides, both passive functions (i.e., power splitting from one to several channels) and active functions (i.e., modulation) can be performed on the light. The primary materials used in the commercial market are glass or fused silica for passive devices and lithium niobate for active devices. Again, there is interest in moving further along, from the microscopic (submicrometer) level to the nanoscale. A closely related area, currently in the early R&D stage, is that of photonic IC devices, where a variety of semiconductor optoelectronic devices are monolithically integrated and interconnected with waveguides. Planar lightguide circuits using silica-based optical waveguides are fabricated on silicon or on a silica substrate by a combination of flame hydrolysis deposition and reactive ion etching [260]. Figure 5.1 depicts an illustrative example of integrated technology.

Related to optoelectronic devices, an example of electronics-enhancing photonic functionality is the integration of all the driver and control functions on a compact silicon-based PLC switch matrix [257, 261]. A more forward-looking example of electronics-enhancing photonic functionality is the incorporation of a 2D photonic crystal (discussed later) layer on silicon electronics as a parallel optical routing plane. Photonic crystal-based PLCs can be fabricated on a SOI platform and integrated with silicon electronics [262]. Wavelength-division multiplexing technology has already facilitated an increase in transmission capacity and an improvement in flexibility in broadband optical telecommunication networks; additional advances are being sought in this area by network operators. The wavelength multiplexer/demultiplexer subsystem (chip) is one of the key constituent components in WDM networks. Various material systems, including AlInP (aluminum indium phosphide) alloys, lithium niobate polymer, and silica (glass), have been used to implement wavelength multiplexers around the 1550-nm wavelength. The SOI technology (more on this in Chapter 6) has been shown to be a promising technology for guided-wave photonic devices operating in this infrared region. A number of SOI guided-wave optical devices and circuits with high performance have already been demonstrated [263]. The recent emergence of SOI-CMOS integrated circuits ensures the availability of high-quality, low-cost substrates; these circuits are based on low-power, high-speed electronic technology. In addition, silicon germanium heterostructures permit the realization of silicon-based photodetectors within the 1200–1600-nm infrared region. It follows that SOI technology offers potential for cost-effective integration of a monolithic multiwavelength optical receiver system, including the wavelength demultiplexer, photodetectors, and electronic circuitry [263].

5. For the general realm of *nanotechnology* and *nanostructures,* microstructured photonic crystal fibers are an example of new technology being investigated. Nanophotonic devices hold the promise for applications in high-speed optical switching in telecommunication networks as well as in high-sensitivity chemical sensing, to list just a few. The development of artificially microstructured materials known as photonic crystals has resulted in new opportunities. Photonic crystals have newly discovered optical properties that allow both passive and active optical components to be realized within an ultrasmall (but *not* nanoscale) volume; photonic crystal structures have periodic index of refraction. The PCFs can be tailored by design to exhibit new transmission properties and enhanced operation. They exploit unusual properties of the new photonic crystal materials to deliver previouslyunattainable optical performance [264]. They enable novel applications in nonlinear optics that are of interest for telecom applications, for example, pulse compression, soliton formation, and wavelength conversion [257]. Focusing on nanophotonics proper, there is currently sustained research on metallic nanostructures (specifically, plasmonic devices) that "concentrate light"; related to this, scientists studying the way light interacts with (metallic) nanostructures are now "throwing out their old optics textbooks and brushing up their quantum mechanics instead" [265]. Nanophotonic circuits are planned to be used in repeaters, multiplexers, switches, and routers for high-throughput fiber-optic communication systems, e.g., for cost-effective fiber-to-the-home (FTTH) applications. The topic of photonic nanostructures is discussed in more detail later.

Table 5.2 depicts some related recent trends [257]. The sections that follow immediately below focus on nanotechnology per se.

5.3 BASIC NANOPHOTONICS

The previous section looked at photonics from a general perspective and at some of the areas of ongoing research and development. We now concentrate more specifically on nanophotonics. Nanoscale photonic structures can manipulate light in a useful manner. Many (if not most) of the phenomena that are being studied in nanooptics at this time have been already topics of research for a number of prior years (even decades), but the application of scanning probe techniques and single-molecule detection, along

TABLE 5.2 Some Recent Trends in Optics

High-speed optoelectronics	Asymmetric twin-waveguide technology is a platform for heterogeneous integration without regrowth. Monolithic integration of semiconductor optical amplifier and a high-speed waveguide p–i–n photodetector using asymmetric twin-guide technology.	[266]
	Fully integratable 1550-nm wavelength continuously tunable asymmetric twin-waveguide distributed Bragg reflector laser.	[267]
	Ultrafast photonic devices with quantum and nanostructures shown to be required for optical communication beyond the 100-Gbps range.	[268]
	Semiconductor optical waveguide switch based on coupled quantum wells.	
	40-Gbps operation of quantum dot SOA made possible by the ultrafast recovery time of QDs, potentially expandable to 160 Gbps.	
Long-wavelength VCSELs	Long-wavelength VCSELs (1310–1600-nm-wavelength range) have become available. These can better match the fiber's transmission properties to provide high-performance, low-cost transmitters for optical networks with a longer reach. Currently longer reaches are achieved only through the use of DFB lasers and electroabsorptive modulated lasers (EMLs); in the near term, some of these applications will be replaced by long-wavelength VCSELs and FP lasers.	[257]
	High-performance 1310-nm VCSELs with low threshold (1–2 mA), high power (2–7 mW), high efficiency (30–40%), high operating temperature (120°C), and high modulation performance (12 Gbps).	[269, 270]
Tbps optical links	Low-cost approaches for Tbps capacity based on CWDM are being developed.	[271]
High-power fiber lasers and optical amplifiers	The output power of fiber lasers has increased with improvements in double-clad fiber design and improved pump diode sources (multiemitter diodes and diode stacks). Cladding-pumped fibers have revolutionized fiber lasers by increasing output power from <1 W using traditional core pumping to more than 100 W single mode and to 1000 W multimode at 1100 nm.	[272, 273, 274]

New applications for photonic crystal fibers	Photonic crystal fibers are a new kind of optical fiber that has microscopic holes running along the length of the fiber and/or a hollow core in the center of the light guide, which give the fiber advantages for transmitting information. By controlling the size and placement of the holes, it is possible to control the physical properties of the light as it propagates through the fiber and/or the hollow core (light travels through air or a vacuum more efficiently than through glass or plastic). Photonic crystals can be employed in the manufacturing of holey fibers.	[272]
	First holey-fiber laser with Er-doped core, achieved lasing at 1535 nm with relatively high output and efficiency.	
	Improved all-silica single-mode photonic crystal fibers.	[275]
	Improved photonic crystal fibers with small effective area and larger nonlinearities; facilitate nonlinear optical processing—e.g., supercontinuum generation, optical 2R regeneration, and optical code division multiple access (CDMA)—using shorter fibers and lower optical power levels.	[275]
	Supercontinuum generation of 400–1600 nm "white light."	[276, 277]
	Soliton generation with high peak power.	[278]
	Microfluidics applications with photonic crystal fibers.	[279]
Active micro- and nanostructure devices	Surface-emitting single-defect photonic crystal lasers.	[280]
	Polarization-controlled single-mode photonic crystal VCSEL using asymmetric holes to break the symmetry for single-polarization-mode operation.	[281]
	Microdisk lasers: ring laser coupled to optical bus, demonstrating that it can be integrated with other components.	[282]
Quantum communications	There has been a steady improvement in the techniques for quantum communications (i.e., the use of the constraints set by quantum mechanics to the measurement of two conjugate observables to device methods for secure communications).	[283]

with the "revival" of high-resolution optical microscopy in the 1990s (e.g., NSOM/SNOM—see Appendix F), have added new impetus of late. Recent developments in micro- and nanophotonic materials and devices includes the areas listed in Table 5.3; nanophotonics areas that we cover here include the following:

- Photonic crystals devices and lasers (making use of "photonic bandgaps")
- Plasmonic devices (supporting the concentration of optical energy)
- Nanointegrated photoelectronics
- Other new technological areas/opportunities

We provide below a brief overview on each of these topics followed by a more inclusive discussion. In some cases a single molecule or nanoparticle is itself the subject of the ongoing research; in some cases the single molecule or nanoparticle is an active source of light; and in yet other cases, the single molecule or nanoparticle is a subwavelength detector of the electromagnetic radiation [284].

5.3.1 Photonic Crystals

Photonic crystals are a new class of materials (expected to enter commercial applications in the 2006–2008 time frame) that make control over light propagation possible: By appropriately designing a photonic crystal, one can manipulate the manner light propagates inside it [285]. Photonic crystals operate on the principle of refraction (bending) of the light as it travels from one material to another; refraction can be used to block specific wavelengths of light [286, 287]. Photonic crystals have the property that they can enhance the light–matter interaction by orders of magnitude, enabling the devices' optical properties to be controlled externally [251]. Photonic crystals have been the subject of intensive studies in the past decade. In the late 1980s, researchers applied concepts that had been developed for engineering the bandgaps of semiconductors to the field of optics; the research predicted that just as electrons may be manipulated by the crystalline structure of a semiconductor, light also may be confined by periodic patterns in glass [288]. Photonic crystals came into existence as part of a program to modify spontaneous emission rates in optical microcavities by modifying the optical mode density [289]. Photonic crystals open up gaps in the optical spectrum via spatially and spectrally overlapping Bragg planes; these electromagnetic bandgaps can be used to connect optical modes in laser cavities. In high-contrast dielectric systems, only a few lattice periods may be necessary to connect an electromagnetic mode; a lattice period has a length scale on the order of one-half of the optical wavelength [290].

A point defect in a photonic crystal results in a resonance in the bandgap and confinement of light to a region smaller than one wavelength. A linear array of such point defects creates a photonic bandgap waveguide and the combination of such waveguides is expected, in turn, to produce a wealth of optical elements such as beam splitters and interferometers. Nevertheless, realization of high-quality two-dimensional photonic crystals with depth of numerous wavelengths remains a nontrivial problem;

TABLE 5.3 Recent Developments in Micro- and Nanophotonic Materials

Photonic crystals	Structures that provide means to manipulate, confine, and control light in one, two, or three dimensions of space; 1D, 2D, or 3D devices with ordered variations in refractive index, more specifically with periodically varying indexes of refraction. The devices are constructed of ultrathin layers of nonconducting material that reflect various wavelengths of light. Photonic crystals are highly engineered material with superior optical properties; may be used to develop optical circuits. These device are designed to create a bandgap structure with forbidden regions and allowed energies that can select or confine electromagnetic waves. They are periodic dielectric or metallodielectric (nano)structures that are designed to affect the propagation of electromagnetic waves in the same way as the periodic potential in a semiconductor crystal affects the electron motion by defining allowed and forbidden electronic energy bands. They can be thought of as “optical analogs” to electronic semiconductors. The periodically varying index of refraction permits the control of the propagation of photons inside the crystals, similar to the manner by which electrons are excited in a semiconductor crystal [291]. The absence of allowed propagating electromagnetic modes inside the structures, in a range of wavelengths called a photonic bandgap, gives rise to distinct optical phenomena such as inhibition of spontaneous emission, high-reflecting omnidirectional mirrors, and low-loss waveguiding among others [292].
Integrated photonic circuits (microphotonic ICs)	Next-generation devices that could eventually replace the CMOS technology below the 5–10-nm frontier will likely be based on new technologies: Optoelectronic and quantum properties of nanometer-scale devices will probably be the basis of alternative technologies [293]. These are circuits where light is guided in optical waveguides that are rendered on a planar substrate. For example, passive devices such as optical splitters and multiplexers can be made by proper materials design and engineering.
Photonic crystal fibers	Microstructured fibers; one of the first commercial products based on 2D periodic photonic crystals. They are fibers that use a nanoscale structure to confine light with radically different characteristics compared to conventional optical fiber for applications in nonlinear devices, guiding exotic wavelengths, among others [292].
Plasmonic devices	Ultrasmall metal structures of various shapes that capture and manipulate light. Plasmonics is an emerging field of optics aimed at the study of light at the nanometer scale; the goal of plasmonics is to develop new optical components and systems that are the same size as today’s smallest ICs and that could ultimately be integrated with electronics on the same chip [265, 294]. Plasmonic devices make use of optical properties of metallic nanostructures. Nanoscale objects can amplify and focus light via a mechanism based on plasmons; plasmons are ripples of waves in the plasma (ocean) of electrons flowing across the surface of metallic nanostructures. The type of plasmon that exists on a surface is related to its geometric structure (e.g., curvature of a nanoscale gold sphere or a nanosized pore in metallic foil). When light of a specific frequency strikes a plasmon that oscillates at a compatible frequency, the energy from

TABLE 5.3 (*Continued*)

	the light is absorbed by the plasmon, converted into electrical energy that propagates through the nanostructure, and eventually converted back to light [265, 294].
Superprism effects	The superprism phenomenon relates to the very large angular dispersion experienced by a light beam when entering a photonic crystal. This arises from the anisotropy of the photonic band structure that can be present even in systems without a complete photonic bandgap [295]. These effects can be exploited for sensing and filtering applications.
Subwavelength phenomena and plasmonic excitations	There is research interest in surface electromagnetic waves and the extraordinary light transmittance (EOT) through an optically thick metal film that is perforated with subwavelength-size holes. EOT was first discussed in the late 1990s and has been intensively investigated since [296]. These phenomena and properties in nanoengineered structures can be used as integrated elements in various optoelectronic and photonic devices, including optical computers [297]. In the optical and infrared spectral ranges, the excitation of the electron density coupled to the electromagnetic field results in a surface plasmon polariton (SPP) traveling on the metal surface. At the metal–air interface, the SPP is a wave, with the direction of the magnetic field parallel to the metal surface. In the direction perpendicular to the interface, SPPs exponentially decay in both media. The SPP can propagate not only on the metal surface but also on the surface of artificial electromagnetic crystals, for example, on wire-mesh crystals. Since the SPP propagation includes rearrangement of the electron density, its speed is less than the speed of light; as a result, the SPP cannot be excited by an electromagnetic wave impinging on a perfectly flat metal surface. The situation, however, changes when the film is modulated: When one of the spatial periods of the modulation coincides with the wavelength of the SPP, the latter can be excited by a normally incident electromagnetic wave [297].
Metamaterials	Artificial (new) types of materials with electromagnetic properties not found in nature. Electromagnetic and multifunctional artificial materials, created in order to comply with certain specifications. New designer materials. Designing new materials with otherwise unattainable properties is one of the promises of nanotechnology. Composites offering a range of magnetic properties that cannot be secured using known naturally occurring materials. The new composites are constructed using nanotechnology to build tiny circuits on a plate made of quartz [298].
Photonic bandgap (PBG)	An optical effect of nanochannel structured optical materials (that can be used in miniturized optoelectronic devices) that relates to spectral regions inhibiting photons from traveling through the structured materials. PBG systems are expected to be of importance in the future for fast optical communication and optical computers. Some researchers claim that future devices based on PBG structures will be as important as transistors for electronic-based devices today; this is because PBG allows the manipulation of photons as much as transistors allow the manipulation of electrons [299].

most recent efforts have concentrated on very thin photonic crystals surrounded by media of lower dielectric constants [284].

Photonic crystals facilitate a plethora of possible applications (e.g., lasers, antennas, millimeter-wave devices, and efficient solar cell photocatalytic processes); additionally, they give rise to interesting new physics (e.g., including, but not limited to, cavity electrodynamics and localization) [300]. Optical components that can be built with photonic crystals include microlasers, wavelength division multiplexers, filters, modulators, waveguides (specifically, PCFs), couplers, and lenses. Photonic crystals have also been considered for optical switches and logic gates; photonic crystal-based components and functionalities are integrable as dense photonic ICs or can be employed as a photonic overlay on electronics for optical communication applications. More details are provided in Section 5.4.

5.3.2 Photonic Crystal Fibers

The focus of telecommunications is transmitting pulses of light at high signaling rates and in an efficient manner. Optical fibers and most microstructured optical fibers have relied on the transparency of silica to provide low transmission losses. Research aimed at reducing the power needed to transmit light signals over long spans of optical fiber has studied the manufacturing of fiber cables utilizing photonic crystal, which, as noted, is an artificial patterned structure consisting of a mix of materials or a material and air [286]. Photonic bandgap structures are periodic dielectric structures that preclude propagation of electromagnetic waves in a certain frequency range. The PBG-based fibers (e.g., PCFs) present the opportunity for minimizing the interaction between the propagating waves and the material, thus allowing for the use of materials that themselves do not have a high intrinsic transparency.

Key to the realization of this transmission improvement opportunity is the establishment of fibers with large PBGs [301]. Designs with high-refractive-index contrast lead to large PBGs and omnidirectional reflectivity; the large PBGs result in very short electromagnetic penetration depths within the layer structure, significantly reducing radiation and absorption losses while increasing signal robustness [302]. Photonic crystal fiber can be constructed by forming patterns of holes along the length of a fiber; the photonic crystal fiber keeps light confined to a hollow core (typically 15 μm in diameter) surrounded by a structure of glass and air. In contrast, traditional fiber strand is constructed from solid glass or plastic surrounded by a reflective coating (e.g., see Fig. 5.2). Photonic crystal is better than traditional reflective coating at keeping light from scattering or being absorbed, which keeps signals stronger over longer distances [287].

Recently researchers have developed prototypes that are 100 times more efficient at carrying the 1550-nm-wavelength light widely used in telecommunications than previous photonic crystal prototypes were able to (attenuation being improved from 1000 to 13 dB/km). More work remains to be done, however, since existing commercial fibers have attenuation levels of around 0.5 dB/km (e.g., see Fig. 5.3); otherwise, repeaters are required at very short distances, which is costly and undesirable from a plant design perspective. Researchers are also working to develop techniques

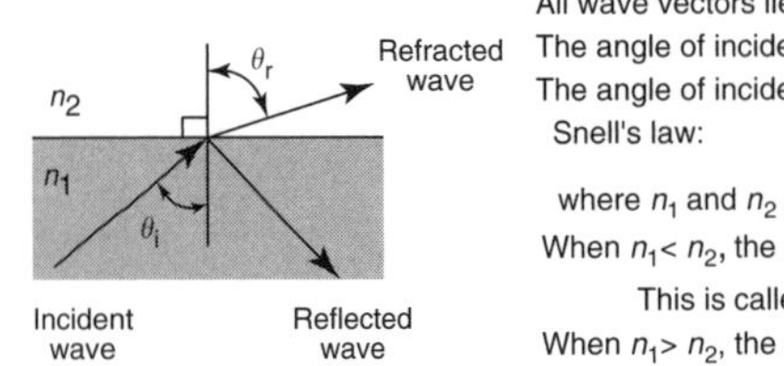

FIGURE 5.2 Internal reflections (applicable to fiber).

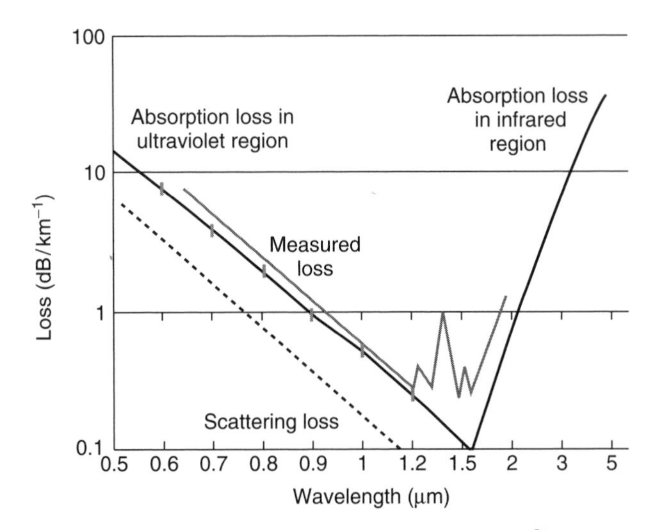

FIGURE 5.3 Basic performance of a telecom fiber.

to ensure that the photonic crystal fiber does not affect signal properties, such as polarization. The manufacturing goal is to make photonic crystal fiber structures that are consistent enough to efficiently channel light over useful distances. The challenge to producing the more efficient fiber is keeping the structure consistent throughout the length of the fiber [242, 287]. The material costs of the new fiber are lower than those of existing fiber lines, but the cost of manufacturing the new fiber could be higher, at least initially. More details are provided in Section 5.5.

5.3.3 Photonic Crystal Lasers

Photonic crystal lasers are lasers where the resonant cavity is formed by a periodic dielectric constant; the dielectric constant is periodic in at least two dimensions [290]. In some ways, they are similar to distributed feedback lasers and to vertical cavity surface-emitting lasers.

5.3.4 Plasmonics

Plasmonics is an emerging field of optics aimed at the study of light at the nanometer scale; as noted, these dimensions are far smaller than a wavelength of light (e.g., 1550 nm), smaller than today's smallest electronic devices [265]. Plasmons are coherent oscillations of the conduction electrons of the metal against the static positive background of the metal ion cores [142]. Some nanostructures act as superlenses, capturing specific wavelengths of light, focusing the light to ultrasmall spots at high intensities, and converting some electrical energy back into light that is reflected away [265, 294]. These metallic nanoscale structures are being investigated for their strong localization via the plasmon resonance; it is expected that these structures can be used to guide light over extended distances with lateral dimensions much less than the wavelength [250]. This research is expected to allow scientists and engineers to design new optical materials and devices "from the bottom up" using metal particles of specifically tailored shapes [294]. The further integration of optical devices (also discussed next) will require the fabrication of waveguides for electromagnetic energy below the diffraction limit of light. Research shows that arrays of closely spaced metal nanoparticles can be used for this purpose. Coupling between adjacent particles sets up coupled plasmon modes that give rise to coherent propagation of energy along the array. Studies have shown that one can obtain group velocities of energy transport that exceed 0.1 c along straight arrays and that energy transmission and switching through chain networks such as corners and T structures are possible and efficient. These plasmon waveguides and switches could be the smallest devices with optical functionality [303]. More details are provided in Section 5.6.

5.3.5 Integration

As already noted, integration is a desirable goal in order to support miniaturization, improved power consumption profiles, reliability, and functionality. Interest exists to develop an optical switching capability based on silicon in a planar geometry. We already discussed this concept in Section 5.2 at the mesoscopic scale, but we revisit it here at the nanoscale. The miniaturization of optical devices to dimensions comparable to VLSI-(very large scale integration) based electronic is a major goal of current research efforts in optoelectronics, photonics, and semiconductor manufacturing [142]. There is major interest in developing optical ICs with high complexity and advanced functionality. Large-index contrast waveguides are seen as the building blocks for future integrated optics systems. In many approaches, high-index contrast structures are employed resulting in photonic wires either for conventional index guiding waveguides or in photonic bandgap structures. The goal of the commercialization initiatives is to achieve low-cost mass-produced photonic structures. To that end, observers believe that an evolutionary path to complex index guiding devices is closer to market than the more revolutionary approach of photonic crystals [304]. In both cases the number of functional elements of a given chip area could be enhanced by several orders of magnitude, eventually resulting in VLSI photonics. Optical microresonators are a promising basic building block for filtering, amplification, modulation, and switching (this is discussed later). Active functions can be obtained by monolithic integration or a hybrid approach using materials with thermo-, electro-, and optooptic properties and materials with optical gain.

As just noted, some researchers believe that plasmon waveguides can facilitate the goal of integration. A high level of integration of optical components enabling the fabrication of all-optical chips for computing and sensing requires a confinement of the guided optical modes to small dimensions along with the ability to route energy around sharp corners. Technologies that are advancing the art (and the science) of fabrication of integrated optical components include planar waveguides, optical fibers, and photonic crystals, all of which can confine and guide electromagnetic energy in spatial dimensions in the sub-micrometer range. Technologies based on plasmonics can drive the minituarization to the nanoscale. "Optics-on-chip" is a platform that is free from conventional microelectronics limitations. As mentioned in the introduction of this chapter, up to now there has been somewhat limited progress in this area, and most photonic devices are still constructed as discrete elements; this is mainly due to the fact that most of the work on active (or tunable) devices is nonintegratable with present-day microelectronics, given the fact that silicon has been considered primarily as a passive material. Recently, however, it has been shown that optical properties of silicon (e.g., for optical switching) can be controlled externally using photon absorption for carrier injection. Photonic crystals can enhance the electromagnetic field and nonlinearities by orders of magnitude and, hence, can be useful in this context [251]. A number of new photonic-level technologies are evolving; as an example, researchers are developing techniques for making photonic microchips, including ways to guide and bend light in air or a vacuum, to switch a beam of light on and off, and to connect nanophotonic chips to optical fiber [305, 305a].

5.3.6 New Technologies

More advanced technologies deal with quantum-level phenomena, such as quantum entanglement. Quantum entanglement is expected to be useful for quantum information processing. Quantum theory is more than a (radical) departure from classical physics: It also offers new possibilities in communications and computing. Quantum theory is nonlocal: It predicts entanglement between distant systems leading to correlations that cannot be explained by any theory, based only on local variables, as demonstrated by Bell inequality. The state of a two-particle system is said to be entangled when its quantum mechanical wave function cannot be factorized into two single-particle wave functions (wave functions are discussed in Appendix D) [306]. Relevant experiments have been found to be in agreement with quantum theory; hence, the physics community faces a "strange" world view: In theory, everything is entangled; in practice, however, decoherence makes it impossible to reveal this entanglement. Entanglement of photons or of material particles (electrons, atoms, ions, etc.) has received attention in the recent past. Experimental realization of quantum entanglement is relatively easy for photons: A starting photon can spontaneously split into a pair of entangled photons inside a nonlinear crystal. Quantum entanglement has recently gained interest and research attention because this phenomenon can be key to massively parallel quantum information processing. The general idea related to computing is that entanglement provides means to carry out tasks that are either traditionally impossible (e.g., quantum cryptography) or would require significantly more steps to perform on a classical computer (e.g., searching a database, factorization) [307]. This topic is fairly advanced and is not pursued further at this juncture; Appendix G provides a discussion from a quantum computing perspective.

Another trend is related to QD lasers and semiconductor optical amplifiers (QDs were introduced in Chapter 4). Researchers have demonstrated quantum dot semiconductor optical amplifiers (SOAs) with saturation power comparable to the best results obtained for bulk and quantum well devices; quantum dot SOAs offer a wider gain bandwidth, higher saturation power, very fast recovery time, and a very small linewidth enhancement factor [254, 307a]. The unique properties of quantum dots may also lead to higher speed for all-optical signal processing applications [268, 308].

5.3.7 Instrumentation

While there is a large engineering effort in the synthesis of nanostructures (clusters, particles, tubes, layers, biomaterials, self-assembled systems), there is also a need for techniques with which to interact with these nanostructures and to probe their physical properties at the nanoscale. As an example, near-field optical techniques can be applied to probe complex semiconductor nanostructures as well as individual protein molecules [250] (we briefly describe this technique in Section 5.6.2). However, there are many other techniques, for example, based on X-ray, electron beam, ion beam, nuclear magnetics, and atomic force principles, as covered in Appendix F.

Next, we provide more extensive discussion on the technologies identified in Section 5.3.

5.4 PHOTONIC CRYSTALS

5.4.1 Overview

We expand here on the concepts that we introduced in the previous sections. Photonic crystals are optical materials that allow for controlling and manipulating the flow of light. Photonic crystals are composed of a regular arrangement of a dielectric material that shows strong interaction with light; any material exhibiting spatial periodicity in refractive index is a photonic crystal [252, 253]. Photonic crystals are materials with repeating patterns spaced very close to one another, with separations between the patterns comparable to the wavelengths of light. When light falls on such a patterned material, the photons of light interact with it, and with proper design of the patterns, it is possible to control and manipulate the propagation of light within the material [309]. Because the physical phenomenon in photonic crystals is based on diffraction, the periodicity of the photonic crystal structure needs to be in the same dimensional scale as the wavelength of the electromagnetic waves of interest. The more the contrast in refractive index, the better the optical properties of the photonic crystal [285]. Photonic crystals are considered to be *microphotonics* structures, but they are generally discussed in the context of nanophotonics.

Photonic crystals are artificially created, multidimensionally periodic structures. Photonic crystals are 2D or 3D ordered structures composed from submicrometer-sized objects. An example of a photonic crystal is opal, the gemstone: The opalescence is a photonic crystal phenomenon based on Bragg diffraction of light on the crystal's lattice planes. In general, photonic crystals are periodic dielectric structures that control the propagation of light. The dielectric structures have lattice parameters on the order of the wavelength of light. When these structures are periodic, that is, when the refractive index exhibits 2D or 3D modulation effects, the structures are "crystallinelike" and demonstrate "Bragg" diffraction. This is associated with the opening of a "bandgap" in the photonic Brillouin zone, namely, there is a range of frequencies at which the propagation of electromagnetic waves is forbidden [299]. Photonic bandgaps give rise to new possibilities for the design of optical switches, wavelength-selective mirrors (Bragg mirrors), lossless reflectors, and lasers.

The propagation of light in photonic crystals has a strong similarity to the wave propagation of a conduction electron in a crystalline solid; hence, the dispersion relations of light can be described with band structures in a Brillouin zone in reciprocal space. Photonic crystals are being pursued to obtain an optical range of frequencies—a PBG—for which propagation is forbidden in all directions simultaneously, much like the energy gap in a semiconductor. This PBG has important quantum optical consequences, such as the inhibition or enhancement of radiative processes like spontaneous emission. Furthermore, point defects can act as small cavities that may form the basis for efficient miniature light sources and for novel solid-state quantum electrodynamics (QED) experiments in the strong coupling regime [310] (see Appendices D and E for a discussion of QED). Photonic crystals represent a new frontier in fundamental aspects of quantum and nonlinear optics [18].

The basic form of a photonic crystal is a 1D periodic structure such as a multilayer film. Electromagnetic wave propagation in such systems was first studied by Rayleigh in the late nineteenth century, when he demonstrated that any such 1D system has a bandgap. The 1D periodic systems eventually appeared in applications ranging from DFB lasers to reflective coatings. The 2D periodic optical structures, without bandgaps, received initial study in the 1970s and 1980s. The possibility of 2D and 3D periodic crystals with 2D and 3D bandgaps was suggested a century after Rayleigh: In 1987 Eli Yablonovitch and Sajeev John independently published articles describing these constructs [311, 312].

Figure 5.4 depicts how the periodicity defines dimensionality of photonic crystals. When the index of refraction changes periodically along one direction only, then the material is a 1D photonic crystal. A 2D photonic crystal has refractive index varying periodically along two directions. In dielectric structures with a 3D periodicity,

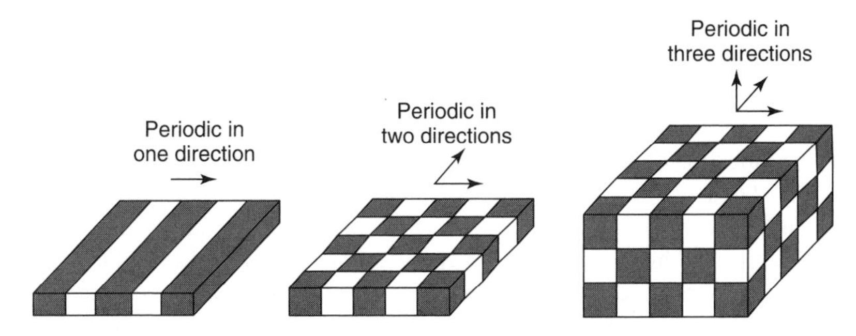

FIGURE 5.4 1D, 2D, and 3D PCs. Different shades represent materials of different values of refractive index.

there are no propagation modes in any direction for a range of frequencies, giving rise to a complete photonic bandgap [285].

5.4.2 Applicability of Technology

The PBG structures represent a new class of dielectric materials that allow guiding and manipulating the flow of light on the scale of the wavelength of light; PBG materials consist of a periodic arrangement of dielectric elements (e.g., hollow cylinders in a dielectric host material with high refractive index) with a lattice constant comparable to the wavelength of light [18]. Similar to the forbidden energy range (bandgap) for electrons in a semiconductor, PBG materials can present a bandgap for the energy spectrum of photons; it follows that the surface of the PBG materials can act like a perfect dielectric mirror for light with incidence from all directions. In turn, this allows one to capture (localize) light in nanoresonators and guide light in two and three dimensions with radii of curvature that were previously inaccessible. In turn, again, this enables new dimensions in terms of miniaturization of photonic devices—microphotonics will be able to evolve to nanophotonics [18].

Most of the work in this field has focused on the manipulation of the photonic bandgap, or the forbidden states for the photons. A photonic crystal with a full photonic bandgap is like a photonic insulator in which photons are not allowed to propagate. One can utilize the allowed states for photons and employ photonic crystals as novel photonic conductors [285]. Periodic structures, including gratings and 3D arrays, have many possible optoelectronic applications. Photonic crystal applications include LEDs, optical fiber, nanoscopic lasers, photonic ICs, radio frequency antennas and reflectors, and pigments.

Two-dimensional periodic photonic crystals already have reached a level where integrated-device applications are in sight; 3D photonic crystals had not yet entered commercialization as of press time (manufacturability considerations require further resolution). Initial commercial products involving 2D periodic photonic crystals are already available in the form of photonic crystal fibers that use a nanoscale structure to confine light with distinct characteristics (compared to conventional optical fiber) for applications in nonlinear devices [292]. There is also interest in higher refractive index material to develop 1D photonic bandgap structures that operate in the near-infrared frequency region (850–1550 nm) (as well as in the visible region). An omnidirectional reflector consisting of alternating layers of tin sulfide and silica (lower refractive index material) can be fabricated with a characteristic length scale on the order of 100 nm. These kinds of materials lead to several new possibilities that include efficient reflectors, high-frequency waveguides for communications and power delivery and high-Q cavities [313] (Q is a factor that measures the quality of a cavity—it is often obtained experimentally—See Glossary for more information).

In electronic microcircuits, electrical currents are guided by thin metal wires; electrons are bound within the cross section of the wire by the work function of the metal; as a result, electrical currents follow the path prescribed by the wire without escaping to the background. The predicament is different for optical waves: Light in an optical fiber can easily escape into the background electromagnetic modes of empty space if the fiber is bent or distorted on a microscopic scale [314]. Photonic crystals with complete photonic bandgaps eliminate this problem by removing all the background electromagnetic modes over the relevant band of frequencies. Light paths can be created inside a PBG material in the form of engineered waveguide channels to realize a number of microoptical functional elements such as straight waveguides [315], sharp bends [316], beam splitters, and tiny interferometers. Similarly, isolated defects in a PBG material act as microresonators and may be used to couple waveguiding structures to realize even more complex functional elements such as add–drop filters [317].

The PBG material can act as a tailored quantum electrodynamic vacuum [318]. The quantum optical and the nonlinear optical properties of atoms, molecules, and artificial atoms (e.g., semiconductor quantum islands or quantum dots) in such dielectric environments are distinctly different from usual free space. The inhibition or strong modification of spontaneous emission in a controlled fashion is one example of distinct properties. At the same time, this new physics allows to design novel photonic devices such as ultralow-threshold lasers and/or ultracompact and ultrafast nonlinear optical switching elements. The controlled incorporation of "defects" can lead to allowed energies within the forbidden gap, which then act like extremely narrow optical filters. There is research underway on linear and nonlinear optical propagation of light (wavelength range of 1300 and 1550 nm) in PBG materials and photonic devices made of them. The evolution of light pulses, "photonic bits," is of obvious relevance for applications in terms of telecommunication [18].

The controlled generation of "hot carriers" in the semiconducting constituents of a photonic crystal may provide a novel avenue to realizing tunable photonic band structure [319]. Similar considerations apply to the recently proposed photonic crystal structures that employ metal-coated spheres [320, 321].

There are other effects related to photonic crystals that may be of interest for future engineering applications. For example, while linear wave propagation is absent in a PBG, nonlinear effects can still occur. When one of the materials exhibits an intensity-dependent refractive index, certain high-intensity "light bullets" (solitary waves) can pass through the material even at frequencies within the gap [18, 322].

This effect may serve as the basis for ultrafast nonlinear switching devices. Conversely, outside a PBG, the photonic band structure may exhibit flat bands over extended frequency ranges, giving rise to extremely low group velocities [18, 323]. In addition, the corresponding photonic mode structure may exhibit strong resonant field enhancements in certain spatial regions [18, 324]. Hence, when a nonlinear photonic crystal is appropriately designed, these effects will lead to very large effective nonlinearities for propagating pulses and a number of nonlinear conversion effects such as second harmonic generation may be realized with increased efficiencies. The potential of nonlinear optics in PBG materials is only now beginning to be explored: Undesired broadening of "optical bits" due to group velocity dispersion could be avoided or reversed by soliton effects, optical bits could be switched or routed, and so on [18] (nonlinear optics are revisited in Section 5.7.1).

5.4.3 Fabrication

The necessary precision in terms of fabrication is on the order of 10 nm; this manufacturing scale is somewhat of a challenge at this time, at least in terms of per-unit costs and yields [18]. Photonic crystals take advantage of the nanofabrication processes that are available to pattern the dielectric function at the subwavelength scale; this patterning ability presents the possibility of designing the electromagnetic modes of photonic devices in microscopic detail [290]. The fabrication of 2D and 3D photonic crystals for near-infrared and visible frequencies remains a difficult undertaking at this time [314]. "Top down" nanotechnology fabrication methods to grow photonic crystals are not inexpensive; hence, researchers have sought approaches based on self-assembled structures starting from colloidal crystals. The strategy used by some researchers is to manufacture 2D PBG structures via self-assembly of functional magnetic colloidal particles as building blocks: Under certain conditions self-assembly of identical particles results in "photonic crystals" [325]. Some researchers use magnetic interactions; this approach is promising for PBG applications since these interactions (and the bandgap) can be tuned by the application of an external magnetic field [299].

One-dimensional photonic crystals can be constructed from a variety of organic and biopolymers, which can be dissolved or melted, by templating the solution-cast or injection-molded materials in porous silicon or porous silicon dioxide multilayer (rugate dielectric mirror) structures [326]. Synthesis of materials using nanostructured templates has emerged as a useful and versatile technique to generate ordered nanostructures [327]. Templates consisting of microporous membranes [328, 329], zeolites [330], and crystalline colloidal arrays [331] have been used to construct fairly elaborate electronic, mechanical, or optical structures. Porous Si is a viable candidate for use as a template because the porosity and average pore size can be tuned by adjusting the electrochemical preparation conditions that allow the construction of photonic crystals, dielectric mirrors, microcavities, and other optical structures.

Besides the necessity for large-scale microstructuring of materials with precision on submicrometer-length scales, there are only a limited number of materials with sufficiently high index of refraction such as Si, Ge, GaAs (gallium arsenide), GaP (gallium phosphide), InP (indium phosphide), and SnS_2. Because of their semiconducting nature, these materials exhibit a certain amount of frequency-dependent absorption and associated frequency-dependent variations in the index of refraction at near-infrared and visible frequencies [314]. (For example, it may be possible that multiple Bragg scattering in photonic crystals could lead to a resonant enhancement of otherwise negligible amounts of absorption in the bulk material and, as a consequence, may compromise the performance of photonic-crystal-based devices).

The information presented above forms the rough outline of the general operation of photonic crystals and PGB materials and its general applicability. The next section looks at some of the specific applications.

5.5 TELECOM APPLICATIONS OF PHOTONIC CRYSTALS

5.5.1 Quantum Cascade Lasers

Quantum cascade (QC) lasers are relatively new devices that exploit photonic crystal technology (as we saw in the previous section, photonic crystals are highly engineered materials with useful optical properties). The QC lasers are a class of high-performance semiconductor lasers invented in the mid-1990s. They are constructed by stacking many ultrathin atomic layers of standard semiconductor materials (such as those used in photonics) on top of one another. By varying the thickness of the layers, it is possible to select the particular wavelength at which a QC laser will emit light, allowing engineers to custom design a laser. These lasers are compact, rugged, and powerful [309].

When an electric current flows through a QC laser, electrons cascade down an energy "staircase," and every time an electron hits a step, a photon of infrared light is emitted. The emitted photons are reflected back and forth inside the semiconductor resonator that contains the electronic cascade, stimulating the emission of other photons. This amplification process enables high output power from a small device. In the decade since their invention, QC lasers have proved to be convenient light sources and are commercially available [309]. Applications also include free-space optics operating at 10 μm transmission, rather than the usual 1-μm range.

Standard QC lasers emit light from the edges (they cannot emit laser light through the surface of the device). Recently, a surface-emitting QC laser was developed by using the precise light-controlling qualities of a photonic crystal to create a QC laser (using electron beam lithography) that emits photons perpendicular to the semiconductor layers, resulting in a laser that emits light through its surface. The laser is "only" 50,000 nm across; this is quite a bit larger than the nanoscale, but it is compact by standard considerations. Compact lasers enable large arrays of devices to be produced on a single chip, each with its own designed emission properties. Such lasers-on-chips, if fabricated in the future, may lead to new possibilities for optical communications as well as other optoelectronics and sensing technologies [309].

5.5.2 Photonic Crystal Fibers

Standard silica-based optical fiber is a strand of glass that has a cladding 125 μm in diameter with an inner core that is 10 μm diameter. The inner core has a refractive index higher than the surrounding cladding. This enables the core to trap and guide the light down the waveguide by total internal reflection; light propagates down the waveguide with relatively little loss (in the range of 0.5 dB/km). At this time the principal source of optical loss is intrinsic scattering of light from the silica due to the molecular structure of the glass itself. The injection of the optical amplifier in the signal path has supported long-transmission circuits without requiring electrical termination and remodulation, but this has also introduced a new limitation, namely, dispersion (i.e., pulse broadening.)

To sustain bandwidth-carrying capacity in a single-fiber strand, one must address loss, dispersion, and other constraints in the fiber. There have been a number attempts to do this in the past (e.g., using new materials such as fluoride fibers), but these initiatives have not progressed to fruition. As briefly discussed in Section 5.4.2, photonic crystals offer the possibility of novel optical devices, such as filters, attenuators, polarization controllers, wavelength converters, and other in-fiber photonic devices that manipulate light rather than just transporting or filtering it. Recently, researchers have studied microstructured fibers where the light is confined by air holes in the fiber rather than being confined by the slight index differential (contrast) of conventional fibers. This provides a very large contrast. This placement of longitudinal airholes around the optical core affords new capabilities regarding optical properties (e.g., nonlinearity resulting from extremely tight light confinement and very low dispersion) [288]. The high-refractive-index contrast between glass and air holes permits light (under the right engineering circumstances) to be guided in air, rather than being taken into the high-index glass. These fibers can support very high information-carrying capacity because of the ability to multiplex across a wide bandwidth.

The idea of using periodically spaced air holes to guide light longitudinally was introduced in the mid-1990s and, although true bandgap operation was not achieved at that time, a lot of research ensued, including but not limited to [332, 333, 334, 335, 336, 337]. The first air core fiber was actually demonstrated in 1998, with propagation over structures about 1 m in length; manufacturing progress is being made, as was already noted earlier in the chapter. The holes in the fibers can also be filled with other materials, such as polymers, liquids, or liquid crystals.

If these kinds of fibers can be manufactured in such a manner to achieve low loss, they would become practical for telecom applications. It has been demonstrated that in theory attenuation can be many orders of magnitude less than silica fibers. However, other practical challenges exist: For example, the fiber must be fabricated with enough dimensional accuracy over long lengths (one needs to maintain bandgap operation over long lengths); low-loss splicing techniques need to be developed; and so on.

A resonant structure may be used to create as perfect a reflector as possible at optical frequencies (creating a Bragg reflection). A resonant structure can be created with a stack of dielectric films with alternating high and low refractive index.

Constructive interference between reflections from each layer results in the inability of the light to pass through the structure, giving rise to (near) perfect reflection. A sheet of this material can be rolled into a hollow tube to confine and guide light down the tube. Optical confinement by Bragg resonance can also be achieved in a 2D crystal structure.

5.5.3 Superprism Effect in Photonic Crystal

The superprism[5] phenomenon in photonic crystals is an effect that utilizes the allowed states for photons to manipulate the propagation of light. Polychromatic light incident at an angle onto one of the surfaces of a prism is dispersed within the prism; that is, light rays of different wavelengths propagate at different angles in the prism. Rays exiting the prism have a wavelength-dependent propagation angle that is due to the prism geometry. Conventional prisms rely on material dispersion; because the change in refractive index with wavelength is rather weak for transparent materials, the obtainable dispersion is limited. On the other hand, photonic crystal structures can be used to obtain much higher spatial dispersion. As a consequence of the wavelength-scale feature sizes of photonic crystals, these structures exhibit a behavior that is distinct from that of bulk materials. Wavelength regimes with high dispersion have been observed in theory and in practice (experiments) for 1D, 2D, and 3D photonic crystals. Considering the fact that these artificial structures exhibit much higher dispersion than the material dispersion of conventional prisms, this phenomenon has been termed the superprism effect [337a].

For certain wavelengths, light propagation inside a photonic crystal is significantly different from that in ordinary crystals. In an ordinary prism, splitting of light depends on the wavelength dependence of the refractive index; as noted, this dependence is small and the light propagation angle is determined by Snell's law. In a photonic crystal, band structure anisotropy determines the internal propagation direction and, consequently, the large angular dispersion. Due to this anisotropy, the propagation direction of light inside a photonic crystal can be an extremely sensitive function of parameters such as the wavelength or the incident angle. This effect, known as the superprism phenomenon, is observed at high frequencies, where anisotropy in photonic band structure is strongest and effects such as negative refraction and birefringence are expected.

The basis for the superprism phenomenon is anisotropy in the photonic band structure, a feature that is strongly present at frequencies near the photonic bandgap. The effect is very sensitive to the particular choice of incident angle relative to the orientation of the photonic crystal as well as sensitive to the incident wavelength. Calculations show a wavelength sensitivity of 14 degrees/nm for an input wavelength of around 1300 nm; calculations at constant wavelength show a change of 8° in the internal propagation angle for a 1° change in the input angle; these sensitivities increase at lower wavelengths. However, there is a very narrow window of operation for superprism effect to occur.

[5]This section is based on [285].

Since the superprism effect is basically an extraordinary angle-sensitive and wavelength-sensitive light propagation, a primary field of application of this effect is the area of integrated optics. Wavelength-dependent propagation can be exploited for WDM devices, whereas angle-sensitive propagation can be utilized for beam-steering and waveguiding. The mechanism of superprism can be applied to a variety of optical modulation devices. The advantage of using photonic crystal for these devices is that the device dimensions can be reduced given that the sensitivity is greater due to large angular deviation of the light beam. There also are applications to an optical sensing technique; this exploits the extreme sensitivity of the diffraction angle to the material properties, such as the refractive index contrast.

5.6 PLASMONICS

5.6.1 Study of Light at the Nanoscale

While interesting, PCs and PCFs discussed in the previous sections are still microlevel constructs. Plasmonics, on the other hand, is an emerging field of optics aimed at the study of light at the nanometer scale. The goal of plasmonics is to develop new optical components and systems that are the same size as today's smallest integrated circuits and that could ultimately be integrated with electronics on the same chip [265, 294]. The properties of metallic nanostructures and their near-field properties and applications are currently receiving increased research attention [338]. There is interest in the manipulation and focusing of optical fields by metal structures with subwavelength features. Advanced nanofabrication techniques that realize subwavelength metal structures, the development of computational methods to analyze their electromagnetic properties, and the observation of new phenomena such as surface-enhanced Raman scattering have all contributed to this resurgence of interest and to the development of a new field newly dubbed "plasmonics" [339]. An entirely new technology is emerging to manipulate optical information on a scale smaller than the wavelength of light, where optical energy is stored in plasmon states of materials [252, 253]. Researchers see "intriguing" properties of small metal and dielectric nanostructures from a physical and materials point of view.

New classes of metal–dielectric nanostructured materials with applications in photonics and optoelectronics are emerging: Metal nanostructures are capable of supporting various plasmon modes, which can result in high local fields (see Section 5.6.2) and, in turn, in major improvement of optical responses. These plasmonic nanostructures act like nanoantennas accumulating and building up the electromagnetic energy in small nanometer-scale areas [340, 341, 342, 343]. Applications envisioned for these nanomaterails typically entail arrays of closely spaced metal nanoparticles (arrays of closely spaced metal rods): Arrays of nanoscale metal structures can be constructed to create plasmon waveguides that can be smaller than a wavelength of light.

Plasmons can be thought of as "ripples of waves" in the "ocean" of electrons flowing across the surface of metallic nanostructures. Plasmonic devices are ultrasmall metal structures of various shapes that capture and manipulate light; they make use of optical properties of metallic nanostructures. The field of plasmonics has existed for only a few years, but it is now attracting a lot of research interest (and funding). The fact that light interacts with nanostructures is *prima facia* at odds with traditional optics, which for over a century postulated that light waves could not interact with anything smaller than their own wavelengths [265, 294]. Recently, however, it has been shown that nanoscale objects can amplify and focus light. References [344, 345] and [345a through 345t] comprise a short bibliography of recent papers on the topic. Table 5.4 lists areas of research and applications within this field [338].

Individual noble metal nanoparticles strongly interact with visible light at their dipole surface plasmon frequency due to the excitation of a collective electron

TABLE 5.4 Areas and Applications of Plasmonics

Nanofabrication of plasmonic architectures: structures and devices	Nanoparticle plasmonic structures
	Chemical fabrication
	Lithographic and nanopatterning (nonlithographic) fabrication
	Materials fabrication
	Biomimetic and bioinspired fabrication
	Integration of nanophotonic materials and devices with microphotonics (e.g., photonic crystal devices and other monolithic integrated photonic technologies)
Plasmonic properties and characterization	Spectroscopies (both spectral and time resolved)
	Local probes, nanooptics, and near-field phenomena
	Microscopies
	Nonlinear optical properties.
Plasmonic phenomena and effects	Surface plasmon optics in thin films
	Surface-enhanced Raman scattering
	Surface-enhanced nonlinear spectroscopy
	Fluorescence enhancement
	QED effects in plasmonic nanomaterials
	Metallic arrays and plasmonic bandgap materials
	Extraordinary transmission/diffractive/refractive phenomena
	Low-frequency plasmons and their applications
	Left-handed plasmonic materials
Plasmonics and plasmonic nanophotonics applications and devices	Sensors
	Waveguides
	Arrays
	Devices
	Plasmonic nanocircuits
	Plasmonic sensors
	Nanocrystal and nanowire lasers and LEDs
	Micromirrors, microcavities

motion (a so-called plasmon) inside the metal particle [303, 346]. The surface of the nanoparticle confines the conduction electrons inside the particle and sets up an effective restoring force, leading to resonant behavior at the dipole surface plasmon frequency. Figure 5.5 shows the flow of electromagnetic energy around a single spherical metal nanoparticle at two different excitation frequencies. When the frequency of the light is far from the intrinsic plasmon resonance of the metal nanoparticle (Fig. 5.5*a*), the energy flow is only slightly perturbed. At the plasmon resonance

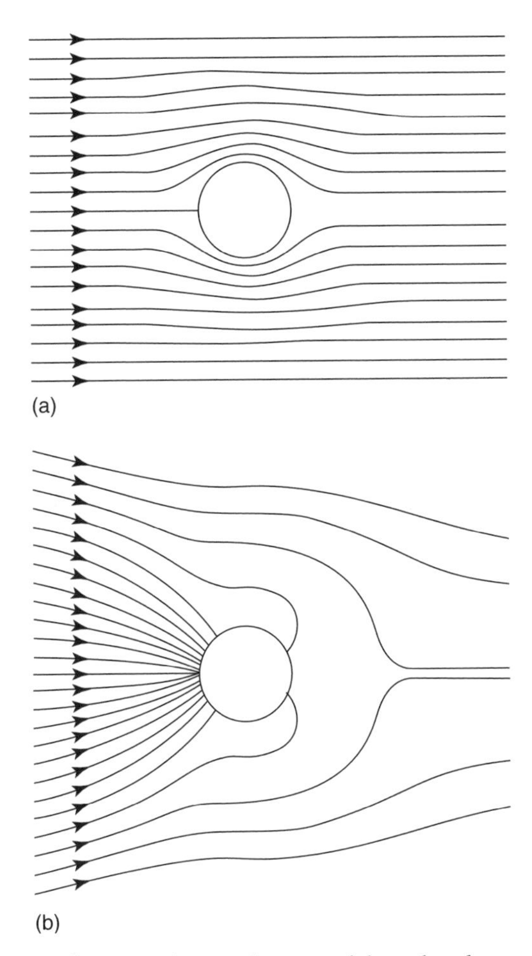

FIGURE 5.5 Energy flux around a metal nanoparticle under plane wave excitation. (*a*) When the excitation occurs far from the plasmon resonance frequency, the energy flow is only slightly perturbed. (*b*) When the excitation occurs at the plasmon frequency, the energy flow is directed toward the particle. This resonant field enhancement is a key element of plasmon waveguides. (From C. F. Bohren and D. R. Huffman, *Absorption and Scattering of Light by Small Particles*, © Wiley, New York, 1983; by permission of John Wiley & Sons, Inc. [356]).

frequency, the strong polarization of the particle effectively draws energy into the particle (Fig. 5.5*b*). This effect can be observed as a strong enhancement in the scattering cross section in optical extinction measurements [303, 346]. The dipole surface plasmon resonance is most pronounced for particles much smaller than the wavelength of the exciting light, since in this case all conduction electrons of the particle are excited in phase. The resonance frequency is determined by the particle material, the shape of the particle, and the refractive index of the surrounding host. Surface plasmons can be efficiently excited in the noble metals gold, silver, and copper due to their free-electron-like behavior; for these metals the plasmon resonance occurs in the visible range of light in a variety of hosts.

Integrated optics faces the fundamental limitation that, for the guiding, modulation, and amplification of light, in a useful manner, structures are needed that have dimensions comparable to the wavelength of light. Recently, it was shown theoretically that this problem can be avoided by transporting electromagnetic energy along linear chains of closely spaced metal nanoparticles. This transport relies on the near-field electrodynamic interaction between metal particles, resulting in coupled plasmon modes [347]. Waveguides based on the principle of total internal reflection (e.g., optical fibers) do not allow for the guiding of light around sharp corners with a bending radius considerably smaller than the wavelength of light λ [142, 348]. Engineering of defect modes in photonic crystals has enabled the fabrication of defect waveguides with (more) complex guiding geometries [142, 349, 350]. The integration of active devices such as defect mode lasers into photonic crystals is expected to advance the creation of optical chips [351]. The size and density of optical devices employing these technologies, however, are restricted by the diffraction limit $\lambda/2n$ of light. This imposes a lower size limit of a few hundred nanometers on the optical mode size; hence, a size mismatch between highly integrated electronic devices with lateral dimensions of a few tens of nanometers and optical guiding components persists and needs to be overcome [142]. The diffraction limit for the guiding of electromagnetic energy can be overcome if the optical mode is converted into a nonradiating mode that can be confined to lateral dimensions smaller than the diffraction limit [142]. Examples of approaches of the latter kind that have been the focus of research over the last few years are surface plasmon–polaritons in metals. Whereas plasmons in bulk metal do not couple to light fields, a two-dimensional metal surface can sustain plasmons if excited by light either via evanescent prism coupling or with the help of surface corrugations to ensure momentum matching [142, 352]. Such surface plasmons propagate as coherent electron oscillations parallel to the metal surface and decay evanescently perpendicular to it. Thus, the electromagnetic energy is confined to dimensions below the diffraction limit perpendicular to the metal surface. Corrugations can further act as light-scattering centers for surface plasmons, allowing for the fabrication of interesting optical devices such as an all-optical transistor [142, 353].

The new field of nanoplasmonics has applications to photonic devices. Plasmonics research provides a model that describes how ultrasmall metal structures of various shapes capture and manipulate light. Metal nanoparticles possess plasmon modes whose strong local fields may be useful for precise, photoinitiated processes in nanostructured arrays [354]. The equations that describe the frequencies of the plasmons in nanoparticles are similar, but not identical, to the quantum theory equations that describe the energies of electrons in atoms and molecules (e.g., Appendix D); this means that nanophotonic structures can be modeled in quantum-theoretic terms, rather than in classical optics terms. The type of plasmon that exists on a surface is directly related to the surface's topology, for example, a nanosized pore in metallic foil or the curvature of a nanoscale gold sphere. When light of a defined frequency interacts with a plasmon that oscillates at a compatible frequency, the energy from the light is captured by the plasmon, converted into electrical energy that propagates through the nanostructure, and ultimately reconverted back to light [265, 294].

Dielectric materials have modest refractive indices and, as a result, the range wave vectors achievable at optical frequencies have held back the scaling of dielectric optical components. Fortunately metallic, metallodielectric, and metal/semiconductor hybrid photonic structures have the capability to significantly alter this situation [249]. These structures support surface and interface plasmons that offer the ability to localize, extract, and enhance electromagnetic fields at metallodielectric interfaces and in nanostructures. The unusual dispersion properties of metals near the plasmon resonance enables excitation of surface modes and resonant modes in nanostructures that access a very large range of wave vectors over a narrow frequency range. This feature constitutes a critical design principle for light localization below the free-space wavelength and opens the path to truly nanophotonic and plasmonic optical devices. Thus, truly nanophotonic components and systems based on such materials now appear to be viable [249].

Another nanoscale arrangement that can sustain surface plasmons are metal nanoparticles [355, 356]. In metal nanoparticles, the 3D confinement of the electrons leads to well-defined surface plasmon resonances at specific frequencies. It has been established that light at the surface plasmon resonance frequencies interacts strongly with metal particles and excites a collective motion of the conduction electrons, or plasmon [357]. These resonance frequencies are typically in the visible or infrared part of the spectrum for gold and silver nanoparticles. For particles with a diameter much smaller than the wavelength λ of the exciting light, plasmon excitations produce an oscillating electric dipole field resulting in a resonantly enhanced nonpropagating electromagnetic near field (see Section 5.6.2) close to the particle surface [142]. It has also been found that near-field interactions between closely spaced metal nanoparticles in regular 1D particle arrays can lead to the coherent propagation of electromagnetic energy along the arrays with lateral mode sizes below the diffraction limit [358]. For Au and Ag nanoparticles in air, group velocities for energy transport higher than the saturated electron velocities in semiconductors and energy decay lengths of a couple of hundred nanometers have been predicted. Furthermore, it has been suggested that these so-called plasmon waveguides can guide electromagnetic energy around sharp corners and T structures, and an all-optical modulator based on interference operating below the diffraction limit may be possible [359].

Metal nanoparticles can be fabricated using a variety of tools, including electron beam lithography [360], colloidal synthesis [361], self-assembly [362], and ion irradiation [363]. Ion beams show the potential for large-volume fabrication of anisotropically distributed metal nanoparticles [354][6].

In particular, organized arrays of nanoparticles can be manufactured. An understanding of the interactions between nanoparticles in an array is important for the development of nanoscale photonic devices. Interactions between nanoparticles serve as communication mechanisms in nanoscale environments which are well below the scale of the conventional semiconductor technology in use today. Fundamental mechanisms of communication can involve transfer of energy in the form of photons, charge, or spin [354].

Applications of Plasmonics

Applications of the new nanostructured materials include [340] (i) plasmonic bandgap materials, (ii) light-gated optical transmitters that allow one to control photons with photons and to develop photonic nanocircuits, (iii) plasmon-enhanced QED control of spontaneous emission for surface-enhanced coherent spectroscopies, (iv) plasmon-enhanced nonlinear photodetectors and other optoelectronic devices, and (v) novel "left-handed" plasmonic nanomaterials that have a negative refractive index and appear to offer untapped potential for photonics (e.g., "superlenses.")

At the practical level, the concept of surface plasmons is applicable to thin films, nanowires, and clusters. Photonic and plasmonic applications range from energy guiding and storage to imaging and sensing. Applications include but are not limited to chemical sensing where chemicals in quantities as small as a single molecule can be detected; this has applicability, for example, to homeland security and medicine. Applications such as waveguiding below the diffraction limit, sensing and Raman spectroscopy, and integration of metal nanostructures with dielectric microoptics in the near infrared are also being investigated [252, 253]. Geometrically ordered metal nanostructures, such as periodic arrays of metal nanoparticles, arrays of holes in metal films, and metal nanowire meshes are being evaluated for tunable optical responses with frequency-, polarization-, and angle-selective enhancements [340].

Geometrically ordered metal nanostructure materials may also be developed as robust photonic crystals with large and scaleable bandgaps due to their negative dielectric permittivities. Researchers have designed and fabricated periodic metal nanostructures as photonic devices for guiding light with unprecedented density and performance [340]. Instead of making use of extended surfaces, a confinement of energy-guiding surface plasmon modes can be achieved using metal nanowires [142]. In nanowires, the confinement of the electrons in two dimensions leads to well-defined dipole surface plasmon resonances when the lateral dimensions of the wire are much smaller than the wavelength of the exiting light.

[6] The dipolar coupling between metal nanoparticles can also be utilized as a model for the study of other structures such as quantum dot chains [365], magnetic nanoparticle arrays [366], and coupled-resonator optical waveguides [367].

Nanoparticle Waveguides using Plasmonics

In recent years, there has been significant progress in the miniaturization of optical devices. As discussed earlier in the chapter, planar waveguides and photonic crystals are technologies that are enabling advancements in integrated optical components [303, 368, 369]. Still, the size and density of optical devices employing these technologies is limited by the diffraction limit of light, which imposes a lower size limit on the guided light mode of about a few hundred nanometers [303, 368]. Another limitation is the typical guiding geometry: While photonic crystals allow for guiding geometries such as 90° corners, planar waveguides are limited in their geometry because of radiation leakage at sharp bends [368, 369]. Scaling optical devices down to the ultimate limits for the fabrication of highly integrated nanophotonic devices and circuits requires electromagnetic energy to be guided on a scale below the diffraction limit and that information can be guided around large corners (bending radius much less than the wavelength of light).

The further integration of optical devices will, therefore, require the fabrication of waveguides for electromagnetic energy below the diffraction limit of light. Recently, new methods for the guiding of electromagnetic energy have been advanced that allows for a further reduction of the device size to below the diffraction limit in a variety of geometries [303, 370, 371, 372, 373]. It has been shown that electromagnetic energy can be guided in a coherent fashion via arrays of closely spaced metal nanoparticles based on near-field coupling. As discussed earlier in this section, there is now theoretical and experimental work underway on metal nanoparticle waveguides: Research shows that arrays of closely spaced metal nanoparticles can be used for integration and miniaturization purposes [303]. Coupling between adjacent particles sets up coupled plasmon modes that give rise to coherent propagation of energy along the array; plasmon-based waveguides and switches currently are the smallest devices with optical functionality. The basic underlying phenomena supporting nanoparticle waveguides are plasmon resonance and near-field coupling, discussed in the previous paragraphs [303].

The strong interaction of individual metal nanoparticles with light can be used to fabricate waveguides (a.k.a. plasmon waveguides) if energy can be transferred between nanoparticles. The dipole field resulting from a plasmon oscillation in a single-metal nanoparticle can induce a plasmon oscillation in a closely spaced neighboring particle due to near-field electrodynamic interactions [370]. The finding that ordered arrays of closely spaced noble metal particles show a collective behavior under broad beam illumination supports such an interaction scheme [303, 374].

When metal nanoparticles are spaced closely together (separation a few tens of nanometers), the strongly distance-dependent near-field term in the expansion of the electric dipole interaction dominates. The interaction strength and the relative phase of the electric field in neighboring particles are both polarization and frequency dependent. This interaction leads to coherent modes with a wave vector **k** along the nanoparticle array. Calculations for 50-nm silver spheres with a center-to-center distance of 75 nm show energy propagation velocities of about 10% of the speed of light; this is 10 times faster than the saturated velocity of electrons in typical semiconductor

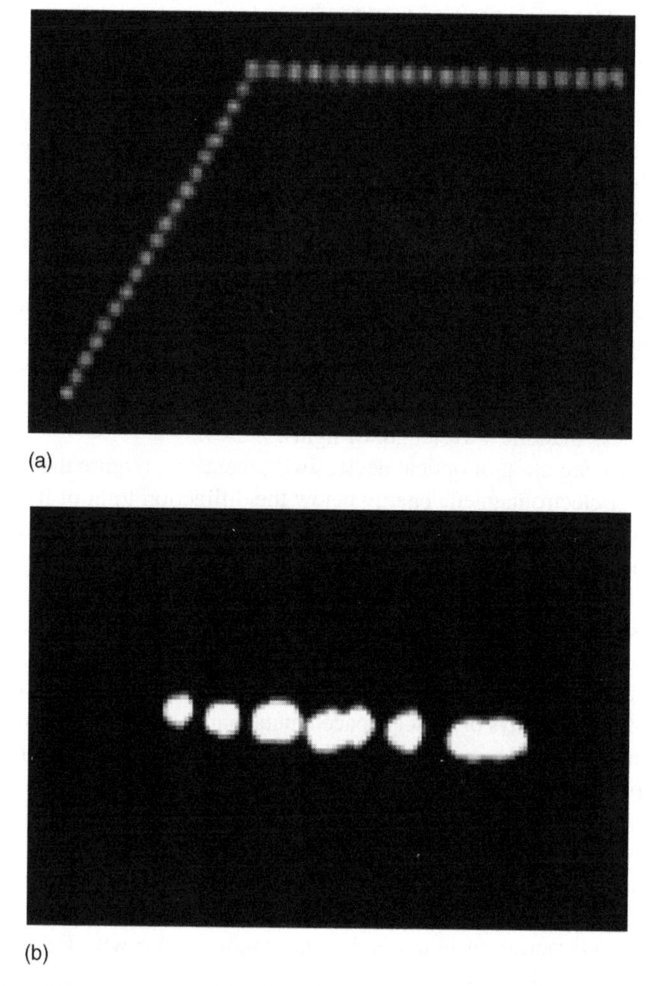

FIGURE 5.6 Plasmon waveguides: (*a*) SEM image of a 120° corner in a plasmon waveguide, fabricated using EBL (Au dots are ~50 nm in diameter and spaced by ~75 nm center to center). (*b*) Straight plasmon waveguide made using 30 nm diameter colloidal Au nanoparticles; the particles were assembled on a straight line using an atomic force microscope in contact mode, and then imaged in noncontact mode. (Courtesy: S. A. Maier, et al., *Plasmonics—A Route to Nanoscale Optical Devices*, © Wiley-VCH, 2001 [303b]).

devices [303]. These nanoparticles act like a light waveguide. Figure 5.6 depicts plasmon waveguides [303].

There are number of approaches for fabrication of nanoscale plasmon waveguides with optical functionality [303a]. The fabrication method should produce a narrow size distribution of the individual particles; additionally, a regular and uniform particle spacing is critical for the transport properties because of the strong distance dependence of the electromagnetic near field. Gold nanoparticles with diameters between 30 and 50 nm can be used as building blocks for plasmon waveguides [303, 346]. Two fabrication techniques have been used: electron beam lithography (EBL) and atomic force microscopy (AFM). Electron beam lithography provides good size and distance control of the nanoparticles constituting the waveguides. A second method of fabrication uses manipulation of randomly deposited nanoparticles using the tip of an atomic force microscope; in this instance, the control of the particle spacing is limited by the spatial resolution of the atomic force microscope.

5.6.2 Physics of the Near-Field

Since plasmonics deals with the near field, a brief discussion of this topic follows; the topic, naturally, is of interest on its own merit. Over the last decade, extensive research on and exploitation of the different kinds of near fields existing spontaneously or induced artificially in immediate proximity to the surface of materials (or at the interface of two materials) have resulted in a plethora of new technical developments. For example, these technologies permit advances in nanolithography and nanoinspection of semiconductor materials: Because near-field optics permits the interaction of light and matter with a resolution of nanometers, manipulation of matter at this scale becomes possible [193]. This section provides a very summarized view of the physics of the near field and is completely based on and abstracted from an excellent paper by Girard et al. [375].

The concept of the near field is not restricted to specific areas in physics but actually covers numerous domains of contemporary physics (electronics, photonics, interatomic forces, phononics, etc.). The theory mainly concerns phenomena involving evanescent fields (electronic density surface wave, evanescent light, local electrostatic and magnetic fields, etc.) and/or localized interatomic or molecular interactions. In fact, the practical exploitation of these *waves* and local *interactions* was latent for a long time in physics, until the emergence and the success of local probe-based methods [scanning tunneling microscopy (STM), SFM, SNOM]. Although near-field physics was a well-established research area before the mid-1970s, its actual and systematic investigation began in the early 1980s with the invention of the scanning tunneling microscope. Within a few years of this important discovery there is now an explosion of new experimental devices able to explore and measure many different kinds of near fields (electronic, photonic, acoustic, force, etc.). At this time, various theoretical approaches and powerful numerical methods well suited to near-field physics are described in the literature.

It has long been known that the surface limiting a solid body locally modifies the physical properties of many materials (dielectric, metal, or semiconductor). In other words, the symmetry loss generated by the presence of an interface produces specific surface phenomena that have been identified in the past (spontaneous polarization, electronic work function, electronic surface states, surface polaritons, surface-enhanced optical properties, etc.). The near field can be defined as the extension outside a given material of the field existing inside this material. Basically, it results from the linear, homogeneous, and isotropic properties of the *space–time* that impose a continuous variation of field amplitudes and energies across the interfaces. In most cases, the amplitude of the near field decays very rapidly along the direction perpendicular to the interface, giving rise to the so-called evanescent wave character of the near field.

In optics, the symmetry reduction occurring in the vicinity of an interface can enhance some hyperpolarizabilities initially absent in the bulk materials. This has been used for surface second-harmonic generation at the metal–air interface. In the vicinity of a metal–vacuum interface, the electron density distribution tails off exponentially into the vacuum and exhibits Friedel oscillations on the metal side. A long list of similar effects extensively described in the *surface science literature* arise due to the existence of this near-field zone. In this context, surfaces can also be considered as a privileged place to generate, guide, manipulate, and detect evanescent waves.

A given field $F(r)$ lying in a spatial region A always presents a continuous extension inside an adjoining domain B. This proposition is true whatever the change between the physical properties of the two regions A and B may be. In well-defined conditions, this leads to the occurrence of a more or less rapid decay of the field $F(\mathbf{r})$ inside the domain B. One can distinguish two important categories of such interfacial near fields:

1. The first corresponds to spontaneous near fields produced in B from a permanently established field in A. For example, permanent electric fields in immediate proximity to an ionic crystal belong to this category. This is also the case of the wave functions of electrons that tail off the surface of a metal.
2. The second class gathers together surface near fields that can only be produced by applying an external excitation (photon and electron beams impinging on a surface). Both optical near fields and surface plasmon–polaritons excited at a solid interface provide good illustrations of this category. These phenomena have a special interest because they can be manipulated at will by an external operator.

Four different kinds of near fields are common (among other): the electrostatic surface field, the optical near field, the fluctuating electromagnetic field, and the electronic evanescent wave function near metallic surfaces. See Table 5.5.

Specifically, near-field principles can be applied to microscopy. The wave nature of light causes it to diffract, which in turn limits the spatial resolution of a microscope using near-field methods. Typically, the minimum detectable separation of two light scatterers for a given optical system is the Rayleigh criterion (see Appendix F). This limits traditional light microscopy to a resolution of 200–300 nm at best (some state-of-the-art photolithographic systems achieve 100 nm resolution by using vacuum ultraviolet light [193]). New far-field methods, also in conjunction with scanning tips, can extend resolution to 10 nm. Figure 5.7 depicts the classical optical method with the near-field method (e.g., see [193, 376]).

Of late there have been advancements in scanning near-field optical microscopy techniques that allow near-field optical imaging using laser-illuminated metal tips. Aperture scanning near-field microscopy is a technique that allows for nanoscale

TABLE 5.5 Four Different Kinds of Near Fields

Electrostatic surface fields	A simple example of permanent electric near field can be found close to the surface of ionic or metal–oxide crystals (NaCl, LiF, MgO, . . .).
Optical near fields	Optical nonfluctuating near fields are not permanent and consequently must be generated by an external light source. The simplest method consists of illuminating the surface of a sample by external reflection. In this case, the structure of the electromagnetic field above the sample critically depends on the incident angle. This effect is particularly important outside the Brewster angle, where the field intensity tends to be modulated by the interferences between incident and reflected waves. The physics of optical evanescent waves [OEWs (which is the central concept used in near-field optics (NFO) instrumentation] has been familiar in traditional optics for a long time: The analysis of the skin depth effect at metallic surfaces about a century ago was probably the first recognition of the existence of evanescent electromagnetic waves.
Electromagnetic fluctuating near field	This less conventional class of surface near fields has a considerable impact in local probe-based experiments. It concerns the fluctuating electromagnetic field existing spontaneously near the surface of any material. Historically, as early as 1930, London showed that the quantum mechanical fluctuations between two neutral atoms or molecules (devoid of any permanent multipole moments and separated by a distance R) could give rise to a force that varies as R^{-7}. Two decades later this concept was generalized by Lifshitz in order to derive a complete scheme able to grasp the origin of van der Waals dispersion forces between solid bodies.
Electronic wave function at a metal surface	The metal–vacuum interface (the surface charge density near metal) can be described with the free-electron Sommerfeld approximation (FESA) where the ground-state properties of the electron gas are obtained by filling up the conduction band with N free electrons obeying a Fermi–Dirac distribution. This free-electron scheme can be completed by applying the density functional method inside a "jellium" environment in which the ion cores are smeared out into a uniform positive background truncated by the surface. The electron charge profile near various metal surfaces was calculated with this technique by Lang and Kohn. It shows up the splitting between delocalized electronic charges and positive jellium into the vacuum side of the interface. In the metal, it exhibits the well-known Friedel oscillations, which have the characteristic wavelength π/K_F (K_F is the Fermi wave vector). One can calculate the permanent and *probabilistic* near-field component of the electronic wave function associated with the FESA electrons. The exponential nature of the evanescent wave function provides the opportunity for a uniquely sensitive form of microscopy. Exploitation of this simple electronic decay law began with the invention of the STM by Binnig and Rohrer in 1981; since then, exploitation of this effect has enabled many original studies at subnanoscale resolution to be undertaken.

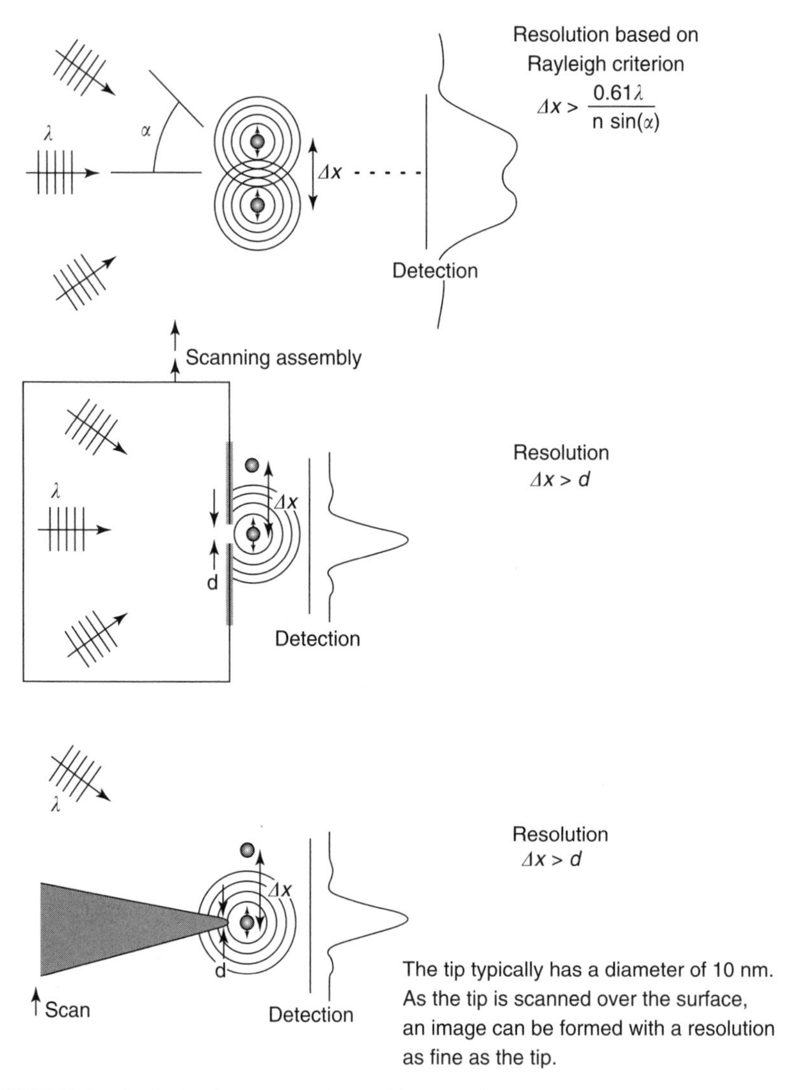

FIGURE 5.7 Optical microscopy enhanced by near-field techniques: (top) traditional optical microscopy, (middle) near-field approach, (bottom) system using a scanning tip (field-enhanced microscopy).

resolution. Aperture scanning functions by scanning a small aperture over the object: Light can only pass through the aperture, and so this size determines the resolution of the system (see Fig. 5.7, middle). This technique is typically implemented by tapering a fiber optic to a narrow point and coating all but the tip with metal. However, the amount of light that can be transmitted by a small aperture poses a limit on how small the aperture can be before no photons get though. Studies show that when the aperture is 100 nm, only one photon in 10,000 makes it through; when it reaches 50 nm, only one photon in 100 million makes it through; and when it reaches 10 nm, only one photon in 1 trillion makes it through. Additionally, the input power cannot be increased arbitrarily because a major fraction of the power is absorbed in the coating, so that increasing the input power above approximately 10 mW will destroy the coating. To address this issue, instead of using a small aperture, one can use a metal tip to provide a local excitation. If a sharp metal tip is placed in the focus of a laser beam, an effect called local field enhancement will cause the electric field to become approximately 1000 times stronger. This enhancement is localized to the tip, which has a typical diameter of 10 nm. As this tip is scanned over the surface (as depicted in Fig. 5.7, bottom), an image can be formed with a resolution as fine as the tip. The tip typically has a diameter of 10 nm. As the tip is scanned over the surface, an image can be formed with a resolution as fine as the tip [193].

5.7 ADVANCED TOPICS

In this section we briefly look at nonlinear optics, microresonators, and quantum optics. These topics have applicability to nanophotonics.

5.7.1 Nonlinear Optics

In this subsection we look briefly at area of nonlinear optics. Only the simplest concepts are introduced; the mathematics can quickly become fairly complex.

When a beam of light is launched into a material, it causes the charges of the atoms to oscillate. Figure 5.8 depicts three basic mechanisms of response:

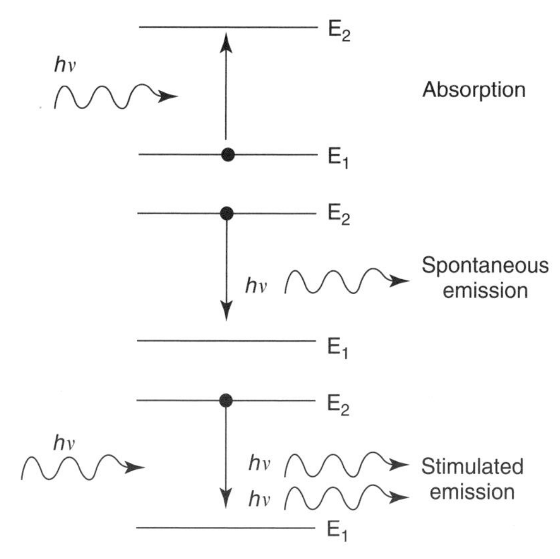

FIGURE 5.8 Photonic interactions with matter.

- *Absorption:* An atom or molecule has many different energy levels. When the atom or molecule absorbs a photon, its energy is increased by an amount equal to the energy of the photon. The atom or molecule then enters an excited state. Hence, absorption is the process by which the energy of a photon is taken up by another entity, for example, by an atom that makes a transition between two electronic energy levels; the photon is destroyed in the process.
- *Spontaneous emission:* This is the process by which matter may lose energy, resulting in the creation of a photon.
- *Stimulated emission:* This is the process by which, when perturbed by a photon, matter may lose energy, resulting in the creation of another photon. The perturbing photon is *not* destroyed in the process and the second photon is created with the same frequency and phase as the original. Stimulated emission is a quantum mechanical phenomenon. The process can be conceived as *optical amplification*, and it forms the basis of the laser (these mechanisms are used, for example, in EDFAs).

In a linear material the amount of charge displacement is proportional to the instantaneous magnitude of the electric field. The charges oscillate at the same frequency as the frequency of the incident light. Either the oscillating charges radiate light at that frequency or the energy is transferred into nonradiative modes that result in material heating (or other energy transfer mechanisms). Generally, the radiated light travels in the same direction as the incident light beam: The light is effectively "bound" to the material; the light excites charges that reradiate light that excites charges; and so on. As a result, the light travels through the material at a lower speed than it does in vacuum. If the motion of some of the charges within the material decays without giving off light, some of the light intensity is lost from the incident beam by scattering and absorbance. The absorbance is defined as the ratio of light exiting a material to the light incident into the material divided by the material thickness. Both the absorbance and refractive index (ratio of speed of light in vacuum to the speed of light in the material) are linear optical properties of a material for low-intensity incident light [377].

In a nonlinear optical material, the displacement of charge from its equilibrium value is a nonlinear function of the electric field (e.g., see Fig. 5.9). All materials when exposed to a high enough light intensity show a nonlinear response. Nonlinearity in optics occurs when the electromagnetic wave is large enough such that the medium responds not only at the fundamental driving frequency but also at higher harmonics. For small forces, the displacement of the charge is small and is approximated by a harmonic potential. When the displacement away from equilibrium is large, the harmonic approximation breaks down and the force is no longer a linear function of the displacement. When the charges in a molecule are bound by a harmonic potential, the induced dipole moment is linear in the applied field. The response of a molecule is "nonlinear" if the charges are bound to the molecule by a nonharmonic potential. In this case, the dipole moment of the molecule is a nonlinear function of applied electric field. More generally, if a nonlinear molecule is exposed to light, the time-dependent induced dipole moment is a nonlinear function of the time-dependent electric field [377].

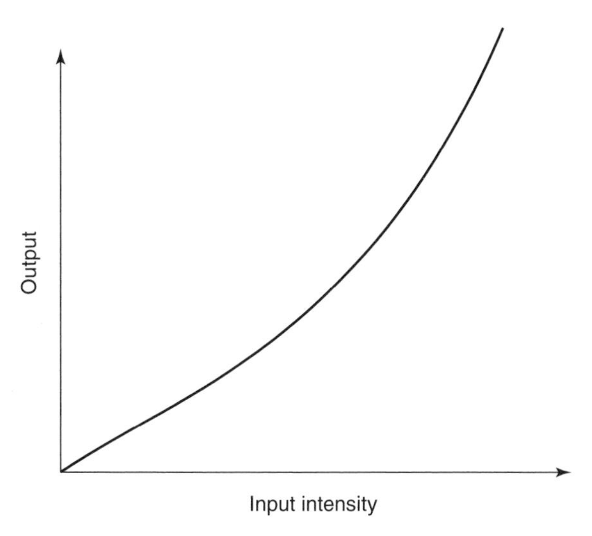

FIGURE 5.9 Nonlinearity.

When the intensity of the incident light to a material system increases to a large value the response of medium is no longer linear. The response of an optical medium to the incident electromagnetic field is the induced dipole moments inside the medium [378]. Typically it takes fields greater than of 10^5 V/m to observe most nonlinear optical phenomena. These optical fields are easily generated by lasers: The coherence of laser light makes it possible to observe many nonlinear phenomena; when the molecules in the material respond coherently to the laser light, their combined effect can be detected even for the weakest nonlinear effects [377]. Nonlinear optics as a field started by the discovery of second-harmonic generation around the time the laser was emerging. While nonlinear optical phenomena can be formulated by Maxwell's and Schrödinger's equations, it was not until the advent of the laser that most nonlinear optical phenomena could be tested; since the invention of the laser researchers have indeed confirmed a large number of nonlinear optical effects and have applied them to many practical uses. Nonlinear optical phenomena can be described in terms of higher order susceptibilities. Various specific nonlinear phenomena include electrooptic modulation, acoustooptic modulation, harmonic generation and frequency conversion, stimulated Raman and Brillouin scattering and amplification, parametric oscillation and amplification, self-phase modulation, soliton propagation, and photorefractive effects. Supportive technologies include nanocomposites, quantum well and quantum dot devices, and photonic bandgap crystals (see Table 5.6 [379]).

5.7.2 Confinement and Microresonators

A cavity resonator utilizes resonance to amplify a wave. Electronic and optical confinement in semiconductor heterostructures enjoy a number of advantages for device applications, including increased wavelength and temperature range and

TABLE 5.6 Synthesis of Nonlinear Optics and Nanophotonics

Topic	Concepts/Phenomena
Maxwell's equations and nonlinear polarizability	Maxwell's equations including higher order polarization response
Susceptibility tensors and symmetry properties	Higher order susceptibilities and symmetry properties
Coupled wave equations	Derivation of coupled wave equations in the slowly varying amplitude approximation
Electrooptic and acoustooptic modulation	Design and application of electrooptic modulators and acoustooptic modulators
Harmonic generation and frequency mixing	Design of frequency doubling and frequency-mixing crystals, phase-matching angles, beam walkoff, and conversion efficiency
Parametric oscillation and amplification	Parametric amplification, angle tuning, efficiency, parametric oscillation
Raman and Brillouin scattering	Spontaneous Raman and Brillouin scattering, stimulated Raman and Brillouin scattering and amplification, application to fiber-optic systems
Self-focusing and self-phase modulation	Intensity-dependent refractive index, self-focusing threshold intensity and length, self-phase modulation, and spectral broadening
Nonlinear effects in fibers	Dispersion, self-phase modulation, solitons, modulation instability, cross-phase modulation, and four-wave mixing
Nanophotonics	Nanocomposites, quantum well and quantum dot devices, photonic bandgap crystals
Phase conjugation, optical bistability, and photorefractive effects	Phase conjugation using four-wave mixing, optical bistability in a Fabry–Perot geometry, photorefractive effect, and application to image amplification and phase conjugation

higher speed. A typical cavity has interior surfaces that reflect one type of wave (specifically, a center frequency f_0). When a wave that is resonant with the cavity enters, it bounces back and forth within the cavity with low loss; as additional wave energy enters the cavity, it combines with and reinforces the standing wave that is created, increasing its overall intensity. Hence, resonant systems respond to frequencies close to the natural frequency f_0 much more strongly than they respond to other frequencies. Examples of cavity resonators include the tube in a microwave oven, a laser cavity, and a tube of an organ.

To expand on the laser example, the components of a typical laser include the following: (i) an energy source (usually referred to as the pump or pump source); (ii) some matter that amplifies light ("active medium," "laser medium"); and (iii) some device that traps the light around the space filled with the medium ("cavity," a system of mirrors forming an optical resonator, or an electrophotonic equivalent). The feedback mechanism is needed because the light would otherwise escape from the medium. An optical cavity such as a Fabry–Perot resonator achieves this feedback because the light can travel back and forth between two mirrors a large number of times.

In order to drive an amplifying medium, one has to "pump" energy into it. The gain medium is effectively a converter between the pump energy and the light emission. The factor Q is the measure of the quality of the resonator; often, the conversion efficiency is low, with values in the range 10–50% being considered "large." In optics, Q is a measure of how much light from the gain medium of the laser is fed back into itself by the resonator; in other words, the optical Q of a resonant cavity is the ratio of energy stored to energy dissipated in the cavity. This concept can, in fact, be used to create "pulsed" signals: if the Q of a laser's cavity is abruptly changed, the laser can be induced to emit a pulse of light; this technique is known as Q-switching. The basis of Q-switching is the utilization of some kind of device or mechanism that can alter the Q of the optical resonator (cavity) of the laser.

Since Purcell's discovery in the 1940s that the spontaneous emission rate of an atom can be changed by placing the atom inside a cavity, a number of interesting and useful results relating to cavities have been discovered, in what is now called cavity quantum electrodynamics. In particular, high-Q cavities supporting a single mode of the electromagnetic field have proved to be excellent tools to study some counterintuitive consequences of quantum mechanics. Cavity effects, however, are limited by damping: Real cavities are not made of perfectlyreflective mirrors and the radiation inside a real cavity eventually decays because of losses [380, 381].

While confinement effects in one dimension (quantum wells and planar microcavities) have been understood and exploited for a number of years, recent progress has made it possible to tailor the bandgap and the refractive index in three dimensions, opening new avenues for research and applications [382]. Electronic confinement in 3D nanoislands (quantum dots) leads to complete energy quantization and provides a solid-state equivalent to atomic physics. Optical confinement in 3D microcavities also allows complete control of spontaneous emission. This may lead to the control of spontaneous emission at the single-photon level.

Microresonators are components that are expected to be used for photonic ICs. Specifically, microresonators are miniature components that enable frequency-selective coupling between waveguides. As an example, a microresonator can completely extract the resonant wavelength from the input port and reroute it to the dropped port. It follows that a number of devices operating on a fixed wavelength, such as add–drop filters, switches, and demultiplexers, can be built. Microresonators are very small and facetless cavities where literally thousands of components can be integrated into a single photonic chip by coupling the microresonator microdisks to the same bus waveguide. Circuits containing laser sources, detectors, switches, routers, and multiplexers are envisioned. The structure typically is 10 μm in size and is grown in a single growth process: a 1-μm-thick p-doped (InP) disk-cladding layer followed by a 0.4-μm-thick intrinsic disk core layer [383].

Optical ring resonators (e.g., based on GaAs–AlGaAs and GaInAsP–InP) are promising building blocks for future all-optical signal processing and photonic logic circuits. Their versatility allows the fabrication of ultracompact multiplexers–demultiplexers, optical channel dropping filters, laser amplifiers, and OADM logic gates (to name a few), which are expected to enable large-scale monolithic integration for optics [384]. One is interested in optical field distribution, resonance wavelength, and the finesse of circular ring and disk microresonators. The microresonator can be considered either as a circularly bent waveguide or as a resonant structure with complex eigenfrequency. Very strong dependence of the finesse of 3D microresonators on their cross-sectional refractive index profile exists [385].

We conclude this section by providing some basic information on traditional (optical) quantum wells (QWs), which are active elements found in optoelectronic semiconductor devices. The term "well" refers to a semiconductor region that is (processed) grown to possess a lower energy, so that it acts as a trap for electrons and holes (electrons and holes gravitate towards their lowest possible energy positions). They are referred to as "quantum" wells because these semiconductor regions are only a few atomic layers thick; in turn, this means that their properties are governed by quantum mechanics, allowing only specific energies and bandgaps. Because QW structures are very thin, they can be modified easily [386]. For example, a 980-nm pump laser array uses QW structures to provide more efficient and higher power light output. The QW structures improve the laser performance by forcing electrons from the n material and holes from the p material to be in the same small volume, thereby optimizing the strength of the recombination process (which results in photon emission). By closely controlling the thickness of the quantum layer, the wavelength of the emitted light can be controlled.

Quantum well intermixing (QWI) is a manufacturing technique that allows the properties of a semiconductor material to be modified so that multiple optical communication functions can be integrated on a monolithic chip. This is achieved by depositing additional layers and then applying heat—exciting the atoms and thereby causing intermixing with surrounding materials. By careful choice of capping layers, it is possible to selectively intermix the quantum wells across a wafer, thus allowing a single chip to perform various optical functions [386]. The QWI principles might be used to modify the absorption spectrum (e.g., allowing data modulators), or filtering, for selecting and switching particular data transmission wavelengths.

5.7.3 Quantum Optics

Quantum optics is the science concerned with the applications of the quantum theory of optics; that is, optics defined in terms of the quanta of radiant energy, or photons [34]. In particular, this field aims at understanding how virtual photons are involved in molecular binding, understanding van der Waals and Casimir forces on the nanoscale, and understanding interactions with semiconductor wave functions (particles that do not have a permanent existence are called virtual particles; virtual particles always come in pairs: a particle and antiparticle—these mutually annihilate within an extremely short time—see Glossary). Considerable research work in underway in this field. In this section we limit our short discussion on quantum optics to possible applications.

Compared with free propagating light, the optical near field is enriched by virtual photons. These photons are similar to particles responsible for molecular binding (van der Waals/Casimir forces) and, consequently, represent promising mechanisms for selective probing of atomic structures. The consideration of virtual photons in the field of quantum optics is expected to enlarge the range of fundamental new discoveries and applications [193].

Applications of quantum optics phenomena span optical imaging, communication, materials, biology, and devices for quantum computing, to list a just a few. Researchers are now exploring methods where signals can optically interact with semiconductor nanostructures on length scales smaller than the extent of their quantum wave functions. Probing and manipulating these wave functions might open up applications in optical switching based on quantum logic [250]. The exploitation of quantum effects for technological applications is one of the most obvious driving forces behind the current miniaturization in optoelectronics. The recent rapid advances are due in large part to the industry's newly acquired ability to measure and manipulate individual structures on the nanoscale (scanning probe techniques, optical tweezers, high-resolution electron microscopes, etc.) [193].

The energy of light lies in the range of electronic and vibrational transitions in matter; therefore, the interaction of light with matter renders unique information about the structural and dynamical properties of matter. These spectroscopic capabilities are of importance for the study of biological and solid-state nanostructures: One can apply near-field optical techniques to probe complex semiconductor nanostructures as well as individual protein molecules. Probing and manipulating these wave functions could open up applications such as data storage and optical switching based on quantum logic [193]. In particular, there is interest in understanding the interaction of a QD with an optical near field.

There is ongoing interest in developing quantum computers. Generation of single quantum particles is rooted at the core of modern physics. Cold neutral atoms have internal states that can store quantum bits (qubits); in turn, these bits can be used in computing and information storage applications [387]. At this time all methods supporting quantum information and quantum computation involve interferometry through the definition of qubits and the entangled bit (ebit). Most of the successful and reliable applications of these concepts have been obtained so far in the field of quantum optics with the generation of single photons and entangled photon pairs and their application to quantum information processes [388].

We revisit this topic in Chapter 6.

5.7.4 Superlenses

Along another line of investigation, surface plasmons can be exploited to develop what have been called superlenses. Researchers at the University of California at Berkeley recently announced the development of an optical superlens based on a thin (35-nm thick) layer of silver; the lens has a negative refractive index. This lens can be used to image structures with a resolution that is about one sixth the wavelength of light, thus overcoming the so-called diffraction limit [388a]. Applications of these new superlenses are expected to include detailed biomedical imaging in real-time and in vivo, optical lithography to make higher density electronic circuits, and faster fiberoptic communications.

Conventional lenses have positive-refractive-index; they create images by capturing the light waves emitted by an object and then bending them. Materials negative refractive index bend the light in the opposite direction to an ordinary material. The idea of "superlens" was advanced three decades ago after Russian physicist Victor Veselago first speculated that negative index materials could exist [388b].

It turns out that objects also emit "evanescent" waves that contain information at very small scales about the object. These waves are more difficult to measure because they decay exponentially and, so, do not reach the image plane—a threshold in optics known as the diffraction limit. In 2000 John Pendry of The Imperial College in London suggested that a material with a negative refractive index could capture and "refocus" evanescent waves. In such a superlens, electromagnetic waves that reach the surface of a negative refraction lens excite a collective movement of surface waves ("surface plasmons"), such as electric oscillations. This process enhances and recovers the evanescent waves.

In 2003, the University of California group showed that optical evanescent waves could indeed be enhanced as they passed through a silver superlens. With recent follow-up work they imaged objects as small as 40-nm across (in contrast current optical microscopes can only resolve objects down to around 400-nm, which is about one tenth the diameter of a red blood cell). This work provides a new imaging method that can beat the optical diffraction limit.

5.8 CONCLUSION

The field of nanophotonics is vast and is growing rapidly. In this chapter we only surveyed the most well-developed areas that comprise the field. In particular, we examined photonic crystals, photonic crystal fibers, and plasmonics. References to integrated optoelectronics were made.

CHAPTER 6

Nanoelectronics

Nanoelectronics is the science related to the design of nanoscale devices that have electronic properties, such as transistor, switching, amplifying, tunneling, and/or logical relay capabilities. Silicon-based semiconductor technology has advanced at exponential rates in both performance and functionality over a period of nearly 50 years. As might be expected, there is a desire at this juncture to continue to decrease gate sizes and increase their intrinsic functionality. The need for smaller and faster electronic devices has given life in the recent past to the new field of nanoelectronics. Researchers and developers are interested in nanoelectronic properties of materials, for the purpose of communication, computation, storage, or control. Given the steady advances in nanoelectronics in recent years, the possibility now exists that current microelectronics could be eclipsed within a decade by the promise of quantum effect devices.

As noted, a trend has existed for several decades toward ultra-large-scale integration and miniaturization of electronic components, with "classical" methods already having crossed the 100-nm range in the early 2000s and crossing the 50-nm range as of press time. Future computational systems will likely consist of superdense, superfast, and very small logic devices [237]. Nanoscale researchers already work with prototype electronic circuits as small as 10 nm. Microelectronics has seen major improvements in gate density in recent decades, and nanoelectronics is simply perceived as the next step in that miniaturization tradition. Currently, the majority of microprocessors are constructed from silicon semiconductor transistors patterned and carved by light beams through photolithographic techniques; new manufacturing tecquniques are being sought.

Many (but not all) nanoelectronic devices now being studied still rely on electrical charges being transferred between points on a device or circuit. Typical devices of interest include reduced-size silicon transistors, single-electron transistors (SETs), resonant tunneling diodes (RTDs), magnetic spin-based devices (spintronics, or, more specifically, spin nanoelectronics), and molecular devices. SETs are devices that have a switching capability controlled by the removal or addition of a single electron (also, these devices allow a single electron to be transported at any one

Nanotechnology Applications to Telecommunications and Networking, By Daniel Minoli

time). Tunneling refers to the ability of using the quantum wave properties of an electron to allow transmission through a thin voltage-potential barrier. Spin nanoelectronics refers to the utilization of the electron's spin for storage or computation.

This chapter focuses on the field of nanoelectronics. It provides an overview of key emerging technologies and commercial opportunities thereof, particularly in the networking and computing environments. After the introductory section that follows (Section 6.1), the chapter is comprised of two parts. The first part (Section 6.2) provides a brief overview of the field; the second part (Section 6.3) covers a handful of special topics. Appendix F expands on nanoinstrumentation while Appendix G expands on the computing application; both of these topics are discussed in preliminary fashion in the chapter itself.

6.1 INTRODUCTION

6.1.1 Recent Past

The microelectronics discipline started in the late 1940s. The first bipolar transistor was demonstrated at Bell Telephone Laboratories in 1948. As discussed in Chapter 2, a transistor is a multilayer device that can switch and/or amplify a signal. The integrated circuit (IC) was developed in 1959, and the field-effect transistor (FET) became available in 1960. A lot of progress has been made since; in particular, a lot of progress has been made since the late 1960s, which were characterized by medium-scale integration technology and/or by the kind of discrete-level electronics with which this researcher started out his career, as illustrated in the cover of this book (the device on the desk is an example of discrete electronics).

The average physical size of microelectronic components and memory devices has been decreasing monotonically over time. Since the mid-1960s, the microelectronics advancements have followed the empirical rule of Moore's law, which we already introduced in Chapter 1. In the mid-1960s Gordon Moore made his now well-known observation that an exponential growth in the number of transistors per integrated circuit could be observed in manufactured components and then predicted that this trend would continue [389]. In fact, through technological advances, Moore's law, the doubling of the number of transistors that can be packed in an IC every 18 months or so, has remained fairly accurate during the past 40 years and still holds true today. At the macrolevel, the following evolutions have been observed over time:

- *Small-scale integration* (SSI): The chip contains of a number of transistors, but not hundreds.
- *Medium-scale integration* (MSI): The chip contains hundreds of transistors, but not thousands.
- *Large-scale integration* (LSI): The chip contains thousands of transistors, but less than 100,000.
- *Very large scale integration* (VLSI): The processor or chip has on the order of 100,000 or more transistors, but not over a million.

- *Ultra large scale integration* (ULSI): The processor or chip has over one million transistors, but less than 1 billion.[1]
- *Giga scale integration—our term* (GSI): the processor or chip has over one billion transistors, but less than 1 trillion.
- *Tera-scale integration—our term* (TSI): The processor or chip has over one trillion (10^{12}) transistors, but less than 1 quadrillion (10^{15}).

Saving space and reducing unit cost are not the only drivers: A useful by-product of miniaturization is that the switching speed increases; this is because miniaturization shrinks the size of the gate and the more narrow the gate is, the faster the transistors can turn electrical streams on and off. Observers (such as Intel Corporation) expect that Moore's law will continue to hold at least through the end of this decade and into the early part of the next decade [25]. See Figure 6.1. At this juncture one can place around 100 million transistors *per square centimeter* and a density of one billion transistors per square centimeter is expected to be reached in the 2008–2009 time frame. Since the 1960s, the success of microelectronics is based on the fact that miniaturization of electronic devices allows for large-scale integration of complex electronic systems and higher data-processing rates, while reducing the energy dissipation and improving the reliability of the overall system.

Another way of stating Moore's law is that the *size of a transistor gate* is reduced in half every 18 months. As noted, there are several advantages with this downscaling: the smaller the gate the less power it consumes, the larger the number of transistors that can be integrated onto one silicon chip, and the faster the transistor can switch state. Fortunately, the economics of fabrication (cost-per-chip) have been favorable to the overall per-unit cost as the size is reduced, although the cost of the required manufacturing plant (factory) has increased substantially. This trend is now continuing toward the nanometer regime, down to the manipulation of single atoms [18].

As a point of reference regarding recent trends, at the beginning of this decade the gate length for metal–oxide–semiconductor field-effect transistors (MOSFETs) that were typical of a computer chip set (e.g., Pentium) was around 100 nm and the thickness of the oxide was around 2 nm; approximately 15–20 million transistors could be placed on a microprocessor *chip*; densities for memory chips were better by an order of magnitude (refer again to Fig. 6.1). By press time the density had just about quadrupled compared with these numbers for both microprocessors and dynamic random-access memory (DRAM) memory chips: PowerPC and Pentium chips had in the range of 50–75 million transistors.

While extrapolations of Moore's law as shown in Figure 6.1 suggests that CMOS chips may be scaled to 20-nm transistor gate lengths by the middle of the next decade, still there are significant technical manufacturing requirements that have to be met to allow the scaling of silicon to continue to this level of miniaturization. Many of these requirements have no present or known solutions, and, as a result, the

[1]A 20-cm silicon wafer now contains more electronic components than there are people on Earth.

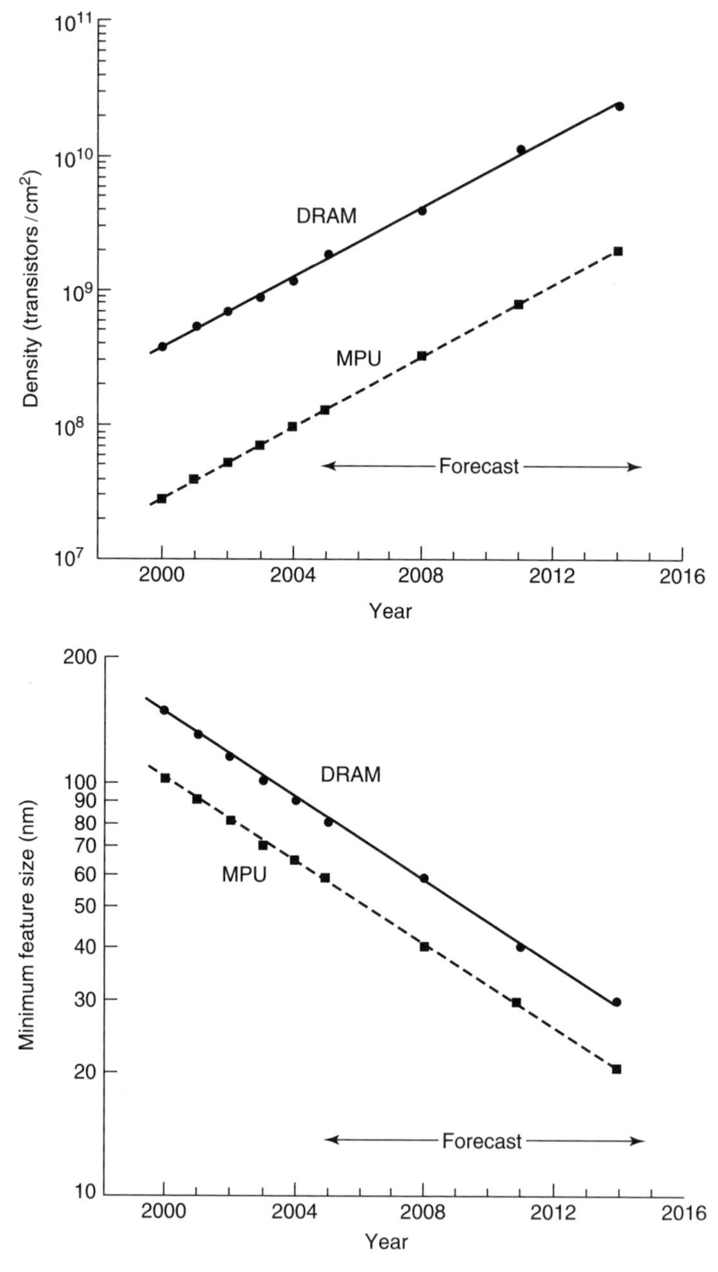

FIGURE 6.1 Trends in power and size of electronic devices.

ability to scale CMOS down to 20 nm is not necessarily guaranteed. In the scaling process all the parts of the transistor need to be reduced in an attempt to keep all the electric fields in the device constant; in reality, this cannot be achieved in short-channel devices and a number of problems arise, as depicted in Table 6.1 [26]. The main challenges lie in the area of the gate dielectric, gate electrodes, substrate and device structure, and device interconnects [23]. Also, as MOSFETs are shrunk, they become sensitive to the fine structure of the random distribution of dopants (random doping fluctuations) in the devices and not simply sensitive to the average or gross distribution (e.g., see [390]).

Furthermore, as hinted above, a by-product of Moore's law is that to increase the yield and reduce the cost per transistor on a chip by scaling down the transistor size, the cost of the semiconductor fabrication equipment ("the factory") doubles every 4 years. While the cost of a DRAM bit was around 4 microcents/bit at press time and the cost of a transistor in a microprocessor was around 60 microcents/transistor, the cost of a semiconductor fabrication plant easily already reached into the $2.5 billion range (some prognosticate that by the year 2012, a single fabrication plants could cost up to $30 billion [391].) It follows that at some juncture in the future, these plants could become too expensive to build, and a point could be reached where no acceptable financial return for the investment can be secured. Because of these considerations, some observers in the semiconductor industry predict that the economics of manufacturing may eventually place a hold on Moore's law before physical limitations actually come into play [26]. It remains to be seen if these predictions hold true.

6.1.2 The Present and Its Challenges

As a point of calibration, looking at the state-of-the-art, at press time an SRAM chip had a capacity of 52 megabits and packed 330 million transistors onto the surface of each chip; that compares with about 50 million transistors on a Pentium 4 processor. Historically, processor performance has been a key driving force behind semiconductor technology innovations: a 30% dimension reduction delivers a twofold increase in transistor density along with a 50% increase in device speed primarily because of the shorter carrier transit [23]. As an illustrative example, a press time announcement by AMD stated that the company had developed a double-gate transistor using industry-standard technology that is smaller than any yet created; as is typically the case with any miniaturization initiative, many obstacles were overcome to realize the 100-million-transistor processor: it required innovations in materials, equipment, maskmaking, and process technology, combined with advancements in design and testing. Figure 6.2 depicts one example of a 100-million-transistor chip. The main challenges in reaching higher levels of integration include not only finer lithography and etch processes, but also vertical scaling of junctions and gate dielectrics (to optimize transistor performance), and advanced interconnects (to minimize RC time-constant delay.) The perceived limit of optical lithography due to diffraction has been continually pushed along with deeper ultraviolet lithography. The resulting gate-level depth-of-focus problem has been accommodated by the adoption of both shallow trench isolation and by chemical mechanical planarization at all levels [23].

TABLE 6.1 Partial List of Scaling Problems for Traditional Silicon-Based Semiconductors

Drain-induced barrier lowering (DIBL)	The p-region (see Chapter 2) produces a barrier between the source and drain n-type contacts. As the gatelength is reduced, the source and drain n-regions deplete out carriers in the p-typeregion, and the barrier between the source and drain is reduced. A practical solution is to increase the p-type dopant density under the gate, but this can only be achieved up to a certain level: Eventually, as the dimensions shrink, electrons can quantum mechanically tunnel through the barrier between the source and drain. It follows that the transistor can never be switched off and it ceases to be a switch (gate lengths of 8 nm have been produced but due to the DIBL, the gain or amplification of the transistor is much lower than that predicted by direct extrapolation from larger gate length transistors).
Quantum mechanical tunneling	Electrons can quantum mechanically tunnel through the gate oxide when the gate becomes ultrathin. It follows that the oxide can only be scaled down in thickness to about 0.8 nm—this value is very close tocurrent manufacturing achievements. One possible way around this is to identify a new material for the gate oxide with a (much) higher dielectric constant. A number of new materials (e.g.,Ta_2O_5) are being researched, but as of yet no commercially manufacturable products have emerged.
Optical lithography fabrication	Rayleigh resolution criteria for optics states that the linewidth that can be achieved is proportional to the wave-length of the light and inversely proportional to the numerical aperture of the lens in the optical system. For the 100-nm gate length transistors, a wavelength of light of 248 nm is used on photosensitive resists that become more soluable when exposed to photons and can be selectively dissolved away. By shining the light through a mask with transparent and opaque regions, the pattern on the mask can be replicated in the resist, and this can be etched into the underlying material. There are only a limited number of known radiation sources with wave-lengths below 248 nm and the appropriate lens and resists either do not exist or require substantial development before they can be used. The radiation at these wavelengths is close to X-rays and therefore the photons are highly energetic; the lenses in the systems are very expensive (in the range of $10 million) and must be replaced on a routine basis as the energetic radiation damages the lens materials. The reduction in wavelength is not guaranteed. One possible method is to use phase-shifting technology. By incorporating sections on the mask that rotate the phase of the photons, the light interferes, and it is possible to outperform the Rayleigh resolution criteria. Intel has demonstrated 30-nm gate length transistors and CMOS static random-access memory (SRAM) cells using phase-shifting technology with 248-nm wavelength resists and optical lithography systems; this demonstrates that optical lithog-raphy with workarounds can produce the required linewidths to at least 2011, although other problems have yet to be resolved. Extreme ultraviolet systems could be used. Also, while electrons can be used instead of photons, almost all electron beam lithography systems are serial in nature and, therefore, are much slower that the parallel lithographic processes.

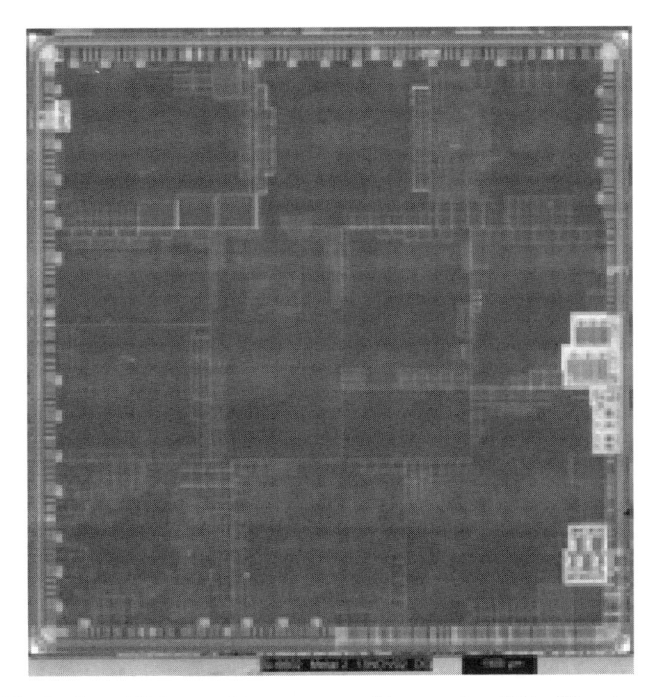

FIGURE 6.2 Radeon 9700, a graphic processor with more than 100 million transistors measuring 14.8 × 14.8 mm. (Courtesy: ATI Technologies, Markham, Canada).

These advances could lead to the placement of 1 billion transistors on chips currently holding 100 million transistors (as noted earlier, based on Moore's law, the 1-billion-transistor processor should be in commercial production around 2008). Around press time, Intel also conceptualized a 1-billion-transistor processor, containing four Intel Itanium 2 cores and a shared cache memory; fabricated at the 65-nm technology node, it would use a gate length of 30 nm and an equivalent oxide thickness of ~8 nm [23, 27a]. Research on productizing 45 nm semiconductor technology was advancing well by press time, and planners were setting their eyes on 32 nm approaches. High-*k* and low-*k* materials, improved interconnects, metal gates, and additional layers, such as strained silicon on germanium, comprise the basic approaches of these advancements [391a, 391b].

As we noted in Chapter 2, the fundamental building block of a microprocessor is the FET. FETs act as basic switches. Specifically, when a voltage of the right level is applied to the gate electrode, it induces charge along the channel; the channel then carries current between the source and the drain, turning the switch on. With sufficiently small gates, these transistors can switch on and off at gigahertz rates [19].

The chip-making process traditionally begins with a large crystal of silicon. With this manufacturing technique one grows the large crystal by starting with a crystal from a small seed crystal that is immersed in a bath of molten silicon. This process yields a cylindrical ingot from which many thin wafers are then produced. For the sub-0.1-μm gate sizes, these single-crystal ingots, however, are no longer adequate because they have a relatively large number of "defects." Defects basically are dislocations in the atomic lattice that degrade the silicon's ability to conduct electrons. It follows that manufacturers now proceed by depositing a thin, defect-free layer of single-crystal silicon on top of each wafer by exposing it to a gas containing silicon. An even better technique is to employ the silicon-on-insulator (SOI) approach. SOI involves placing a thin layer of insulating oxide below the surface of the wafer in order to lower the capacitance between the transistors and the underlying silicon substrate. Capacitance introduces delays in the form of the RC constant and drains power: SOI topologies can increase the switching rate of transistors by up to 30% (this gain is equivalent to what one gets in moving one generation ahead in feature size, per Moore's law.) A specific approach is to bombard the silicon material with oxygen ions that then implant themselves with atoms in the wafer, and in turn form a layer of silicon dioxide. One drawback of this approach is that the oxygen implantation process is relatively slow, which makes it costly. Hence, manufacturers tend to reserve SOI methods for their high-end microprocesors, which command higher prices and make the process financially viable (faster new methods are emerging of late).

Against the backdrop of the press time figures-of-merit listed above and the manufacturing techniques just described, we make note that the demise of Moore's law as the scaling driver for integrated circuitry has been predicted a number of times in the past 15 years; so far, this prediction has been proven to be premature. For example, in 1988 it was predicted that devices with a feature size of 100 nm would no longer function in a useful manner due to fundamental physical limitations; by contrast, today work for 65- and 45-nm device nodes is well underway [229].

Nonetheless, we are at a near-point of inflection: CMOS approaches will be stretched for perhaps another decade, but nanoelectronics solutions are being sought to take over where CMOS will leave off. Successful IC development beyond the 45-nm feature sizes is not guaranteed a priori because of fundamental limits imposed by the basic laws of physics (as we have seen in Table 6.1). As the cross section of conventional electrical wires reaches the mean free path for electronic scattering, surface scattering from boundaries of ultranarrow conductors inhibits electronic conduction and becomes a serious roadblock to additional miniaturization at a fundamental level. In the case of copper metalization schemes, this problem is predicted to emerge as dimensions approach the mean free path for electron scattering in copper, namely, 39 nm [229].

Morphological imperfections and finite-size effects tend to significantly increase electrical dissipation as feature sizes decrease. Figure 6.3 displays a model of copper line resistivity as a function of linewidth [392]; the figure plots the individual detrimental effects on resistivity as a function of linewidth due to surface scattering, grain boundary scattering, and surface roughness. Grain boundary and surface roughness scattering are caused by morphological issues and can be addressed through the use of epitaxial or single-crystal copper to mitigate grain boundary scattering effects; other processes include the incorporation of surface pretreatment protocols combined with tightly controlled film growth methodologies to produce narrow metal lines with atomically smooth surfaces. On the other hand, surface scattering is a fundamental quantum theory problem for which there are no known solutions within the scope of conventional electrical conductivity. The most promising solutions to the scaling

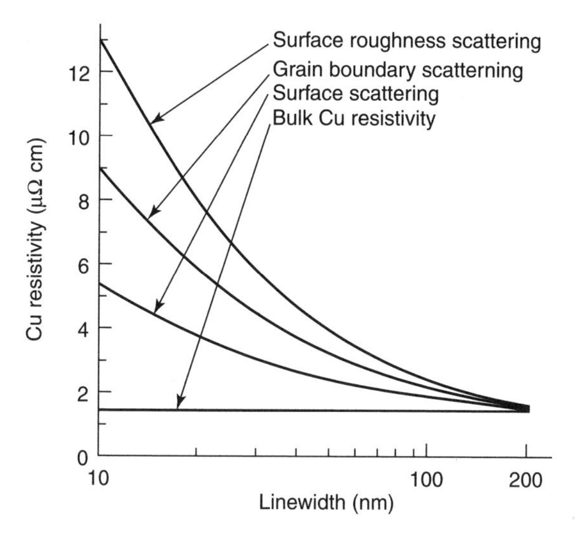

FIGURE 6.3 Theoretical predictions for the increase in copper resistivity as a function of thickness under various effects.

issues are in the context of nanotechnology; potential solutions involve, among others, novel conductivity mechanisms such as ballistic or scatterless electron transport phenomena [229].

6.1.3 Future

Getting There

With CMOS being stretched to the "limit" as we discussed in the previous section, it is advantageous to start looking at other technologies to sustain the miniaturization process. As we noted, the exponential growth of the number of transistors per IC predicted by Moore's law is accompanied by a steady reduction of the size of each individual transistor. For the year 2016, the International Technology Roadmap for Semiconductors predicts a physical gate length of 9 nm for both logic and memory applications. Efforts were under way in Europe and elsewhere to demonstrate the feasibility of 45-nm CMOS logic technology in 2005, while simultaneously starting research activities for the next-generation 32- and 22-nm technology nodes; specifically the objectives were to achieve a demonstration of feasibility of a 45-nm CMOS logic as early as 2005, a first full CMOS process integration in 2007, and a demonstration of feasibility of a 32-nm CMOS logic process as early as 2007. A project of this kind was considered ambitious at press time: These smaller nodes were considered at the limits of present-day technologies, and the move to these gate lengths may require the use of new nanotechnologies, such as silicon nanowires and carbon nanotubes. As part of these efforts, researchers were exploring introduction of the necessary changes in the materials, process modules, device architectures, multilevel metallization structures, and all related characterization, test modeling, and simulation technologies to keep the scaling trends viable [393].

It is worth noting that for decades, each IC generation could be derived from the previous one by simple scaling of device geometry and voltages, the only limiting factor at those stages being manufacturing technology. But scaling eventually will reduce the extensions of semiconductor devices to sizes at which some of the assumptions underlying scaling break down; one example of this relates to the *locality of transport parameters* [394]. An important question is "How small can electronic devices be made before the classical laws of physics prohibit them from operating?" To answer this question one needs to bring together several scientific disciplines in the nanoscience field: condensed-matter physics, solid-state electronics, chemistry, materials science, and electrical engineering [22]. Examples for new nanoelectronic devices and applications of interest include single-electron transistors for use as ultra-high resolution charge sensors and memory devices, and molecular electronics for the realization of circuit functions [18].

Some researchers take an optimistic view in reference to the outlook for miniaturization using silicon methods. From 1960 to 2000, the energy transfer associated with a binary switching transition—the canonical digital computing operation—decreased by approximately 5 orders of magnitude, and the number of transistors per chip increased by approximately 9 orders of magnitude. But such exponential advances must eventually come to a halt imposed by a hierarchy of physical limits. The five levels of this hierarchy are defined as [395]: fundamental, material, device, circuit, and system. (Perhaps fortunately) an analysis of the key limits of each of these levels reveals that silicon technology has the theoretical potential to achieve the TSI goal of more than 1 trillion transistors per chip, with critical device dimensions or channel lengths in the 10-nm range. This potential represents more than a 3-decade increase in the number of transistors per chip and more than a one-decade reduction in minimum transistor feature size compared with the state-of-the-art in the early 2000s. Researchers see limited TSI on a massive scale as being feasible assuming the development and economical mass production of *double-gate* MOSFETs with gate oxide thickness of approximately 1 nm, silicon channel thickness of approximately 3 nm, and channel length of approximately 10 nm. The development of interconnecting wires for these transistors presents a major current challenge to the achievement of nanoelectronics involving TSI [395].

Other researchers take a less optimistic view in reference to traditional silicon methods. Technical and economic difficulties in further miniaturizing silicon-based transistors with present fabrication technologies have motivated a strong effort to develop alternative electronic devices, including devices based on single molecules and/or nanotubes. Carbon nanotubes have already been successfully used in the lab for nanometer-sized devices such as diodes, transistors, and random-access memory cells [396].

As noted earlier in the chapter, a large amount of money has been invested in the semiconductor industry in order to consistently shrink and improve our semiconductor electronics. However, this shrinking of components cannot continue for the long haul [4]. Furthermore, the direct shrinking of circuits predicted by the Moore law may not be the most economical method for the future. We have already noted that as transistors such as the MOSFET, (one of the primary components used in integrated circuits) are made smaller, both the properties and manufacturing expense change with the scale. Currently, ultraviolet light is used to develop the silicon circuits with a lateral resolution around 200 nm (the wavelength of ultraviolet light). As the circuits shrink below 50–100 nm, new fabrication methods must be created, resulting in increasing costs. Furthermore, once the circuit size reaches only a few nanometers, quantum effects such as tunneling begin to become important, which support a different behavior than at greater physical dimensions. Thus, novel methods for computer chip fabrication have been and are being intensely sought by microchip manufactures [4].

More cost-effective methods of manufacturing microchips may gradually replace multibillion-dollar foundries with table-top devices [397]. According to a number of observers, while the industry's technical innovation may continue on an aggressive path, the economic limitations of the future may curtail some technologies from reaching their true market potential [398]. Some of these economics problems were already encountered for the 90-nm design node, where one saw implementation problems and considerably higher costs that originally predicted; this will only become magnified in the future, as the industry searches to find the financial support required to continue the migration downward, to the 65-, 45-, 32-, and 22-nm nodes. Figure 6.4, shows the increase in new product development costs by design rule [398]. When observers refer to the costs of future devices, either above or below the 45-nm node, they are talking about total development costs, from design inception to final silicon, and that cost has been increasing exponentially as design rules shrink. With the single exception of standard products, such as microprocessors and memory, volume requirements for devices manufactured per design, have dropped considerably from just ten years ago. In large part, this relates to both the rapid increase in product diversity and development costs, always coupled with time to market [398].

What's There When We Get There

As we set out for a possible transition, recent research in the field of nanoscale electronics has focused on two fundamental issues: (a) the operating principles of small-scale devices and (b) schemes that lead to their realization and eventual integration

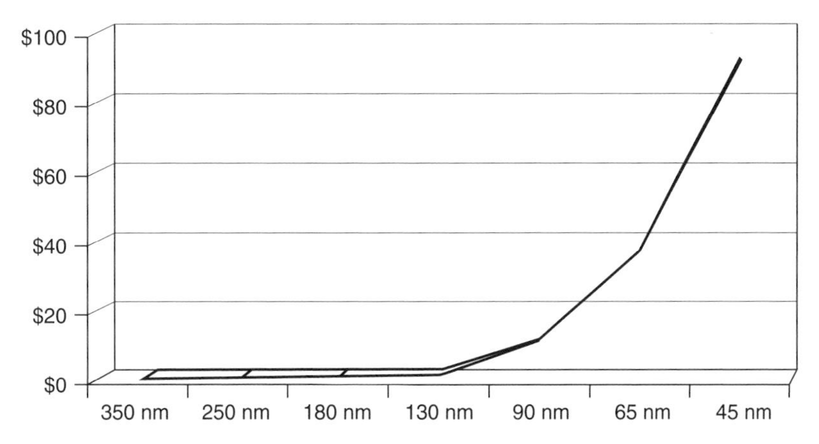

FIGURE 6.4 Nonrecurring engineering by design rule category (US$ in millions). (Source: In-Stat / MDR, 05/04).

TABLE 6.2 Current Research Area in Nanoelectronics (Partial List)

Fabrication and characterization of metallic nanostructures	Controlled generation of metallic nanostructures with the electrochemical atomic force microscope
	Electrochemical growth of nanostructures
	Metalloid clusters: Correlation between properties and topology within the cluster core
	Preparation and characterization of metallic nanostructures
Electron transport in nanostructures	Tunneling in solid-state systems on femtosecond time scales
	Quantum effects in single-electron devices
	Atomic-scale point contacts by electrochemical deposition
	Ballistic 2-DES-hybrid (two-dimensional electron gas hybrid) nanostructures
	Interaction and coherence in disordered conductors
	Quasiclassical theory of superconducting and ferromagnetic hybrid devices
	Experimental investigation of electron transport in nanostructures
	Surface acoustic waves
	Transport through normal conducting nanostructures: effect of electron–electron interaction
	Transport through nanostructures subject to time-dependent fields
	Correlation effects in nanostructure devices
Nanoelectronic devices	Solid-state realizations of quantum networks
	Spintronics
	Quantum information devices

into useful circuits [397]. Some areas of interest in nanotechnology include the following [208] (also see Table 6.2): (i) nanoengineering and nanodesign—nanomachines, nanoCAD, nanodevices, nanoscale logic circuits; (ii) nanoelectronics—molecular and computational nanoelectronics, nanodevices, electronic states, quantum dots, nanowires; (iii) nanomagnetism—magnetic properties of nanostructures and nanostructured materials; and (iv) quantum computers—theoretical aspects, computational methods for simulating quantum computers, devices, and algorithms. Nanoscientists working from the bottom up are now attempting to create a new understanding and structure from the dynamics of the basic materials and their molecules (the physical disciplines); those working from the top down seek to improve existing devices, such as transistors, and to make them smaller (engineering disciplines) [22].

Nanoscale metal and semiconductor particles exhibit a transition between quantum and bulk properties. When electrons become confined in the mesoscopic regime, they display quantum mechanical behavior. A mesoscopic scale is not as small as a single atom, but small enough so that properties are significantly different from those in a microscopic piece of a material. Specifically, nanoscale electronic components are governed by the quantum theory: with decreasing particle size, bulk properties are lost as the continuum of electronic states becomes discrete (the quantum size effect) and as the fraction of surface atoms becomes large. The electronic and magnetic properties of metallic nanoparticles and nanoclusters show new characteristics that can be utilized in novel applications in areas that range from nonlinear optical switching and catalysis to high-density information storage [399]. In order to understand the properties of IC at the nanoscale, one needs to understand quantum wires, quantum transistors, quantum resistors, and other novel circuit elements. One challenge in nanoelectronics is finding ways to position and attach nanowires to the tiny molecular-scale components. Progress is being made in this arena. For example, researchers have already announced the ability to grow nanowires between electrodes that were created using common patterning techniques; because the nanowires grow and connect automatically, the method promises to provide a relatively inexpensive way to mass-produce nanoelectronics [400].

Quantum devices that hold promise for future systems include ballistic electron transport, Coulomb blockade devices, resonant tuneling diodes (RTD), quantum dots, electron-wave coupling devices, and (hypothetical) nano-MOSFETs operating in the quantum's coherent transport limit.

The following devices and systems (among others) have applicability to nanoelectronics (some call this the "nanoscale toolbox"):

- *Silicon nanoelectronics* (e.g., double-gate MOSFET, nano-MOSFET). Some newer CMOS transistor architecture for 30- to 50-nm environments (e.g., double-gate and ultrathin-body (UTB) MOSFETs) could see deployment in the next few years, based on the ability of these newer architectures to achieve higher performance compared with traditional CMOS approaches; these approaches offer paths to further scaling, while one is waiting for full-fledged nanoelectronic devices. Double-gate transistors allow twice the drive current, with an inherent coupling between the gates and channel that makes the design more scalable; in UTB-SOI, power consumption is reduced along with leakage current (see Fig. 6.5) [401].
- *Carbon nanoelectronics*. For example, some researchers are exploring ways to build ultrasmall electronic devices out of atom-thick carbon nanotubes and have incorporated them into a new kind of field-effect transistor.

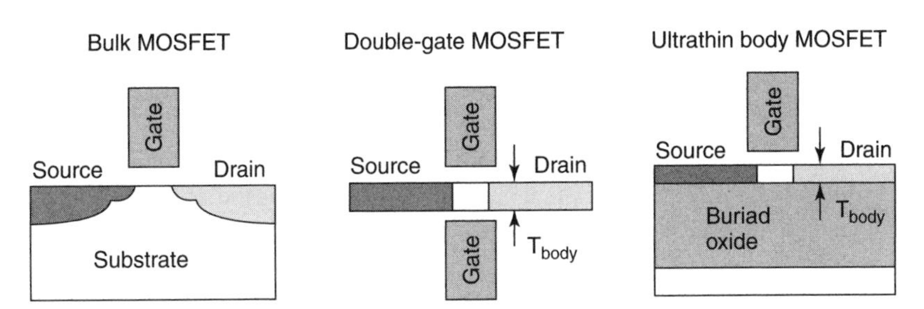

FIGURE 6.5 Schematic of traditional, double gate, and UTB MOSFETs.

- *Nanowires and nanocontacts.* Among the many potential building blocks within this nanoscale toolbox, nanowires are considered one of the key components because they can be used as interconnects and other functional devices in nanoelectronics [402]. For example, field-effect transistors can be constructed from semiconducting indium oxide nanowires grown directly out of and vertical to a substrate's surface; this approach not only reduces the area taken up by individual transistors, but could potentially make it easier for them to be connected together to form complex circuits [403].
- *Ballistic magnetoresistance (MR).* Ballistic transport refers to the transport of electrons in a medium where the electrical resistivity due to the scattering by the atoms, molecules, or impurities is negligible or altogether absent. It also refers to the motion of electrons in ultrasmall (highly confined) regions in semiconductor structures, at very high electric field with velocities much higher than their equilibrium thermal velocity. This phenomenon allows ultrafast devices to be developed [39]. As noted, ballistic electrons are not subjected to scattering, and, therefore, they can move with high velocity. Ballistic transport is determined by electronic structure of the semiconductor and is different for different semiconductors.
- *Single-electron systems.* These are devices operating at the quantum/nanoscale that have switching properties controlled by the removal or injection of a single electron; also, a device through which only one electron can be transported at a time.
- *RTD (resonant tunneling diodes)-based devices.* These are devices operating at the quantum/nanoscale making use of tunneling. Resonant tunneling occurs when one of the QW (quantum well) bound states is mono-energetic with the input electrode Fermi level; peaks in the electrical current as a function of bias voltage are observed in the current–voltage characteristics [404]. These devices have the potential for a number of high-speed electronic applications including terrahertz oscillators and logic circuits with switching speeds as low as 2 ps at room temperature.
- *Josephson arrays.* A Josephson junction is an electronic circuit operating at temperatures approaching absolute zero (0 K) and capable of switching at very high speeds. The device makes use of the phenomenon of superconductivity, which we discussed in Chapter 3. A Josephson junction is comprised of two superconductors, separated by a nonsuperconducting layer; the nonsuperconducting layer is so thin that electrons can cross (tunnel) through the barrier under certain conditions. The movement of electrons across the barrier is known as Josephson tunneling (when a voltage is applied, the current stops flowing thru the barrier). A Josephson interferometer is comprised of two or more junctions joined by superconducting paths. Josephson junctions are utilized in highly sensitive microwave detectors magnetometers (the Josephson effect is influenced by magnetic fields in the proximity, a capability that allows the Josephson junction to be utilized in devices that measure extremely weak magnetic fields—e.g., subtle changes in the human body's electromagnetic energy field), and

superconducting quantum interference devices (SQUIDs). A SQUID is a device utilized to measure extremely weak signals.

- *Spintronics.* These are devices that rely on an electron's spin to perform their functions (conventional electronic devices rely on the movement of electric charges, ignoring the spin carried on each electron).
- *Molecular nanoelectronics.* This area deals with nanoelectronics based on the nanometer scale building blocks such as organic molecules, nanoparticles, nanocrystals, nanotubes, and nanowires [405].
- *DNA nanoelectronics* (where DNA is developed to transport electrical current as efficiently as a semiconductor.)
- *Neuromorphic nanoelectronics* (neurons are the cells that comprise the nervous systems of human beings and animals); the goal here is for neuromorphs (silicon-based neurons) to be designed with certain lifelike characteristics such that networks of neuromorphs could be constructed to emulate the functions of biological nervous systems.

Some of these technologies are more promising that others; the more promising ones are discussed in the next section. In the sections that follow a number of topics are expanded upon.

6.2 OVERVIEW OF BASIC NANOELECTRONIC TECHNOLOGIES

As implied above, the exponential scaling of standard silicon-based technology will eventually come to an end: while the scaling is expected to continue through this decade and into the early part of the next decade, alternative technologies may be required (and are being sought) to sustain further improvements. Nanoelectronics, and in particular nanoelectronice devices such as *single-electron devices*, *quantum mechanical tunnel devices*, and *spin nanolectronics*, among others, may, in due course, be able to continue the minituarization march. The subsections that follow covering these technologies (Sections 6.2.1 through 6.2.7) are loosely based on [26].

6.2.1 Single-Electron Devices

In spite of the fact that CMOS transistors now have small gate lengths (say 50–100 nm), the number of electrons that are employed in a switching operation is still large (tens of thousands). One approach (this being the basic philosophy of single-electron transistors [SETs]) is to reduce the flow to the point where only one electron is used; this implies that the energy required to switch is much lower than what is needed at present. SET devices could become the potential successors to the conventional technology employed to manufacture MOSFETs. SETs make use of the smallest unit of electrical charge, the electron, to represent bits of information. While electron tunneling in MOSFETs limits their smallest usable fabrication scale, this same phenomenon in SET devices may prove useful in reaching the next miniaturization level.

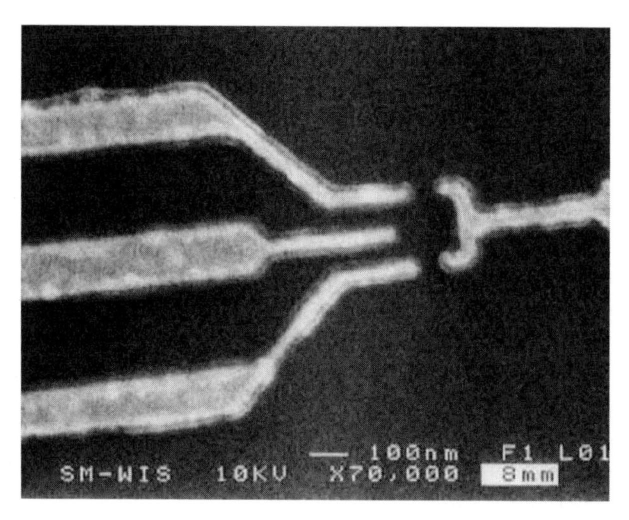

FIGURE 6.6 Example of single-electron transistor. (Courtesy: Quantum-Effect Devices Group of the Research Laboratory of Electronics, Massachusetts Institute of Technology).

Nanoscale semiconductor SET devices consist of a semiconductor quantum dot (QD) held between two metallic leads (see Fig. 6.6 for an example [406]). The metallic SET was developed in the late 1980s and is considered one of the simplest types of artificial atoms. A metallic SET consists of a metal particle isolated from its leads by two tunnel junctions (which are similar to diodes) and capacitively coupled to a common gate electrode. The tunnel junctions create a Coulomb island, which the electrons can enter only by tunneling through one of the insulators. Coulomb repulsion prohibits more than one extra electron at a time on the island (near the gate). Hence, electronic circuits can be made to pump or count electrons one at a time. Because a SET electrical resistance is highly sensitive to the electrical fields from nearby charges, it can easily detect not only single electrons, but also charges as small as 1% of an electron's electrical field. The current as a function of bias across the tunnel barriers can also be measured in order to observe the so-called Coulomb staircase, a stepwise increase of current as electrons are added to the metal particle [22]. Some of the technical issues that researchers seek to overcome include problems associated with quantum charge fluctuations and the SET's sensitivity to microwave radiation. Because SETs exhibit extremely low power consumption, reduced device dimensions, excellent current control, and low noise, they promise to lead to innovative electronic devices [22]. Applications of SETs include specialized metric scale applications (e.g., current standards and precision electrometers); high-density neural networks, high-density computer memory, and computer systems.

As noted, a SET consists of a metal island a few hundred nanometers across, coupled to two metal leads via tunnel barriers. At temperatures below 1 K, no current can pass through the island when one applies low-voltage biases. This effect is known

TABLE 6.3 Comparison of Microelectronic Devices and Nanoelectronics Devices

	Conventional Memory		Quantum Dot Memory			
				Nanoflash		
	DRAM	Flash	SET	Multidot	Single Dot	Yano-type
Read time	~6 ns	~6 ns	1 ns	~10 ns	~10 ns	~20 μs
Write time	~6 ns	1 ms	1 ns	~100 ns	<1 μs	~10 μs
Erase time	<1 ns	~1 ms	<1 ns	~1 ms	<1 ms	~10 μs
Retention time	250 ms	~10 years	~1 s	~1 week	~5 s	~1 day
Endurance cycles	Infinite	10^6	Infinite	10^9	10^8	10^7
Operating voltage	1.5 V	10 V	1 V	5 V	10 V	15 V
Voltage for state inversion	0.2 V	3.3 V	<0.1 V	0.65 V	0.1 V	0.5 V
Electron number to write bit	10^4	250	1 (excluding no. to change gate potential)	10^3	1 (excluding no. to change gate potential)	2 (excluding no. to change gate potential)
Cell size	8.5 F^2/bit	~9F^2/bit	9–12 F^2/bit	9F^2/bit	9F^2/bit	2F^2/bit

Courtesy: D. J. Paul, Cavendish Laboratory.

as the Coulomb blockade, which is the result of the repulsive electron–electron interactions on the island. What is of interest is the fact that the current through the island can be accurately controlled down to a single electron. SETs are also realized in semiconductor devices, where their behavior is characterized as a QD. As we saw in Chapter 4, QDs are nanometer-sized human-made boxes that control the flow of atoms by selectively holding or releasing them [22]. There are several types of SETs. SETs can be based on *Coulomb blockade*, *miniature flash memory*, or *Yano-type memory*. Table 6.3 summaries the experimental results in SETs with the present production memory of DRAM and flash produced using CMOS processing lines.

Coulomb blockade involves a small island between two electrodes. For room temperature operation of such devices the island must be less than 10 nm in diameter. When the island is small enough and contains N electrons, then an energy gap opens up between the energy of the last (Nth) electron and the first empty electron state ($N + 1$th). When this energy gap is larger than the thermal energy in the system, then electrons cannot quantum mechanically tunnel through the system (this is because the only free states that electrons may tunnel onto the island are above the energy of the electrons in the electrodes); if, however, a gate is used to electrostatically move the islands energy states with respect to the electrodes, then the $N + 1$th

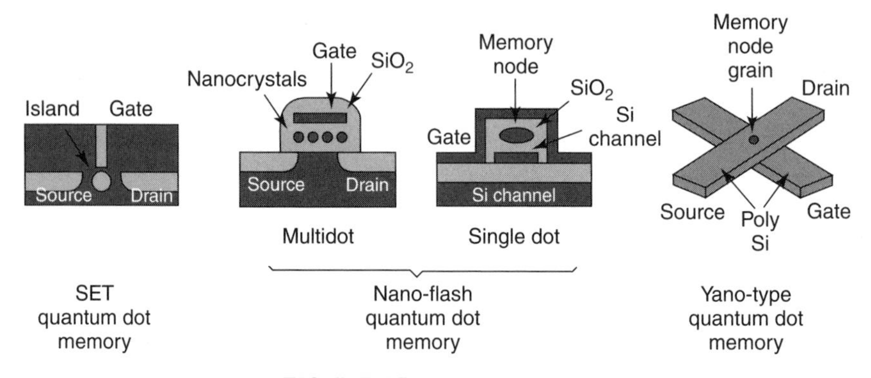

FIGURE 6.7 Memory devices.

electron free state can be moved below that of the left electrode, allowing electrons to quantum mechanically tunnel through the island one at a time. A number of implementations of such transistors have been achieved using a variety of materials; for example, silicon and silicon dioxide. SETs are more likely to be employed for memory applications because of the fact that they have no gain. It is more difficult to create logic circuits because one needs the gain in a transistor or logic device to overcome the losses in the circuit and interconnects; hence, it remains an open issue if one is able to scale these kinds of circuits to the levels of present CMOS microprocessors.

Miniature flash memory is a miniature version of the conventional CMOS flash memory. Here the addition of a single electron to the memory node results in a substantial change to the electron current through the transistor channel. Another approach is to use a number of silicon nanocrystals as nodes in the oxide rather than just one; this approach has the advantage of being more robust to single-electron fluctuations in the system. Figure 6.7 depicts a number of configurations.

Yano-type memory has been demonstrated by K. Yano at Hitachi. Yano-type memory involves the fabrication of two crossed poly-Si wires using standard CMOS fabrication lines. This type of memory device uses the grains as memory nodes. The ability to mass produce such memory devices with the required control of properties may, however, be challenging.

6.2.2 Quantum Mechanical Tunnel Devices

Quantum mechnanical tunnel devices make use of the electron's ability to tunnel through thin barriers when the electron wave function can penetrate through the barrier. The barrier thickness must be less than about 10 nm to allow a useful amount of tunneling current. There are two main classes of devices that use quantum mechanical tunneling to produce a negative differential resistance current–voltage characteristic that may be put to use in a number of circuit applications: *resonant tunneling diodes* (RTDs) and *Esaki diodes*.

An RTD is a device with two electrodes (a "left" electrode and a "right" electrode) with two tunnel barriers between the electrodes. The (electronic) quantum well (QW) that is created by the confinement of the electron wave function between the two barriers generates a discrete set of allowed electron energy states in the quantum well. When (and only when) an electron from the "left" electrode has an energy that corresponds to the allowed state in the quantum well such electron can quantum mechanically tunnel through the two barriers and through the quantum well, and reach the "right" electrode. A peak forms in the current when the two electrodes are aligned after the application of a voltage between the electrodes; once an additional voltage is applied, the electrons do not have allowed states in the quantum well to tunnel through, and the current drops. The best results for RTD devices are obtained when using InP/InGaAs/InAlAs substances, but given that silicon is the dominant material in electronics and it is substantially cheaper to manufacture than some of the other substances, there has been interest in a silicon-based system. So far, however, silicon-based diodes have shown poor peak current densities and peak-to-valley current ratios. Typical RTD performance is summarized in Table 6.4. To date tunnel diodes represent one of the few quantum device technology that has successfully demonstrated circuit operation at room temperature. For example, one has seen InP-based tunneling SRAM memory cells using two RTDs and one FET; generic logic can change functionality between NAND, NOR, or NOT by changing the relative sizes of the transistors. Circuit operation up to 12 GHz has been claimed. Double-current peak RTDs have also been used to demonstrate multivalued logic that allows a substantial reduction in device count for logic chips. RTD devices have been used to demonstrate a large number of circuits that either provide higher speed or lower power dissipation than conventional CMOS or III-V FET technologies. At present,

TABLE 6.4 RTD Parameters

Parameter	High-Speed InP RTD Logic	Predicted High-Speed Logic	Low-Power InP RTD Memory	Predicted Low-Power Memory	Nanometer Scaled RTDs	SiGe RTD	SiGe Interband Tunnel Diode
Peak-to-valley current ratio	4	3	2	3	3	3.3	5.45
Peak current density (A/cm^2)	40k	10k	0.16	0.1	10k	25k	8k
Minimum feature size	2 μm	200 nm	500 nm	200 nm	50 nm	14 μm	14 μm
Peak voltage	0.35	0.16	0.20	0.20	0.20	1.5	0.28
Maximum clocking frequency	12.5 GHz	6.25 GHz	592 kHz	56.8 MHz	6.25 GHz	NA	NA
Time constant	0.02 ns	0.04 ns	422 ns	4.4 ns	0.04 ns	~0.02 ns	~0.31 ns

Courtesy: D. J. Paul, Cavendish Laboratory.

all these demonstrators are in III-V technology and no silicon RTD or Esaki diode circuits have been demonstrated (a drawback of the III-V system is that since the tunneling current depends exponentially on the barrier thickness, a difference of as little as two atoms across a wafer can have a substantial change in the tunneling current; this limits the yield of the circuits).

The Esaki diode (described by Leo Esaki in 1958) is an interband device where electrons tunnel from the conduction band to hole states in the valence band. The devices produced in the beginning were fabricated by diffusing one species of doping to create the p-i-n diode; this results in poor uniformity and poor control of the performance. Recent advances through the use of epitaxial growth has resulted in better performance results and good uniformity.

6.2.3 Spin Nanoelectronics (Spintronics)

The spin of an electron may be used for memory storage or logical computing. Hard discs in computers provide an example of how spins may be used to store information; this is done through the measurement of a ferromagnetically polarized particle. Spintronics is at a relatively mature stage of research and may (soon) allow both memory and logic functions in much smaller and robust systems than Si CMOS.

The basic operation of a spin device relies on the ferromagnetic material having a larger number of one spin close to the Fermi energy or surface: if the ferromagnetic material is put into a circuit, then only the down spins can be transported through the system. Hence, switches can be built by designing structures with different ferromagnetic and normal metallic layers. By having two ferromagnetic layers placed on either side of a metal, if both ferromagnets are polarized in the same direction, then the majority spins in the system will have a low-resistance path through the device; if, however, the ferromagnets have opposite polarization, then both the majority and minority spins encounter a high-resistance path.

This kind of effect can be used to produce a spin tunnel device. If two ferromagentic contacts are placed on either side of a metal (or semiconductor), then when a bias is applied to the system electrons of one particular spin will be injected into the metallic layer. If the second contact to the metal also has the same spin polarization as the first, then the electrons can pass into the second contact and the circuit has a low resistance. If, however, the second contact is polarized opposite to that of the first contact, no free states exists close to the Fermi energy surface for the electrons to "tunnel into" and, so, no current can flow in the system. If the metal is replaced by a semiconductor, then the application of an electric field to the semiconductor will cause a rotation of the spin of the electrons as they are transported across the semiconductor layer, and the device can be switched on and off using the electric field from a gate. These effects can be used to form a number of different spintronic devices, including switches and spin-polarized filters.

The most successful magnet nanoelectronic device to date is the magnetic random-access memory (MRAM). It corresponds to a GMR element that is written and read using the magnetic fields generated by currents in the word and bit lines.

A number of demonstrations have been produced with memory capacities above 1 MB and access times of about 50 ns using 250-nm dimensions. These devices can be easily scaled to smaller sizes, where their performance is predicted to increase substantially. To date there have only been a few demonstrations of *spin injection into semiconductors*. Specifically, a spin-polarized light-emitting diode has been demonstrated where the output photons are polarized depending on the spin of the electron and holes in the system; significant spin injection can only be achieved from magnetic semiconductors into normal semiconductors (results from metallic ferromagnets into semiconductors has to date been poor).

6.2.4 Molecular Nanoelectronics

In the last couple of decades a lot of attention has focused on the research of molecules as a potential future component in electronic devices; this research is motivated, as covered earlier, by the possible economical and physical limits of the minimization of the current solid-state electronic devices, including, among a host of other issues, the increasingly complex and expensive lithography processes, the doping fluctuations, and the short inversion-channel effect [237]. Molecular electronics is a potentially interesting alternative to traditional silicon-based semiconductors because the proposed molecular electronic structures occupy an area less than thousand times of the area currently used in solid-state semiconductor integrated circuits [237]. Key structures of interest include: (i) molecules as wires/switches and (ii) rectifiers, diodes, rectifying diodes, and resonant tunneling diodes.

Molecular nanoelectronics was proposed by Aviram and Ratner in 1974 as a way to produce a rectifier from organic molecules. The first example of a single molecular electronic device appeared in 1990. Several approaches have been studied in the past few years. The major challenge relates to the difficulty of making individual electrical contacts to molecules, these being only be a few nanometers in size. The development of the scanning tunneling microscope (STM) enabled the first advancements in this field and has remained one of the major tools for characterizing single molecules. Some of the first demonstrations of electronic properties of single molecules by Purdue University included Coulomb blockade and Coulomb staircase. A second set of experiments at IBM have demonstarted a STM tip deforming a C_{60} buckyball; the resulting mechanical deformation modifies the resonance tunneling bands of the molecule and produces electromechanical amplification. While these devices demonstrate functionality that may be used in circuits, the scaling to the level of present CMOS circuits appears to be out-of-reach at this time.

Another approach to molecular electronics is the use of organic molecules. Yale and South Carolina Universities have reported the conduction through a benzene molecule attached to two gold electrodes using thiol groups to bind the molecule to the gold. Benzene rings have delocalized π electrons out of the plane of the molecule through which electrons can be transported when an appropriate bias is applied across the molecule. Carbon–carbon double and triple bonds provide similar orbitals out of the plane and, therefore, combinations of these polyphenylene molecules create conducting wires now known as Tour wires (named after James Tour). Molecules can then be designed with conducting sections. Molecular RTDs operating at room temperature have been demonstrated. Mitre Corporation has also investigated possible architectures using organic molecules. A number of designs have been advanced based on diode logic using the Tour wires and diodes. AND, OR, and XOR gate designs have been shown, along with an adder. The major problem with such organic systems is that the conductivity is relatively poor through the interconnecting Tour wires (the RC time constants of most of the devices is likely to be relatively large, and, unless better conductors or architectures for which the performance does not depend on resistance can be found, the organic systems will always be much slower than silicon).

To be able to utilize molecules in future electronics, entire structures will have to be scaled down, including the wires connecting each component. Therefore, it is essential to develop "molecular wires." An interesting group of molecules are the ones that consist of chains of aromatic benzene rings, such as polyphenylenes, polyporphyrines, and polythiophenes. Several researchers have shown that molecules of this type conduct current and that they are capable of switching small currents [407, 408]. Aromatic-based molecules bonded with multiply bonded groups (such as ethenyl, —HC=CH—, or ethynyl, —C=C—) are also conductive. Because of this capability, triply bonded ethynyl or acetylenic linkages can be inserted as spacers between rings in the Tour wire; spacers are needed to eliminate steric interference that can affect the extent of p-orbital overlap between neighboring rings, thereby reducing conductiveness [237].

Carbon nanotubes (discussed in Chapter 4) have also been used to conduct electrons and demonstrate reasonable conductivities for their size. Both SWNTs and MWNTs can be created that may potentially reduce resistance. Conductivities as high as 2000 Sm^{-1}, which corresponds to a resistance of 200 $\Omega/\mu m$, have been measured, substantially better than organic molecules. A transistor has been demonstrated, although the gate was a silicon substrate and the carbon nanotube had been placed across two metal electrodes fabricated on top of the thermally oxidized silicon substrate. Metal-nanotube rectifiers have also been demonstrated. To date, no switch or three-terminal devices have been produced, which are basic requirements for most logical architectures.

Some researchers have demonstrated measurements on nanocrystals; for example, CdSe nanocrystals can be prepared with sizes down to about 2 nm and with attached linker molecules that may bind to numerous surfaces, including gold. A single-electron transistor has been demonstrated where an oxidized silicon substrate is used as a gate to a CdSe nanocrystal placed between two gold electrodes on the SiO_2 surface.

There have also been efforts at measuring and achieving conductivity in different types of DNA. Results range from highly insulating behavior to semiconducting and even to metallic behavior. The best results to date correspond to Pd nanoparticles attached to a DNA strand that has demonstrated a conductivity of 100 Sm^{-1}. There are a large number of other approaches discussed in the literature. To date, however, there has not been any demonstration of an all-molecular transistor or any structure with gain.

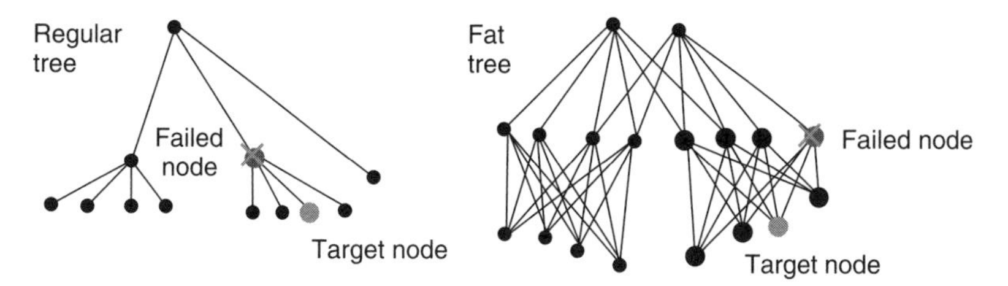

FIGURE 6.8 Example of fat-tree design.

6.2.5 Fault-Tolerant Designs

One of the key reasons for the high cost of CMOS fabrication plants is that the top-down architecture required from lithographic fabrication techniques is not defect tolerant: Any transistor or interconnect that fails on the circuit potentially destroys the entire chip. Some redundancy can be built in the design, but this approach comes at some added expense. This problem is more severe for applications-specific integrated circuits (ASICs) because each application has a different circuit design, unlike the mass repetition of memory or microprocessor designs. Testing represents a substantial cost, around two-thirds of the total chip development/manufacturing cost: A new circuit design must be tested to check that no mistakes in circuit design or mask fabrication have occurred. With up to 20 mask levels in some chips and with up to 300 million transistors in designs, the potential for errors at some point in the design and fabrication process is nontrivial.

Nanoelectronic devices still rely on elements that have no intrinsic fault tolerance. Hence, the question is whether an architecture could be found where not all the transistors or interconnects are required to function correctly in order for the operation of the complete chip to still be viable. As implied by the discussion of the previous paragraph, this could reduce costs considerably.

One of the few defect-tolerant architectures that has appeared in the recent past is the Teramac[2] architecture. This concept relies on the ability to have a large number of interconnects in a system (e.g., see Fig. 6.8), so that some path may always be found around defects and nonfunctional parts of a circuit. It is similar, in a way, to the concept of a *crossbar switch, Clos network*, *Benes network*, or *fat tree* (these are designs of nonblocking networks, networks for which a connection from sources to destinations can always be accomplished without blocking). Before the system is run, a map of all the defects is found and the system is configured so that the defects can be circumvented. Therefore, the Teramac approach is to build a cheap computer that is allowed to have defects, find the defects, configure the resources with software, compile the program, and then run the computer. Teramac (also called a Custom Configurable Computer) is a prototype supercomputer developed at HP. While the architecture of the prototype is built on conventional electronic

[2]The name Teramac derives from the word *tera* which means *trillion* and *mac* from multiple architecture computer.

methods, its principles and approaches are novel. Teramac consists of 864 identical chips comprised of field-programmable gate arrays (FPGAs). Each FPGA contains a large quantity of computation units and a flexible connection network, called LUTs and Crossbar, respectively. All the LUTs are identical for their physical structure and can implement different logic function; hence they do not consist of digital logical component like an AND gate, but rather with memory. Teramac differs from conventional computer architecture because its architecture is tolerant to defects. The key property of Teramac is its software-changeable architecture. It has been estimated that there were 200,000 defects in the HP prototype computer, but it still works, and in fact, it can run (in some of its configurations) 100 times faster than a single-processor workstation. One can use code or instruction to change the logic composition of Teramac [409, 410, 411, 412, 413, 414, 415, 416, 417, 418, 419].

6.2.6 Quantum Cellular Automata

The quantum cellular automata (QCA) provides an interconnectless method of information exchange and computation. QCA has received a lot of attention of late. The basic concept relies on four (or five) quantum dots being produced with only two electrons in the system (see Fig. 6.9). The Coulomb repulsion of the electrons forces them to occupy diagonally opposite sites and, therefore, the arrangement dictates whether the cell is in a "1" or "0" state. As long as the electrons are forced to remain in this cell, then other cells can be placed beside the first cell. If a field is applied to change the arrangement of the electrons in the first cell, then this information is passed to neighbor cells and, therefore, is transported through the system. By appropriate arrangements, logic calculations can also be achieved. So far only a single cell has been demonstrated experimentally (no logic gates have been demonstrated). Calculations suggest that silicon QDs of less than 2 nm size are required to produce cells that may operate at room temperature. There is some belief that molecular self-assembled QD systems may be the ideal system to produce QCAs, but additional work is required. The QCA is an interesting concept (being an interconnectless architecture) but substantial problems exist. Calculations have shown that QCAs may

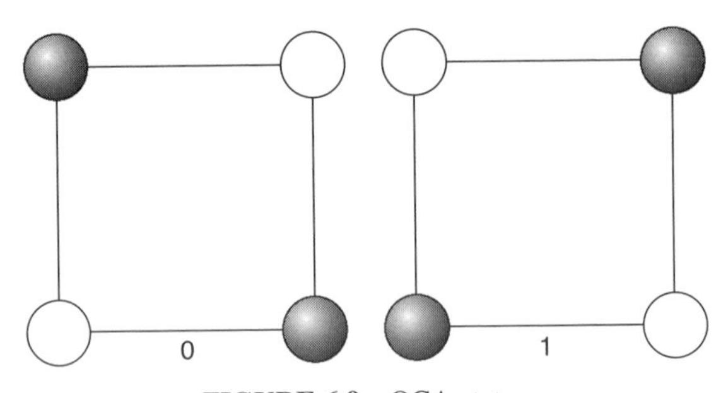

FIGURE 6.9 QCA states.

be much slower than conventional CMOS circuits at relatively small integration densities.

6.2.7 Quantum Computing

Quantum computing (QuC[3]) is a massively parallel architecture at a level that is impossible in any classical architecture. In QuC, information is encoded in "quantum bits (qubits)"; but in contrast to classical logical bits, which can be in either state 1 or state 0, quantum bits exist as a combination (a linear superposition) of two quantum logic states, represented as |1> and |0>. It is because of this capacity for parallel processing that a quantum computer is able to perform calculations much faster than classical computers.

QuC uses the interference properties of entangled quantum mechanical particles to allow each bit of quantum information (called a qubit) in the computer to be intimately linked to every other qubit in the system. The quantum theory mechanism for QuC is the Einstein–Podolsky–Rosen paradox and the Bell inequality: If two particles are entangled and taken to the opposite ends of the universe, then quantum theory determines that if one measures the properties of one particle, one automatically then knows the properties of the other since they are intimately linked. The state of one after a measurement automatically defines the state of the other, this is as if information had traveled instantaneously across the distance of the universe, which is impossible from Einstein's special relatively theory. An example is the spin of a photon where if one photon has a right-hand polarization then the other photon after entanglement must have a left-hand polarization through quantum theory.

From a physical point-of-view a bit is a two-state system. In quantum mechanics, if a bit has two distinguishable states, then it may also exist in a superposition of these two states, called a qubit. If each of these superpositions is entangled, so that there is a further superposition of each qubit, then the state of each qubit depends on the states of every other qubit. Gate operations on these qubits provide a time evolution of the system where every entangled particle is affected. As a result, a massive parallel computation results. To produce a quantum computer, there are five criteria, called the DiVincenzo checklist [420], that must be adhered to:

1. A scalable physical system with well-characterized qubits.
2. Initialization of simple fiducial states (initialization of the qubits to a well-known state). Basically this is setting the qubits into similar quantum states before any calculation can proceed. An example is setting all the spins in one system to be identical.
3. Long relevant decoherence times, much longer than the gate operation. As the decoherence time dictates the length of time the qubits can be entangled without loss of any information, any computation must be finished before the qubits loose information.

[3] The term QuC refers in the rest of the chapter to either quantum computing or quantum computer, based on the context.

4. A universal set of quantum gates. For a quantum computer, two types of gate operation are required to produce a universal computer that may be designed and programmed to complete any computation task. These are the single- and two-qubit operations; from these a computer for factorization, database searching, or quantum system simulation may be built and operated. These two gate operations can be achieved by different techniques that depend on the specific two-state system used for each qubit.
5. A qubit-specific measurement capability. Once a calculation is complete, the information must be outputed from the qubits.

Satisfying these criteria is a major technical challenge, since this entails reconciling two conflicting requirements: the need to access the qubits so one can initialize, manipulate, and read out their state, but at the same time, the qubits must be highly isolated from the environment, so they remain coherent for a long time [421].

In a quantum computer, the quantum bits first have to be controlled individually in order to initialize the quantum register in which information is stored; then, a controllable interaction between the quantum bits must be established so that the quantum states become entangled. "Entanglement" is the key to a quantum computer's ability to operate: In effect, a rigid coupling is introduced between the qubits, which can then no longer be considered individually, but are affected simultaneously by a calculational operation. R&D efforts are now aimed at searching for a physical system that could form a reliable, controllable quantum bit [422].

The QuC is presently at an (immature) experimental stage, but if many of the problems can be overcome, the potential power of the technique will allow many numerically intensive calculations to be completed that are presently impossible with classical computers. By press time only a few systems with up to 7 qubits have been demonstrated. For most applications such as factorization of large numbers for cryptography, 30,000 qubits or so are required, although even 50 qubits will quickly solve problems that are very slow on classical computers. Ion traps and nuclear magnetic resonance systems have so far demonstrated the largest number of qubit entanglements. For large integration, however, solid-state quantum computers provide the best platforms. Examples for the two-state systems required for qubits include semiconductor QDs, superconducting device, the nuclear spins of donors in silicon, the electron spins of donors in SiGe heterostructures, and the use of surface acoustic waves with low-dimensional structures in GaAs heterostructures. These concepts are revisited in the next section and in Appendix G.

6.3 ADDITIONAL DETAILS ON NANOELECTRONIC SYSTEMS

In the subsections that follow a brief survey of some research and engineering topics is provided to expand some of the nanoelectronics concepts introduced earlier at various points in the chapter. Specifically, we focus on quantum dots, quantum wires, quantum computing, fabrication, and instrumentation.

6.3.1 Quantum Dots and Quantum Wires

Quantum Dots

As we noted in Chapter 4, QDs are small electrically conducting cavities, typically less than 1000 nm in diameter, that contain from one to a few thousand electrons. They are "tiny" clusters of semiconductor material. Because of the constrained volume, the electron energies within the QD are quantized; it follows that the behavior of the QD is somewhere between that of an atom and that of a classical macroscopic object [31]. Single QDs are referred to as *artificial atoms* and an array of coupled QDs is called an *artificial crystal* [423]. QDs are called artificial atoms because the charge carriers in these systems (electrons or holes) can only occupy a restricted set of energy levels, just as it is the case for the electrons in an atom (QDs occupy well-defined, discrete quantum states) [422, 424]. QDs can be considered "designer" atoms since their electronic properties can be controlled via the synthetic method used to prepare the dots [425]. Two QDs can be connected to form an *artificial molecule* and depending on the strength of the interdot coupling (which supports quantum mechanical tunneling of electrons between the dots), the two dots can form *ionic* or *covalent* bonds. In the ionic bond case, the electrons are localized on individual dots, and in the covalent bond case, the electrons are delocalized over both dots [424]. The goal of the research in this area is to create useful electronic and optical nanomaterials that are quantum mechanically engineered by tailoring the shape, size, composition, and position of various QDs [426].

Basic mechanisms of operation depend on the size of the QD and on the nature of the ligands used to prevent coalescence of the dots. The ligands control how closely the dots can be packed, and, in turn, the strength of the coupling between adjacent QDs. An important parameter in this context is the energy cost of adding an electron to a dot: Because of the large size of the dots, the Coulomb repulsion of the added electron is low (the energy required to remove or add an electron to the QD is determined by the size of the dot). Unlike most ordinary atoms, QDs have a high capacity for accommodating an additional electron [425].

Concerning the physics of QDs, adding even a single electron to such a system requires a significant amount of extra energy, as noted, because of the repulsion between the negatively charged electrons as they are forced into a smaller volume. One result of this, called the Coulomb blockade, is to make possible a greater control of the number of electrons in a QD, that is, researchers can tune the number of electrons by manipulating input energy [426]. It should be noted that the optical excitation of a semiconductor leads to the creation of a quasi-particle known as an *exciton*—a negatively charged electron bound together with a positively charged "hole." In contrast to the Coulomb blockade resulting from electrical injection of electrons into a QD, such dots remain neutrally charged following optical excitation, and the QD exciton may possess useful properties [426].

An electron in a QD can be described by a quantum wave function that is similar to that used for an electron in a single atom, although the energy of the electron in the QD is spread in a coordinated way (spread "coherently") over the lattice of atomic nuclei. The electronic wave functions of QDs are often labeled with atomic notation, but QDs are, in reality, solid-state nanostructures that can be tailored into different shapes [426].

Recent progress in semiconductor manufacturing makes it feasible to tune the physical properties of QDs in a controllable manner. Quantum effects in artificial crystals are useful in nanotechnology devices [423]. When an individual molecule, a nanocrystal, a nanotube, or a lithographically defined QD is attached to metallic electrodes via tunnel barriers, electron transport is dominated by single-electron charging and energy-level quantization. As the coupling to the electrodes increases, higher-order tunneling and correlated electron motion give rise to new phenomena, including the Kondo resonance (see Glossary) [427].

Electroluminescence is also supportable with QDs. Single-photon sources have recently been demonstrated using a variety of devices, including molecules, mesoscopic quantum wells, color centers, trapped ions, and semiconductor QDs [428]. The fabrication of high-efficiency organic light-emitting diodes has been announced in the literature [429]. The light is generated through fluorescence, as electrons make transitions between orbital states of π-conjugated organic molecules (as seen in Chapter 3, the π bond arises from the overlap of the 2p orbitals of electrons in carbon atoms). In addition to having high quantum efficiency for electron-to-photon conversion, π-conjugated molecules in organic LEDs have the advantage of color tunability, so that they can be used to build full-color displays based on red–green–blue (RGB) emitters [429]. Organic LEDs permit robust fabrication technique and high performance; this coupled with the luminescent properties of nanocrystals offers prospects for practical devices, for example, color displays for mobile telephones. QDs that emit light are also expected to form the basis of a new generation of lasers [426].

A number of applications for single-photon sources have been advanced in the field of quantum information, but most—including linear-optical quantum computation—also require consecutive photons to have identical wave packets. For a source based on a single quantum emitter, the emitter must be excited in a rapid and/or deterministic way and must interact in a rather limited way with its surrounding environment. Most proposed applications for single-photon sources in the field of quantum information (with the notable exception of quantum cryptography) involve two-photon interference. Such applications include quantum teleportation, postselective production of polarization-entangled photons, and linear-optics quantum computation [428].

Semiconductor QDs have been synthesized, opening up the possibility of implementing qubits in a solid-state environment (the original proposals for quantum computers were based on atomic systems, such as atoms held in traps, where the qubits formed by two energy levels between which an atomic electron can make transitions) [422]. In fact, QDs offer a number of two-level systems, based on charge or spin (or both); an example of one such two-level system is a coupled electron–hole pair—an exciton. The absence (equivalent to the state $|0\rangle$) and presence (state $|1\rangle$) of an exciton in a semiconductor QD could represent the two levels of a quantum bit [197, 422].

The following quote summarizes rather well the opportunities offered by QDs:

> Quantum dots have great flexibility because their properties can be artificially engineered, but this comes at a price. Nature has given us atoms; scientists must make QDs. Further advances in this exciting field of science and technology will depend heavily on the creativity of physicists, chemists, and materials scientists who make these tiny structures. [426]

Quantum/Nanowires

As noted in Chapter 4 there is interest in developing nanoscale-level conductors (wires) that can be employed to construct the next generation of computer and memory chips. The technical and economic difficulties in further miniaturizing silicon-based transistors with the present fabrication technologies that we discussed in the previous sections have motivated sustained efforts to develop alternative electronic devices, based, for example, on single molecules. Recently, carbon nanotubes have been successfully used for nanometer-sized devices such as diodes, transistors, and random-access memory cells. These nanotube devices are usually very long compared to silicon-based transistors [430, 431].

Currently, wires in the 30- to 50-nm range are being studied; however, one is ultimately interested in much smaller systems, for example, systems based on carbon nanotubes with diameters as small as 1 nm. As already discussed, carbon nanotubes combine physical strength, true nanoscale dimensions, and flexible electronic properties. They can behave either as conducting metals or as semiconductors, depending on how carbon atoms are arranged on the wall(s); as a result, they afford the possibility for use as components in electronic devices that are even smaller than those available today [75].

As an example of work under way which we alluded to in the previous section, researchers have already incorporated a semiconducting nanotube as a component in experimental FETs. In this environment the nanotube is grown on a surface of silicon dioxide with metal electrodes evaporated on the nanotube's surface serving as the device's electron source and drain; a layer of silicon fabricated under the silicon dioxide serves as the transistor's gate [75] (it is important in this process that the gate be properly coupled with the rest of the device, otherwise excess power is required to turn the device from off to on).

A first step relates to the potential use of new nanomaterials for *interconnect applications* on chips. There are two approaches to nanoscale (molecular) interconnects (see Table 6.5 [229]): (i) organic approach and (ii) inorganic approach. The use of carbon nanotubes and/or molecular crystal wires as interconnects, however, remains somewhat of a future technology in the context of commercial-level applications because of significant material processing and integration challenges. Processing challenges are triggered by the nature of the self-assembly techniques that are commonly used to form nanoscale organic systems [229].

As we noted in Chapter 4, current self-assembly methods are not as of yet sufficiently reliable and reproducible for the deposition of individual nanotubes and wires on chips; furthermore, self-assembly fabrication techniques are, in general,

TABLE 6.5 Nanoscale Interconnects

Organic interconnects	Entail the fabrication and integration of molecular crystal wire infrastructures comprising two building blocks: (a) charge carrier groups and (b) direct-attachment functionality groups. The charge carrier groups consist of conjugated polymeric systems or organic conductors. Here overlapped conjugated electron states (known as π-π* in adjacent carbon atoms) serve as coherent electron carrier states to enable ballistic electron transport via delocalized p-electron molecular orbitals.
Inorganic interconnects	Entail the use of metal ion-based organometallic groups as charge carrier entities, with the overlap of atomic or molecular orbitals between adjacent metal ions providing 1D "metallic" conduction. Reaction of customized metal salts at specific surface sites ensures directed (selective) attachment of metal ions with large ionic radii or organometallic groups on selected surfaces.

incompatible with prevailing semiconductor processing techniques [229]. Integration complexities are caused by the parametric variances between the chemical, mechanical, and thermal properties of organic systems and their metallic counterparts, which gives rise to significant interfacing challenges. These challenges are further compounded by the physics of electron propagation across the interfaces between organic systems and conventional metals, including suboptimal ohmic contact with metal leads [229].

While the field of interconnect nanotechnology is just starting to develop, preliminary theoretical and experimental results point to its extensive potential to manipulate matter at the atomic-length scale, leading to the formation of nanometer-scale IC building blocks that employ novel signal propagation schemes; these structures and devices may lead to the deployment of ICs at terahertz speeds [229].

Along another avenue of research, resistance-free current flow is expected to offer manufacturers in the electronics industry key functional improvements. Resistance-free current flow is usually associated with superconductivity. Electrical resistance arises because charge carriers (electrons and holes) collide with imperfections (e.g., impurities and dislocations) in the material they are traveling through. In theory, a perfectly crystalline conductor offers no resistance, yet experiments undertaken in the late 1990s and early 2000s failed to validate this and virtually defect-free wires have resistances of several kilo-ohms; however, of late, researchers have shown for the first time that resistance does vanish in a small but perfectly formed wire [432]. Ballistic quantum wires offer an alternative to superconductivity: These nanoscale structures are almost completely free of the defects that give rise to resistance in traditional wire conductors and the wire is termed *ballistic* because the electrons can travel its entire length before encountering a defect (at press time results using a layer of gallium arsenide on a sliver of aluminium gallium arsenide GaAs/AlGaAs were being announced [432].)

Conductive polymers also are of interest for a plethora of research and commercial applications. New growth methods were emerging at press time, pointing to increased versatility of these polymers [433]. Conducting polymers are materials that possess the electrical properties of metals, yet retain the mechanical properties of polymers. A new chain-growth method allows scientists to "cap" each conducting polymer with chemical groups that link to other structural polymers; with this method researchers can form highly conductive nanowire sheets. Variations in the chemical cap also allow conducting polymer strands to adhere directly to metal, silicon, or other industrially important templates used in devices such as transistors. Conducting polymers have a number of potential applications, for example, for dissipating static electrical charges that build up on coated floors for use in disposable radio frequency identification (RFID) tags [433].

Researchers have also been looking at methods for dividing a semiconductor nanotube into multiple QDs with lengths of about 10 nm by inserting Gd-C_{82} endohedral fullerenes. Techniques such as these can be used for fabricating an array of QDs that can be used for nanoelectronics and nanooptoelectronics [430].

In conclusion, major progress is expected in the near term in the QD and quantum wire areas, thereby facilitating additional nanoelectronic advances.

6.3.2 Quantum Computing

As noted in the opening pages of this chapter, over the past half-a-century computer technologies have continued to advance in terms of reduced physical dimensions, improved density, extended feature set, increased speed, and increased memory storage; current lithographic techniques can place tens of millions of transistors per cubic centimeter on a silicon chip. Technology is continuously being improved, but, as already stated, a limit is expected to be reached in the next few years. To work around these limitations, atomic-scale computer devices may be developed within the next decade or so. In addition to single-electron transistors, two promising alternatives to traditional computers are molecular computing and quantum computing/computers (QuC). When logic gates are comprised of only a few atoms, computers may be able to manipulate the quantum states of subatomic particles, atoms, and molecules, to perform the basic logic operations of computing [307]. QuCs are expected to be much faster than existing supercomputers because they operate on all the particles and, so, on all the possible coded numbers simultaneously. Much progress has been made during the last few years, and this technology has been shown to be a potential feasible replacement for semiconductor chips [4]. Expanding on the introduction on the topic provided in Section 6.2, this section provides a short overview of QuC; a more inclusive treatment of QuC is included in Appendix G.

Quantum computing seeks to write, store, process, and read information on the quantum level. Qubits are the basic units of QuC: In this environment physical particles can be used as qubits; more specifically, qubits can be equated to the individual atoms or subatomic particles [434]. This is not a trivial task because of the complex nature of quantum mechanical systems; for example, the laws of quantum theory involve unintuitive principles such as superposition and entanglement. However, these complexities also offer novel opportunities: Taking advantage of superposition means that a qubit of information can be used in several computations at the same time; taking advantage of entanglement means that the information can be processed over long distances without the extant requirement of actual wires [4].

Traditional computer techniques involve the representation of information as a sequence of bits, that is, a sequence of binary digits (0's or 1's). Each such bit is realized physically through a macroscopic physical system, for example, magnetization on a hard disk, an electrical voltage level in a RAM cell, and the like. Traditional computation can be described as a physical process consisting of a sequence of systematic manipulations on an input group of bits [307]. The underlying concept associated with QuC is to utilize quantum theory systems to represent individual physical bits of information (i.e., microscopic physical systems that obey the laws of quantum mechanics, such as individual atoms, electrons, or photons; for example, because particles spin in one of two directions, the state can be used to represent 1's and 0's of binary computing) [307].

So far, researchers have been able to build quantum computers of only a few qubits, but that has been adequate to confirm the basic principles of QuC. Although the practical importance of present-day quantum processors (e.g., state-of-the-art 7-qubit computers available experimentally at press time) may be limited in themselves; they are considered to represent an important step toward larger-scale QuC. Lateral QDs (see Fig. 6.10), as one example, are well-suited for QuC applications since all the tunnel barriers can be easily controlled using electrostatic gates [186] (researchers have been able of late to isolate a single electron in lateral quantum dot structures; until recently, this had only been achieved in vertical dots [188]). The actual realization of a QuC is challenging because it requires the ability to initialize, coherently manipulate, and measure individual quantum systems [421].

Two qubits represent every possible two-bit number (00, 01, 10, 11). Each additional qubit increases the number of possible numbers exponentially: n qubits represent 2^n possible numbers (e.g., 20 qubits represent numbers from 0 to more than 1 million, 30 qubits represent numbers from 0 to more than 1 billion, and 40 qubits represent numbers from 0 to more than 1 trillion).

Electron spins isolated in QDs and located in a magnetic field provide a useful two-level system. A QuC acts on a set of particles by influencing the probabilities of

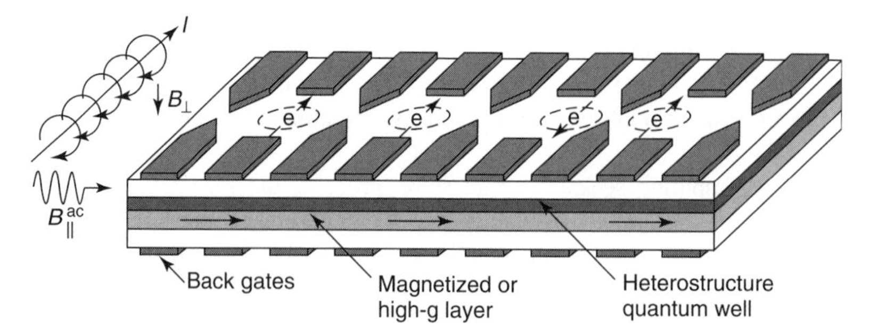

FIGURE 6.10 Schematic diagram of the Loss and DiVincenzo computer. (Courtesy: D. Loss and D. P. DiVincenzo).

the particles' spins (posing the problem) so that when the particles leave their quantum state the resulting spins represent a specific number (getting the answer) [434]. In the quantum state, a particle has a probability of spinning in either direction when it leaves the quantum state; this implies that a qubit represents both 0 and 1 at the same time. Qubits achieve an additional state (called a superposition) beyond the traditional logical state 0 or the logical state 1, with a state that is a blend of the states representing the classical bits 0 or 1, respectively. This feature enables the input for QuC as a superposition of many classically different inputs and calculations on all the inputs to be performed simultaneously (quantum parallelism). Another mechanism is that two or more qubits can be in an entangled state, which essentially means that neither of the involved qubits can be said to be in a well-defined state of its own and that all the information about the qubits is encoded in their joint state(s); important physical applications of entanglement are quantum teleportation and quantum cryptography [307].

As we noted, in classical computers, information is stored and processed in the forms of bits, which must always be either 0 or 1. Any physical system with two possible states can be used to represent a bit, including two-level quantum systems. Hence, one can encode information into the spin of an electron (up/down), the polarization of a photon (horizontal/vertical), two energy levels of an atom (ground/excited state), and so forth [421]. Interestingly, quantum theory predicts that such systems can also exist in a "superposition" of the two basis states,

$$|\psi> = \cos\theta\,|0> + e^{i\varphi}\sin\theta\,|1>$$

effectively, this means a qubit can be *0 and 1 at the same time*[4] (it follows that one can represent qubits by a point on a sphere, as shown in Fig. 6.11). In general, an n-qubit computer can perform 2^n *computations at the same time*. This suggests that the computational capability of a quantum computer increases exponentially with its size (compared to only linearly for a classical computer) [421].

The emerging field of quantum information processing (QIP), which makes use of QuC mechanisms, has the potential to revolutionize information processing/IT and to provide new methods for securing, processing, storing, retrieving, transmitting, and displaying information. Through the exploitation of quantum phenomena, a QuC is anticipated to be orders of magnitude more powerful than traditional computers (assuming that these QuCs can be realized into commercially available products). More specifically, a QuC should be able to perform many computations much faster than a traditional computer; QuCs are anticipated to achieve functionality that may result in the ability to tackle qualitatively different computational problems that

[4]When a qubit is measured, the result will always be either 0 or 1, even if it was both 0 *and* 1 before the measurement. Thus, after 2^N parallel computations, only one of the corresponding output values will be measured; while QuCs cannot give us the output value for all possible input values at the same time, special quantum algorithms are able to exploit quantum parallelism to solve relevant problems in exponentially fewer steps than is possible classically. The best-known quantum algorithm is Peter Shor's 1994 algorithm for efficiently finding the prime factors of large integer numbers; another algorithm, invented by Lov Grover in 1996, offers a quadratic speedup for unstructured search problems [421].

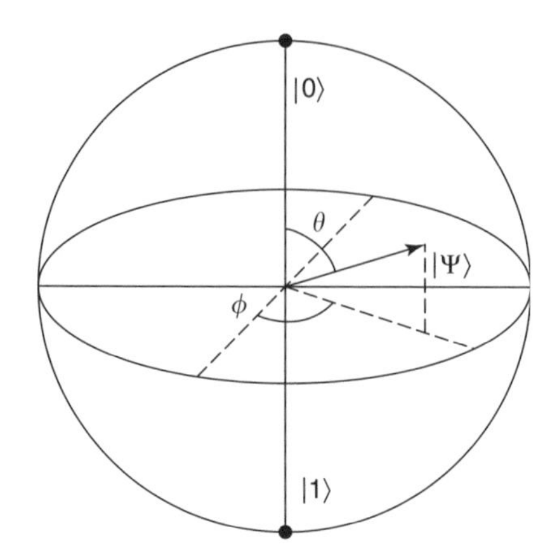

FIGURE 6.11 Qubits representation on a sphere.

are too complex for traditional computers. A significant expectation is that a QuC will be able to effectively factorize large integers into their primes, posing a threat to widely used encryption methods [307]. Quantum cryptography is considered to be the most advanced application of QIP; studies have suggested that quantum cryptography can provide absolutely secure (new) encryption methods.

There are decoherence roadblocks for QIP as discussed next [435]. QIP relies to a large extent upon the ability to ensure and control unitary evolution of an array of coupled qubits for long periods of time. There are a number of physical effects that act against this coherent evolution. These include interaction of the qubits with a larger environment, unwanted or uncontrolled interactions between qubits, and imperfections in applied unitary transformations. The latter can be either systematic or random and can also give rise to additional unitary errors. The term *decoherence* referred originally explicitly to errors that arise in the wave function phase, that is, to decay of off-diagonal terms in the density matrix. This decay of phase is basis-set dependent; it also does not constitute the only source of loss of unitarity. At this juncture, decoherence is, therefore, more generally understood in the field of QIP to refer to all manifestations of loss of unitarity in the qubit state time evolution. It includes [435]: (i) explicit loss of coherence, (ii) dissipative or energy relaxation effects, and (iii) leakage out of the qubit state space.

The most general objective QuC research is to develop novel systems and techniques for information processing, transmission, and security by exploiting the properties of quantum mechanical operations. Specific objectives (currently) include the items listed in Table 6.6 [307]. Some specific research efforts underway in the early 2000s included the items listed in Table 6.7 [307].

6.3.3 Fabrication Methods and Techniques for Nanoelectronics

This section builds on the discussion of Chapter 4, but with a more direct focus on nanoelectronics.

TABLE 6.6 Partial List of Generic QuC Research and Development

Development of an elementary scalable quantum processor	The current state of development is not far enough along to identify a specific technology that could uniquely achieve the properties, performance, and other design factors required for full-fledged QuC. Existing approaches and technologies all have limitations. Hence, the identification of appropriate new technologies and systems that are practical and scalable is critical to advancement. The developments and demonstration of technologies and systems that exhibit fewer limitations than current approaches are desirable (new technologies include nuclear magnetic resonance, ion traps, cavity quantum QED, quantum dots, etc.). The feasibility and acceptance of a successful technology depends on several factors and requirements, including: (i) the ability to control decoherence and perform fault-tolerant operations, (ii) scaling properties in terms of number of qubits, time per gate, and physical size, and (iii) the likelihood of becoming a low-cost, low-power technology with the potential to be commercialized in appropriate applications.
Quantum algorithm development for grand challenges	Up to now there have been few quantum algorithms that tackle problems of practical significance. Research to advance the development of "killer applications" for quantum computers is needed to demonstrate their potential and justify future investment (see Appendix G for a more inclusive discussion). The automated discovery of new quantum algorithms is also an area of research.
Longer distance and secure quantum communication	Significant challenge exists in scaling quantum communication protocols over distance and to demonstrate the compatibility of quantum communications with traditional telecom infrastructure, such as optical fibers. It is also desirable to expand (i) quantum key distribution for longer distance applications, (ii) multiparty quantum key distribution, and (iii) free-space key distribution. The development of quantum key repeaters is considered another important enabling factor for the development and implementation of underlying concepts and technologies.
Development of novel applications	Range of applications currently conceived is limited, and the full potential of the technologies and concepts are not fully understood at this time. Current concepts and associated applications are seen as early demonstrations and are based on quantum systems with only a few qubits. There are fundamental challenges including the following: (i) devising suitable methodologies for harnessing decoherence (these would include, among others, error correction algorithms, redundancy techniques, and fault-tolerant architectures); (ii) quantum information storage and retrieval, including associated error control; (iii) readout of intermediate and final results by measurement techniques (the efficient readout of information from a quantum system remains one of the key challenges); and (iv) building a quantum toolbox (this being a set of design components, development of components, and engineering processes that perform specific quantum tasks that can be used as building blocks of a QuC system—examples of these components are light sources with controlled fluctuations for photon-based QuCs; nonlinear optical fibers for guided photon-pair generation and coupled semiconductor QDs for single-spin or single-photon applications; examples of quantum engineering processes include quantum nondestructive measurement techniques; production and use of correlated optical solutions and quantum interferometry techniques).

TABLE 6.7 Some Specific Areas of QuC Research and Development

Area	Description
Atomic chips for quantum information research	Development of novel approaches to QIP utilizing neutral-atom manipulation with integrated microdevices, such as: (i) tools of quantum optics, long coherence times and high-precision neutral atom manipulation for a quantum component in information processing and (ii) technology to integrate optical and electronic elements, in conjunction with a classical approach to atom manipulation.
Entanglement in QuC and communication	This research is focused on quantum mechanical entanglement as a key enabling technological component for QIP. The main objectives of the research are to advance the theory of bi-partite entanglement and to develop a theory of multipartite entanglement, which would ultimately establish the foundation for future technological applications of quantum mechanics in information processing and communication. Ways must be found for entanglement principles to be carried through to implementation for practical communication and computation applications. Advances in optimal coding/decoding in quantum channel communication, suppression of decoherence in QuC and communication, scalable quantum computer, and scalable communication line repeaters are sought.
Enabling technologies for quantum information systems	There has been a lack of progress in developing technologies that will be required for practical application of new concepts. Research seeks to address this problem by demonstrating research prototypes utilizing a range of key enabling technologies for quantum communication systems, including single-photon detectors, integrated interferometers, and customized photon data acquisition PC cards (utilize new technologies to demonstrate PC-based quantum key distribution systems with enhanced performance in terms of bit rate, distance, and improved practicality).
Novel algorithms and many-body implementations	Goals of this line of research are to develop new quantum algorithms (algorithms for difficult problems in combinatorics) and to investigate the theoretical limits of their implementation in devices based on realistic physical structures. A study of microscopic-complex regularities related to the physical interactions and symmetry mirroring of the system's invariance with respect to specific state-space transformations is needed.

Quantum algorithms and information processing	Based on current concepts for QuC and the development of concrete quantum algorithms, the possibility exists that several computational problems might be solved more efficiently by exploiting the principles of quantum mechanics in the computational process. The objectives of this research are to better understand this technological opportunity and exploit it fully; topics include: (i) quantum algorithms and complexity, (ii) cryptography, (iii) fault-tolerant computation, and (iv) information and communication theory. Anticipated results include new quantum polynomial time algorithms and few-qubit applications, complexity theory-based quantum cryptographic protocols, better QuC codes, self-testing/correcting quantum programs, advanced information theories, communication complexity, and the Kolmogorov complexity in the quantum domain.
QuC device using doped fullerenes	There is research underway to establish the possibility of constructing a solid-state spin-based QuC using doped fullerenes. Current concepts and designs require doping of single atoms with ultrahigh position and precision, while maintaining a regular silicon lattice structure. Some have looked at a variation whereby the qubit resides in the nuclear spin of a single dopant atom that is either attached or trapped inside a fullerene cage.
QuC and transfer with single atoms and photons	There are efforts to study decoherence and entanglement of a variety of quantum systems (atoms, ions, and cavities), with a view to the practical realization of scalable quantum registers for use in QuC and quantum communication. Goals of this research include the demonstration of quantum gates using single ions and strings of ions trapped in different types of ion traps, the demonstration of trapping and cooling of single atoms for use in novel quantum gates, the realization of novel quantum devices using techniques based on cavity QED, and the demonstration of elementary quantum communications by bringing together aspects from the above areas.
Long-distance photonic quantum communication	There are research efforts to scale secure quantum communication over longer distances, to demonstrate novel applications, and to identify and transfer spin-off applications to industry. Physical resources to be studied that include entangled quantum states having no classical counterpart. There is interest in novel sources for direct generation of entangled photon states in electrically pumped structures, on diode-laser-pumped nonlinear optical crystals, and on detectors or multiphoton states. Basic quantum information building blocks, such as teleportation and entanglement swapping, need to be developed, and field demonstrations will be used to validate the technology at 700–800 nm (free space and optical fibers) and at 1300 and 1550 nm (telecom fibers and systems).

TABLE 6.7 *(Continued)*

Quantum information with continuous variables	There is a need for the development, fabrication, and demonstration of quantum communication with continuous variables, such as the quadrature amplitudes of quantized electromagnetic fields. The advantage of using these highly excited quantum systems involving many photons in a light field over single-particle quantum systems is expected to result in high optical data rates and simple processing tools based upon techniques used in telecommunications. The objectives of this line of research are the transfer and storage of continuous quantum information encoded on light and matter systems and the implementation of quantum cryptography using quantum continuous variables.
Solid-state sources for single photon	There is a need to undertake the development of solid-state single-photon sources and their use in QIP by addressing the following issues: (i) The light source should involve a single dipole, so that upon emission it produces a single photon. Such a single emitter intrinsically emits radiation almost isotropically. To remedy this, (ii) the emitter should be embedded in an optical microcavity designed to increase the spontaneous emission in a restricted number of modes, which can be collected with high efficiency for further use, e.g., by coupling into an optical fiber. And (iii) in order to assess its performance, the source that will reach fully operational status must be able to be integrated in an operating quantum information system.
Semiconductor-based quantum information devices	Researchers are looking for semiconductor-based implementation of QuC devices directed at a specific class of novel nanostructures: single and coupled semiconductor QDs. The possibility of large-scale integration of these structures makes them ideal candidates for the practical solid-state implementation of massive QuC. A strategy for their implementation as potential quantum-computing devices is still lacking.
Superconducting qubits	There is interest to demonstrate the feasibility of quantum information processors using scalable solid-state low-T_c Josephson junction nanotechnology. Efforts were underway for manufacturing systems of quantum logic gates by developing Josephson junction, single-electron and semiconductor-based implementation of quantum information devices technologies to achieve initiation, processing, and readout of superconducting qubits information.

Top-Down Method

As we discussed earlier in this chapter, miniaturization of silicon electronics is now being intensely pursued at the commercial level, although fundamental limits of lithography may prevent current techniques from reaching the deep nanometer regime (1–25 nm) for highly integrated devices [136, 436, 437, 438, 439, 440]. Devices at the (true) nanoscale (e.g., 1 nm) or at the atomic scale (e.g., 0.2 nm) are about two to three orders of magnitude smaller than the scale of electronic devices fabricated by current standard optical lithographic means [18]. As we have discussed throughout this text, at the (true) nanoscale a system's dynamics are governed by the quantum-theoretic nature of the electron: Quantization of electric charge and quantum coherence effects influence the transport and behavior of electrons; at the same time, these novel transport properties afford new opportunities [18]. Electronic circuits as small as 10 nm are expected to be developed in support of the next generation of electronic technology. Devices on this scale are miniaturized to a degree that can only be manufactured with novel lithographic techniques and can be observed only with special instrumentation, such as electron microscopes, atomic force microscopes, and so on.

Up to the present time the reduction in scale has been made possible by photolithographic procedures that use photochemical patterning followed by selective etching to carve microscopic structures into semiconductor wafers [441]. However, as discussed elsewhere in this book, this approach has fundamental limits of resolution, imposed by the optics of the patterning beams: At present, a resolution of around 0.10–0.05 μm for commercial photolithography remains the baseline, although progress is being made on a continual basis. Electron-beam lithography, extreme ultraviolet (EUV) lithography, and X-ray lithography offer finer resolution (as briefly discussed at the end of Chapter 2); the development of new lithographic techniques—such as electron-beam lithography, chemical or electrochemical lateral structuring, and scanning probe microscopy (SPM) techniques—enables the manufacturing of structures of only a few nanometers in size, as well as the manipulation of individual atoms [18]. However, currently these techniques operate at greater cost so that they are not, as of yet, routine industrial techniques. Nonetheless, if current trends continue, the scale of miniaturization will approach in just a couple of decades the diameter of large molecules (a few nanometers, as covered in Chapters 2 and 3); once this happens, completely new technologies will be needed [441].

Intel (as an illustrative example of a well-known IC producer) was manufacturing microprocessors using 90-nm masks in 2003 and was expected to use 65-nm processes by 2005 (the mask set for a chip must be ready at least a year before production starts); the company has also reportedly fabricated its first masks geared for the 32-nm node, which was expected to be based on EUV lithography techniques, making it the first company to put EUV onto the production line by 2007 (Intel was planning to use EUV masks starting with the 45-nm nodes and then progressing to the 32-nm nodes) [442].

In the late 1990s microprocessor and memory chip vendors pushed the lithography tools to the point that the wavelength of light they utilized in manufacturing exceeded the minimum feature size of the transistors. Unfortunately, optical wavelength scaling did not keep up with the rate of feature size scaling [442]. One way to correct this is to use larger lenses, but lenses can only be extended up to a point to keep aberrations to a minimum. The maximum numerical aperture is 1 and the largest lenses today already have a numerical aperture of 0.8 [442]. The practical way to keep up the (desired) process technology cycle is to "keep pushing" the lithography tools no matter how "fuzzy" the image becomes and then find ways to correct the image. This approach is known as "pushing the $K1$," also known as the "process complexity" or "lithography aggressiveness" factor. $K1$ is the product of device feature size and lens size divided by the wavelength of the light source. The goal, therefore, is to lower the $K1$ factor. The theoretical limit for $K1$ is 0.25 (as an illustrative example, Intel claimed at press time that it had achieved a value of 0.4 [442].)

The problem with fixing an image is that it makes the process of preparing the masks much more challenging. For example, a 90-nm device needs 200 GB of data to describe a mask with up to 25 layers, which equates to about 1 trillion pixels; finding and repairing a defect in such a mask is technically daunting [442]. Another approach is to use EUV masks, as mentioned above (with EUV one is less worried about the $K1$ factor and the effects of optical proximity correction, because the wavelength from the light source is below the device feature size).

Immersion lithography unfolded quickly in the early 2000s, in a concentrated effort to get it ready for volume production; EUV lithography is another area that has seen industry cooperation of late. Nanoimprint lithography can also be used to mold structures into a thin polymeric film; like immersion lithography and EUV lithography, nanoimprint lithography is seen as a possible solution for work at the 32-nm node. Researchers on nanoimprint lithography in the mid-1990s and although the technology is now moving from research labs to industry, it is still in its infancy [443].

Advanced instrumentation is needed to facilitate manufacturing of a number of nanostructrures. Typically, advanced instrumentation provides high-intensity sources of radiation for increased sensitivity and resolution. On the atomic level, synchrotron "light" (this being a continuous spectrum of electromagnetic radiation ranging from infrared to X-rays) is well suited to surface studies; but various other experimental methods have been developed that involve diffraction, microscopy, and spectroscopy, enabling researchers to get a closeup look at materials and their properties. Methods (such as atomic layer epitaxy, molecular-beam epitaxy, and the Langmuir–Blodgett technique) have been recently developed to fabricate thin 2D nanostructures with an accuracy of a single atomic layer; however, the challenge arises when developers and researchers look to create 3D structures smaller than 100 nm [406].

For example, to enable the generation and emission of photons, it is typically necessary to develop fabrication processes that give rise to structures with very thin semiconductor layers (these thin layers are called epitaxial layers.) These growth mechanisms are based on top-down methods on top of bulk semiconductor wafers[5] (e.g., gallium aluminum arsenide may be grown on a gallium arsenide substrate). Fitting very thin layers of a semiconductor material between layers of a different

[5]Epitaxial layers are typically grown on flat surfaces, but new methods are being explored to grow layers on nonplanar structures and, hence, add features such as ridges or channels, etched into the surface of semiconducting devices.

composition enables scientists to control the heterostructure's bandgap; bandgap engineering applied to silicon and germanium and III-V compound semiconductors supports the development of semiconductor materials with new properties [406].

The most precise method to grow epitaxial layers on a semiconducting substrate is molecular-beam epitaxy. This technique uses a beam of atoms (or molecules) emitted from a common source; after traveling across a vacuum, the beam strikes a heated crystal surface and forms an epitaxial layer with the same crystal structure as the substrate. Another deposition method is atomic layer epitaxy, which is a digitally controlled layer-by-layer technique used to produce thin material layers with atomic accuracy. Here, each layer that is formed is the result of a saturating surface reaction; the thin films are virtually free of defects and have near-perfect step coverage, which is required in submicron semiconductor technology. The surface control achieved in atomic layer epitaxy produces thin films with bulk density and acceptable uniformity on large-area substrates. In the future, atomic layer epitaxy may be a valuable tool for nanotechnology, especially when combined with appropriate patterning and micromachining techniques [406].

Bottom-Up Method

As discussed in Chapter 4, an alternative to the top-down techniques of conventional semiconductor engineering is the bottom-up approach that builds nanostructures from molecular components. In a bottom-up approach, rather than building devices and patterning materials by "reduction" (i.e., carving them from larger, monolithic blocks), one can handle the manufacturing by synthesis, that is, putting the structures together molecule-by-molecule and/or atom-by-atom. The methods of standard chemical synthesis provide one option, but by exploiting self-assembly and self-organization, one can attain the same ends in a spontaneous, preprogrammed, and in a less labor-intensive manner [441]. Fundamental physical constraints and economics are expected to limit continued miniaturization in electronics by conventional top-down manufacturing during the next one or two decades (as noted earlier); this predicament has motivated efforts to search for new strategies to meet expected computing demands of the future [444]. Rather than milling down from the macroscopic level using tools of greater and greater precision (and probably cost), researchers and manufacturers seek to build nanoconstructs from the bottom up, starting with chemical systems [11].

Bottom-up approaches, where the functional electronic structures are assembled from well-defined nanoscale building blocks (such as carbon nanotubes, molecules, and/or semiconductor nanowires), have the potential to go beyond the limits of top-down manufacturing [445]. The process of self-assembly is a coordinated action of independent entities under distributed control that produces a larger structure or achieves a group effect [406]. The use of nanoscale structures as building blocks for self-assembled structures could, in theory, ultimately eliminate conventional and costly fabrication factories. One-dimensional structures, such as nanowires and carbon nanotubes, are expected to be basic building blocks for nanoelectronics because they can function both as devices and as the wires that access them [136, 436, 437, 438, 439, 440]. As noted elsewhere, nanotubes can be used for field-effect transistors, low-temperature single-electron transistors, intramolecular metal–semiconductor diodes, intermolecular-crossed nanotube–nanotube diodes, and an inverter[6] ([444]). However, the wide-scale use of nanotube building blocks is currently limited at the practical level because the selective growth and/or assembly of semiconducting or metallic nanotubes is not currently possible [136, 436, 437, 438, 439, 440]. Successful implementation of a building-block approach for the assembly of nanodevices and device arrays requires that the electronic properties of the blocks to be defined and controlled. Researchers recently demonstrated that the carrier type (electrons, n-type; holes, p-type) and carrier concentration in single-crystal silicon nanowires could be controlled during growth [136].

In building nanostructures and nanoelectronic devices, chemical self-assembly has of late become an important factor in constructing supramolecular nanostructures with useful electrical properties: new self-assembly techniques hold promise for going beyond the limits of current top-down technology. One example is in the field of X-ray nanolithography: Although it can generate patterns on the submicron scale, it is not the best method for accurately manipulating structures that are less than approximately 30 nm wide. Hence, the challenge now is for scientists to fully understand the self-assembly process—how it is achieved, how it can be controlled, and how it can be effectively applied to strategies for nanofabrication [406].

Nature utilizes self-assembling materials for nanostructures as the components for living cells and it assembles different materials into a variety of useful composites at the cellular level. In the (recent) past scientists successfully caused materials to self-assemble into microscopic structures such as layered films and liquid-crystal phases; unfortunately, these structures lacked the complexity of the natural composites [406]. Also, up to now, human attempts have been limited to building self-assembling nanoscale materials a few atoms or molecules at a time.

One approach for designing nanoscale materials is to organize molecular constituents into assemblies that perform complex functions. A critical factor in this context is the ability to control the spatial arrangement of the molecular components, especially if intermolecular energy- and charge-transfer processes are at the core of the function of the material [446]. Spatial organization of molecular components can be achieved through the design of covalently bonded supramolecules where molecular subunits are linked together in such a manner that their relative geometries (i.e., separations and orientations) are well defined. These structures are demanding to synthesize, but they offer an ample degree of control over engineering parameters. Work in this are is expected to continue in the future.

6.3.4 Microscopy Tools for Nanoelectronics

As noted earlier in this section, nanotechnology requires the use of new fabrication and probing techniques. Classical methods utilized to analyze the structure and composition of structures have included the optical microscope, X-ray diffraction, infrared spectroscopy, and ultraviolet spectroscopy. Appendix F provides a description of a number of microscopy tools; this subsection introduces the field, loosely based on observations and material from [406].

[6]Inverters represent a key component for logic operations.

Propitiously, new microscopy techniques are being added on a continual basis; in fact, the development of new tools and techniques for nanostructure research has become a science in itself. Specifically, scanning probe microscopy has created a revolution in microscopy, with applications ranging from condensed matter physics to biology. The first scanning probe microscope, the scanning tunneling microscope, was invented by G. Binnig and H. Rohrer in the 1980s, and this invention has been the engine of technological advancements [31].

Electron beams have played a significant role in semiconductor technology for the past quarter century. Electron microscopy can obtain nearly atomic resolution of a material's atomic arrangement and chemical composition. This technique requires a clean sample that meets ultrahigh-vacuum standards in order to provide surface characterizations such as reconstruction and phase transitions. Scanning electron microscopy (SEM) is performed by scanning a focused probe across the surface of the material. Secondary electrons emitted from the sample are typically detected by a photomultiplier system, the output of which is used to modulate the brightness of a monitor synchronized with the electron-beam scan. The more electrons a particular region emits, the brighter its image. Scanning transmission electron microscopy (STEM) has made possible new imaging techniques by using inelastically scattered electrons, emitted X-rays, and other forms of an elastically scattered beam. In scanning transmission electron microscopy, the electron beam is rasterized across the surface of a sample in a similar manner to scanning electron microscopy, however, the sample is a thin section and the diffraction contrast image is collected on a solid-state detector.

Early electron-beam machines used a raster-scanned beam spot to write patterns onto electron-sensitive polymer resist materials. At this juncture, electron-beam lithography is employed to make the smallest components on silicon substrates and is the most effective method of creating patterns on substrates such as photomasks and X-ray masks. A scanning-electron-beam lithography system can be used to direct-write onto device substrates, to make photomasks and X-ray masks, and to develop a new technique called spatial-phase-locked electron-beam lithography, which will improve the writing precision of the current electron-beam lithography system.

All scanning probe microscopy systems are based on the interaction between a submicroscopic probe and the surface of the material under consideration; what differentiates various scanning probe microscopy technologies is the nature of the interaction and the mechanism by which the interaction is monitored. The scanning probe microscope is an offshoot of the scanning tunneling microscopes. The scanning probe microscope is used to study the surface properties of materials from the atomic to the micron level; it can also be used in the 3D imaging of structures, where it has a large dynamic range that encompasses the domains of both the optical and electron microscopes. The scanning probe microscope's probe is atomically sharp and it typically scans a material's surface at a distance of a few angstroms or nanometers. Not only does it allow probing under various conditions (such as air, gas, liquid, and vacuum), but it also permits the selective manipulation of single atoms on a solid surface; thus, it has the ability to measure many physical properties.

Scanning probe microscopes have no lenses. In lieu of a lense, a "probe" tip is brought in close proximity of the specimen surface, and the interaction of the tip with the region of the specimen immediately next to it is measured. The type of interaction measured effectively defines the type of scanning probe microscopy: when the interaction measured is the force between atoms at the end of the tip and atoms in the specimen, the technique is called *atomic force microscopy* (AFM); when the quantum mechanical tunneling current is measured, the technique is called *scanning tunneling microscopy* (STM). These two techniques, atomic force microscopy and scanning tunneling microscopy have been the progenitors of a variety of scanning probe microscopy techniques that have emerged in recent years [31].

Both scanning and electron microscopies are used to study the geometric structure of substances. Scanning tunneling microscopy can determine the distance between the probe and the surface under study by exploiting the quantum tunneling effect. Held near the surface of a material, the stylus probe generates an electric current into the surface; by making several passes over the surface, the location of the electron orbitals can be determined and the rate at which electrons tunnel quantum mechanically from the surface to the probe can be measured. With this information, a graphic of individual atoms can be formed; moving the probe up and down over the surface produces a 3D image. Scanning tunneling microscopy allows both the topographical and electrical properties of a material to be studied. Scanning tunneling microscopy also permits the manipulation of atoms on the material's surface. In scanning tunneling microscopy, electrons quantum mechanically "tunnel" between the tip and the surface of the sample. This tunneling process is sensitive to overlaps between the electronic wave functions of the tip and of the sample (the tunneling depends exponentially on the separation). The scanning tunneling microscope makes use of this marked sensitivity to distance. In actual use, the tip is scanned across the surface while a feedback circuit continuously adjusts the height of the tip above the sample to maintain a constant tunneling current. The recorded trajectory of the tip creates an image that maps the electronic wave functions at the surface, revealing the atomic landscape in crisp detail [244].

In another type of 3D atomic resolution microscopy, the atomic force microscope exploits various interatomic forces that occur when two objects are brought within nanometers of each other. Similar in operation to the scanning tunneling microscopy, the atomic force microscope also creates three-dimensional images. An atomic force microscope operates when the sharp silicon-tip probe contacts the surface, which causes a repulsive force, and/or when the probe is a few nanometers away, which results in an attractive force. Atomic force microscopy produces a topographic map of the sample as the probe moves over the sample surface. Unlike most other scanning probe microscopy technologies, atomic force microscopy is not dependent on the electrical conductivity of the product being scanned, therefore atomic force microscopy can be used in ambient air or in a liquid environment. The basic atomic force microscope is composed of a stylus-cantilever probe attached to the probe stage, a laser focused on the cantilever, a photodiode sensor (recording light reflected from the cantilever), a digital translator recorder, and, a data processor and monitor [447]. Atomic force microscopy differs from other scanning probe microscopy technologies because the probe makes physical (albeit gentle) contact with the sample. The crux of this technology is the probe, which is composed of a surface-contacting stylus attached to an elastic cantilever mounted on a probe stage. As the probe is dragged across the sample, the stylus moves up and down in response to surface features; this vertical movement is reflected in the bending of the cantilever, and the movement is measured as changes in the light intensity from a laser beam bouncing off the cantilever and recorded by a photodiode sensor. By optically monitoring the cantilever motion it is possible to detect extremely small chemical, electrostatic, or magnetic forces that are only a fraction of those required to break a single chemical bond or to change the direction of magnetization of a small magnetic grain [244]. The data from the photodiode is translated into digital form, processed by specialized software on a computer, and then visualized as a topological 3D shape [447].

Atomic force microscopy is currently the most widely used scanning probe microscopy technique [244]. The atomic force microscope can be used to investigate contact and hardness on the atomic scale. As discussed, it uses a feedback loop to control the distance between the sample and a probe tip at the end of a cantilever arm. Rather than a tunneling current, an atomic force microscope monitors an optical signal as feedback to measure the level of deflection. Thus, both attractive and repulsive interactions of the tip and sample can be monitored. As the microscope tip approaches the surface, attractive forces are first exerted on the tip by the surface and can be measured to as small a value as 10^{-9} N. Upon contact with the surface, further motion of the tip results in repulsive forces between the tip and the sample. This procedure is capable of producing loads that overlap the forces encountered in macroscopic mechanical measurements [448]. Atomic force microscope techniques are also used to fabricate microelectromechanical systems (MEMS) in order to build an array of many atomic force microscopy tips on a single chip.

Both scanning tunneling microscopes and atomic force microscopes can be used in resist-based lithographic processes. However, scanning tunneling microscopes are limited to the study of metal surfaces while atomic force microscopes can be used on both nonmetallic and metallic surfaces.

The expectation is that this discipline will continue to advance over time and be a valuable tool for nanoelectronics.

6.3.5 Microelectromechanical Systems and Microoptoelectromechanical Systems Applications

This section briefly looks at microelectromechanical systems (MEMS) and microoptoelectromechanical systems (MOEMS—optical MEMS).

The MEMS technique have been studied for a couple of decades now. Micromirror arrays can be utilized in a variety of areas from optical displays, scanners and communication switches, to maskless lithography and optical spectroscopy. Ink-jet printers represent a major use of micromachined integrated electromechanical systems. MEMS-based accelerometers/actuators, used as sensors for deploying automobile air bags, are also in wide use [449]. MEMS fabrication uses planar processing technologies (similar to technologies used in the manufacturing of electronic integrated circuits) to simultaneously "machine" large numbers of relatively simple mechanical devices in an integrated manner.

The MOEMS technology has made significant progress of late. MOEMS are being used in components for telecom equipments, in addition to the deployment into projection display systems and adaptive optics. Optical switching is possible with the aid of MEMS/MOEMS-based micromirrors, which deflect the input optical signal into desired output port directly [250]. In the conventional design of these arrays, mirrors are mounted on tiltable cantilevers. However, in high-frequency applications a phased-mirror approach would be more useful; also nonmechanical systems would be of interest [250]. Many of the devices in practical use today are made with silicon-based fabrication technology because of the well-developed methods created for use by the microelectronics industry [449]. Typical dimensions of MEMS devices are in the several micrometers to hundreds of micrometers range. Nanoelectromechanical systems (NEMS) are characterized by small dimensions, where the dimensions are relevant for the function of the devices. Critical feature sizes may be from hundreds to a few nanometers. New physical properties, resulting from the small dimensions, may dominate the operation of the devices, and new fabrication approaches may be required to make them [449].

The NEMS systems with dimensions in the "deep submicron" mostly operate in their resonant modes; in this size regime, NEMS enjoy extremely high resonance frequencies, diminished active masses, and acceptable force constants; the quality (Q) factors of resonance are in the range, significantly higher than those of electrical resonant circuits [450] (see Glossary for more information on Q). These attributes collectively make NEMS suitable for a variety of applications such as ultrafast actuators, sensors, and high-frequency signal-processing components.

According to some industry observers, NEMS are among the most promising manifestations of the emerging field of nanotechnology [450]. Mechanical devices are shrinking in thickness and width to reduce mass, increase resonant frequency, and lower the force constants of these systems. Advances in the field include improvements in fabrication processes and new methods for actuating and detecting motion at the nanoscale. Lithographic approaches are capable of creating freestanding objects in silicon and other materials, with thickness and lateral dimensions down to about 20 nm. Similar processes can make channels or pores of comparable dimensions, approaching the molecular scale. This allows access to a new experimental regime and suggests new applications in sensing and molecular interactions [449].

6.4 CONCLUSION

In this chapter we examined basic nanoelectronic technologies that are now emerging. These technologies promise to address eventual limits in miniaturization imposed by existing semiconductor-based approaches. Among other technologies that we highlighted in the chapter, the following appear to be the most promising for the medium term (5 to 10 years): Silicon nanoelectronics (e.g., double-gate MOSFET, nano-MOSFET); carbon nanoelectronics; single electron systems; resonant tunneling diodes; Josephson arrays; spintronics; and nanowires and nanocontacts. Practitioners should track developments in these areas for near-tem opportunities for applications in telecommunications and computing.

APPENDIX A

Historical Developments Related to Atomic Theory and Some Additional Perspectives

This appendix provides a brief historical perspective on developments in physics and chemistry and is based in part on [81].

The word *atom* is derived from the Greek word *atomos*, meaning *indivisible* (we now know that atoms are, indeed, "divisible"). Democritus (460–370 BC) advanced philosophically the theory that matter is composed of fundamentally "indivisible particles," called "atomos." It took about 2200 years for any further significant developments. While Isaac Newton in the 17th century thought that matter was comprised of particles, it was John Dalton who formally postulated in 1802–1803 that everything we see/know is made from atoms. Dalton's atomic theory was based on the following assumptions:

1. Each element is composed of very small particles called atoms.
2. All atoms of a given element are identical; the atoms of different elements are different and have different properties (including different masses).
3. Atoms of an element are not changed into different types of atoms by chemical reactions; atoms are neither created nor destroyed in chemical reactions.
4. Compounds are formed when atoms of more than one element combine; a given compound always has the same relative number and kind of atoms.

Dalton postulated that atoms are the *basic building blocks of matter*; they are the smallest units of an element (atoms are the smallest particle of an element that retains the chemical properties of that element). Furthermore he postulated that:

- An *element* is composed of only one kind of atom.
- In *compounds* the atoms of two or more elements combine in definite arrangements.

Nanotechnology Applications to Telecommunications and Networking, By Daniel Minoli

- *Mixtures* do not involve the specific interactions between elements found in compounds, and the elements that comprise the mixture can be of varying ratios.

Simple "laws" of chemical combination that were known at the time of Dalton were:

1. The *law of constant composition* (in a given compound the relative number and kind of atoms are constant)
2. The *law of conservation of mass* (the total mass of materials present after a chemical reaction is the same as the total mass before the reaction)

Dalton used these laws to derive another "law"—the *law of multiple proportions*, which states that if two elements can combine to form more than one compound, then the ratios of the relative masses of each element that can combine can be represented by characteristically small whole numbers.

Dmitri Mendeleev's first periodic table in 1869 helped establish the view, prevalent throughout the late-19th century, that matter was comprised of atoms.

Additional developments came with an understanding of the behavior of moving charge in a magnetic field:

- A charged particle moving though a magnetic field feels a force *perpendicular* to the plane described by the velocity vector and magnetic field vector.
- This force deflects the moving charged particle according to the "right-hand rule" (based on a positive charge).
- A negative charge will be deflected in the *opposite* direction.

Electrical discharge through partially evacuated tubes produce radiation. This radiation originates from the *negative* electrode, known as the cathode.

- "Rays" travel toward or are attracted to the positive electrode (anode).
- Rays are not directly visible but can be detected by their ability to cause other materials to glow, or fluoresce.
- Rays travel in a straight line.
- The path of the rays can be "bent" by the influence of magnetic or electrical fields.
- A metal plate in the path of the "cathode rays" acquires a negative charge.
- The cathode rays produced by cathodes of different materials appear to have the same properties.

These observations indicated that the cathode ray radiation was composed of *negatively* charged particles (now known as electrons).

Work by J. J. Thomson established that atoms are composed of light electrons and much heavier protons. Thompson (1897) measured the charge-to-mass ratio for a stream of electrons using a cathode ray tube apparatus at 1.76×10^8 C/g.

- Charged particle stream can be deflected by both an electric charge and by a magnetic field.
- An electric field can be used to compensate for the magnetic deflection—the resulting beam behaves as if it were neutral.
- The required current needed to "neutralize" the magnetic field indicates the charge of the beam.
- The loss of mass of the cathode indicated the "mass" of the stream of electrons.

Thompson determined the charge-to-mass ratio for the electron but was not able to determine the mass of the electron. Robert Millikan (1909) was able to successfully measure the charge on a *single* electron (the Milliken oil drop experiment). This value was determined to be 1.60×10^{-19} C.

Wilhelm Roentgen (1895) discovered that when cathode rays struck certain materials (e.g., copper) a different type of ray was emitted. This new type of ray, called the "X" ray, had the following properties:

- X-rays could pass unimpeded through many objects.
- They were unaffected by magnetic or electric fields.
- They produced an image on photographic plates (i.e., they interacted with silver emulsions like visible light).

Henri Becquerel (1896) was studying materials that would emit light after being exposed to sunlight (i.e., phosphorescent materials). The discovery by Roentgen made Becquerel wonder if the phosphorescent materials might also emit X-rays. Becquerel discovered that uranium-containing minerals produced X-ray radiation (i.e., high-energy photons). Marie and Pierre Curie set about to isolate the radioactive components in the uranium mineral.

Ernest Rutherford validated that the protons are concentrated in a compact nucleus. The nucleus was initially modeled as being composed of protons and confined electrons (this to explain the difference between nuclear charge and mass number), but later it was established that the nucleus is composed of protons and neutrons. Rutherford studied alpha (α) rays, beta (β) rays, and (γ) gamma rays, emitted by certain radioactive substances. He noticed that each behaved differently in response to an electric field:

- β rays were attracted to the anode.
- α rays were attracted to the cathode.
- γ rays were not affected by the electric field.

The α and β rays were composed of (charged) particles and the γ ray was high-energy radiation (photons) similar to X-rays. β Particles are high-speed electrons (charge $= -1$). α Particles are the positively charged core of the helium atom (charge $= +2$).

In 1900, Thompson advanced a model of the atom (also known as the "plum pudding" model of the atom) as follows:

- The atom consists of a sphere of positive charge within which was buried negatively charged electrons.

Rutherford model of the atom (1910) had the following highlights:

- Most of the mass of the atom, and all its positive charges, reside in a very small dense centrally located region called the "nucleus."
- Most of the total volume of the atom is empty space within which the negatively charged electrons move around the nucleus.

The nuclear model proposed by Rutherford conceives the atom as a heavy, positively charged nucleus, around which much lighter, negatively charged electrons circulate, much like planets in the solar system. This model is, however, completely unsustainable from the standpoint of classical electromagnetic theory because an accelerating electron (circular motion represents an acceleration) should radiate away its energy. In fact, a hydrogen atom should exist for no longer than 5×10^{-11} s, time enough for the electron's death spiral into the nucleus [451].

Rutherford (1919) discovered protons (positively charged particles in the nucleus), and Chadwick (1932) discovered neutrons (neutral charge particles in the nucleus).

Bohr considered an electron in a circular orbit of radius r around the proton. Using Newton's second law and other assumptions, Bohr showed that the allowed orbital radii are then given by:

$$r_n = n^2 a_0$$

where

$$a_0 \equiv \frac{\hbar^2}{me^2} = 5.29 \times 10^{-11}\ \mathrm{m} = 0.529\ \text{Å}$$

which is known as the Bohr radius. The corresponding energy is

$$E_n = -\frac{e^2}{2a_0 n^2} = -\frac{me^4}{2\hbar^2 n^2} \qquad n = 1, 2 \ldots$$

Rydberg's formula can now be deduced from the Bohr model. The Bohr model can be readily extended to hydrogenlike ions. De Broglie's proposal that electrons can have wavelike properties was inspired by the Bohr atomic model (this topic is treated in Appendix D). Wilson (1915) and Sommerfeld (1916) generalized Bohr's formula for the allowed orbitals. Hence, the Bohr model was an important first step in the historical development of qantum theory. It introduced the quantization of atomic energy levels and gave quantitative agreement with the atomic hydrogen spectrum. With the Sommerfeld–Wilson generalization, it accounted as well for the degeneracy of hydrogen energy levels. Nonetheless, it had flaws that required a new formulation [451].

Quantum theory was postulated around this time. The *Schrödinger equation* was developed by the Austrian physicist Erwin Schrödinger in 1925. This equation describes the time dependence of quantum mechanical systems; it is central to the theory of quantum mechanics, and it fulfills a role analogous to Newton's second law in classical mechanics.

Many of the subatomic particles (also known as subnuclear particles or elementary particles) were discovered in the 50 years that followed. As we noted in Chapters 2 and 3, subatomic particles are smaller than an atom; these particles include electrons, protons, and neutrons, as well as particles produced by radiative and scattering processes, such as photons, neutrinos, and muons (protons and neutrons are actually composite particles, made up of quarks). During the 1950s and 1960s, a relatively large number/variety of particles was identified through scattering experiments. To systemitize these composite particles, the standard model was developed during the 1970s. This model posits that large number of particles can be explained as combinations of a (relatively) small number of fundamental particles. The model is currently perceived to be a provisional theory (until a more comprehensive theory is developed), also because it appears that there may be some elementary particles that are not properly described by the model (such as graviton—the hypothetical particle that carries gravitational force).

All elementary particles are either fermions (named after Enrico Fermi) or bosons. Fermions are particles that form totally antisymmetric composite quantum states; they have half-integer spin (fermions are subject to the Pauli exclusion principle and obey Fermi–Dirac statistics). Examples of fermions include electrons, protons, neutrons, and quarks. Fermions are classified into two groups: leptons and quarks; see Table A.1. Focusing on matter, the elementary particles that make up matter are fermions and electrons.

TABLE A.1 Fermions

Leptons, Spin = $\frac{1}{2}$			Quarks, Spin = $\frac{1}{2}$		
Flavor	Mass (GeV/c^2)	Electric Charge	Flavor	Approx. Mass (GeV/c^2)	Electric Charge
ν_e electron neutrino	$<1 \times 10^{-8}$	0	u up	0.003	$\frac{2}{3}$
e electron	0.000511	−1	d down	0.006	$-\frac{1}{3}$
ν_μ muon neutrino	<0.0002	0	c charm	1.3	$\frac{2}{3}$
μ muon	0.106	−1	s strange	0.1	$-\frac{1}{3}$
ν_τ tau neutrino	<0.02	0	t top	175	$\frac{2}{3}$
τ tau	1.7771	−1	b bottom	4.3	$-\frac{1}{3}$

Leptons contain no quarks but are small irreducible particles; leptons include electrons, muons, tauons, and neutrinos. Hadrons, on the other hand, are particles composed of quarks; hadrons include mesons and baryons. Mesons are composed of a normal quark and an antiquark; they are not very stable and have half-lives on the order of nanoseconds. Baryons are composed of three quarks, usually of the up or down variety. Examples of baryons include:

Name	Quarks	Mass (GeV)
Proton	uud	0.938
Neutron	udd	0.940
Λ	uds	1.116
Σ^{+}	uus	1.189
Σ^{0}	uds	1.192
Σ^{-}	dds	1.197
Δ^{2+}	uuu	1.232
Ω^{-}	sss	1.672
Λ_c^{+}	udc	2.273

The three original quarks were up (u), down (d), and strange (s) (additional types are shown in Table A.1). Each quark is a *fermion*. It took some thinking to understand how three similar quarks could coexist in the same state within a baryon because the extension of the Pauli exclusion principle forbids this. The resolution was to propose that *each* quark has three different complementary *color* states that have to be combined to make the composite particle (meson or baryon); the composite particle is *colorless* (white) (in this context *color* has nothing to do with the wavelengths of visible light). Using this *quark model* with *gluon exchange* (gluons are *color changers*; they convert a quark from one color to another when emitted or absorbed), physicists are now able to describe the structure of hadrons. The theory became known as quantum chromodynamics (or QDC), by analogy with quantum electrodynamics (QED), except with the color (Greek *chromos*) focus in place of electric charge.

String theory has also been advanced as a modeling mechanism in particle physics. String theory is a theory of elementary particles based on the idea that the fundamental entities are not pointlike particles but finite lines (strings), or closed loops formed by strings, the strings one-dimensional curves with zero thickness and lengths (or loop diameters) of the order of the Planck length of 10^{-35} m [53]. String theory is a model with fundamental building blocks consisting of one-dimensional extended objects (strings); this is in contrast with the zero-dimensional points (particles) that were the basis of most earlier physics. It follows that string theories are able to avoid problems associated with the presence of pointlike particles in a physical theory. String theories do not just describe strings but other objects as well, including points, membranes, and higher-dimensional objects. String theory has not yet made the kind of predictions that would allow it to be experimentally tested.

This concludes our broad-brush description of the advancements in physics in the 19th and 20th centuries.

APPENDIX B

Brief Introduction to Hilbert Spaces

This appendix provides some basic mathematical formalism that is relevant to quantum theory.

The concept of *Hilbert space* is key in the mathematical formulation of quantum theory.[1] Hilbert spaces were named after David Hilbert, who worked on the concept in the context of integral equations.

Hilbert space is an *inner product space* (see below) that is complete with respect to the norm (see below) defined by the inner product. Hilbert spaces serve to clarify and generalize the concept of Fourier expansion and/or certain linear transformations such as the Fourier transform. In Fourier analysis one can express a given function as an infinite sum of multiples of specified base functions (e.g., sine and cosine terms). This can be studied more abstractly in the context of Hilbert spaces. Hilbert spaces is and infinite-dimensional *vector space*.

Inner product space is a vector space endowed with an inner product, scalar product, or dot product that allows one to talk about angles and lengths of vectors.

We now define a vector space (also called linear space). The concept of a vector space can be defined abstractly, but it is a generalization of the concept of geometrical vectors in three-dimensional space.

A set V is a vector space over a field F (typically the field of real or of complex numbers), if, given an operation *vector addition* defined in V (shown as $\mathbf{v} + \mathbf{w}$ for all $\mathbf{v}$, $\mathbf{w}$ in V), and an operation *scalar multiplication* in V (shown as $a^*\mathbf{v}$ for all $\mathbf{v}$ in V and a in F), the following properties hold for all a, b in F and $\mathbf{u}$, $\mathbf{v}$, and $\mathbf{w}$ in V:

1. V is closed under vector addition: $\mathbf{v} + \mathbf{w}$ belongs to V.
2. Associativity of vector addition in V: $\mathbf{u} + (\mathbf{v} + \mathbf{w}) = (\mathbf{u} + \mathbf{v}) + \mathbf{w}$.
3. Existence of an additive identity element in V: there exists an element $\mathbf{0}$ in V, such that for all elements $\mathbf{v}$ in V, $\mathbf{v} + \mathbf{0} = \mathrm{v}$.
4. Existence of additive inverses in V: for all $\mathbf{v}$ in V, there exists an element $-\mathbf{v}$ in V, such that $\mathbf{v} + (-\mathbf{v}) = 0$.

[1]Some material based on [452].

Nanotechnology Applications to Telecommunications and Networking, By Daniel Minoli

5. Commutativity of vector addition in V: $\mathbf{v}+\mathbf{w}=\mathbf{w}+\mathbf{v}$. (Actually this property can be shown to follow for the others in this list.)
6. V is closed under scalar multiplication: $a^*\mathbf{v}$ belongs to V.
7. Associativity of scalar multiplication in V: $a^*(b^*\mathbf{v})=(ab)^*\mathbf{v}$.
8. Neutrality of 1: if 1 denotes the multiplicative identity of the field F, then $1^*\mathbf{v}=\mathbf{v}$.
9. Distributivity with respect to vector addition: $a^*(\mathbf{v}+\mathbf{w})=a^*\mathbf{v}+a^*\mathbf{w}$.
10. Distributivity with respect to field addition: $(a+b)^*\mathbf{v}=a^*\mathbf{v}+b^*\mathbf{v}$.

The members of a vector space are called *vectors*. Properties 1 through 5 indicate that V is an abelian group under vector addition. Hence, to establish that a set V is a vector space, one must specify a field F, define vector addition and scalar multiplication in V, and determine if V satisfies the above properties over the field F. Examples include: the vector space $\mathbf{R}^n$, over $\mathbf{R}$, with component-wise operations; the set of all continuous real-valued functions on a closed interval; and so on. A group $(G, \#)$ is abelian if "#" is commutative, namely: $g\#h=h\#g$ for all g and h in G.

As stated, an inner product space is a vector space endowed with an inner product, scalar product, or dot product that allows one to talk about angles and lengths of vectors. Inner product spaces are generalizations of Euclidean space. More formally, an inner product space is a real (or complex) vector space V together with a map f: $V\times V\rightarrow F$ where F is either the field of real numbers ($\mathbf{R}$) or the field of complex numbers ($\mathbf{C}$). One often writres $\langle x, y\rangle$ in lieu of $f(x, y)$.

Next we define the inner product. The following axioms must be satisfied to have an inner product:

For all $x\in V$, $\langle x, x\rangle\geq 0$, and $\langle x, x\rangle=0$ if and only if $x=0$.
For all scalars a and for all $x, y, z\in V$, $\langle z, ax+y\rangle=a\langle z, x\rangle+\langle z, y\rangle$.
For all $x, y, z\in V$, $\langle x, y\rangle=\langle y, x\rangle^*$ [if $F=R$, then $\langle x, y\rangle=\langle y, x\rangle$].

where * represents complex conjugation. A function that satisfies to the second and third axioms is called a *sesquilinear operator*; a sesquilinear operator that is *positive* ($\langle x, x\rangle\geq 0$) is called a *semi-inner product*. A function that satisfies all three axioms is called an inner product (contrary to the definition provided here, some researchers require an inner product to be linear in the first and conjugate-linear in the second argument). The inner product allows one to perform many "geometrical" construction that are analogous to those that can be done in spaces of finite dimensions.

Every inner product $\langle .,. \rangle$ on a real (or complex) vector space H has a norm $\|\cdot\|$ defined as follows:

$$\|x\|=\sqrt{\langle x, x\rangle}$$

H is a Hilbert space if it is complete with respect to this norm. *Completeness* here means that any Cauchy sequence of elements of the space converges to an element in the space, in the sense that the norm of differences approaches zero. A Cauchy sequence (named after the French mathematician Augustin Louis Cauchy) is a sequence whose terms become arbitrarily close to each other as the index of the sequence increases. Generally speaking, the terms of the sequence are getting closer together in a manner that suggests that the sequence ought to have a limit (this, however, does not need to be the case). Formally, a sequence $y_1, y_2, y_3, \ldots$ in a metric space (M, d) is called a Cauchy sequence if for every positive real number r, there is an integer N such that for all integers m and n greater than N the distance $d(y_m, y_n)$ is less than r. All finite-dimensional inner product spaces (such as Euclidean space and the dot product operation) are Hilbert spaces.

The elements of Hilbert spaces are also called vectors; they are typically functions or sequences. Every Hilbert space has an *orthonormal basis* (see below), and, hence, by the very definition of a vector space, every element of the Hilbert space can be written in a unique way as a sum of multiples of these base elements. The infinite-dimensional Hilbert spaces are important in the application, particularly of quantum mechanics. Hilbert spaces are "well behaved" and are somewhat similar to finite-dimensional spaces.

In quantum mechanics, a physical system is described by a Hilbert space that contains the "wave functions" that define the possible states of the system (these are described in Appencides D and E).

A typical infinite-dimensional Hilbert space is the spaces $L^2([a, b])$ of square-Lebesgue-integrable functions over the interval [0,1]. The inner product of the two functions f and g is here given by:

$$\langle f, g \rangle = \int f(t)g(t)\, dt$$

Many other examples could be provided.

An orthonormal basis of a (Hilbert) vector space H is defined as follows: a subset B of H with the properties:

1. Every element of B has norm 1, namely, $\langle e, e \rangle = 1$ for all e in B.
2. Every two different elements of B are orthogonal: $\langle e, f \rangle = 0$ for all e, f in B with $e \neq f$.
3. The linear span of B is dense in H.

It follows that every element x of H can be derived as:

$$x = \sum_{b \in B} \langle b, x \rangle b$$

For example, the set {(1,0,0),(0,1,0),(0,0,1)} forms an orthonormal basis of $\mathbf{R}^3$. Another example is the set $\{f_n : n \text{ belongs } \mathbf{Z}\}$ with $f_n(x) = \exp(2\pi inx)$; this set forms an orthonormal basis of the complex space $L^2([0,1])$. Again, $L^2([0, 1])$ is the of square-Lebesgue-integrable functions in the interval [0,1].

Every Hilbert space admits an orthonormal basis (and any two orthonormal bases of the same space have the same cardinality.) A Hilbert space is separable if and only if it admits a countable orthonormal basis. It can be shown that all separable Hilbert spaces are isomorphic. Nearly all Hilbert spaces used in physics are separable; hence, when physisists talk about *the Hilbert space* they mean any separable Hilbert space.

For a Hilbert space H, the continuous linear operators $A: H \to H$ are of particular interest. Such a continuous operator is *bounded* in the sense that it maps bounded sets to bounded sets. This permits one to define its *norm* as:

$$||A|| = \sup_{||x|| \leq 1} ||Ax||$$

In quantum mechanics one also considers linear operators that need not be continuous and that need not be defined on the entire space H.

APPENDIX C

Reference Information

This appendix provides the electronic configuration of the elements, as well as some other information.

Electronic Configuration of the Elements

Num.	Symbol	K	L		M			N				O				P				Q	
1. Period		**1s**	**2s**	**2p**	**3s**	**3p**	**3d**	**4s**	**4p**	**4d**	**4f**	**5s**	**5p**	**5d**	**5f**	**6s**	**6p**	**6d**	**6f**	**7s**	**7p**
1	**H**	1																			
2	**He**	2																			
2. Period		**1s**	**2s**	**2p**	**3s**	**3p**	**3d**	**4s**	**4p**	**4d**	**4f**	**5s**	**5p**	**5d**	**5f**	**6s**	**6p**	**6d**	**6f**	**7s**	**7p**
3	**Li**	2	1																		
4	**Be**	2	2																		
5	**B**	2	2	1																	
6	**C**	2	2	2																	
7	**N**	2	2	3																	
8	**O**	2	2	4																	
9	**F**	2	2	5																	
10	**Ne**	2	2	6																	
3. Period		**1s**	**2s**	**2p**	**3s**	**3p**	**3d**	**4s**	**4p**	**4d**	**4f**	**5s**	**5p**	**5d**	**5f**	**6s**	**6p**	**6d**	**6f**	**7s**	**7p**
11	**Na**	2	2	6	1																
12	**Mg**	2	2	6	2																
13	**Al**	2	2	6	2	1															
14	**Si**	2	2	6	2	2															
15	**P**	2	2	6	2	3															
16	**S**	2	2	6	2	4															
17	**Cl**	2	2	6	2	5															
18	**Ar**	2	2	6	2	6															
4. Period		**1s**	**2s**	**2p**	**3s**	**3p**	**3d**	**4s**	**4p**	**4d**	**4f**	**5s**	**5p**	**5d**	**5f**	**6s**	**6p**	**6d**	**6f**	**7s**	
19	**K**	2	2	6	2	6	—	1													
20	**Ca**	2	2	6	2	6	—	2													
21	**Sc**	2	2	6	2	6	1	2													
22	**Ti**	2	2	6	2	6	2	2													
23	**V**	2	2	6	2	6	3	2													
24	**Cr**	2	2	6	2	6	5	1													

25	**Mn**	2	2	6	2	6	5	2													
26	**Fe**	2	2	6	2	6	6	2													
27	**Co**	2	2	6	2	6	7	2													
28	**Ni**	2	2	6	2	6	8	2													
29	**Cu**	2	2	6	2	6	10	1													
30	**Zn**	2	2	6	2	6	10	2													
31	**Ga**	2	2	6	2	6	10	2	1												
32	**Ge**	2	2	6	2	6	10	2	2												
33	**As**	2	2	6	2	6	10	2	3												
34	**Se**	2	2	6	2	6	10	2	4												
35	**Br**	2	2	6	2	6	10	2	5												
36	**Kr**	2	2	6	2	6	10	2	6												
5. Period		**1s**	**2s**	**2p**	**3s**	**3p**	**3d**	**4s**	**4p**	**4d**	**4f**	**5s**	**5p**	**5d**	**5f**	**6s**	**6p**	**6d**	**6f**	**7s**	**7p**
37	**Rb**	2	2	6	2	6	10	2	6	—	—	1									
38	**Sr**	2	2	6	2	6	10	2	6	—	—	2									
39	**Y**	2	2	6	2	6	10	2	6	1	—	2									
40	**Zr**	2	2	6	2	6	10	2	6	2	—	2									
41	**Nb**	2	2	6	2	6	10	2	6	4	—	1									
42	**Mo**	2	2	6	2	6	10	2	6	5	—	1									
43	**Tc**	2	2	6	2	6	10	2	6	6	—	1									
44	**Ru**	2	2	6	2	6	10	2	6	7	—	1									
45	**Rh**	2	2	6	2	6	10	2	6	8	—	1									
46	**Pd**	2	2	6	2	6	10	2	6	10	—	—									
47	**Ag**	2	2	6	2	6	10	2	6	10	—	1									
48	**Cd**	2	2	6	2	6	10	2	6	10	—	2									
49	**In**	2	2	6	2	6	10	2	6	10	—	2									
50	**Sn**	2	2	6	2	6	10	2	6	10	—	2									
51	**Sb**	2	2	6	2	6	10	2	6	10	—	2	3								
52	**Te**	2	2	6	2	6	10	2	6	10	—	2	4								
53	**I**	2	2	6	2	6	10	2	6	10	—	2	5								
54	**Xe**	2	2	6	2	6	10	2	6	10	—	2	6								

Electronic Configuration of the Elements *(Continued)*

Num.	Symbol	K	L		M			N				O				P				Q	
6. Period		**1s**	**2s**	**2p**	**3s**	**3p**	**3d**	**4s**	**4p**	**4d**	**4f**	**5s**	**5p**	**5d**	**5f**	**6s**	**6p**	**6d**	**6f**	**7s**	**7p**
55	**Cs**	2	2	6	2	6	10	2	6	10	—	2	6	—	—	1					
56	**Ba**	2	2	6	2	6	10	2	6	10	—	2	6	—	—	2					
57	**La**	2	2	6	2	6	10	2	6	10	—	2	6	1	—	2					
58	**Ce**	2	2	6	2	6	10	2	6	10	2	2	6	—	—	2					
59	**Pr**	2	2	6	2	6	10	2	6	10	3	2	6	—	—	2					
60	**Nd**	2	2	6	2	6	10	2	6	10	4	2	6	—	—	2					
61	**Pm**	2	2	6	2	6	10	2	6	10	5	2	6	—	—	2					
62	**Sm**	2	2	6	2	6	10	2	6	10	6	2	6	—	—	2					
63	**Eu**	2	2	6	2	6	10	2	6	10	7	2	6	—	—	2					
64	**Gd**	2	2	6	2	6	10	2	6	10	7	2	6	1	—	2					
65	**Tb**	2	2	6	2	6	10	2	6	10	9	2	6	—	—	2					
66	**Dy**	2	2	6	2	6	10	2	6	10	10	2	6	—	—	2					
67	**Ho**	2	2	6	2	6	10	2	6	10	11	2	6	—	—	2					
68	**Er**	2	2	6	2	6	10	2	6	10	12	2	6	—	—	2					
69	**Tm**	2	2	6	2	6	10	2	6	10	13	2	6	—	—	2					
70	**Yb**	2	2	6	2	6	10	2	6	10	14	2	6	—	—	2					
71	**Lu**	2	2	6	2	6	10	2	6	10	14	2	6	1	—	2					
72	**Hf**	2	2	6	2	6	10	2	6	10	14	2	6	2	—	2					
73	**Ta**	2	2	6	2	6	10	2	6	10	14	2	6	3	—	2					
74	**W**	2	2	6	2	6	10	2	6	10	14	2	6	4	—	2					
75	**Re**	2	2	6	2	6	10	2	6	10	14	2	6	5	—	2					
76	**Os**	2	2	6	2	6	10	2	6	10	14	2	6	6	—	2					
77	**Ir**	2	2	6	2	6	10	2	6	10	14	2	6	7	—	2					
78	**Pt**	2	2	6	2	6	10	2	6	10	14	2	6	9	—	1					
79	**Au**	2	2	6	2	6	10	2	6	10	14	2	6	10	—	1					
80	**Hg**	2	2	6	2	6	10	2	6	10	14	2	6	10	—	2					
81	**Tl**	2	2	6	2	6	10	2	6	10	14	2	6	10	—	2	1				
82	**Pb**	2	2	6	2	6	10	2	6	10	14	2	6	10	—	2	2				
83	**Bi**	2	2	6	2	6	10	2	6	10	14	2	6	10	—	2	3				

84	Po	2	2	6	2	6	10	2	6	10	14	2	6	10	—	2	4				
85	At	2	2	6	2	6	10	2	6	10	14	2	6	10	—	2	5				
86	Rn	2	2	6	2	6	10	2	6	10	14	2	6	10	—	2	6				
7. Period		**1s**	**2s**	**2p**	**3s**	**3p**	**3d**	**4s**	**4p**	**4d**	**4f**	**5s**	**5p**	**5d**	**5f**	**6s**	**6p**	**6d**	**6f**	**7s**	**7p**
87	Fr	2	2	6	2	6	10	2	6	10	14	2	6	10	—	2	6	—	—	1	
88	Ra	2	2	6	2	6	10	2	6	10	14	2	6	10	—	2	6	—	—	2	
89	Ac	2	2	6	2	6	10	2	6	10	14	2	6	10	—	2	6	1	—	2	
90	Th	2	2	6	2	6	10	2	6	10	14	2	6	10	—	2	6	2	—	2	
91	Pa	2	2	6	2	6	10	2	6	10	14	2	6	10	2	2	6	1	—	2	
92	U	2	2	6	2	6	10	2	6	10	14	2	6	10	3	2	6	1	—	2	
93	Np	2	2	6	2	6	10	2	6	10	14	2	6	10	4	2	6	1	—	2	
94	Pu	2	2	6	2	6	10	2	6	10	14	2	6	10	6	2	6	—	—	2	
95	Am	2	2	6	2	6	10	2	6	10	14	2	6	10	7	2	6	—	—	2	
96	Cm	2	2	6	2	6	10	2	6	10	14	2	6	10	7	2	6	1	—	2	
97	Bk	2	2	6	2	6	10	2	6	10	14	2	6	10	9	2	6	—	—	2	
98	Cf	2	2	6	2	6	10	2	6	10	14	2	6	10	10	2	6	—	—	2	
99	Es	2	2	6	2	6	10	2	6	10	14	2	6	10	11	2	6	—	—	2	
100	Fm	2	2	6	2	6	10	2	6	10	14	2	6	10	12	2	6	—	—	2	
101	Md	2	2	6	2	6	10	2	6	10	14	2	6	10	13	2	6	—	—	2	
102	No	2	2	6	2	6	10	2	6	10	14	2	6	10	14	2	6	—	—	2	
103	Lr	2	2	6	2	6	10	2	6	10	14	2	6	10	14	2	6	1	—	2	
104	Rf	2	2	6	2	6	10	2	6	10	14	2	6	10	14	2	6	2	—	2	
105	Db	2	2	6	2	6	10	2	6	10	14	2	6	10	14	2	6	3	—	2	
106	Sg	2	2	6	2	6	10	2	6	10	14	2	6	10	14	2	6	4	—	2	
107	Bh	2	2	6	2	6	10	2	6	10	14	2	6	10	14	2	6	5	—	2	
108	Hs	2	2	6	2	6	10	2	6	10	14	2	6	10	14	2	6	6	—	2	
109	Mt	2	2	6	2	6	10	2	6	10	14	2	6	10	14	2	6	7	—	2	
110	Uun	2	2	6	2	6	10	2	6	10	14	2	6	10	14	2	6	9	—	1	
111	Uuu	2	2	6	2	6	10	2	6	10	14	2	6	10	14	2	6	10	—	1	
112	Uub	2	2	6	2	6	10	2	6	10	14	2	6	10	14	2	6	10	—	2	
114	Uuq	2	2	6	2	6	10	2	6	10	14	2	6	10	14	2	6	10	—	2	2

Some Fundamental Physical Constants

Quantity	Unit	Value 10
Absolute zero	°C	−273.16
Avogadro constant	$1\ mol^{-1}$	6.0221367×10^{23}
Bohr magneton	$J\,T^{-1}$	$9.2740154 \times 10^{-24}$
Bohr radius	m	$5.29177249 \times 10^{-11}$
Boltzmann constant	$J\,K^{-1}$	1.380658×10^{-23}
Classical electron radius	m	$2.81794092 \times 10^{-15}$
Dirac's constant	J s	$1.0545887 \times 10^{-34}$
Elementary charge	C	$1.60217733 \times 10^{-19}$
Electron mass	kg	$9.1093897 \times 10^{-31}$
Electron–proton mass ratio		$5.44617013 \times 10^{-4}$
Electronvolt	J	$1.6021892 \times 10^{-19}$
Faraday constant	$C\,mol^{-1}$	96485.309
Feigenbaum's constant		4.669210609102990
Fine-structure constant		$7.29735308 \times 10^{-3}$
First radiation constant	$W\,m^{2}$	$3.7417749 \times 10^{-16}$
Newtonian constant of gravitation	$N\,m^{2}\,kg^{-2}$	6.67259×10^{-11}
Hydrogen Rydberg number	m^{-1}	1.0967758×10^{7}
Josephson frequency–voltage quotient	$Hz\,V^{-1}$	4.8359767×10^{14}
Loschmidt constant	m^{-3}	2.686763×10^{25}
Magnetic flux quantum	Wb	$2.06783461 \times 10^{-15}$
Molar gas constant	$J\,mol^{-1}\,K^{-1}$	8.314510
Molar Planck constant	$J\,s\,mol^{-1}$	$3.99031323 \times 10^{-10}$
Molar volume of ideal gas	$m^{3}\,mol^{-1}$	0.02241410
Nuclear magneton	$J\,T^{-1}$	$5.0507866 \times 10^{-27}$
Permeability of vacuum	$N\,A^{-2}$	$12.566370614 \times 10^{-7}$
Permittivity of vacuum	$F\,m^{-1}$	$8.854187817 \times 10^{-12}$
Planck constants	J s	$6.6260755 \times 10^{-34}$
Rydberg constant	m^{-1}	10973731.534
Second radiation constant	m K	0.01438769
Speed of light in vacuum	$m\,s^{-1}$	299792458
Stefan–Boltzmann constant	$W\,m^{-2}\,K^{-4}$	5.67051×10^{-8}

Energy Conversion Tables

Multiply # of to obtain # of	by by	to obtain # of Divide # of
Btu	1.0548×10^3	joules (absolute)
Btu	0.25198	kg-cal
Btu	1.0548×10^{10}	ergs
Btu	2.930×10^{-4}	kW-h
Btu/lb	0.556	g-cal/g
eV	1.6021×10^{-12}	ergs
eV	1.6021×10^{-19}	joules (abs)
eV	10^{-3}	keV
eV	10^{-6}	MeV
ergs	10^{-7}	joules (abs)
ergs	6.2418×10^5	MeV
ergs	6.2418×10^{11}	eV
ergs	1.0	dyne-cm
ergs	9.480×10^{-11}	Btu
ergs	7.375×10^{-8}	ft-lb
ergs	2.390×10^{-8}	g-cal
ergs	1.020×10^{-3}	g-cm
g-calories	3.968×10^{-3}	Btu
g-calories	4.186×10^7	ergs
joules (abs). Joule is the unit of work and energy. It is equal to the work done when the point of application of a force of one newton moves in the direction of the force, a distance of one meter.	10^7	ergs
joules (abs)	0.7376	t-lb
joules (abs)	9.480×10^{-4}	Btu
g-cal/g	1.8	Btu/lb
kg-cal	3.968	Btu
kg-cal	3.087×10^3	ft-lb
ft-lb	1.356	joules (abs)
ft-lb	3.239×10^{-4}	kg-cal
kW-h	2.247×10^{19}	MeV
kW-h	3.60×10^{13}	ergs
MeV	1.6021×10^{-6}	ergs

APPENDIX D

Basic Nanotechnology Science—Quantum Physics

We noted in Chapter 2 that one is interested to determine how the electrons are arranged when bound to nuclei to form atoms and molecules. These arrangements ultimately determine the shape and properties of atoms and molecules, and in turn, the shape and properties of materials, including materials of interest in nanoscience. While empirical methods have evolved within the field of chemistry, a more well-rooted theoretical model is desirable; this model is provided by quantum theory. From a scientific point of view, there is no choice but to deal with quantum principles when operating at the atomic and/or nanoscale. Hence, a basic treatment of this topic is provided in this appendix. In our treatment, we alternate between general theory and specific instances (applications) in order to develop a grounding for the discussion. From a nomenclature perspective, we use the term *quantum theory* to encompass all aspects of the discipline; however, in the material that follows, the term *quantum mechanics* could also be used almost interchangeably.

For the casual reader the takeaway from this appendix should be a general appreciation of quantum theory principles. The general reader should not necessarily feel obligated to acquire an in-depth knowledge of this science, but, rather, just the general outline of the mechanics involved.

D.1 PHYSICS DEVELOPMENTS LEADING TO A QUANTUM MODEL

Nanotechnology is concerned with materials and systems whose structures and components exhibit novel and significantly improved physical, chemical, and biological properties, phenomena, and processes due to their nanoscale size [2, 33]. The nanoscale requires operation in the quantum theory realm, which we describe in the sections that follow in some general terms.

Nanotechnology Applications to Telecommunications and Networking, By Daniel Minoli

D.1.1 Experimental Highlights

As hinted in Appendix A, early 20th century attempts to account for the stability of an atomic system of elementary particles using Newtonian mechanics ran into logical obstacles: With this modeling of the atom, all of the electrons would spiral into the nucleus giving rise to emission of light. In turn, it would follow that all matter would collapse to a much smaller volume, the volume occupied by the nuclei; clearly, this is obviously not the case in the real world. Other "complications" also were encountered: Experimental work on electrons supports a perspective that an electron is a particle with a small mass (which we identified in Chapter 2); furthermore, the trajectory of the electron can be detected in a cloud chamber. But in the early part of the 20th century a number of experiments were undertaken that could be interpreted by classical mechanics only if one assumed that electrons possessed a wave motion; for example, a beam of electrons, when passed through a suitable grating, gives a diffraction pattern similar to that obtained in diffraction experiments with light.

Ultimately, these efforts lead to the formulation of quantum theory. A tenet of quantum theory is that certain quantities (e.g., energy, angular momentum, light) can only exist in definable discrete amounts, called *quanta*. The theory is used to describe physical systems that are of atomic dimensions or less. Initially, the theory was developed by Max Planck to explain that radiating bodies emit energy not in a continuous stream but in discrete units (quanta) (the energy being directly proportional to the frequency). Soon the theory was expanded to provide rules with which one can (probabilistically) calculate and predict how matter behaves. Once the system of interest is defined and the interactions among the particles of the system are described, the quantum theory equations are solved to quantify properties of the system [38].

Let us look at electrons passed through a grating. Based on diffraction experiments, in 1923 physicist de Broglie reasoned pragmatically that a relationship should exist between the "particle" and "wave" properties for light: If light is a stream of particles, they must possess momentum. At the same time, de Broglie reasoned that if light is a wave, then it possesses a characteristic frequency ν with wavelength λ, and he then derived the following relationship that is known by his name:

$$\lambda = h/\text{momentum}$$

de Broglie also reasoned that light and electrons might behave in the same way (in 1927 researchers carried out experiments that supported de Broglie's prediction). Consequently, a beam of electrons (with each electron of mass m and with a velocity v, and so, with a momentum mv), should exhibit diffraction effects with an apparent wavelength:

$$\lambda = \frac{h}{mv}$$

Now, let us look at light. Not surprisingly, light can be studied from the viewpoint of waves; for example, diffraction can be easily understood from the perspective of wave motion. However, a number of experiments were undertaken in the early part of the 20th century that could be interpreted by classical mechanics only if it was assumed that light was composed of a stream of particles. Specifically, certain metals emit electrons when they are exposed to a source of light; this is called the photoelectric effect. One finds that [59, 60]:

1. The number of electrons released from the surface increases as the intensity of the light is increased, but the energies of the emitted electrons are independent of the intensity of the light. This implies that light cannot be a wave motion in the classical sense because when light "waves" strike a substance only the number of emitted electrons increases as the intensity is increased; the energy of the most energetic electrons remains constant. This can be explained only if it is assumed that the energy in a beam of light is not transmitted in the manner characteristic of a wave, but rather that the energy comes in discrete bundles or packets (a packet of light energy is called a photon) and that the size of the packet is determined by the frequency of the light.
2. No electrons are emitted from the surface of the metal unless the frequency of the light is greater than a certain minimum value. When electrons are ejected from the surface, they exhibit a range of velocities, from zero up to some maximum value. The energy ε of the electrons with the maximum velocity is found to increase linearly with an increase in the frequency ν of the incident light, namely,

$$\varepsilon = h\nu$$

where the constant $h = 6.62 \times 10^{-27}$ erg s is called Planck's constant.[1]

This equation turned out to be revolutionary to the field of physics because it implies that the energy of a specified frequency of light cannot be varied continuously (as would be the case classically), but, rather, that the energy is available only in discretized quantities. The energy of light is said to be *quantized*, and a photon is one quantum of energy (the constant h determines the size of the light quantum). Planck is credited with postulating in 1900–1901 that energy is not a continuously variable quantity but occurs only in discretized quantities. To the modern computer scientist (or even just to a number theorist), the concept of discretized quantities is not revolutionary at all, but to the physicist of the early part of the 20th century, it was.

A substance in gaseous form emits light when an electrical discharge is passed through it (think of a neon sign); the atoms serve to transform electrical energy into the energy of light. This happens because the electrons flowing through the gas transfer a portion of their energy to the electrons of the atoms that comprise the gas, and when the electrons lose this added energy and return to their normal state, the excess energy is emitted in the form of light. However, when the emitted light is passed through a prism (or a diffraction grating) to separate the light, only certain wavelengths appear in the spectrum: a "line" spectrum rather than a continuous spectrum is obtained when

[1]The quantity can also be expressed as 6.6×10^{-34} Js (one converts from erg s to Js by multiplying by 10^{-7}). Some also use $\hbar = h/2\pi$.

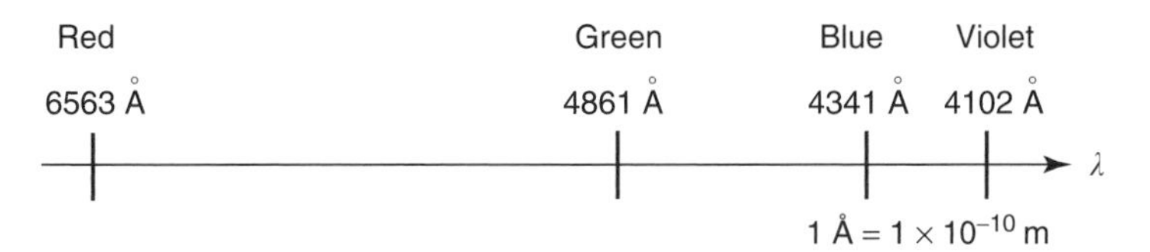

FIGURE D.1 Spectrum of H atoms.

atomic electrons are excited by an electrical discharge (e.g., see Figure D.1, which illustrates the visible spectrum observed for the hydrogen atom). If all energies were permitted for an electron bound to an atom, then all wavelengths (frequencies) should appear in the corresponding spectrum, that is, a continuous spectrum should be observed. The fact that only certain lines appear implies that only certain values for the energy of the electron are permitted. In fact, Rydberg (1890) found that all the lines of the atomic hydrogen spectrum could be described by a single formula:

$$\frac{1}{\lambda} = R\left(\frac{1}{n_1^2} - \frac{1}{n_2^2}\right) \qquad n_1 = 1, 2, 3, \ldots, \quad n_2 > n_1$$

where R, known as the Rydberg constant, has the value $109{,}677\,\mathrm{cm}^{-1}$ for hydrogen.[2] It was Bohr who discovered in 1913 that atomic line spectra could be accounted for if one assumed that the energy of the electron bound to an atom is quantized.

This discussion shows the apparent parallelism between the properties of light and electrons: Both exhibit the wave–particle dualism and the energies of both are quantized. The bottom line of these discoveries is that light behaves in a different way from ordinary particles and/or waves and it requires a special description. At this point in time, we now understand that the bottom line is that "particle" and "wave" are concepts inherited from classical mechanics and that such views lead to logical paradoxes. The conclusion, therefore, is reached that the classical equations of motion are inapplicable to the electron and other subatomic particles; furthermore, a dual description must also be abandoned in favor of a "better" theory.

Another issue comes into play. When attempting to observe an object as small as an electron, we must consider the interaction of an individual photon with an individual electron. It is found experimentally that when a photon is scattered by an electron, the frequency of the emergent photon is lower than it was before the scattering (this is known as the Compton effect). This crystallizes an important point regarding observations at the atomic level: *One cannot make an observation on an entity without at the same time disturbing the entity*; the interaction of the observer with the system he or she is observing can be ignored in classical mechanics where the masses are large, but this is not the case at the atomic level. The information regarding both the position and the momentum of an electron cannot be obtained with unlimited accuracy.

[2]Other elements (even compounds) have line spectra, which can be used as a "fingerprint" to identify the element by spectroscopists; however, no atom other than hydrogen has a simple relation analogous to the Rydberg equation for its spectral frequencies.

Building on all of these results, in 1926 Schrödinger formulated an equation whose role in solving problems in atomic physics parallels the role enjoyed by Newton's equation. It turns out that Schrödinger's equation correctly predict all physical behavior, including the ones we highlighted above. Quantization emerges directly from Schrödinger's equation. Schrödinger's equation forms the basis of quantum theory. It should be noted, however, that only the energies of bound (constrained) systems are quantized; the energies in unconstrained systems do not show quantization.

Below we provide a basic introduction of quantum theory: We discuss the behavior of electrons in atoms and molecules as predicted and interpreted by quantum theory (the section that follows is intended for readers that have interest in advanced concepts).

D.1.2 Basic Mechanisms

Standing Waves in a String Constrained at Both Ends

The properties of waves in a (one-dinensional) string constrained at both ends provide insight on some of the important basic quantum theory properties of atoms and molecules. By empirical observation, waves in a string constrained at both ends exhibit discrete wavelengths:

$$\lambda = 2L/n \qquad n = 1, 2, 3, \ldots$$

because the ends of the string cannot be displaced. Define Ψ as the function that describes the displacement at any point along the wave; the *solutions* to the wave equation, Ψ, are called the wave functions. It follows, by definition, that the waves that are generated by plucking such string constrained at both ends can be described by solving the wave equation for Ψ. The classical mechanics wave equation in three dimensions is [453]

$$\nabla^2 \Psi(x, y, z) = -(2\pi/\lambda)^2 \Psi(x, y, z) \tag{D.1}$$

where ∇^2 is the Laplacian operator:

$$\nabla^2 \equiv \frac{\partial^2}{\partial x^2} + \frac{\partial^2}{\partial y^2} + \frac{\partial^2}{\partial z^2}$$

In this particular case of the string constrained at both ends, Ψ is one dimensional with:

$$\frac{\partial^2 \Psi(x)}{\partial x^2} = -\left(\frac{2\pi}{\lambda}\right)^2 \Psi(x)$$

and Ψ^2 is proportional to the energy density at any point along the wave.

Wavelike Particles (the de Broglie Wavelength)

As noted above, the hypothesis that particles have wavelike properties was developed by de Broglie. Also as noted above, Planck had demonstrated in 1900–1901 that electromagnetic radiation was emitted and absorbed in discrete quanta having energy proportional to the frequency of the radiation $E = h\nu$. Einstein showed in 1905 that the energy of a particle is $E = mc^2$. These expressions taken together ($h\nu = mc^2$) provide the relationship between a photon's momentum and frequency (or wavelength). We get $h\nu/c = mc = p$, or $\nu\lambda = c$, from which $h\nu/\nu\lambda = p$, and finally, $\lambda = h/p$.

This implies, at face value, that electromagnetic radiation has particlelike characteristics (momentum) in addition to its wavelike characteristics (wavelength, diffraction). In the mid-1920s, de Broglie postulated that matter also had a comparable dual nature and proposed that a wavelength can be associated with the momentum of any particle, not just photons. With this assumption one can express λ in terms of energy by reworking the above equations as (with V the potential energy):

$$\lambda = h/p = h/[2m(E - V)]^{1/2} \tag{D.2}$$

Schrödinger's Equation for a One-Electron Atom

The relation between classical waves and de Broglie's particle waves was made by Schrödinger. Schrödinger substituted the de Broglie wavelength for λ [Eq. (D.2)] in the classical wave equation given above [Eq. (D.1)] to adapt the situation to the particle waves. Hence, one obtains the Schrödinger equation as [453]:

$$\nabla^2\Psi = -(2\pi/\{h/[2m(E - V)]^{1/2}\})^2\,\Psi = -[8\pi^2 m/h^2\,(E - V)]\,\Psi$$

This equation can be rearranged in a series of algebraic steps to a more convenient form:

$$[(-h^2/8\pi^2 m)\nabla^2 + V]\Psi = E\Psi$$

and by defining the operator H (the Hamiltonian operator) as follows:

$$H \equiv [(-h^2/8\pi^2 m)\nabla^2 + V]$$

one obtains

$$H\Psi = E\Psi$$

where Ψ is called an eigenfunction and E an eigenvalue (see Glossary for definitions). An eigenfunction is a function such that when an operation is performed on it, the result is the same function times a constant; that constant is known as the eigenvalue. Notice that the left-hand side is an operator (operation) on Ψ. Hence, one cannot just drop the term from both sides as would be the case in simple algebra. H is the Hamiltonian linear operator acting on the state space. The Hamiltonian operator is Hermitian—a continuous linear operator H: $M \rightarrow M$ on a Hilbert space M is called Hermitian or self-adjoint if $(x,Hy) = (Hx,y)$ for all elements x and y of M; in this definition, the parentheses denote the inner product given on M.

The second derivative of Ψ is the rate of change of slope of Ψ at any given point, and, so, it describes the curvature of "wiggliness" of the function. One can deduce that

TABLE D.1 Nomenclature

Ψ: atomic/molecular wave functions
ψ: atomic/molecular spin orbitals
ϕ: one-electron wave functions used to create the ψ

the "wigglier" is Ψ at a given point in space, the greater is the kinetic energy of the electron at that point [453]. The solutions for E are discrete (quantized), having $n = 1, 2, 3, \ldots$ possible values; this is reminiscent of the $n = 1, 2, 3, \ldots$ wavelengths (given by $2L/n$) of a string constrained at both ends we described at the beginning of this section.

As it will become clearer in the sections that follow, Schrödinger's equation supports a probabilistic (rather than deterministic) view of the location of the atomic particles (such as electrons). Table D.1 provides some nomenclature used in the discussion that follows.

D.2 QUANTUM CONCEPTS

In the previous section we introduced the fundamental concepts of quantum theory. Here we expand on these concepts. This section is intended for readers that have interest in advanced concepts; others can skip it, if they so choose. As noted earlier, in its simplest form, quantum theory deals with the behavior of atomic particles in atoms. Because the energy that these particles can have takes on discrete rather than continuous values, their behavior, location, and trajectories are different than would otherwise be the case. At lengths in the range of atomic units, that is 0.52917×10^{-10} m (0.052917×10^{-9} m, or 5% of a nanometer/nanoscale), quantum effects will be evident.

Quantum theory is the underlying framework of many aspects of physics and chemistry. It describes the behavior of physical systems at ultrashort distances. Quantum theory is derived from a small set of basic principles and explains four types of phenomena that classical mechanics and classical electrodynamics is unable to account for: quantization, the uncertainty principle, wave–particle duality, and quantum entanglement.

Besides atoms, quantum theory can also be applied to molecules, which is of interest in nanotechnology. For example, the theory of atoms in molecules (AIM), invented by Richard Bader [59, 60], is gaining acceptance as an objective way to describe molecular structure and to predict key molecular properties using quantum principles. AIM uses the topology of derivatives of the electronic charge density (which is a scalar field in three-dimensional space and a "quantum mechanical observable"), to objectively and unambiguously explain the following (and more) about molecules [454]: (i) bonding (topology or "molecular structure"), (ii) bond types (covalent, ionic, etc.), (iii) bond orders, (iv) physical locations of "lone pairs," (v) physical locations of "unpaired electrons," (vi) electrophilic/nucleophilic reactive sites, and (vii) per-atom electron populations, magnetic properties, and so on. Specifically, AIM provides objective justification using quantum theory principles for the theory of orbitals, the theory of electron pairs, and the valence shell electron pair repulsion model of molecular geometry (in cases where these other theories differ from the theory of atoms in molecules, it is the latter that agrees with experiment [454]). This is in contrast to analyses based on molecular orbitals (MOs). MOs are not quantum mechanical observables. MOs are projections onto a (essentially arbitrary) set of basis functions; these basis functions are then, often, recombined by eye and/or selectively highlighted to fit the data [454]. Most of our treatment here, however, relates to simple atoms (not molecules).

D.2.1 Electron Density of Atoms

This subsection starts the discussion by looking at a "topological" electron density view of the atom. This is related to the electronic structure of atoms mentioned earlier. Then, it is argued that the atom defined in terms of the topology of the electron density is consistent with a quantum view.

Topological Perspective

As discussed in Chapter 2, matter is comprised of atoms. Electrons are one of the many particles in the atom and are of particular interest because they are involved in chemical reactions and, of course, also, make electricity possible. Electrons are distributed in space based on the attractive field exerted by the nucleus or nuclei. The nuclei act as point attractors immersed in a cloud of negative charge. The electron density $\rho(\mathbf{r})$ describes the manner in which the electronic charge is distributed throughout space in the proximity of the nucleus (or nuclei). The electron density is typically expressed in terms of the number of electronic charges per unit volume of space, e^{-}/V (V is usually expressed in atomic units of length cubed—a cube with a length of 0.52917×10^{-10} m, or a_0^3). It follows that one atomic unit of electron density equates to 6.7 electronic charges per cubic Ångstrom. $\rho(\mathbf{r})$ is a measurable quantity and, importantly, it determines the form and the appearance of matter. In the paragraphs that follow, the definition of an atom in terms of the topology of the electron density $\rho(\mathbf{r})$ is formulated.

For illustration purposes, Figure D.2 depicts the spatial distribution of $\rho(\mathbf{r})$ in the plane containing the two carbon and four hydrogen nuclei of the ethene molecule. This density is at a maximum value at the position of each nucleus and decreases rapidly away from these positions. As the reader should be able to infer from the diagram, there are five critical points in this figure [places where $\rho(\mathbf{r})$ has an extremum; that is, a point where $\nabla\rho(\mathbf{r}) = 0$, with $\nabla\rho(\mathbf{r})$ being the gradient]: There is a saddle found between C (carbon) nuclei and four saddle points located between adjacent H (hydrogen) nuclei. These critical points, called *bond critical points*, will be reexamined shortly.

One is interested not only in the density $\rho(\mathbf{r})$ but also the field that is obtained by following the trajectories traced out by the gradient vectors of the density. The gradient vector is a vector that points in the direction of maximum increase in the density. Starting at some point, one determines the gradient of $\rho(\mathbf{r})$; then one takes an infinitesimal imaginary step in the direction indicated by the gradient and recomputes the new gradient to determine, yet again, the new direction. By continuing this process, one plots out a trajectory of $\nabla\rho(\mathbf{r})$. A gradient vector graph is shown Figure D.3 for the

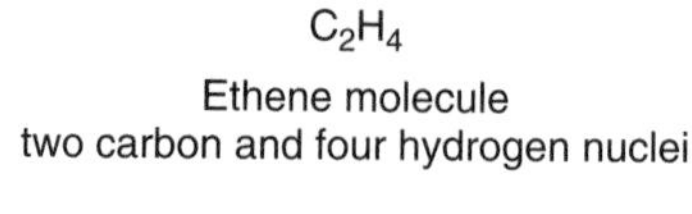

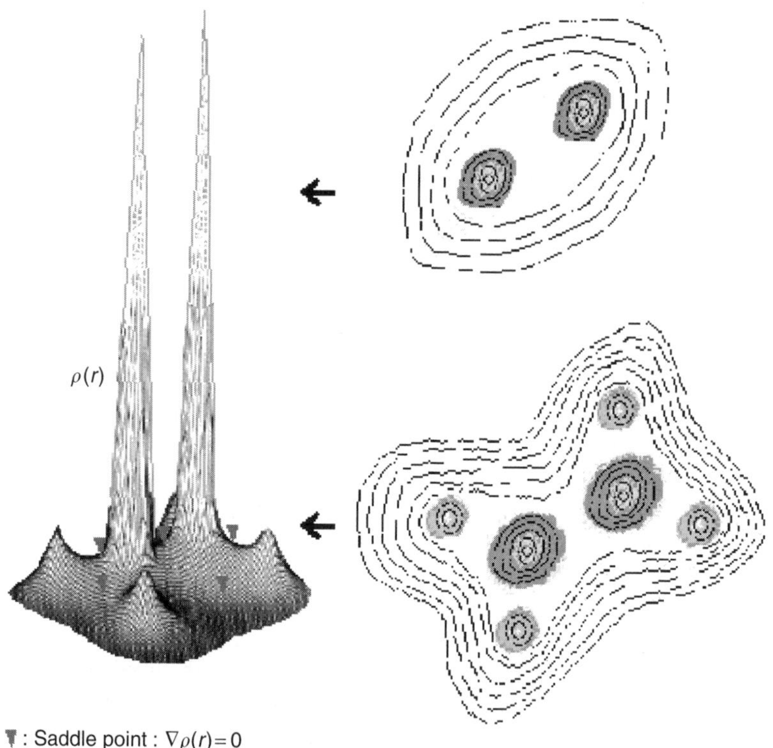

FIGURE D.2 Electron density ρ(*r*) of the ethene molecule (C_2H_4) in the plane containing the carbon and the hydrogen nuclei: (left) three-dimensional graph and (right) planes thru the graph at left.

same plane of the ethene molecule of Figure D.2; each component of the plot represents a trajectory traced out by the vector $\nabla\rho(\mathbf{r})$.

Before continuing, we further define for the reader the concept of gradient. Formally, a gradient is a vector-valued operator ∇ that acts on field ϕ that shows its rate and direction of change.[3] The gradient is represented by the nomenclature $\nabla\phi$ where ∇ is the vector differe operator del, and ϕ is a scalar function. In other words, given a scalar field, the gradient of the field is a vector field, where all vectors point toward the higher values; the magnitude equal to the rate of change of values.[4]

[3]The nomenclature grad(ϕ) is also used by some.

[4]The gradient does not always exist at all points; e.g., it may not exist at discontinuities or where the function or its partial derivative is undefined.

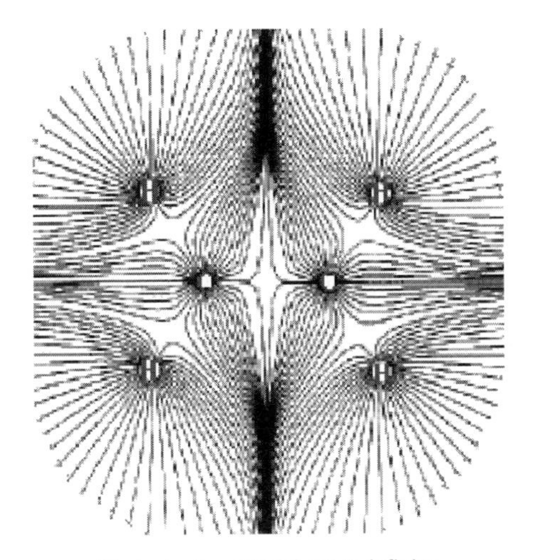

Trajectories extend out to infinity
but are not shown in the diagram

FIGURE D.3 Ethene (C_2H_4) gradient trajectories that terminate at the nuclei of the carbon and hydrogen atoms.

For example, consider a room with a three-dimensional Cartesian axis (x,y,z) anchored at a specified corner of the room. Consider light intensity at any point (x,y,z) as a scalar field $\phi(x,y,z)$. $\phi(x,y,z)$ is a number associated to each point vector (we assume a time-invariant situation). At any given point (x,y,z), the gradient ∇ is a vector that points in the direction of the greatest rate of change and has a magnitude equal to that rate. Figure D.4 illustrates this example but considers only a two-dimensional slice [assume, e.g., that $\phi(x,y)$ is the light intensity off a black-and-white photograph].

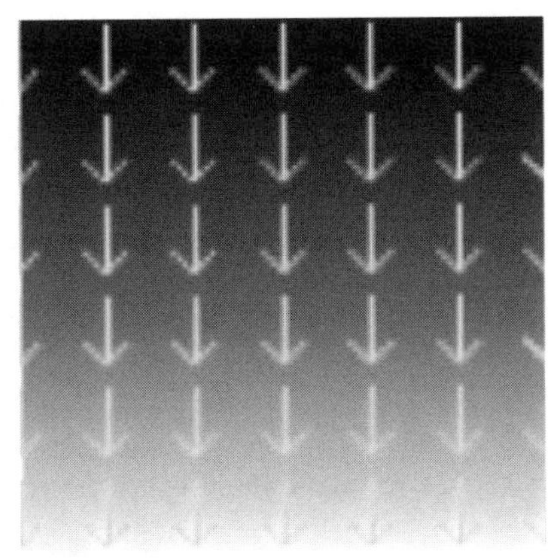

FIGURE D.4 Gradient.

In this example, the scalar field is the light intensity and the gradient is represented by the arrows. In three dimensions, in Cartesian coordinates, the expression expands to

$$\nabla\phi = \begin{pmatrix} \partial\phi/\partial x \\ \partial\phi/\partial y \\ \end{pmatrix}$$

Returning to the discussion of the atom, the nuclei are the *attractors* of the gradient vector field of the electron density $\rho(\mathbf{r})$, and the space of the molecule is partitioned into *basins*. By definition, a basin is the region of space traversed by the trajectories terminating at a given nucleus or attractor. Because a single attractor is associated with each basin, an *atom* can be defined as the union of an attractor and its basin [59, 60].

The next question of interest is "What is a bond?" Figure D.5 is a repeat of Figure D.3, but it includes the trajectories that originate at the *bond critical point*s we called out before $[\nabla\rho(\mathbf{r}) = 0]$ and terminate at an attractor or at infinity. As it can be seen in Figure D.5 (keeping in mind that the figure is only two dimensional), with each critical point there is an associated set of trajectories that start at infinity and terminate at the critical point (only two of which appear in the specific plane shown in the figure, but keeping in mind that there will be a set in each plane—Fig. D.6 attempts to illustrate the three-dimensional physical environment that exists). The trajectories define a surface (called *interatomic surface*) that separates the basins of neighboring atoms.

In the figures for ethane being discussed here, there is a unique pair of trajectories that originate at each critical point and terminate, one each, at the neighboring nuclei (in more complex molecules there could be more than just two neighboring nuclei). The unique pair of trajectories define a line through space. Along this line the electron density is a maximum. Keep in mind that there are two sets of trajectories associated with such a critical point, the set that terminates at the critical point

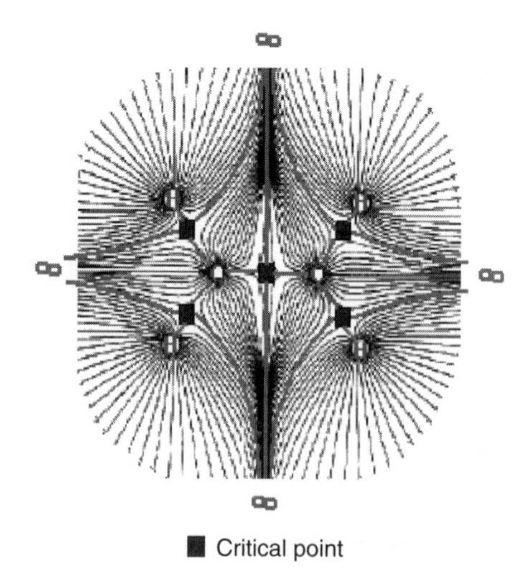

FIGURE D.5 Ethene (C_2H_4) gradient trajectories that originate at critical points and terminate at the attractors and/or infinity.

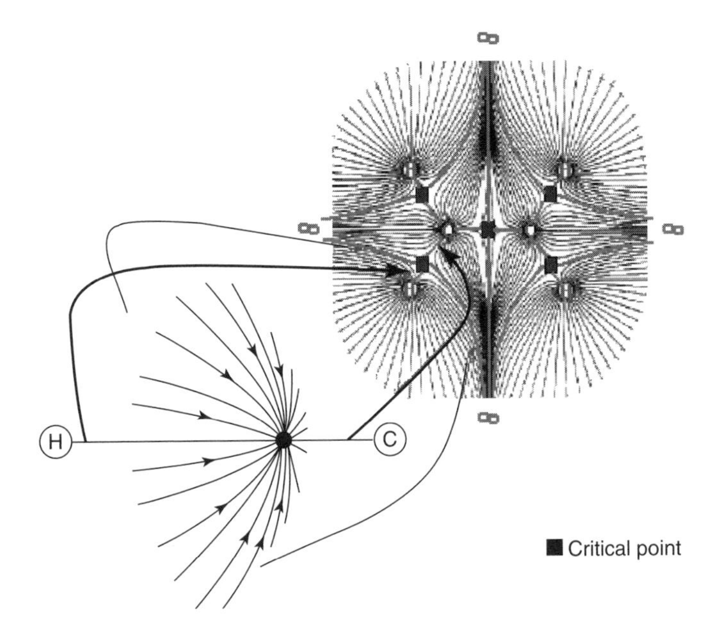

FIGURE D.6 Interatomic surface and bond critical points.

and defines the interatomic surface and the pair that originates there and defines the line of maximum density.

In an equilibrium environment, the line of maximum density is called a *bond path*. The set of bond paths for a given molecule is known as the *molecular graph*. The molecular graph accurately recovers the group of chemical bonds that are assigned on the basis of chemical considerations. Hence, a pair of bonded atoms are linked by a line along which the electron density, the glue of chemistry, is maximally concentrated. A molecular graph and the characteristics of the density at the bond critical points provide a concise summary of the bonding within a molecule or crystal [59, 60]. A small set of examples of molecular structures predicted by the molecular graphs determined by the electron density are shown in Figure D.7 for various carbon-based structures.

Observations shows that the molecular graph is subject to radical changes if the nuclei are somehow impacted via some kind of process akin to that in a chemical reaction (e.g., proximity with select elements along with some catalyst; abrupt or discontinuous changes in temperature, pressure, electrostatic status, etc.). When this occurs, the external agent breaks and/or makes certain of the bonds and alters a preexisting structure into another. With a theoretical machinery at hand, one can describe changes and properly predict outcomes.

Figure D.8, inspired by [59, 60], depicts an example, using the molecule C_6H_6. The top portion of the figure shows the molecular graph for C_6H_6 in a quiescent state, with the gradient vector field maps shown for the symmetry plane containing the C–C bond critical point and the three apical C atoms. When the separation between the two bridge-head nuclei is increased to a critical value via some (electrochemical) process, the bond critical point coalesces with the three critical points in the proximity, resulting in a singularity in $\rho(\mathbf{r})$, as shown in the middle portion of the figure. The singularity is

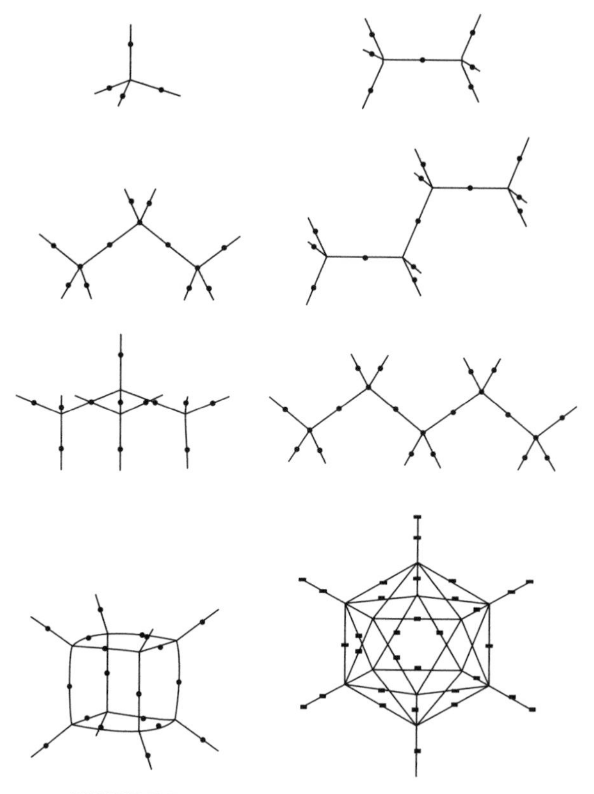

FIGURE D.7 Examples of molecular graphs.

unstable, representing the breaking of the C–C bridge-head bond. Additional separation of the nuclei (as shown in the bottom portion of the figure) causes the nuclei to form a new structure in which the bridge-head carbon atoms are not bonded to one another.

In the subsections that follow, the point is made that the *atom defined in terms of the topology of the electron density is consistent with a quantum view.* Physicists want to demonstrate that the above-suggested topological definitions and the quantum definitions of an atom coincide, so that the fundamental nature of the atom, as the building block of matter, is validated at the theoretical level.

Quantum Perspective

One is interested in what quantum theory predicts about the properties of electrons whose motions are confined to a relatively small space around the nucleus by the attractive force of the nucleus. One is also interested in extending the theory to cover molecules. As we have seen, quantum theory is based on Schrödinger's equation.

The mathematical basis of the quantum theory is in the form of a probability calculus expressed in terms of a Hilbert function space (we briefly discussed Hilbert function space in Appendix B). This approach is more general than earlier classical probability theories because not only does it yield the probabilities for matter to be

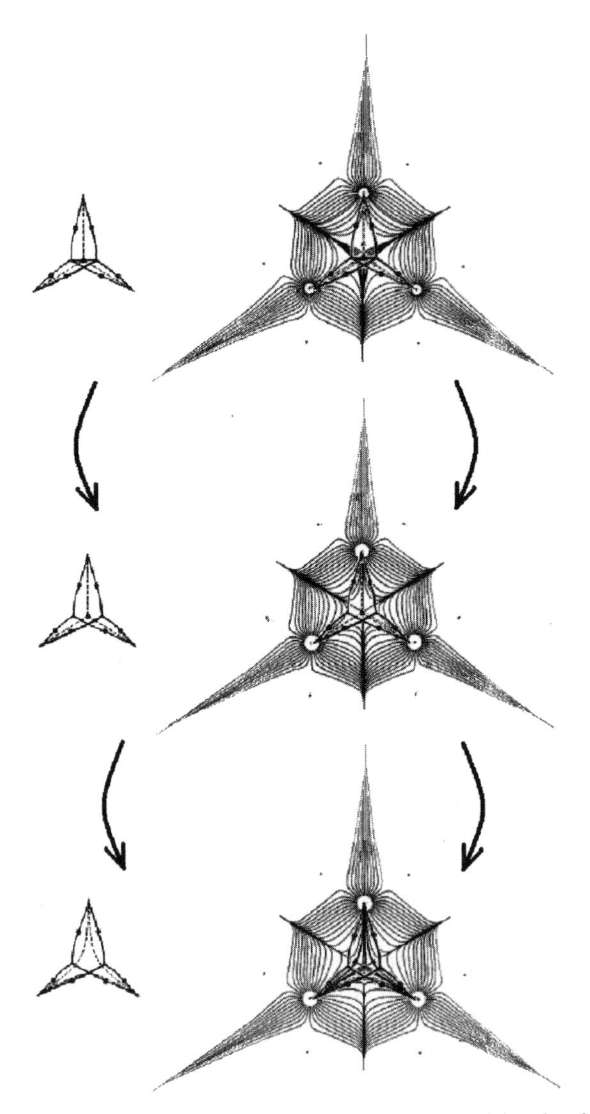

FIGURE D.8 Example depicting changes in structure induced by the dynamics of the nuclei.

in the (infinite multitude of) states of a microsystem, but it also yields the probabilities of transitions between these states. The probabilities of the material system to be (1) in particular states and (2) making transitions between them, are fundamental in the definition of the elementary particle of matter [455].

In classical mechanics the description of the dynamic state of *a particle* at a given time t is based on the specification of *six parameters*, the components of the position $\mathbf{r}(t)$ and linear momentum $\mathbf{p}(t)$ of the particle. All the *dynamical variables* (energy, angular momentum, etc.) *are determined* by the specification of $\mathbf{r}(t)$ and $\mathbf{p}(t)$. *Newton's laws* enable one to calculate $\mathbf{r}(t)$ through the solution of second-order differential equations with respect to time. Consequently, these laws fix the values of $\mathbf{r}(t)$ and $\mathbf{p}(t)$ for any time t when they are known for the initial time [456]. Quantum theory requires a more complex description machinery. In quantum theory, a state is described by the wave function (Ψ functions) also known as probability amplitude. The probability amplitude is a complex-number-valued function of position, that is, a quantity whose value is a definite complex number at any point in space. The probability of finding the particle described by the wave function (e.g., an electron in an atom) at that point is proportional to square of the absolute value of the probability amplitude. The "pictures" of orbitals in chemistry textbooks are a representation of the region within which the wave function gives a high probability of finding an electron for that state [457].

Schrödinger's equation addresses the following questions [458]:

- Where are the electrons/nuclei of an atom or molecule in space? This relates to configuration, conformation, size, shape, and the like.
- Under a given set of conditions, what are the energies of the electrons/nuclei of an atom or molecule? This relates to heat of formation, conformational stability, chemical reactivity, spectral properties, and so forth.

As noted, the *dynamic state of a particle*, at a given time, is characterised by a *wave function* $\Psi(\mathbf{r}, t)$ that contains all the information that is possible to obtain about the particle. The *state* no longer *depends* on six parameters, but *on an infinite number of parameters:* namely, *the values of* the wave function $\Psi(\mathbf{r}, t)$ *at all points* $\mathbf{r}$ in the coordinate space. For the classical concept of *trajectory* (the succession in time of the various states of the classical particle) one must substitute the idea of the *propagation of the wave associated with the particle.* $\Psi(\mathbf{r}, t)$ is interpreted as the *probability amplitude of the particles presence.* Abs$[\Psi(\mathbf{r}, t)]^2$ is interpreted as the probability density of the particle being, at time t, in a volume element d^3r situated at point $\mathbf{r}$. The equation describing the evolution of the wave function $\Psi(\mathbf{r}, t)$ is the *Schrödinger equation.* The *result of a measurement* of an arbitrary dynamic variable must belong to the set of the eigenvalues of the operator representing the dynamic variable. With each eigenvalue it is associated an eigenstate, the eigenfunction of the operator belonging to the particular eigenvalue. Recall that an eigenfunction is a function such that when an operation is performed on it, the result is the same function times a constant; that constant is known as the eigenvalue. If a measurement yields a particular eigenvalue, the corresponding eigenfunction is the wave function of the particle immediately after the measurement. The predictions of the measurement results are only probabilistic: they yield the probability of obtaining a given result in the measurement of a dynamical variable [456].

When a particle of mass m is subjected to the influence of a scalar potential $V(\mathbf{r}, t)$, the Hamiltonian operator of the particle in Schrödinger representation can be expressed as $H = T + V(\mathbf{r}, t)$, where T is the kinetic energy operator of the form [456]; that is,

$$T = \frac{p^2}{2m} = \frac{\hbar^2 k^2}{2m} = -\frac{\hbar^2}{2m}\nabla^2$$

with

$$\hbar = \frac{h}{2\pi}$$

The Schrödinger equation

$$H\Psi(\mathbf{r}, t) = i\hbar\frac{\partial\Psi(\mathbf{r}, t)}{\partial t}$$

that governs the time evolution of the physical system is of first order in t. Here i is the unit imaginary number ($i = \sqrt{-1}$) and H is the Hamiltonian linear operator acting on the state space (which is a Hilbert space.) From this it follows that, given the initial state $\Psi(\mathbf{r}, t)_0$, the final state $\Psi(\mathbf{r}, t)$ at any subsequent time t is determined. There is *no indeterminacy* in the time evolution of a quantum system: indeterminacy appears only when a physical quantity is measured, the state function then undergoing an unpredictable modification; however, between two measurements, the state function evolves in a deterministic manner in accordance with the equation just listed [456]. Also note that the Schrödinger equation is linear and homogenous; its solutions are linearly superposable; this leads to wave effects.

The Hamiltonian H, has two related meanings:

1. In classical mechanics, it is a function that describes the state of a mechanical system in terms of momentum and position variables. Using the Hamiltonian operator, one can develop a reformulation of classical mechanics known as Hamiltonian mechanics.
2. In quantum theory, the Hamiltonian refers to the observable (that which can be observed) related to the total energy of a system. (Since H is a Hermitian operator, the energy is always a real number.)

As noted earlier, Schrödinger's equation can also written as

$$H\Psi = E\Psi$$

in which electrons are considered as wavelike particles whose "waviness" is represented mathematically by a set of wave functions Ψ obtained by solving Schrödinger's equation. E is energy, and H is the Hamiltonian operator. The Hamiltonian operator describes the total energy of the system, but the exact form of H is not provided by the Schrödinger equation and must be independently determined based on the physical properties of the quantum system. The Hamiltonian operator acts on the state space. The universe of all possible states of a system is described by using a complex Hilbert space, and the instantaneous state of a system is described by a unit vector in that space. This state vector represents the probabilities for the outcomes of all possible measurements applied to the system. Because the state of a system is time dependent, the state vector is a function of time. The Schrödinger equation provides a quantitative formulation of the rate of change of the state vector.

The instantaneous state vector at time t can be represented as $|\psi(t)\rangle$ using Dirac's notation. Then, Schrödinger equation is written as:

$$H|\psi(t)\rangle = i\hbar\frac{\partial}{\partial t}|\psi(t)\rangle$$

As just stated, in quantum theory, the physical state of a system may be characterized as a vector in an abstract Hilbert space, and physically observable quantities are characterized as Hermitian operators acting on these vectors. The eigenvectors of H, denoted $\{|a\rangle\}$, provide an orthonormal basis for the Hilbert space. The spectrum of allowed energy levels of the system is specified by the set of eigenvalues, denoted $\{E_a\}$:

$$H|a\rangle = E_a|a\rangle$$

Depending on the system, the energy spectrum may be either continuous or discrete. (Some systems have a discrete spectrum in one range of energies and a continuous energy spectrum in another range—e.g., a finite potential well: It permits bound states with discrete negative energies and free states with continuous positive energies.)

Returning to the establishment of behavior, to follow the evolution of state of the system one has to solve the quantum mechanical equation of motion—the time-dependent Schrödinger equation. Analytical solution exists only for some oversimplified cases. In one dimension one can find the solution for an arbitrary $V(x)$ potential by numerical integration of the time-dependent Schrödinger equation. This is performed such that first the effect of the Hamiltonian on $\Psi(x,t)_0$, the state function at the initial time instant initial t_0, is calculated. This gives the time rate of change of the state function at the initial time instant t_0. From this one gets the change of the state function for the time interval $\delta_t = t - t_0$ and, thus, the state function at the time instant t, $\Psi(x, t)$. By choosing short time intervals and close values of the x coordinate, the method provides a reasonable approximation of the evolution of the state function. To get the evolution of the wave function by the outlined method in two or three dimensions is practically impossible, mainly because of computational time limitations. These limitations can be addressed pragmatically by an efficient numerical techniques [456].

Schrödinger's equation for *molecular systems* is even more complicated and can only be solved approximately. The approximation methods can be categorized as either ab initio or semiempirical. Semiempirical methods use parameters that compensate for neglecting some of the time-consuming mathematical terms in Schrödinger's equation, whereas ab initio methods include all such terms. The parameters used by semiempirical methods can be derived from experimental measurements or by performing ab initio calculations on model systems. Their practical differences are listed below [458]:

- Ab initio
  - Limited to tens of atoms and best performed using a supercomputer
  - Can be applied to organics, organometallics, and molecular fragments (e.g., catalytic components of an enzyme)
  - Vacuum or implicit solvent environment

- Can be used to study ground, transition, and excited states (certain methods)

- Semiempirical
  - Limited to hundreds of atoms
  - Can be applied to organics, organometallics, and small oligomers
  - Can be used to study ground, transition, and excited states (certain methods)

In summary, ab initio (Latin for "first principles") is a quantum mechanical nonparameterized molecular orbital treatment for the description of chemical behavior taking into account nuclei and all electrons. In theory ab initio is the most accurate of the three computational methodologies, these being, as noted, (i) ab initio, (iii) semiempirical all-valence electron methods, and (iii) molecular mechanics (MM), discussed in Appendix E [459, 460, 461].

As a concluding observation here note that to define probability amplitude functions one can start from the general perspective of path integrals. A description follows verbatim from [462]:

> Path integrals date back to work done in the 1920's by the mathematician Norbert Wiener, who gave a rigorous definition of integration over a space of paths. This work offered a way to understand the statistical properties of quantities determined by the path of a randomly moving particle. In the approach developed by Wiener, one considers the likely location of the particle at a number of instants. Thus, its motion is represented by a sequence of "snapshots". The location of the particle in each snapshot is a random variable, with statistical properties that can be determined. One can then attach a probabilistic weight to each possible sequence of particle locations. Though potentially relevant to a variety of problems, for many years Wiener's work was appreciated by few outside the fields of probability theory and measure theory. Moreover, because of the emphasis these disciplines place on rigour, the application of Wiener's ideas to real-world problems was largely neglected.
>
> This situation changed in 1948, when the U.S. physicist Richard Feynman used similar ideas to reformulate the rules of quantum physics. Prior to this date, quantum theory had always been described in the language developed primarily by Heisenberg, Schroedinger, Von Neumann and Dirac. Feynman suggested that, when considering the Quantum Theory of a particle in motion, every conceivable path of the particle could be assigned a certain complex number called the probability amplitude for that path. The probability amplitude for any event could then be obtained by adding together contributions from all paths consistent with that event. (The actual probability of the event is the square of the magnitude of this total amplitude.) A simple argument shows that the contributions from erratic non-Newtonian paths should tend to cancel each other out, so that by far the most important contribution to the total amplitude comes from paths very close to the "classical" path predicted by the deterministic Newtonian theory. The classical theory would therefore be expected to provide a good approximation to the quantum theory in many situations, as indeed is the case.
>
> Feynman gave two postulates for Quantum Theory:
>
> 1. *If an ideal measurement is performed to determine whether a particle has a path lying in a region of space-time, then the probability that the result will be affirmative is the absolute square of a sum of complex contributions, one from each path in the region.*

2. *The paths contribute equally in magnitude, but the phase of their contribution is the classical action (in Planck units); i.e. the time integral of the Lagrangian taken along the path.*

Compared to the previous formulations of Quantum Theory, these two postulates seem very straightforward. It must be acknowledged, however, that there is some vagueness in the phrase "the paths contribute equally in magnitude". This is the problem of choosing the functional measure.

Feynman later generalized his ideas to apply to fields as well as particles. In this case one adds contributions from all possible *histories* of the field, instead of particle paths. By using an elegant diagrammatic notation, Feynman and his followers were again able to reproduce all the important results obtained by the traditional approach. Feynman believed that his formulation of quantum theory was equivalent to the traditional one. In order to convince other physicists of this, much effort went into the rederivation of established results using the new techniques. Perhaps for this reason, people were slow to consider the possibility that the new approach might be *more* powerful in some cases than the traditional one.

It is now apparent that the usefulness of the path integral formulation goes far beyond its ability to reproduce known results. In the first place, many of the techniques developed by Feynman are very easily adapted to a variety of stochastic (rather than quantum) processes. The only major modifications here are that the contribution from each path will be real rather than complex, and that the total from all contributions will represent a probability rather than a probability amplitude. In certain cases this approach has yielded results that have not yet been obtained by any other methods. Path integrals are now used by a growing number of people to describe stochastic phenomena in applications ranging from polymer science and the modelling of chemical reactions to the modeling of interest rates and the pricing of derivative securities.

Wave functions will be discussed more in the material that follows.

D.2.2 Energy Levels

The previous subsection provided a view into the mathematical machinery of quantum theory. In this section we provide some of the simplest (practical) examples of quantum-based thinking.

Atoms are complicated systems because there is motion in three physical dimensions. This is the case even for the simplest of atoms: the hydrogen atom. We look first at an idealized problem, where motion occurs in just one dimension (in this idealized system the electron moves along the x axis only over a line segment of given length L—there is no nucleus present in this very simple example). We look at the results afforded by quantum theory for this simple environment and compare these results with the results that would arise from classical mechanics if indeed these did apply. Later we look briefly at the hydrogen atom with the nucleus and one electron. This material that follows is loosely based on/inspired from references such as [59, 60].

We look at (1) a macrolevel system composed of a mass of 1 g confined to move on a line 1 m in length. We apply classical mechanics to this problem. Then, (2) we compare this with an atomic-level system composed of electron of mass m, which as seen earlier is of the order of magnitude 10^{-28} g, that is confined to move on a line segment (along an imaginary x axis) of length L. L is taken to be the approximate diameter of an atom, which as seen earlier, is of the order of magnitude 10^{-10} m. In either case, it is assumed that when the electron or particle reaches the end of the segment, it is reflected by some force; the electron or particle is said as being in a potential well. We apply quantum theory principles to these two cases and we look at the kinetic energy (energy of motion).

In classical mechanics, through empirical experience, the velocity v can have any possible value from zero upward; classical mechanics predicts that the total kinetic energy is quadratic on v, it being $E = (\frac{1}{2})mv^2$. It follows that since all values for v are permissible, then all values for E are also permissible. Classical mechanics correctly predicts what an experimenter can measure and observe empirically for large particles or objects. For a Newtonian-level particle or object one can establish (measure) the position and velocity of a particle at any given instant of time. It further follows that, given the object's mass, its initial velocity, and a specification of the forces acting on it, one can use classical mechanics equations to predict the exact position and velocity of the particle or object at any future time. Hence, one can determine the trajectory of the particle and calculate it to any degree of accuracy.

The fact that all values for E are permissible in classical mechanics is in contrast with the prediction that quantum theory makes regarding the energy of an electron in a comparable situation. Quantum theory, based on Bohr's explanation of the line spectra that are observed for atoms, predicts that there are only certain specific quantized values of the energy can be achieved by electron confined to move on the one-dimensional line. The equation for the allowed energies as given by quantum theory for the simple system under discussion is

$$E_n = \frac{h^2 n^2}{8mL^2} \qquad n = 1, 2, 3, 4, \ldots$$

or

$$E_n = Kn^2 \qquad n = 1, 2, 3, 4, \ldots \qquad (\text{with } K = h^2/8mL^2)$$

where h is Planck's constant and n (called the quantum number) is an integer that may assume any value from one to infinity. As can be seen from the equations just given, each value of the quantum number n results in a value of E_n (notice that we can express the value of E_n in terms of so many units of K). The lowest value of E (when $n = 1$) is called the ground level, E_1. Notice that since the lowest allowed value of the quantum number is $n = 1$, the energy can never equal zero (it is, in fact, E_1); this means that a confined electron can never be motionless.

Figure D.9 plots the equation for the energy. As can be seen, only discrete values can be taken. Each value (line) in the diagram is called an energy level. We noted above that in a classical system the energy can vary in a continuous manner and can assume

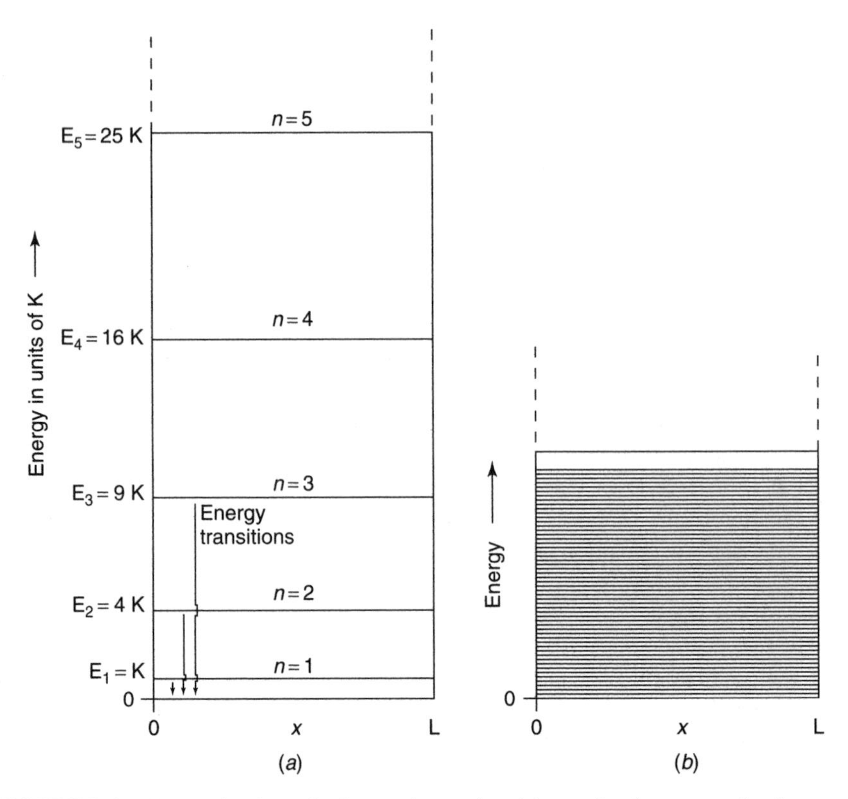

FIGURE D.9 Energy-level graphs for an electron/particle moving in a constrained manner on a line of length *L*. (*a*) Quantum behavior (only the first 5 levels are depicted. Note that a line spectrum (emitted energy) is a direct result of the quantization of energy. (*b*) Classical behavior. The energy can vary in a continuous manner and can assume any value. Diagram shows an "infinite" number of lines with infinitesimally small spacings from one line to the next. Note that a continuum of possible energy values produces a continuous spectrum of frequencies of emitted energy (not shown).

any value. This energy continuum of classical mechanics is supplanted by a discrete set of energy levels in quantum theory. One needs to impart a certain amount of energy to energize an electron from a certain energy state to a higher permissible state. When an electron is imparted sufficient energy to place it in one of the higher energy levels, we know that the electron is "excited." When the electron "falls" back down to the ground level, a photon is be emitted, and the energy ε of the photon is given by $E_n - E_1$. Because $\varepsilon = h\nu$, the frequency of the photon is a discrete value, $\nu = (E_n - E_1)/h$, which is consistent with Bohr's frequency condition. This reinforces that only certain frequencies can be emitted and the spectrum consists of a series of discrete lines.

As deduced from Figure D.9, the spectrum is comprised of a series of lines. The spacings between the spectral lines increase as ν increases. The figure shows pictorially that if the energy was not quantized (and all values were possible), then all increments in energy would be possible and all frequencies would appear in the spectrum.

Let us consider the momentum $p = mv$ of the electron. In the quantum case one obtains, by simple substitutions,

$$E_n = \frac{n^2h^2}{8mL^2} = \frac{1}{2}mv^2 = \frac{p^2}{2m}$$

Solving the quadratic for p one obtains

$$p = \pm\frac{nh}{2L} \qquad n = 1, 2, 3, 4, \ldots$$

The + and − sign are a result of solving a quadratic. The interpretation is that while one knows the magnitude of the momentum, one does not know (and cannot determine) the direction of the motion (if the electron moves from right to left, the sign is negative, and, if it moves left to right, the sign is positive). It follows that the most one can know about the momentum itself is its average value: This is zero because of the equal possibility for motion in either direction on the x axis. Notice in the equation just given that the kinetic energy and the momentum increase as the length of the line L is decreased, and the motion of the electron becomes more confined.

Now look at the issue of position in the quantum environment. We have already hinted at the difficulties that are encountered when one attempts to determine the position of an electron because of the Compton effect. The Compton effect establishes that a portion of the energy of the photon that is "used" in making any observation is transferred to the electron. The result is that the observation acts to measure the electron's position, itself disturbs the electron as we attempt to measure its position. From these observations it follows that quantum theory cannot predict the position of an electron exactly at a deterministic level. A new modeling tool (mathematics) is needed, and this is probability theory. It turns out (as we have already hinted) that in the quantum environment, one is able to provide only a *probability measure* as to where the electron will be found. Fortunately one can indeed validate experimentally these probabilistic measures.

Probabilistic quantities are well-defined concepts. An extensive body of methods and results are available from the discipline of mathematical probability theory. Focusing just on the most rudimentary concepts of probability theory, an example follows to illustrate a key concept. As the example goes, this author has taught about 75 college classes in the past 20 years (a number large enough to achieve statistical significance). Assume that for argument's sake, all classes had 36 students. Consider the experiment: Identify the precise (sitting) location in the Class TM-601, always running Wednesday night, of students in the class that have a birthday in the month of August. Probability theory says that if this experiment was carried out 75 times, and stats kept, that there will be just about 3 students in each class that have an August birthday, when the numbers are averaged over all the 75 runs of the experiment. Hence, with a high degree of certainty one can say that Wednesday nights from 1984 to 2004 in a specified 8 m × 8 m classroom at Stevens Institute of Technology, Hoboken, New Jersey, say with exact coordinates (UMT) 18 581694E-to-18 581695E and 45 10789N-to-45 10790N, that three students satisfying this condition can be located. However, one cannot further define from any theory *exactly where those students will be sitting* within the confines of that classroom. Extrapolating this example, and making reference to predictability, one can predict with good confidence that any Tuesday night of the academic year, in a specific classroom at NYU in New York (or Rutgers in New Jersey, etc.) where a class with 36 students runs routinely, that then one can predict that 3 students born in August can be found, particularly if we looked at this classroom over a period of, say, 10 semesters; but no theory that we know of could predict precisely where those students will be sitting.

Not to imply that this example is a precise analogy with the atom, but just that (a) with probability Theory one can make certain strong statements about certain things, but (b) that one cannot answer every question one can pose about things (e.g., a question about exact position).

Regarding the position of the electron, it turns out that the maximum information that can indeed be obtained both theoretically and experimentally is at the probabilistic level. The probabilities are expressed in terms of a distribution function, say

$P_n(x)$ = {Probability that when the electron is at energy level E_n, it will be at position x on the interval 0 to L}.

Thus, $P_n(x)$ give the probability at a single point. Now consider a very small segment of length Δx (see Fig. D.10). Then the probability that the electron is in that particular small segment Δx is defined by the product of Δx and the value of the probability distribution function $P_n(z)$, where z is the coordinate of the midpoint (on the x axis) for the interval defined by Δx. Namely, the probability is $\Delta x\, P_n(z)$ (is equal to the area of the rectangle, the shaded area in the figure) *with z being infinitesimally close to x.* Eventually, we let Δx approach zero, so that z approaches x.

The probability distribution $P_1(x)$ can be established by designing an experiment that determines whether or not the electron is in one particular segment Δx of the interval (e.g., 0 to $0.01L$, $0.01L$ to $0.02L$, etc., here $\Delta x = 0.01L$), when it is known to be in the quantum level $n = 1$. For each segment on the x axis, true to the "frequentist view of probability," we perform the experiment a large number of times and record the ratio of the number of times the electron is found in a particular segment to the total number of observations made for that segment. For example, an electron is found to be in the segment marked 0 to $0.01L$ in 2 out of 100 observations, or 2% of the time (in the other 98 observations the electron was in one of the other segments). Thus, $P_1(z)\,\Delta x = (0.2/L)\,(0.01L) = 0.02$ or 2% (z being some representative but "choosable" value in the 0 to $0.01L$ interval). A similar set of experiments is made for each of the segments Δx. The limiting case in which the length L is subdivided into a very large number of very small segments ($\Delta x \rightarrow dx$ and $z \rightarrow x$) results in the smooth curve shown in Figure D.11 for $P_1(x)$. Naturally, this experiment is laborious because we have to operate at two levels: One selects a Δx and repeats a large number of experiments (do this throughout the interval); then repeat the entire experiment for a smaller Δx; then repeat the entire experiment for yet a smaller Δx. To make tings more complicated, there is a different probability distribution for each value of E_n, or each quantum level. *Every* quantum level (i.e., every allowed value of the energy) *has associated with it a distinct probability distribution for the electron.* After all the experiments are carried out, one can arrive at a graph like the one shown in Figure D.10.

Let us look at the predictions of quantum theory regarding the position of a bound electron. By examining Figure D.11, we note that for $n = 1$, the first quantum level, the graph of $P_1(x)$ shows the electron will most likely be found at the midpoint. In addition to this, the form of $P_n(x)$ changes with every change in energy. Yet, at every level there are loci where the electrons are most likely to be found.

We noted in the previous section that in classical mechanics, given enough input variables, one can determine the position of a particle in a deterministic fashion at any instant; hence, the concept of a probability distribution is alien to the theory of classical mechanical analysis. However, at the pragmatic level, assuming we were unaware of some input variable (either it was not available or it was too expensive to obtain), we could use the machinery of an a posteriori probabilistic analysis to determine the probability distribution for the particle. This will turn out to be a uniform distribution, namely, $P_{\text{classical}}(x)$ is identical for all values of x and equals $1/L$. Because there are no forces acting on the particle as it traverses the segment, it will be equally likely to be found at any point on the segment; hence, probability will be the same regardless of the energy.

This points out, once again, that there is a conspicuous difference between the classical and the quantum theory environment.

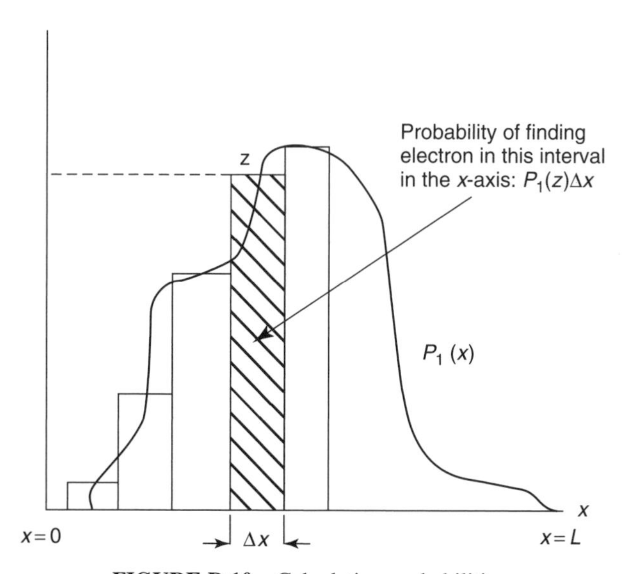

FIGURE D.10 Calculating probabilities.

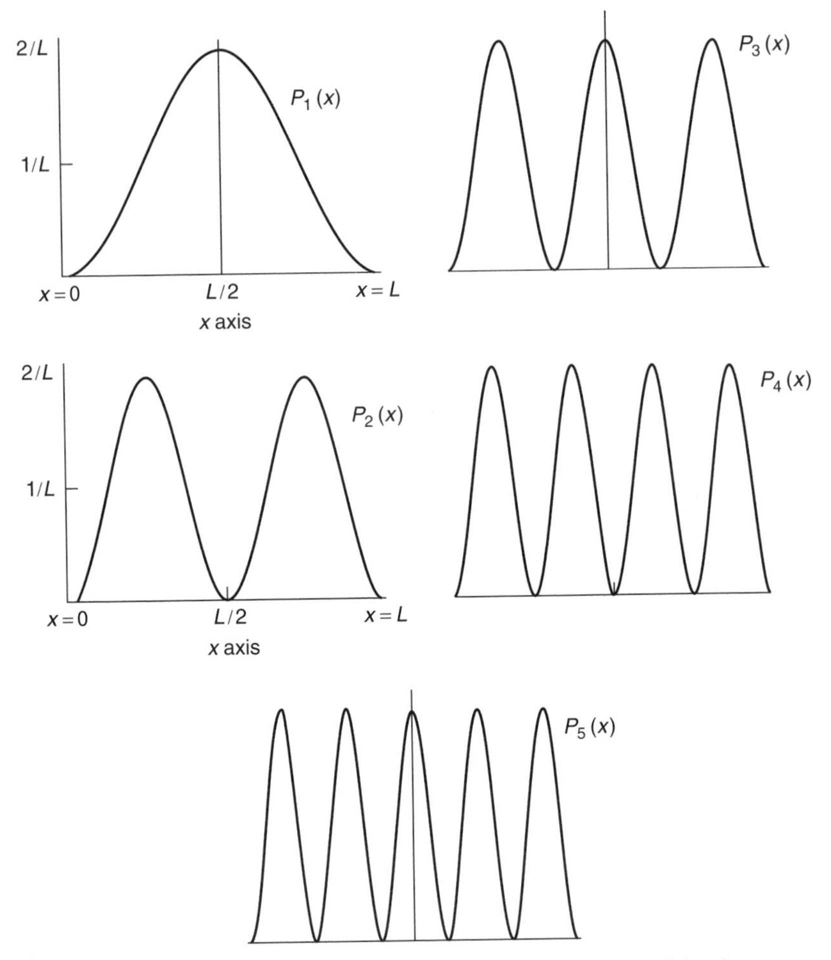

FIGURE D.11 Example of probability distributions for the position of the electron at various energy levels.

D.2.3 Heisenberg's Uncertainty Principle

Next we provide some background that leads to a basic understanding of the Heisenberg uncertainty principle. We are still examining the simple illustrative case of an electron moving along a segment of length L (no nucleus present in this example) (with L equal to a typical atomic dimension of 1×10^{-10} m). The idea is to try to locate an electron that is confined to move on a line. To facilitate the process, we seek to find the electron to within a length that is very small compared to the length of the line. To do this we would use an instrument that injects light that has a wavelength much less than L. The shorter the wavelength of the light that is used to observe the electron, the smaller will be the uncertainty Δx. Say we use light with

$\lambda = (1/1000)L$. It follows that the frequency and energy of a photon with wavelength of $\lambda = (1/1000)L$ are

$$\varepsilon = h\nu = \frac{hc}{\lambda} = \frac{6.6 \times 10^{-27} \times 3 \times 10^{10}}{10^{-11}} = 2.0 \times 10^{-5} \text{ ergs}$$

$$E_1 = \frac{(6.6 \times 10^{-27})^2}{9.1 \times 10^{-28} \times 8 \times 10^{-16}} = 6.0 \times 10^{-11} \text{ ergs} = K$$

Given Compton's effect (the collision of a photon with an electron imparts energy to the electron), this creates a difficulty because the energy of the photon is approximately 3×10^5 times greater than the energy of the electron. It would follow that after the collision the electron will certainly not be in the state $n = 1$, but it will have been excited to some higher levels with $n = 2$ ($E = 4K$) or $n = 3$ ($E = 9K$), and so forth. The conclusion is that if we seek exact information of what the position of the electron is in a given state, we can obtain this information only at the expense of imparting to the electron an amount of energy that disturbs (even destroys) the system. If this experiment was repeated a large number of times and a record kept of the number of times an electron was located in each segment of the line [say, $(1/1000)L$], a probability plot similar to Figure D.11 can be obtained.

A complementary question regarding the position of the electron can be formulated: How much information can be obtained regarding the position of the electron in a given quantum level without destroying that level in the process of making the determination? In answer, obviously, the electron can only accept energy in an amount less than that necessary to excite it to the next quantum level, $n = 2$. The difference in energy between E_2 and E_1 is $3K$; thus, if we wish to leave the electron in a state of known energy and momentum, we must use light whose photons have an energy less than $3K$. Here, however, the wavelength of the light with $\varepsilon = 2K$ turns out to be greater than the length of the line:

$$\lambda = \frac{hc}{\varepsilon} = \frac{6.6 \times 10^{-27} \times 3.0 \times 10^{10}}{12 \times 10^{-11}} = 1.7 \times 10^{-6} \text{ cm}$$

Hence, it follows that the uncertainty in the position of the particle will be of the order of magnitude of, or greater than, L itself; that is to say, in a single experiment the electron will appear to be blurred over the complete length of the segment. Two interpretations that can be formulated regarding the probability distributions:

- Interpret the P_n's as the true probability of finding the electron in a given small segment of the line, using light of very short λ relative to L. The experimental process excites the electron and leaves the electron with an unknown amount of energy and momentum. We have frustrated (changed) the object of our investigation because we now know where it was in a given experiment but not where it will be, in terms of energy or position.

- Interpret the P_n's as instantaneous pictures of the electron when it is bound in a known state. Utilizing a light with a $\lambda = L$, we do not excite the electron, so we leave it in a known energy level. However, in this experimental process the knowledge of the position is uncertain: The photons are scattered from the system and give us the "smeared" distribution P_1 pictured in Figure D.11.

With the second interpretation we must accept the fact that when the electron remains in a given state it is "smeared out" and "appears like" the pictures given for P_n. This distribution is given a particular name; it is called the *electron density distribution*, the electron density that we already discussed above. Here the P_n's represent a charge density distribution that is considered static as long as the electron remains in the nth quantum level. Hence, the P_n functions tell us either (a) the *fraction of time* the electron is at each point on the line for observations employing light of short wavelength, or (b) they tell us the *fraction of the total charge* found at each point on the line (the whole of the charge being spread out) when the observations are made with light of relatively long wavelength [59, 60].

The Heisenberg's uncertainty principle speaks to the *magnitude* of the uncertainties encountered in measurements on the atomic level. This *magnitude* can be derived with quantum mechanical principles. Continuing with the simple case of an electron constrained in a unidimensional interval L: When use of light with $\lambda \sim L$ (which leaves the particle bound in a given quantum level, say $n = 1$), the minimum uncertainty (call it call Δx) in the position of the electron moving on a line obtained in an experiment, is equal to L, the length of the line ($\Delta x = L$). We noted above that the momentum of the electron in the nth quantum level is given by the range $+nh/2L$ to $-nh/2L$; hence, the minimum uncertainty in our knowledge of the momentum is the difference between these two possibilities, or for $n = 1$, h/L (by simple algebra).

The product of the uncertainties in the position and the momentum is

$$\Delta p \, \Delta x = L\frac{h}{L} = h$$

This result is a particular example of a general relationship pertaining to the product of the uncertainties in the momentum and position, known as Heisenberg's uncertainty principle.

Having derived this for a specific case, we now state the general case and its implications: In the general case we have $\Delta p \, \Delta x \geq h$, namely, the product in the uncertainties $\Delta p \, \Delta x$ *equals or exceeds* the value of Planck's constant h. The interpretation is as follows: If we seek to decrease the uncertainty in the position (i.e., make Δx small), there will result a corresponding increase in the uncertainty of the momentum of the electron along the same coordinate, such that the product of the two uncertainties is always equal or greater to Planck's constant. There is no way to defeat Heisenberg's uncertainty principle by an experiment that sets out to decrease the length of the line L because the decrease in Δx obtained by decreasing L is offset by the increase in Δp that accompanies the increased confinement of the electron; again, the product $\Delta x \, \Delta p$ remains unchanged in value.

D.2.4 Motion in Two Dimensions

Next we look briefly at what happens when we allow the electron to move on the $x = y$ plane rather than just along the x axis. We assume that the motion is confined to a length L along each axis.

Here, the motions along the x and y directions are independent of one another. The total energy of the system will be given by the sum of the energy quantum for the motion along the x axis plus the energy quantum for motion along the y axis. Since two dimensions (x and y) are now required to specify the position of the electron, the probability distribution $P_{n,m}(x, y)$ must be plotted in the third dimension; on paper we do this in the form of a contour map. All points in the $x = y$ plane having the same value for the probability distribution $P_{n,m}(x, y)$ are identified by a contour line.

Two quantum numbers are needed (n_x and n_y), one to indicate the amount of energy along each coordinate. One has

$$\begin{aligned} E_{n_x,n_y} &= \frac{h^2}{8mL^2} n_x^2 + \frac{h^2}{8mL^2} n_y^2 \qquad & n_x &= 1, 2, 3, \ldots \\ & & n_y &= 1, 2, 3, \ldots \\ &= \frac{h^2}{8mL^2}(n_x^2 + n_y^2) \end{aligned}$$

Similarly to the unidirectional case, when $n_x = 1$ and $n_y = 1$, the energy $E_{1,1}$ is $h^2/4mL^2$, as was the case before.

Figure D.12 depicts a few of these contours. The values of the contours increase from the outermost to the innermost. For $n_x = n_y = 1$, the electron is most likely to be found at $y = L/2$ and $x = L/2$, when in the level. By way of terminology, a contour map is a graph of the probability or density distribution in a plane; a profile is a graph of the density distribution along a line.

Now for $n_x = 1$, $n_y = 2$ and $n_x = 2$, $n_y = 1$ we have, respectively:

$$E_{1,2} = \frac{5h^2}{8mL^2} \qquad E_{2,1} = E_{1,2}$$

The probability distributions, however, are different as shown in Figure D.12 [e.g., $P_{1,2}(x,y) \neq P_{2,1}\ (x,y)$]. When $n_x = 1$ and $n_y = 2$, there is a zero probability of finding the electron at $y = L/2$, while a slice parallel to the x axis will still be similar to $P_1(x)$. Just the reverse is true for the case $n_x = 2$ and $n_y = 1$. So there are two different arrangements for the distribution of the electron, both of which have the same energy. The energy level is said to be degenerate. The degeneracy of an energy level is equal to the number of distinct probability distribution for the system, all of which belong to this same energy level [59, 60].

D.2.5 Ψ—The Probability Amplitude

In quantum theory, Newton's equations of motion are replaced by Schrödinger's equation. When Newton's laws of motion are applied to a system, we obtain both

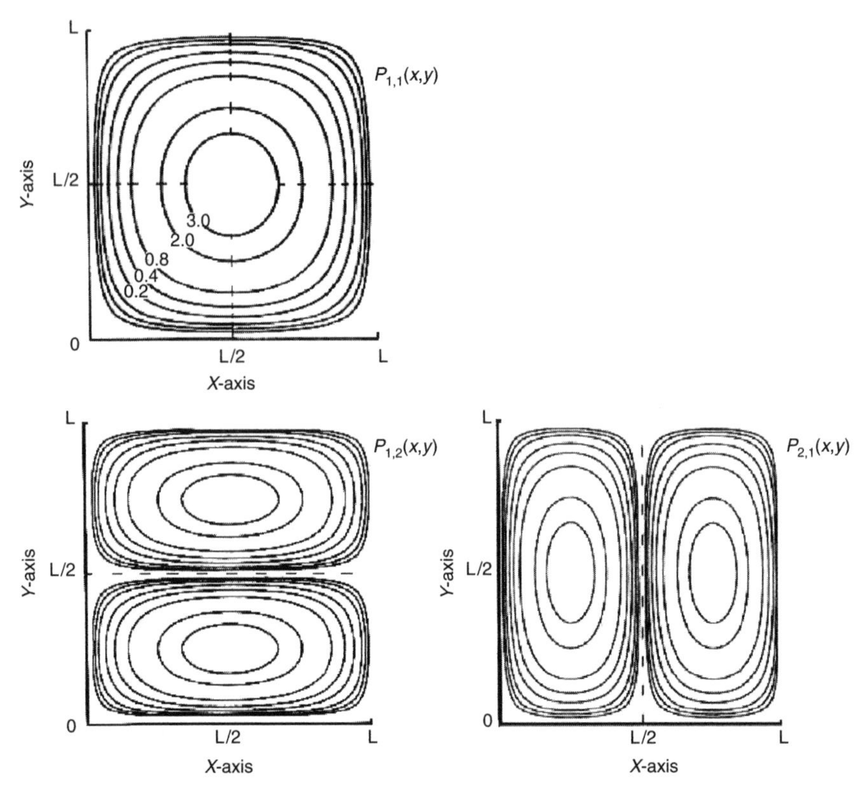

FIGURE D.12 Contour maps of the probability distributions $P_{nx,\ ny}$ (x,y) for an electron moving in the *x*-*y* plane (only a few cases shown). Note: the values shown in the top diagram are units of energy $(4/L^2)$.

the energy and an equation of motion. The equation of motion allows us to calculate the position or coordinates of the system at any instant of time.

Earlier we described the concept of probability amplitudes. For this section we simply define the probability amplitude (or wave function) as a complex-number-valued function of the position of a particle, such that the probability of finding the particle described by the wave function (e.g., an electron in an atom) at that point is proportional to the square of the absolute value of the probability amplitude. The probability amplitudes are functions only of the positional coordinates of the system. Because of the fact that probability amplitudes take on complex values in the complex plane, for every allowed value of the energy, we can have one or more probability amplitudes. For example, the (r, c) tuples $(1, 0i)$, $(0, i)$, $(-1, 0i)$, and $(0, -i)$ all would have a squared norm of 1, where $\|\text{norm}\| = |r^2 + c^2|$. Pictorially, the "graphs" of orbitals in a standard chemistry book are a representation of the region within which the wave function gives a high probability of finding an electron for that state. Ψ is the mathematical function of the positional coordinates.

For a constrained system the amplitudes as well as the energies are determined by one or more quantum numbers. Hence, for every E_n we have one or more Ψ_n's. By squaring the Ψ_n's we can obtain the corresponding P_n's (put another way, the probability distribution itself is obtained by squaring the probability amplitude). Some basic examples are given below to put a sense of concreteness to the Ψ_n's.

By now we have become acclimated that in a quantum environment we have to deal with various probability distribution functions (at various quantum levels) to secure information about the possible position of particles. But we have what might at first appear as a setback: When Schrödinger's equation is solved for a given system, we obtain the energy directly, however, *we do not obtain the probability distribution function* (which, as noted, is the function that contains the information regarding the position of the particle). Instead, the solution of Schrödinger's equation gives only the amplitude of the *probability distribution* function Ψ_n's along with the energy.

To explain the concepts let us look here at amplitude functions Ψ for the simple system of an electron confined to motion on a segment of length L. For any system, Ψ is simply some mathematical function of the positional coordinates. In the example at hand there is a single coordinate x, but in general much more complex situations arise. This is convenient because it allows us to provide a plot of the amplitude functions Ψ based on x, that is, to derive a $\Psi_n(x)$. It turns out that the equations are (no derivation included here)

$$\Psi_n(x) = \sqrt{2/L}\,\sin(n\pi x/L) \qquad n = 1, 2, 3, \ldots$$

Now, as stated, $\Psi_n^2(x)\,\Delta x$, or $P_n(x)\,\Delta x$, is the probability that the electron will be found in some particular small segment of the line Δx. Hence, when the Ψ_n's function are squared, we obtain

$$P_1(x) = (2/L)\,\sin^2(\pi x/L)$$

The first three Ψ_n's are shown plotted in Figure D.13, where we also show the corresponding $P_n\,(x)$. Observe that $\Psi_n(x)$ may be negative for certain values of x and, clearly, wherever $\Psi_n(x)$ crosses the x axis a node appears in the corresponding $P_n(x)$. The probability distribution functions $P_n(x)$ are obtained by squaring $\Psi_n(x)$. It follows that $P_n(x)$ are positive for all values of x, as they should, since they represent a probability.

Obviously,

$$\int_0^L \Psi_n^2(x)\,dx = 1$$

This implies that the probability that the electron is somewhere on the line is unity, that is, a certainty. Also, each Ψ_n must go to zero at $x = 0$ and $x = L$, because the probability of the electron not being on the segment is zero.

By looking at Ψ_n we see that since there is only a single value of the energy for each of the possible Ψ_n functions, only certain discrete values of the energy are

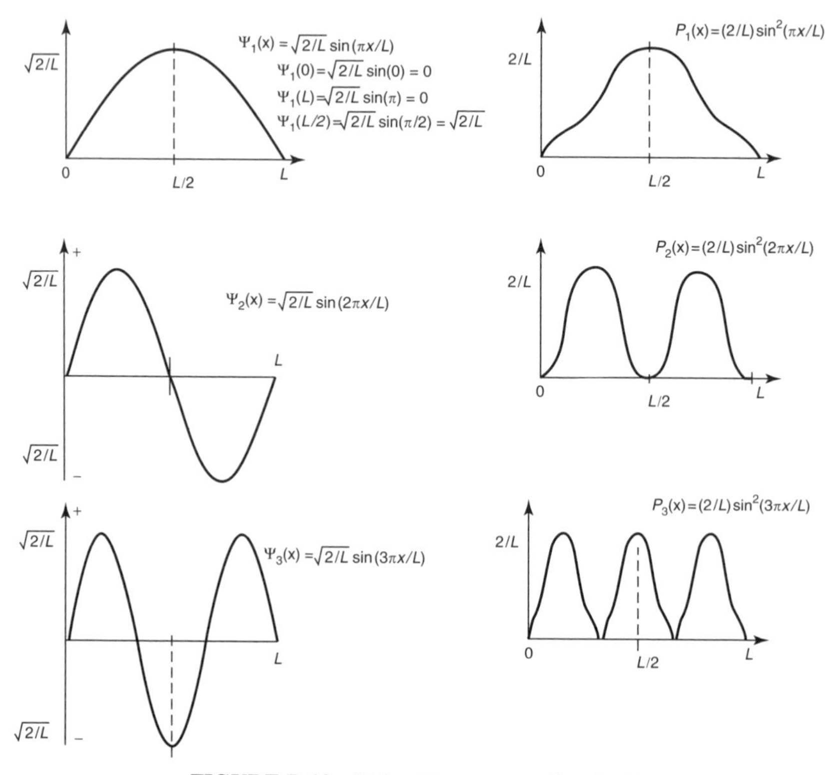

FIGURE D.13 Ψ_n's with corresponding $P_n(x)$.

allowed. The physical restraint of confining the motion to a finite length of line results in the quantization of the energy.[5]

The Ψ_n's are periodic functions of x and have the appearance of a wave (periodicity means that given the value of $\Psi_n(x)$ is repeated as x is increased, that is, for a certain λ, $\Psi_n(x + \lambda) = \Psi_n(x)$ for all x). The wavelength of Ψ_1 is $\lambda_1 = 2L$; the wavelength for Ψ_2, is $\lambda_2 = L$; for $n = 3$ we have $\lambda_3 = (\frac{2}{3})L$, or, more generally, $\lambda_n = 2L/n$. Because of the wavelike nature of the Ψ_n's, quantum physics is also called wave mechanics, and the Ψ_n functions are called wave functions. However, keep in mind that a wave functions Ψ_n themselves have *no physical reality* (e.g., a Ψ_n does not represent the trajectory or path followed by an electron in space). *All physical properties are determined by the square of the wave function* (this being the physically measurable probability distribution). (When one refers to the wavelengths of electrons and other particles, one is talking about the wavelength of the wave function, but one must keep in mind the wavelengths refer only to a property of the amplitude functions and not to the motion of the particle itself.) It turns out, however, that the wavelengths postulated by de Broglie to be associated with the motions of particles equate numerically to the wavelengths of the probability amplitudes or wave functions.

[5]If the segment is made infinitely long and the electron is free and no longer constrained (bound), then solutions for any value of n (integer or otherwise) are possible; in turn, all energies are permissible.

When λ is smaller than the important physical dimensions of the system, quantum effects disappear and the system behaves in a classical fashion; however, when λ is comparable to the physical dimensions of the system, quantum effects predominate. This is made clear by noting that as n increases, the wavelength becomes much smaller than L ($\lambda_n << L$). For *large values of n, say N*, Ψ_N and P_N look like the graphs of Figure D.14. P_n appears to be a "continuous" function of x. Observations of the position of the electron would yield a result for P_N similar to that obtained in the classical case. The condition $\lambda_n << L$ is always true when the system possesses a large amount of energy, that is to say, a high n value.

The value of λ for an electron bound to an atom is $\lambda = 10^{-10}$ m. This is about the same order of magnitude as an atomic diameter. It follows that electrons bound to atoms exhibit quantum effects. In contrast, the wavelength associated with the motion of the mass of 1 g moving on a line segment 1 m in length with a velocity of 0.01 m/s has a $\lambda = 6.6 \times 10^{-29}$ m. This is very short wavelength and quantum effects will not be present.

D.2.6 The Hydrogen Atom—Developing the Hydrogenic Atomic Orbital Concepts

Chemical reactions can be predicted at the practical level by a table of "orbitals," as shown in Table D.2. What theory lies behind this kind of table? In this section we briefly put the machinery that we discussed to work to describe the concept of orbitals. In general, the term *orbital* is used for a wave function that determines the motion of a single electron. If the one-electron wave function is for an atomic system, it is called an atomic orbital. Specifically, the wave functions Ψ_n for the hydrogen atom are called *atomic orbitals*, and they play an important role in all discussions of the electronic structure of atoms.

The hydrogen atom is a physical system that can be treated exactly by quantum theory, and so the wave functions Ψ_n for the hydrogen atom can be derived from the

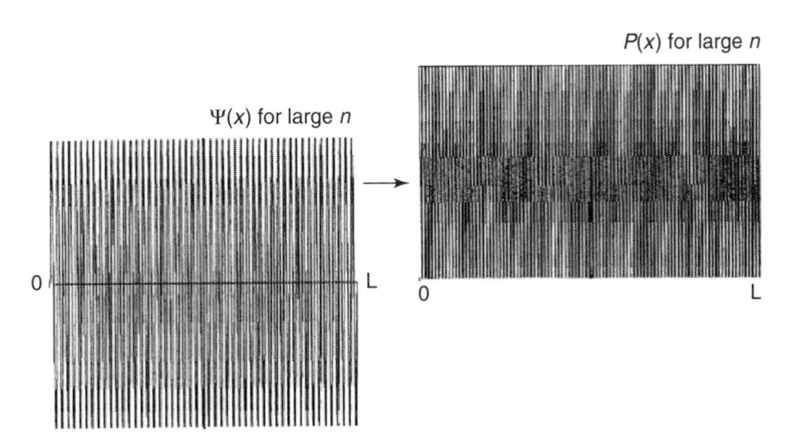

FIGURE D.14 Wave function and corresponding probability distribution for large n.

TABLE D.2 Orbitals

1. Period		**1s**	**2s**	**2p**	**3s**	**3p**	**3d**	**4s**	**4p**	**4d**	**4f**	**5s**	**5p**	**5d**	**5f**	**6s**	**6p**	**6d**	**6f**	**7s**	**7p**
1	**H**	1																			
2	**He**	2																			
2. Period		**1s**	**2s**	**2p**	**3s**	**3p**	**3d**	**4s**	**4p**	**4d**	**4f**	**5s**	**5p**	**5d**	**5f**	**6s**	**6p**	**6d**	**6f**	**7s**	**7p**
3	**Li**	2	1																		
4	**Be**	2	2																		
—																					
3. Period		**1s**	**2s**	**2p**	**3s**	**3p**	**3d**	**4s**	**4p**	**4d**	**4f**	**5s**	**5p**	**5d**	**5f**	**6s**	**6p**	**6d**	**6f**	**7s**	**7p**
11	**Na**	2	2	6	1																
12	**Mg**	2	2	6	2																
—																					
4. Period		**1s**	**2s**	**2p**	**3s**	**3p**	**3d**	**4s**	**4p**	**4d**	**4f**	**5s**	**5p**	**5d**	**5f**	**6s**	**6p**	**6d**	**6f**	**7s**	
19	**K**	2	2	6	2	6	—	1													
20	**Ca**	2	2	6	2	6	—	2													
—																					
5. Period		**1s**	**2s**	**2p**	**3s**	**3p**	**3d**	**4s**	**4p**	**4d**	**4f**	**5s**	**5p**	**5d**	**5f**	**6s**	**6p**	**6d**	**6f**	**7s**	**7p**
37	**Rb**	2	2	6	2	6	10	2	6	—	—	1									
38	**Sr**	2	2	6	2	6	10	2	6	—	—	2									
—																					
6. Period		**1s**	**2s**	**2p**	**3s**	**3p**	**3d**	**4s**	**4p**	**4d**	**4f**	**5s**	**5p**	**5d**	**5f**	**6s**	**6p**	**6d**	**6f**	**7s**	**7p**
55	**Cs**	2	2	6	2	6	10	2	6	10	—	2	6	—	—	1					
56	**Ba**	2	2	6	2	6	10	2	6	10	—	2	6	—	—	2					
—																					
7. Period		**1s**	**2s**	**2p**	**3s**	**3p**	**3d**	**4s**	**4p**	**4d**	**4f**	**5s**	**5p**	**5d**	**5f**	**6s**	**6p**	**6d**	**6f**	**7s**	**7p**
87	**Fr**	2	2	6	2	6	10	2	6	10	14	2	6	10	—	2	6	—	—	1	
88	**Ra**	2	2	6	2	6	10	2	6	10	14	2	6	10	—	2	6	—	—	2	
—																					
114	**Uuq**	2	2	6	2	6	10	2	6	10	14	2	6	10	14	2	6	10	—	2	2

Only certain “orbitals” or “shells” of electron probability densities are permitted. The “shells” are identified by a *principal quantum number n* (quantum numbers arise from the solution of the Schrödinger’s equation). The “subshell within the shell” is identified by a second quantum number *l*. Two additional numbers characterize the states within the subshells. Pauli exlusion principle asserts that only one electron in an atom can have a specified 4-tuple of the four quantum numbers.

Schrödinger equation. The study of the hydrogen atom, however, is more complicated than the example of an electron confined to move on a line segment that was discussed above: Not only does the motion of the electron occur in three dimensions, but there is also a force acting on the electron. This force, the electrostatic force of attraction, is responsible for holding the atom together. We do not study this atom here beyond some basic discussion in order to hint at some of the concepts that are applicable, with appropriate modifications and considerations, to larger atoms and chemistry. In addition to their inherent significance, these solutions suggest prototypes for atomic orbitals used in approximate treatments of complex atoms and molecules [451].

The discussion that follows is relatively terse; the interested reader may consult [59, 60] directly.

We have seen that by solving Schrödinger's equation we can obtain the energy function and one or more wave functions for any given quantum level n. We have seen that knowledge of the wave functions, or probability amplitudes Ψ_n, allows a calculation of the probability distributions for the electron in any given quantum level. As discussed earlier in the chapter, two interpretations exist for P_n:

1. *Interpretation 1* The term P_n represents the fraction of the total electronic charge to be found in a small volume element of space. Here, the measurement procedure designed to detect the position of the electron with an uncertainty significantly less than the diameter of the atom itself will, if repeated a large number of times, result in a diagram such as Figure D.15 for P_1; this measurement procedure uses light of short wavelength, but, in making each of these measurements, the atom will be ionized because the energy of the photons with a wavelength much less than 10^{-10} m is greater than the amount of energy required to ionize the hydrogen atom. With interpretation 1, the electron will be detected close to the nucleus most frequently and the probability of observing it at some distance from the nucleus decreases rapidly with increasing r. Note in Figure D.15 that by the time $r = 2$ is reached, nearly 100% of the probability is already covered.

2. *Interpretation 2* The term P_n represents the instantaneous snapshot of the electron when it is bound in a known state; in this situation, the knowledge of the position is uncertain and one has the "smeared" distribution. Here, if instead of a measurement procedure that uses light of short wavelength (as above), one uses light with a wavelength comparable to the diameter of the atom in the measurement procedure, then the electron will not be excited to another level, but the knowledge of its position will be less precise. So, in these measurement procedures where the electron's energy is not changed, the electron will appear to be "smeared out." Thus, one can interpret P_n as giving the fraction of the total electronic charge to be found in a small volume of space. With interpretation 2 the electron is in a deWnite energy level, hence, it makes sense to refer to the P_n distributions as *electron density distributions*. In summary, the P_n describe the fashion in which the total electronic charge is distributed in space.

Recalling the concept of degeneracy of an energy level from the earlier discussion as being equal to the number of distinct probability distributions for a system, all of which belong to the same energy level, one finds that, for the hydrogen atom for every

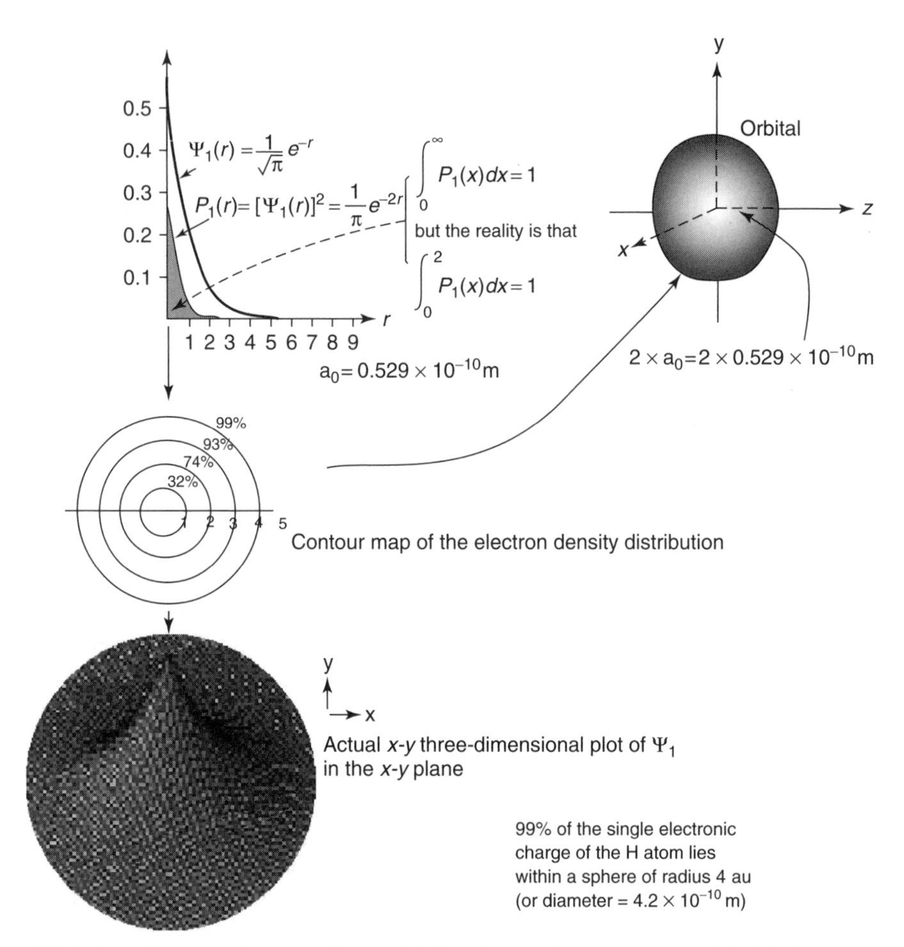

FIGURE D.15 Wave function Ψ_1, probability distribution P_1, and 1s orbital for H atoms: 99% of the single electronic charge of the H atom lies within a sphere of radius 4 au (or diameter = 4.2×10^{-10} m).

value of the energy E_n, there is a degeneracy equal to n^2. Hence, for $n = 1$, there is exactly one atomic orbital and one electron density distribution. For every quantum number n there are n^2 possible ways in which the electronic charge may be distributed in three-dimensional space and still possess the same value for the energy. It turns out that for every value of the quantum number, one of the possible atomic orbitals has a spherical electron density distribution that can be represented by circular contours (the angular momentum is independent of direction). For a given value of $n > 1$, the other atomic orbitals exhibit a directional dependence and give rise to density distributions that are not spherical but are concentrated along certain axes or in certain planes.

When $n = 1$, the wave function (and, hence, the probability function) depend only on the distance r between the electron and the nucleus, as seen in Figure D.15. As seen in the figure, the electron density of a 1s orbital is at a maximum at the nucleus.

For $n = 2$, there are four different atomic orbitals and four different electron density distributions, all of which possess the same value for the energy, E_2. One is spherical and is known as a 2s orbital, as shown in Figure D.16.

Now we focus on the other orbitals. For the hydrogen atom, the angular dependence of the atomic orbitals and the shapes of the contours of the corresponding electron density distributions are connected with the angular momentum possessed by the electron. In fact, the angular momentum plays a major role in the electronic structure of atoms. The angular momentum vector remains constant as long as the plane and radius of the orbit remain unchanged and the speed (angular velocity) of the electron in the orbit is constant. It can be shown that the *magnitude* of the angular momentum may assume only those values given by:

$$M = \sqrt{l(l+1)}(h/2\pi) \qquad l = 0, 1, 2, 3, \ldots, n-1$$

For the State n = 1 Already Covered, l Can Only Have the Value of Zero. For $n = 2$, l may equal 0 or 1, and so on. Clearly, when $l = 0$ the angular momentum of the electron is zero. The atomic orbitals that describe the states of zero angular momentum are called s orbitals (we have already used this nomenclature in the preceding paragraphs). The s orbitals are identified by giving the value of the principal quantum number n, and they are referred to as the 1s, 2s, 3s, ... atomic orbitals. All s orbitals predict spherical density distributions for the electron. While many orbitals

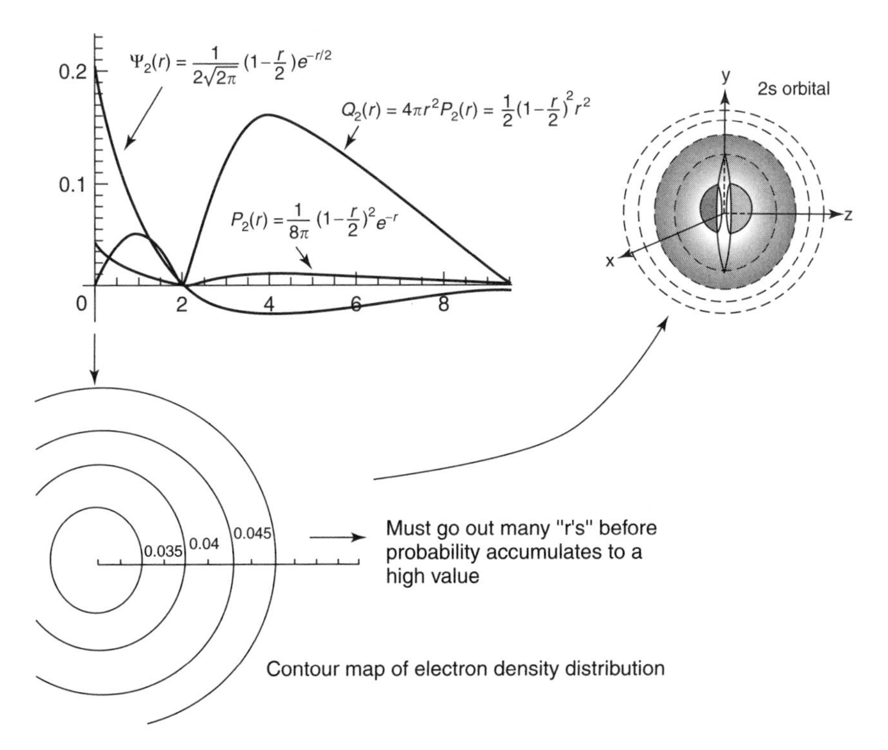

FIGURE D.16 Wave function Ψ_2, probability distribution P_2, and the 2s orbital for H atoms.

TABLE D.3 Number of Electron States in Some "Shells" and "Subshells"

Principal Quantum Number n	Second Quantum Number l	Subshell	Number of Distinct States	Number of Electrons per Subshell	Number of Electrons per Shell
1	0	s	1	2	2
2	0	s	1	2	8
2	1	p	3	6	
3	0	s	1	2	18
3	1	p	3	6	
3	2	d	5	10	
4	0	s	1	2	32
4	1	p	3	6	
4	2	d	5	10	
4	3	f	7	24	

exist, for the purpose of practical chemistry only certain "outer orbitals" are important. As seen in Table D.4, for H this is the 1s orbital; for Na it is the 3s orbital; for K it is the 4s orbital, and so forth. (See Appendix C for a full list.) Table D.3 summarizes some of the discussion that follows; also see Figure D.17.

For $n = 1$ and $l = 0$ in the Hydrogen Atom, One Obtains the 1*s Orbital.* This orbital, and all s orbitals in general, exhibit spherical density distributions for the electron. Figure D.18 shows the radial distribution functions $Q_n(r) = 4\pi r^2 P_n = 4\pi r^2[\Psi_n(r)]^2$

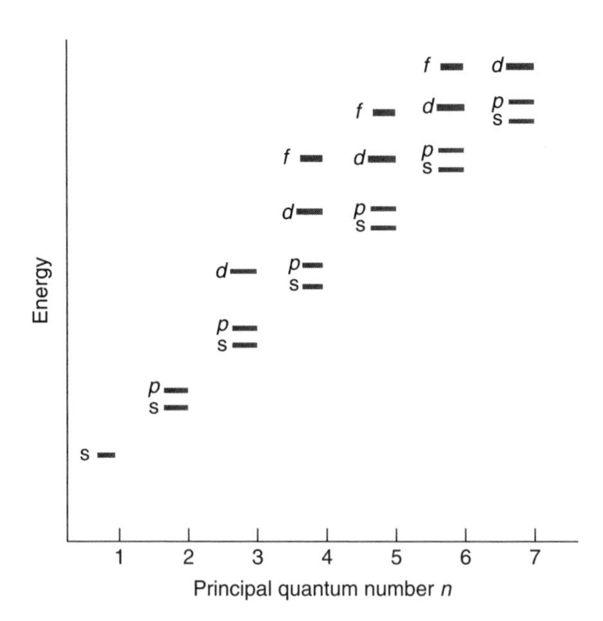

FIGURE D.17 Energy levels and orbitals: energy level sequence: 1s, 2s, 2p, 3s, 3p, 4s, 3d, 4p, 5s, 4d, 5p, etc.

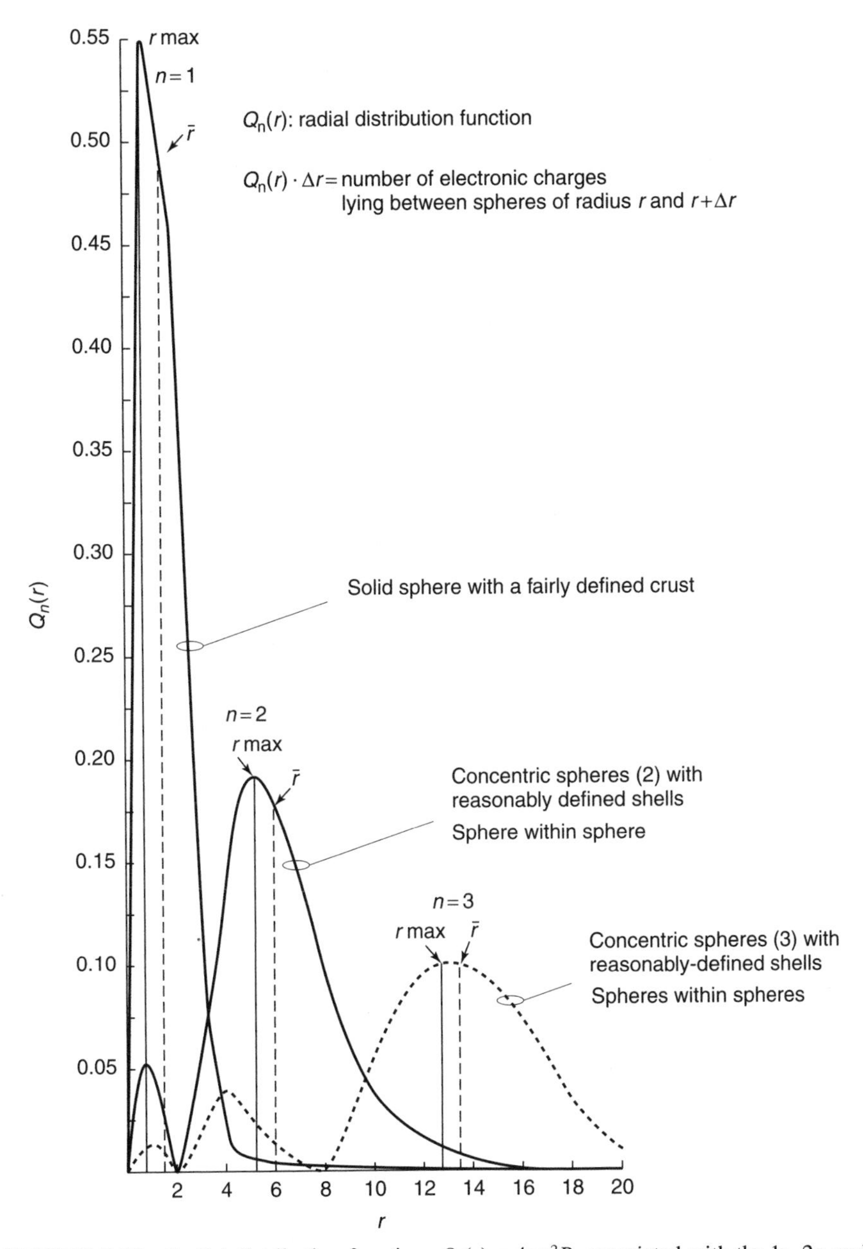

FIGURE D.18 Radial distribution functions $Q_n(r) = 4\pi r^2 P_n$ associated with the 1s, 2s, and 3s density distributions.

when the electron is in a 1s, 2s, or 3s.[6] $Q_n(r)$ *represents the probability density within the entire shell of radius r*. Note how the intrinsic character of the density distributions change as the value of n is increased. For $n = 1$ one effectively has a sphere with a fairly defined "crust": the electron could be more or less around the crust layer of the sphere (refer back to Fig. D.14). For $n = 2$ or 3 we have concentric spheres: Spheres of different radii are permitted (refer back to Fig. D.14). Some other observations are as follows: as n increases the average value of r increases (this is consistent with the fact that the energy of the electron also increases as n increases). The increased energy results in the electron being, in a probabilistic fashion, pulled further away from what would be the attractive force of the nucleus. This, in turn, implies that the electron could leave the atom all together during a chemical reaction.

When the electron have a nonzero angular momentum, the density distributions are not spherical: for each value of l, the electron density distribution has a characteristic shape (as stated, only electrons in s orbitals with zero angular momentum give rise to spherical density distributions and exhibit a charge density at the position of the nucleus). Figure D.19 depicts some contours (but not the three-dimensional shape). In the figure positive values for the contours of the orbitals are indicated by solid lines and negative values are indicated by dashed lines.

When $l = 1$ (clearly with $n > 1$), the Orbitals are Called p Orbitals. When $l = 1$, the orbital and its electron density are concentrated along a ray in space, as depicted in Figure D.19. One has three wave functions:

$$\Psi_{210}(r,\theta,\phi) = \frac{1}{4\sqrt{2\pi}} r e^{-r/2} \cos\theta$$

and

$$\Psi_{21\pm1}(r,\theta,\phi) = \mp \frac{1}{4\sqrt{2\pi}} r e^{-r/2} \sin\theta \, e^{\pm i\phi}$$

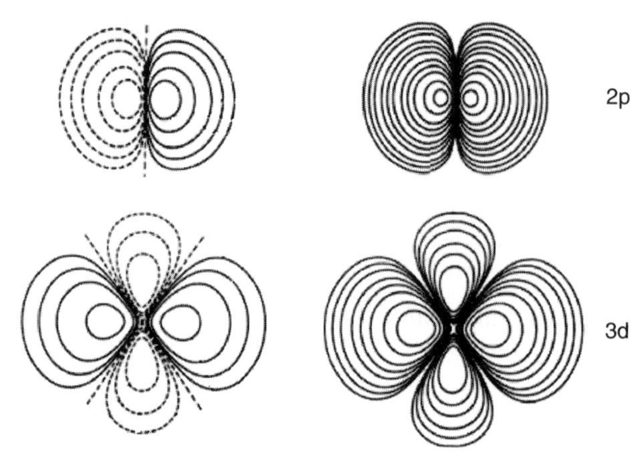

FIGURE D.19 Contour maps of the 2p and 3d atomic orbitals.

[6]Some also use the nomenclature $D_n(r)$.

The function Ψ_{210} is real and contains the factor $r\cos\Theta$, that is equal to the Cartesian variable z. In chemical applications, this is designated as a $2p_z$ orbital:

$$\Psi_{2p_z} = \frac{1}{4\sqrt{2\pi}} z e^{-r/2}$$

A contour plot is shown in Figure D.20. Note that this function is cylindrically symmetrical about the z axis with a node in the $x = y$ plane. Consider a plane that is perpendicular to the axis of the orbital and passes through the nucleus. The $2p$ wave function is positive in value on one side and negative in value on the other side of said plane. The orbital has a node in this plane, which means that an electron in a 2p orbital does not have any electronic charge density at the nucleus. An identical graph for the 2p density distribution is obtained for any plane that contains to the axis of the orbital. Hence, in three dimensions the electron density would appear to be concentrated in two lobes as depicted in Figure D.20. The $\Psi_{21\pm1}$ are complex functions and not as easy to represent graphically.

When $l = 2$, the Orbitals are Called d Orbitals. Figure D.19 shows that the electron density is zero at the nucleus and two nodes in the orbital. Notice that as the angular momentum of the electron increases, the density distribution becomes concentrated along a ray or in a plane in space. Figure D.21 depicts the electron density functions (wave functions) in terms of the first few orbitals.

We close the discussion by providing the most general formulation of the theory, as follows. For an electron in the field of a nucleus of charge $+Ze$, the Schrödinger equation can be written as:

$$\left\{-\frac{\hbar}{2m}\nabla^2 - \frac{Ze^2}{r}\right\}\Psi(\mathbf{r}) = E\Psi(\mathbf{r})$$

In atomic units where length is measured in bohrs and energy in hartrees:

$$a_0 = \frac{\hbar^2}{me^2} = 5.29 \times 10^{-11}\ \text{m} \equiv 1\ \text{bohr}$$

$$\frac{e^2}{a_0} = 4.358 \times 10^{-18}\ \text{J} = 27.211\ \text{eV} \equiv 1\ \text{hartree}$$

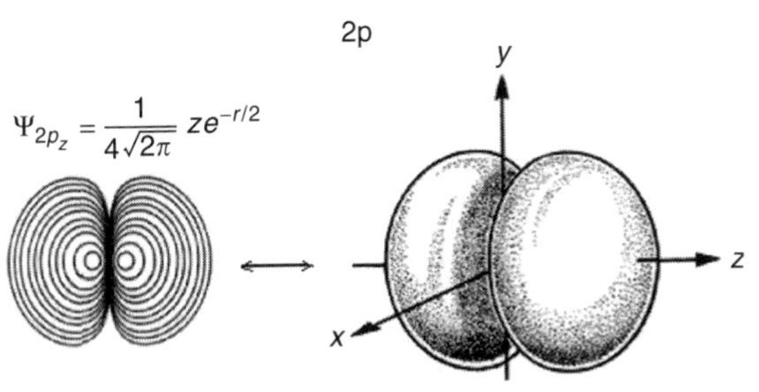

FIGURE D.20 Three-dimensional rendering of the 2p electron density distribution.

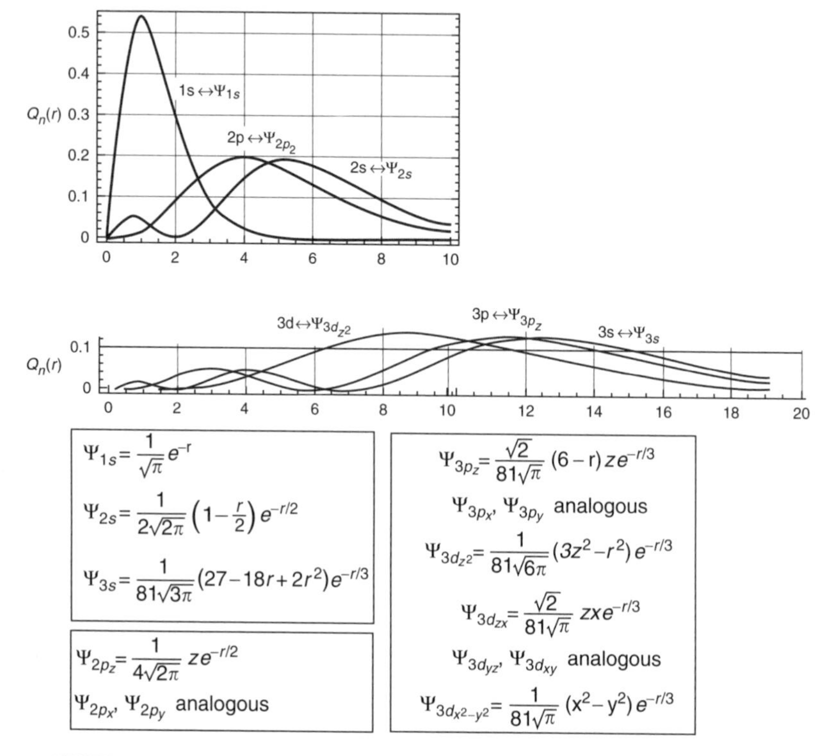

FIGURE D.21 First few electron density distribution functions for hydrogen.

one gets:

$$\left\{ -\frac{1}{2}\nabla^2 - \frac{Z}{r} \right\} \Psi(\mathbf{r}) = E\Psi(\mathbf{r})$$

(since the potential energy is spherically symmetrical, it is advantageous to work with the problem in spherical polar coordinates r; Θ; Ψ). These equations give rise to what are called hydrogenic atomic orbitals, as depicted in Figure D.22, which also summarizes various diagrams presented in this section. Said orbitals have applications to the structure of atoms and molecules.

D.2.7 Formal Application of Theory

So far, we have discussed many of the quantum theory concepts. Here we apply in a formal manner these concepts to the a one-electron atom, a multielectron atom, and molecules.

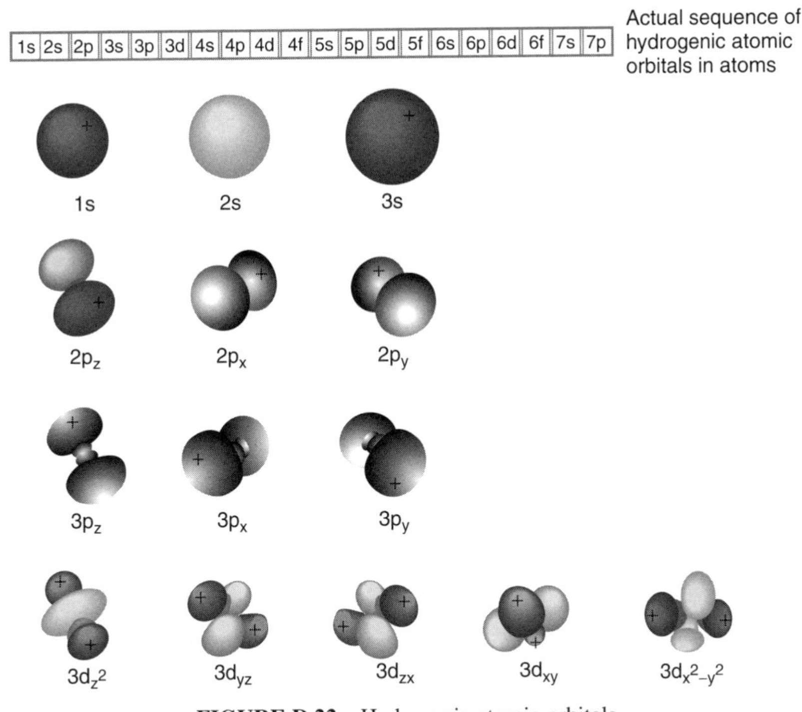

FIGURE D.22 Hydrogenic atomic orbitals.

Application to One-Electron Atom

We have already discussed this situation in the previous subsections, but we reapply the formalism here. The electrostatic potential energy, V, of one charged particle in the field of another is given by:

$$V = q_1 q_2 / r_{12}$$

where q is the charge on each particle and $\mathbf{r}$ is the distance between them. The potential energy between an electron ($q_1 = -e$) and a nucleus ($q_2 = +Ze$) can be written as:

$$V = -Ze^2/r$$

Substituting for V in the wave equation gives [453]:

$$[(-h^2/8\pi^2 m_e)\nabla^2 - Ze^2/r]\Psi = E\Psi$$

Each Ψ and its corresponding energy relate to a single electron bound to a nucleus of charge $+Ze$. The Ψ are called one-electron orbitals (also called hydrogenlike atomic orbitals). The orbitals are obtained analytically by solving Schrödinger's equation in closed form. Ψ^2 provides the probability density of finding an electron at a given point in space. Inspection of the solutions shows that each solution function has a radial term containing r, and some solution functions have angular terms. The orbitals without angular components (the s orbitals) are spherical. All other orbitals have spatial directionality, and the corresponding wave functions have angular terms. Notice that Ψ contains an imaginary component; on the other hand, Ψ^2 is real valued; Ψ^2 can be expressed as $\Psi^* \Psi$, where Ψ^* is the complex conjugate of Ψ. Also Ψ can be symmetric or antisymmetric: The superposition of *all* wave functions of an atom is spherically symmetric because there is no predisposed direction in an atom [453].

Schrödinger's Equation for a Multielectron Atom

The Hamiltonian for an atom with k electrons is given by looking at the electron state, the interaction between the electron and the nucleus, and the electron-to-electron interactions:

$$H = \left[\left(\frac{-h^2}{8\pi^2 m_e} \right) \sum_{i=1}^{k} \nabla^2 - \sum_{i=1}^{k} \frac{Z}{r_i} \right] + \sum_{i=1}^{k-1} \sum_{j=1+i}^{k-1} \frac{1}{r_{ij}} \tag{D.3}$$

The Schrödinger equation for a multielectron atom can be solved numerically. The electron-to-electron interactions (also called "electron correlation") are difficult to be included as an explicit term in the Hamiltonian, but the effect on Ψ can be accounted for by looking at how each electron interacts with an *average* of the nucleus and all other electrons (this is called the self-consistent field approximation). Keep in mind that multielectron wave functions are influenced by nuclear–electron attraction and this tends to spatially contract the electron density distribution toward the nucleus; however, the effect of electron–electron repulsion tends to make the orbitals larger and more diffuse [453].

Example

Consider a two-electron atom such as helium and ignore the electron-to-electron interactions. Then Eq. (D.3) becomes

$$H \cong -(\tfrac{1}{2})\nabla_1^2 - (2/r_1) - (\tfrac{1}{2})\nabla_2^2 - (2/r_2)$$

Schrödinger's equation $H\Psi = E\Psi$ can then be *approximated* as [453]:

$$[h(1) + h(2)]\Psi = (\varepsilon_i + \varepsilon_j)\Psi$$

where

$$h(1) = -(\tfrac{1}{2})\nabla_1^2 - (2/r_1)$$
$$h(2) = -(\tfrac{1}{2})\nabla_2^2 - (2/r_2)$$

and where ε_i and ε_j are the one-electron energies ($\varepsilon_i + \varepsilon_j = E$ with E the total energy). Ψ is composed of a combination of one-electron wave functions, ϕ; the arguments of the wave functions denote the positions of each electron. Notice that the probability of finding either electron at a given position (i.e., ϕ^2) does not depend on the position of the other electron (since the electron–electron interactions are neglected in this "simplified" approach). Electrons can exist in two possible states called "spin" states designated as α and β.

The multielectron wave function must take into consideration the fact that electrons are indistinguishable, and therefore interchanging electron position assignments in a wave function cannot lead to a different wave function. Hence, one obtains the following [453]:

$$\psi_s = \phi_i(1)\,\phi_j(2) + \phi_i(2)\,\phi_j(1) = \phi_i(2)\,\phi_j(1) + \phi_i(1)\,\phi_j(2) \quad \text{(symmetric function)}$$

and

$$\psi_a = \phi_i(1)\,\phi_j(2) - \phi_i(2)\phi_j(1) = -[\phi_i(2)\,\phi_j(1) - \phi_i(1)\,\phi_j(2)] \quad \text{(asymmetric function)}$$

where $\phi_i(1)$ and $\phi_j(2)$ are the one-electron wave functions (e.g., *1*s, 2s, etc.). ψ_s and ψ_a are called "space" functions because they depend on the spatial positions of the electrons. In practice, only the antisymmetric form is physically meaningful.

The above-mentioned spin states are represented by spin functions, ω_s and ω_a. Spin functions must be symmetric or antisymmetric with respect to the interchange of electron state assignments. For a two-electron system, the possible spin functions are

$$\omega_a = \{\alpha(1)\,\beta(2) - \alpha(2)\,\beta(1)\}$$

$$\omega_a = \{\alpha(1)\,\beta(2) + \alpha(2)\,\beta(1)\}$$

$$\omega_s = \{\alpha(1)\,\alpha(2)\}$$

$$\omega_s = \{\beta(1)\,\beta(2)\}$$

A wave function can be taken as being composed of a space function (ψ_s or ψ_a) and a spin function (ω_s or ω_a) (recall that only the antisymmetric form is physically meaningful). Namely,

$$\Psi_a = \psi_s\omega_a$$

with

$$\omega_a = \{\alpha(1)\,\beta(2) - \alpha(2)\,\beta(1)\}, \text{ or}$$

$$\omega_a = \{\alpha(1)\,\beta(2) + \alpha(2)\,\beta(1)\}$$

and

$$\Psi_a = \psi_a\omega_s$$

with

$$\omega_s = \{\alpha(1)\,\alpha(2)\}, \text{ or}$$

$$\omega_s = \{\beta(1)\,\beta(2)\}$$

There are four possible antisymmetric atomic spin orbitals for a *1*s(1) 2s(2) configuration:

$$\Psi_1 = \psi_s\,\omega_a$$
$$\Psi_2 = \psi_a\,\omega_s$$
$$\Psi_3 = \psi_s\,\omega_a$$
$$\Psi_4 = \psi_a\,\omega_s$$

or

$$\Psi_1 = \{\phi_1(1)\,\phi_2(2) + \phi_1(2)\,\phi_2(1)\}\{\alpha(1)\,\beta(2) - \alpha(2)\beta(1)\}$$
$$\Psi_2 = \{\phi_1(1)\,\phi_2(2) - \phi_1(2)\,\phi_2(1)\}\{\alpha(1)\,\alpha(2)\}$$
$$\Psi_3 = \{\phi_1(1)\,\phi_2(2) + \phi_1(2)\,\phi_2(1)\}\{\alpha(1)\,\beta(2) + \alpha(2)\,\beta(1)\}$$
$$\Psi_4 = \{\phi_1(1)\,\phi_2(2) - \phi_1(2)\,\phi_2(1)\}\{\beta(1)\,\beta(2)\}$$

Notice that the Ψ's are *approximate* wave functions for the entire atom because they are obtained by "mixing" one-electron spin orbitals.

The effects of electron–electron interactions cannot be completely ignored: Ψ and E cannot correctly predict atomic properties without somehow accounting for electron correlation. As noted the nuclear–electron attraction tends to spatially contract the electron density distribution toward the nucleus while the effect of electron–electron repulsion makes the orbitals larger and more diffuse. One approach to tackle the overemphasized degree of contraction when using the approximations described above is to implicitly reduce the effects of nuclear–electron attraction.

Schrödinger Equation for a Molecule

The Hamiltonian for a molecule with N atoms and k electrons can be approximated by considering the nuclei to be stationary relative to the electron motions (this is called the Born–Oppenheimer approximation). In this approach one looks at the energy of the electron, the electron-to-nucleus interaction, the electron-to-electron interaction, and the nucleus-to-nucleus interaction. One then has [453]

$$H = \left[(-h^2/8\pi^2 m_e)\sum_{i=1}^{k}\nabla^2 - \sum_{j=1}^{N}\sum_{i=1}^{k}\frac{Z_j}{r_{ji}}\right] + \sum_{i=1}^{k-1}\sum_{l=1+i}^{k}\frac{1}{r_{il}} + \sum_{j=1}^{N-1}\sum_{m=j+1}^{N} Z_j\,Z_m/R_{jm}$$

Just as for multielectron atoms, approximate *molecular* wave functions (molecular spin orbitals) can be developed using a set of one-electron wave functions. The major distinction between the various molecular orbital calculation methods relates to their inclusion (or lack of inclusion) of electron-to-electron interaction (electron correlation.)

The set of one-electron wave functions used to build molecular orbital wave functions is called the *basis set* (the hydrogenlike wave functions modified for electron correlation not used in general because they lead to mathematical complexities).

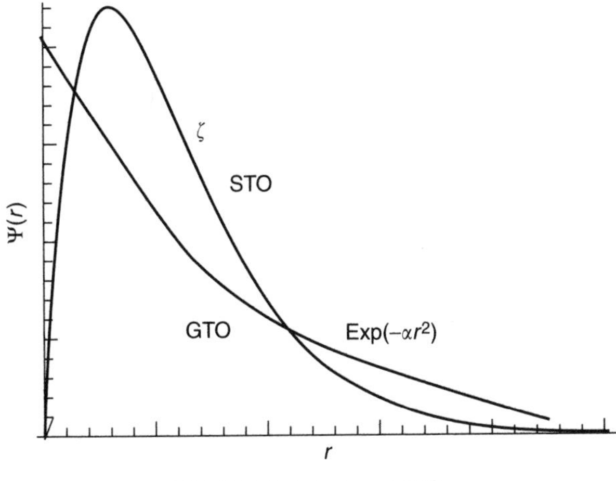

FIGURE D.23 STO/GTO.

Instead, wave functions are used where the radial terms, $\Psi(r)$, are simplified. The most common such wave functions are (see Fig. D.23):

1. Slater-type orbitals (STO):

$$\Psi(r) = r^{[n-1]}\exp[-(Z-s)r/n]$$

   where s is a screening constant and $Z - s = \zeta$ is taken as an adjustable parameter.
2. Gaussian-type orbitals (GTO):

$$\Psi(r) = \exp(-\alpha r^2)$$

   where α is a curve-fitting constant used to approximate an STO.

The mathematics for solving the integrals of the wave equation using STOs are demanding; hence, the STO screening constants are calculated for small model molecules using rigorous methods and then generalized for use with actual molecules of interest. Gaussian orbitals are mathematically simpler than STOs but are less accurate; the one-electron orbitals can be built by combining sets of Gaussian functions that approximate each STO. The issue of molecular modeling is revisited in Appendix E.

D.3 OTHER TOPICS

D.3.1 Field Theory

This section provides a short introduction and some observations on field theory. A field is an entity that acts as intermediary in interactions between particles, that is distributed over part or all of space, whose properties are functions of space coordinates, and (except for static fields) also functions of time [31]. Examples of fields include (but are not limited to) the electric and magnetic fields that envelop electric charges and magnets. Put differently, a field is a domain or region throughout which a force may be exerted: Fields are used to describe the situation where two bodies separated in space exert a force on each other (the alternative to using a field-based view is to postulate that physical influences can be transmitted through empty space without any material or physical mechanism). In a field-based formulation, instead of postulating that body A directly exerts a force on body B, it is assumed that body A (the source) creates a field in every direction around it, and body B (the detector) experiences the field that exists at its position.

There is also a quantum mechanical analog of field concept in which the function of space and time is replaced by an operator at each point in space–time [31]. Quantum field theory addresses the understanding of the quantum mechanical interaction of fields and elementary particles. Quantum field theory aspires to be the most fundamental of sciences. When quantum field theory is applied to the study of electromagnetism, it is called quantum electrodynamics (QED). When quantum field theory is applied to the study of the strong interactions between quarks and between protons, neutrons, and other baryons and mesons, it is called quantum chromodynamics (QCD). QED is currently the best theory available for describing effects of the electromagnetic force. The term *quantum field theory* is used interchangeably with *high-energy physics* and *particle physics*, this based on the fact that the experimental support for this theory comes from a wide array of experiments involving high-energy beams of particles.

The use of QED has proven useful in describing the interaction of light with matter. One of QED's key features is its gauge symmetry. Gauge symmetry implies that when independent changes to local field values are made at different points in space, the equations of quantum electrodynamics are not altered (symmetry is ensured only if the quantum description of a charged particle contains an electromagnetic field with its gauge boson). The underlying mathematical calculations, however, are complex (these are carried out using Feynman diagrams that represent possible variations of interactions and provide a shorthand for precise mathematical formulation). QCD has a mathematical mechanism similar to that of QED.

The following observations of interest are made by Roman Jackiw [463]:

Our present-day theory for fundamental processes in nature (including descriptions of elementary particles and forces) is perceived to be successful: Experimental data confirms theoretical prediction, and where accurate calculations and experiments are attainable, agreement is achieved to six or seven significant figures. No experiment has thus far contradicted our understanding of the gravitational interactions as described by Einstein's general relativity, nor of the strong nuclear interactions, nor of the electromagnetic and radioactivity-producing weak interactions that are now collected into the Glashow–Weinberg–Salam standard model. The strong and electroweak theories make use of a quantum mechanical description, whereas classical physics suffices to account for all known gravitational phenomena. The theoretical structure within which this success has been achieved is local field theory, which offers physicists a variety of applications. Local field theory is a language with which physical processes are discussed and it provides a model for fundamental physical reality, as described by our theories of strong, electroweak, and gravitational processes. No other framework currently exists in which one can calculate so many phenomena with such ease and accuracy.

Field theory first arose from a mathematical account of the propagation of fluids. Field theory was used in the discussion within classical physics of Faraday–Maxwell electromagnetism and soon thereafter of Einstein's gravity theory. Schrödinger's wave mechanics became a bridge between classical and quantum field theory: The quantum mechanical wave function is also a local field, which when "second" quantized gives rise to a true quantum field theory (but a nonrelativistic one). Quantization of electromagnetic waves produced the first relativistic quantum field theory, which when supplemented by the quantized Dirac field gave us quantum electrodynamics, whose further generalization to matrices of fields—the Yang–Mills construction—is the present-day standard model of elementary particles. This development carries with it an extrapolation over enormous scales: Initial applications were at microscopic distances or at energies of a few electron volts, whereas contemporary studies of elementary particles involve 10^{11} electron volts or short distances of 10^{-16} cm. The "quantization" procedure, which extended classical field theory's range of validity, consists of expanding a classical field in normal modes and taking each mode to be a quantal oscillator. Field theoretic ideas also reach for the cosmos through the development of the "inflationary scenario"—a speculative analysis of the early universe. Additionally, quantum field theories provide effective descriptions of many-body, condensed-matter physics. Here the excitations are not elementary particles and fundamental interactions are not probed, but the collective phenomena that are described by many-body field theory exhibit many interesting effects, which in turn have been recognized as important for elementary particle theory.

But in spite of these successes, today there is little confidence that field theory will advance our understanding of nature at its fundamental workings, beyond what has been achieved. Although in principle all observed phenomena can be explained by present-day field theory (in terms of the quantal standard model for particle physics, perhaps slightly extended to incorporate massive neutrinos, and the classical Newton–Einstein model for gravity), these accounts are still imperfect. The particle physics model requires a list of ad hoc inputs that give rise to conceptual, general questions such as: Why is the dimensionality of space–time four? Why are there two types of elementary particles (bosons and fermions)? What determines the number of species of these particles? The standard model also leaves us with specific technical questions: What fixes the matrix structure, various mass parameters, mixing angles, and coupling strengths that must be specified for concrete prediction? Moreover, classical gravity theory has not been integrated into the quantum field description of nongravitational forces, again because of conceptual and technical obstacles: Quantum theory makes use of a fixed space–time, so it is unclear how to quantize classical gravity, which allows space–time to fluctuate; even if this is ignored, quantizing the metric tensor of Einstein's theory produces a quantum field theory beset by infinities that cannot be controlled [463].

But these shortcomings are actually symptoms of a deeper lack of understanding that has to do with symmetry and symmetry breaking. In summary [31, 463]:

1. Present-day theory for fundamental processes (i.e., descriptions of elementary particles and forces) is phenomenally successful. Experimental data confirms theoretical prediction, and where accurate calculation and experiments are attainable, agreement is achieved to six or seven figures. Two examples: (a) The helium atom ground-state energy (Rydbergs) is experimentally measured as -5.8071394 and theoretically calculated as -5.8071380. (b) The muon magnetic dipole moment is experimentally measured as 2.00233184600 and theoretically calculated as 2.00233183478.
2. The theoretical structure within which this success has been achieved is local-field theory, which offers a wide variety of applications, and which provides a model for fundamental physical reality as described by our theories of strong, electroweak, and gravitational processes. No other framework exists in which one can calculate so many phenomena with such ease and accuracy.
3. But is spite of these successes, today there is little confidence that field theory will advance our understanding of nature at its fundamental workings beyond what has already been achieved. Although in principle all observed phenomena can be explained by present-day field theory, these accounts are still imperfect, requiring ad hoc inputs. Moreover, because of conceptual and technical obstacles, classical gravity theory has not been integrated into the quantum field description of non-gravitational forces: Quantizing the metric tensor of Einstein's theory produces a quantum field theory beset by infinities that apparently cannot be controlled.
4. These shortcomings are actually symptoms of a deeper lack of understanding concerning symmetry and symmetry breaking. Physicists are happy in the belief that nature in its fundamental workings is essentially simple, but observed physical phenomena rarely exhibit overwhelming regularity. Therefore, at the very same time that we construct a physical theory with intrinsic symmetry, we must find a way to break the symmetry in physical consequences of the model.
5. These problems have produced a theoretical impasse for over two decades, and in the absence of new experiments to channel theoretical speculation, some physicists have concluded that it will not be possible to make progress on these questions within field theory, and they have turned to a new structure, string theory. In field theory, the quantized excitations are point particles with point interactions, and this gives rise to the infinities. In string theory, the excitations are extended objects—strings—with nonlocal interactions; there are no infinities in string theory, and that enormous defect of field theory is absent.
6. Yet in spite of its positive features, until now string theory has provided a framework rather than a definite structure, and a precise derivation of the standard model has yet to be given. On previous occasions when it appeared that quantum field theory was incapable of advancing our understanding of fundamental physics, new ideas and new approaches to the subject dispelled the pessimism. Today we do not know whether the impasse within field theory is due

to a failure of imagination or whether indeed we have to present fundamental physical laws in a new framework, thereby replacing the field theoretic one, which has served us well for over 100 years.

D.3.2 String Theory

In particle physics, string theory is a theory of elementary particles based on the idea that the fundamental entities are not pointlike particles but finite lines (strings), or closed loops formed by strings; the strings one-dimensional curves with zero thickness and lengths (or loop diameters) of the order of the Planck length of 10^{-35} m [31, 53]. String theory is a model with fundamental building blocks consisting of one-dimensional extended objects (strings); this is in contrast with the zero-dimensional points (particles) that were the basis of most earlier physics. It follows that string theories are able to avoid problems associated with the presence of pointlike particles in a physical theory. String theories do not just describe strings but other objects as well, including points, membranes, and higher-dimensional objects. String theory has not yet made the kind of predictions that would allow it to be experimentally tested.

String theory elementary particle physics aims to give a consistent description of the elementary particles and their interactions. As discussed throughtout this text, experiments indicate the existence of four independent forces: the electromagnetic, the weak, the strong, and the gravitational force. The first three forces are described by the so-called standard model. The gravitational force is described by Einstein's general relativity. The standard model is a quantum theory of elementary particles; general relativity is a classical theory. String theory is a proposal to unify gravitation with the other forces in a single quantum theory. In string theory it is assumed that particles manifest themselves as the different vibration modes of a stringlike structure. String theory is still in a premature stage. Although many interesting results have been obtained, much work remains to be done to achieve the unification mentioned above. The investigation of string theory has led to many unexpected connections with other parts of theoretical physics and mathematics [465].

D.3.3 A Few Words About the Current Understanding of Physics

Modeling the world is a complex task. Many models have been developed in the past half-century, including the standard model and string theory. No final unifying theory has yet been developed. On this topic, the following points are made by Frank Wilczek on limitations of theoretical physics may be of interest [31, 53]:

1. There have been extraordinary triumphs of physicists using analysis and synthesis or, alternatively, reductionism, to account for the behavior of matter and the structure of the universe as a whole. However, there has been quite a lot that physicists might once have hoped to derive or explain based on fundamental principles, for which that hope now seems dubious or forlorn. One important limitation concerns the lack of a principle that could lead to a unique choice

among different seemingly possible solutions of the fundamental equations and could select out the universe we actually observe.

2. It is very instructive to consider the corresponding problem for atoms and matter. Like the classical equations governing planetary systems, the equations of quantum mechanics for electrons in a complex atom allow all kinds of solutions. In fact, the quantum equations allow even more freedom of choice in the initial conditions than the classical equations do. The wave functions for N particles live in a much larger space than the particles do: They inhabit a full-bodied $3N$-dimensional configuration space, as opposed to $2N$ copies of three-dimensional space. (For example, the quantum description of the state of two particles requires a wave function that depends on 6 variables, whereas the classical description requires 12 numbers, namely their positions and velocities.)
3. Yet the atoms we observe are always described by the same solutions—otherwise we would not be able to do stellar spectroscopy, or even chemistry. Why? A proper answer involves combining insights from quantum field theory, mathematics, and some cosmology. Quantum field theory tells us that electrons—unlike planets—are all rigorously the same. Then the mathematics of the Schrödinger equation, or its refinements, tells us that the low-energy spectrum is discrete, which is to say that if our atom has only a small amount of energy, then only a few solutions are available for its wave function.
4. But because energy is conserved, this explanation begs another question: What made the energy small in the first place? Well, the atoms we study are not closed systems; they can emit and absorb radiation. So the question becomes: Why do they emit more often than they absorb, and thereby settle down into a low-energy state? That is where cosmology comes in. The expanding universe is quite an effective heat sink. In excited atoms, energy radiated as photons eventually leaks into the vast interstellar spaces and redshifts away. By way of contrast, a planetary system has no comparably efficient way to lose energy—gravitational radiation is ridiculously feeble—and it cannot relax.
5. So one selection principle that applies to many important cases is to choose solutions with low energy. In the same spirit, when the residual energy cannot be neglected, one should choose thermal equilibrium solutions. This is appropriate for systems and degrees of freedom that have relaxed but is not appropriate in general. The selection procedure that dominates the literature of high-energy physics and string theory is energy based. It is traditional to identify the lowest-energy solution with the physical vacuum, and to model the content of the world using low-energy excitations above that state. For the solution to count as successful, the excitations should include at least the particles of the standard model.

APPENDIX E

Mechanical Molecular Models and Quantum Aspects of Chemistry

E.1 MECHANICAL MOLECULAR MODELS

In this section (as well as in the one that follows) we discuss methodologies to model the behavior of molecules.

As we saw earlier in Chapter 3, chemistry has empirical roots. Chemistry is generally based on the hypothesis that a molecule is a collection of atoms linked by a network of bonds (this being known as the *molecular structure hypothesis*). This hypothesis arose in the 19th century on the basis of experimental considerations. The molecular structure hypothesis has continued to serve as the modeling tool for ordering and classifying the observations of chemistry: In practical chemistry one has an understanding of the science based on a classification scheme that is powerful; however, because of its empirical nature, it is also limited. The problem with this hypothesis has been that it has not been directly aligned at the conceptual level to quantum theory—this topic is addressed in the next section.

We already noted in Appendix D that three methods exist to undertake this modeling: (i) ab initio, (ii) semiempirical all-valence electron methods, and (iii) "mechanical" molecular model, also known as molecular mechanics (MM). MM was developed in order to be able to describe molecular structures and properties in a pragmatic manner, particularly for molecules containing thousands of atoms. Molecular mechanics methods are based on working assumptions such as the following: atomlike particles are spherical (with radii obtained from measurements or theory) and have a net charge (with charge obtained from theory); nuclei and electrons are lumped into atomlike particles; and, interactions[1] (which determine the spatial distribution of atomlike particles and their energies) are modeled based on springs and classical potentials, where the physics of spring deformation is used to describe the bonds' ability to stretch, bend, and twist. Nonbonded atoms (when they are more than two

[1]Interactions are easiest to describe analytically when atoms are considered as spheres of characteristic radii.

Nanotechnology Applications to Telecommunications and Networking, By Daniel Minoli

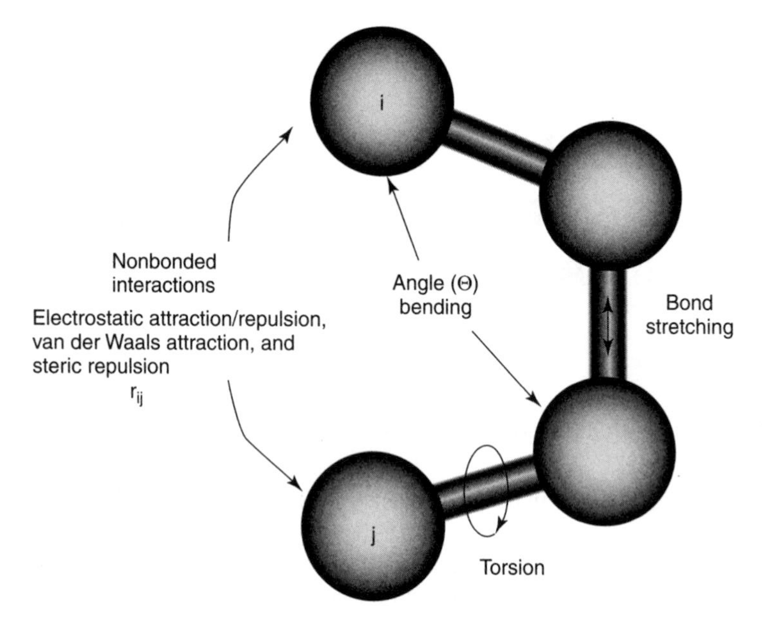

FIGURE E.1 Mechanical molecular models.

bonds apart) interact through electrostatic attraction/repulsion, van der Waals attraction, and steric repulsion (steric repulsion refers to the repulsion when two groups of atoms or molecules are very close together and "try" to occupy the same space.) See Figure E.1. The goal of MM is to predict the energy of any given configuration of a molecule; however, only differences in energy between two or more configurations have meaning (the absolute values per se have no meaning) [465].

A simple MM energy equation is:

$$\begin{aligned}\text{Energy} &= \text{stretching energy} + \text{bending energy} \\ &\quad + \text{torsion energy} + \text{nonbonded interaction energy}\end{aligned}$$

This equation along with the parameters used to describe the behavior of different kinds of atoms and bonds is called a force field.[2]

Stretching Energy

The stretching energy equation is based on Hooke's law:

$$E = \sum_{\text{bonds}} k_b \, (r - r_0)^2$$

[2]A number of different force fields have been developed over the years, in order to improve the accuracy of the mechanical model. For example, some force fields include additional energy terms that describe other deformations; other force fields account for coupling between bending and stretching in adjacent bonds. Clearly, the mathematical formulation of the energy terms varies from force field to force field.

where k_b determines the stiffness of the bond spring, while r_0 defines its equilibrium length; this equation estimates the energy associated with vibrations about the equilibrium bond length. Specific k_b and r_0 values are assigned to each pair of bonded atoms based on their types (e.g., C–C, H–O, etc.). Note that this equation is a parabola with a minimum (of zero) at $r = r_0$. The model is more accurate in the vicinity of this point (the model becomes less useful as a bond is stretched toward the point of dissociation) [465].

Bending Energy

The bending energy equation is also based on Hooke's law, where the k_θ parameter controls the stiffness of the angle spring, while θ_0 defines its equilibrium angle:

$$E = \sum_{\text{angles}} k_\theta (\theta - \theta_0)^2$$

Note that this equation of a parabola with a minimum (of zero) at $\theta = \theta_0$. Specific parameter values for angle bending are assigned to each bonded triplet of atoms based on their types (e.g., C–C–C, C–O–C, etc.). The effect of the k_b and k_θ parameters is to broaden the parabola or make it tighter; for example, the larger the value of k, the more energy is required to deform an angle (or bond) from its equilibrium value.

Torsion Energy

The torsion energy is modeled by a simple periodic function:

$$E = \sum_{\text{torsions}} A\,[1 + \cos(n\tau - \phi)]$$

The values of the parameters are determined from curve fitting: A establishes the amplitude of the curve, n determines its periodicity (e.g., 120° for a CH_3–CH_3 bond), and ϕ shifts the curve along the rotation angle axis (τ). Specific parameter values for torsional rotation are assigned to each bonded quartet of atoms based on their types (e.g., C–C–C–C, H–C–C–H, etc.). The torsion energy in molecular mechanics is primarily used to correct the remaining energy terms to make the total energy agree with experiments or with rigorous quantum mechanical calculation, rather than to represent a physical process [465]. See Figure E.2 for an example.

Nonbonded Energy

The nonbonded energy accounts for repulsion, van der Waals attraction, and electrostatic interactions and is computed as the pairwise sum of the energies of all possible interacting nonbonded atoms i and j:

$$E = \underbrace{\sum_i \sum_j \frac{q_i q_j}{r_{ij}}}_{\text{Electrostatic term}} + \underbrace{\sum_i \sum_j \frac{-A_{ij}}{r_{ij}^6} + \frac{B_{ij}}{r_{ij}^{12}}}_{\text{van der Waals term}}$$

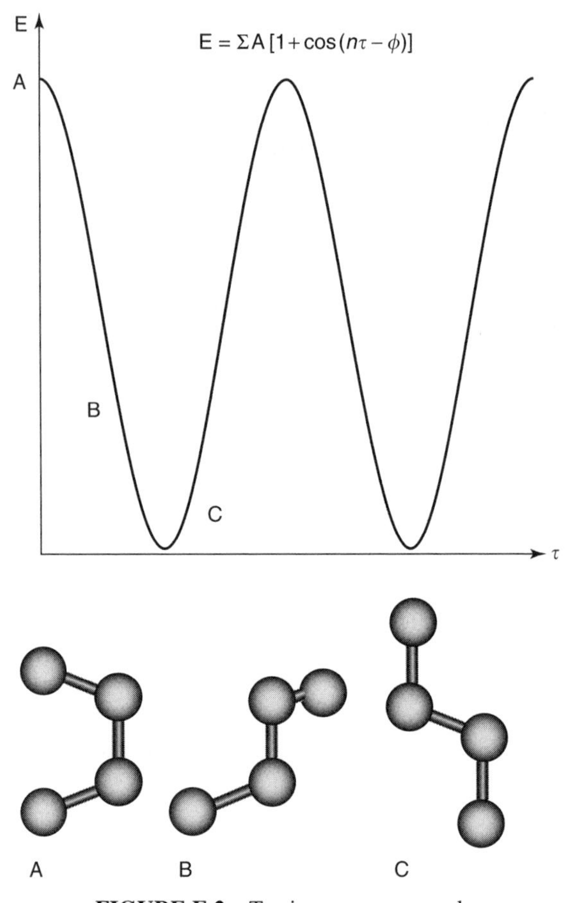

FIGURE E.2 Torsion energy example.

The *A* and *B* parameters control the depth and position (interatomic distance) of the potential energy well for a given pair of nonbonded interacting atoms (e.g., C:C, O:C, etc.). Specifically, *A* determines the degree of "stickiness" of the van der Waals attraction, and *B* determines the degree of "hardness" of the atoms. The *A* parameter can be calculated from quantum theory or can be obtained from atomic polarizability measurements; the *B* parameter is derived from crystallographic measurements and is intended to reproduce observed average contact distances between different kinds of atoms in crystals of various molecules [465].

As can be seen from the equation just given, the nonbonded energy van der Waals attraction applies at short range and rapidly dies off as the interacting atoms move apart by a few angstroms; the decay follows a power of -6 (also see Fig. E.3). Repulsion applies when the distance between interacting atoms becomes even slightly less than the sum of their contact radii; repulsion is modeled by an equation that decays following a power of -12. The energy term that describes attraction/repulsion provides for a smooth transition between these two modalities.

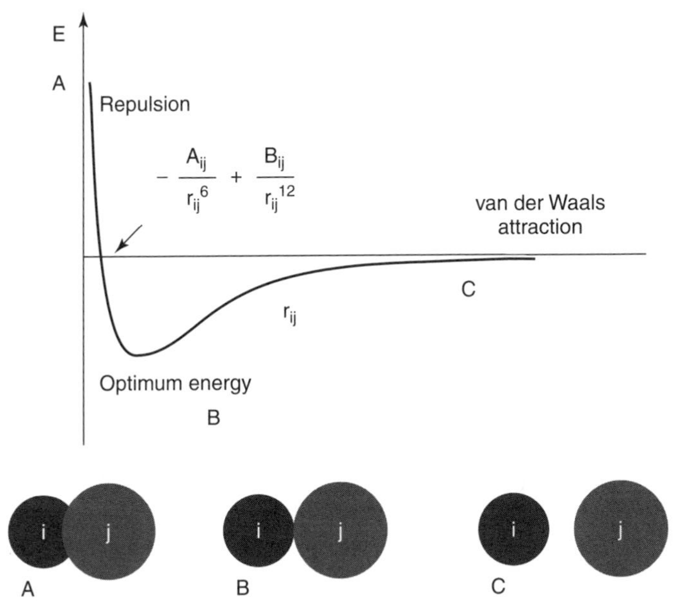

FIGURE E.3 Nonbonded energy example.

The electrostatic term is modeled using a Coulombic potential: This energy is a function of the charge on the nonbonded atoms, their interatomic distance, and a molecular dielectric term that accounts for the attenuation of electrostatic interaction by the environment (the molecular dielectric is typically set to a constant value between 1.0 and 5.0) [465].

E.2 QUANTUM CHEMISTRY/LINEAR COMBINATION OF ATOMIC ORBITALS

In this section we apply some of the quantum theory principles discussed in Chapter 3 to the multiatom (molecule) environment typical of a chemical reaction applications. This section can be skipped by the average reader.

In recent years there has been an interest in using quantum theory to support chemistry. Quantum chemistry is the application of quantum theory to the study of chemical phenomena; it is part of the discipline of quantum theory and is based on Schrödinger's equation. As we discussed in Appendix D, quantum theory is the physics that governs the motions of the nuclei and electrons that make up the atoms and the bonds. Here we add some additional observations, as they apply to multiatom molecules.

In the 1950s, Richard Feynman and Julian Schwinger provided a reformulation of physics that enables one to pose and answer the questions, "What is an atom in a molecule and how does one predict its properties?" Bader [59, 60] demonstrated that this new formulation of physics,[3] when applied to the observed topology of the distribution of electronic charge in real space, yields a unique partitioning of some total system into a set of bounded spatial regions; the form and properties of the groups so defined faithfully recover the characteristics ascribed to the atoms and functional groups of chemistry. By establishing this association, the molecular structure hypothesis discussed in the previous section is freed from its empirical constraints, and the predictive machinery of quantum theory can be incorporated into the resulting theory—a theory of atoms in molecules and crystals [59, 60]. Bader's theory recovers the central operational concepts of the molecular structure hypothesis, that of a functional grouping of atoms with an additive and characteristic set of properties, together with a definition of the bonds that link the atoms and impart the structure. The theory quantifies and provides the physical understanding of the existing concepts of chemistry, and at the same time it makes possible new applications of theory. Specifically, these new applications enables one to perform on a computer, in a manner directly paralleling experiment, everything that can now be done in the laboratory, but more quickly and more efficiently, by linking together the functional groups of theory. These applications include the design and synthesis of new molecules and new materials with specific desirable properties.

As we saw in Appendix D, Schrödinger's equation for molecular systems can only be solved approximately. The approximation methods can be categorized as either semiempirical or ab initio. Semiempirical methods use parameters that compensate for neglecting some of the time-consuming mathematical terms in Schrödinger's equation; ab initio methods include all such terms. The parameters used by semiempirical methods can be derived from experimental measurements or by performing ab initio calculations on model systems. A basic understanding of how approximate molecular wave equations are constructed and solved is essential to the understanding of quantum chemistry. Quantum chemistry is highly mathematical in nature.

E.2.1 Linear Combination of Atomic Orbitals Approach

A well-accepted method for building molecular spin orbitals is the linear combination of atomic orbitals (LCAO) method. With this method the molecular orbitals (ψ_i) are expressed as a set of one-electron orbitals (ϕ) centered on each nucleus; namely,

$$\psi_i = \sum_{j=1}^{m} c_{ij}\phi_j$$

where m is the size of the basis set and the c_{ij} are coefficients that determine the weight of each one-electron wave function in each molecular orbital i. The goal of

[3]Feynman and Schwinger have developed the new formulation of quantum mechanics based upon the classical principle of least action. The generalization of the action principle, as contained in his *principle of stationary action*, in addition to determining the field equation, yields a variational derivation of Heisenberg's equation of motion. With this one to ask and answer questions that could not be answered using the Hamiltonian-based approach to quantum mechanics.

the LCAO-based methods is to find the c_{ij} values that best approximate the actual wave function ψ_i.

The c_{ij} can be calculated in a number of ways, all of which are based on the method of *linear variation*; specifically, according to the *variational theorem*, one needs to find c_{ij} values that give rise to the lowest possible energy (minimum energy) for a given choice of Hamiltonian and basis set [453]. The average energy $<E>$ is obtained from the following expression:

$$<E> = \frac{\int (c_1{}^*\phi_1{}^* + c_2{}^*\phi_2{}^* + \cdots) H (c_1\phi_1 + c_2\phi_2 + \cdots)\, d\tau}{\int (c_1{}^*\phi_1{}^* + c_2{}^*\phi_2{}^* + \cdots)\,(c_1\phi_1 + c_2\phi_2 + \cdots)\, d\tau}$$

where $d\tau$ is an infinitesimal "volume" of space and spin. As usual, the minimum energy with respect to each c_{ij} is obtained when $d<E>/dc_j = 0$. Hence, differentiating the equation for $<E>$ with respect to the LCAO coefficients for ψ_1 leads to [453]:

$$\begin{aligned} & c_1\left[\int \phi_1 H \phi_1\, d\tau - <E> \int \phi_1\phi_1\, d\tau\right] \\ & + c_2\left[\int \phi_1 H \phi_2\, d\tau - <E> \int \phi_1\phi_2\, d\tau\right] \\ & \cdots \\ & + c_n\left[\int \phi_1 H \phi_n\, d\tau - <E> \int \phi_1\phi_n\, d\tau\right] = 0 \end{aligned}$$

If one defines

$$H_{kl} \equiv \int \phi_k H \phi_1\, d\tau \qquad \text{(call this the Coulomb integral)}$$

and

$$S_{kl} \equiv \int \phi_k \phi_1\, d\tau \qquad \text{(call this the overlap integral)}$$

then the derivative expression $d<E>/dc_j = 0$ can be rewritten as:

$$c_1(H_{i1} - <E>S_{i1}) + c_2(H_{i2} - <E>S_{i2}) + \cdots + c_n(H_{in} - <E>S_{in}) = 0$$

where n is the number of one-electron wave functions in the basis set of ψ_i.

This set of equations (one for each ψ_i) can be used to solve for $<E>$ via the method of determinants. As an example, the determinant for the case of H_2^+ (one electron) is written as:

$$\begin{vmatrix} H_{11} - <E>S_{11} & H_{12} - <E>S_{12} \\ H_{21} - <E>S_{21} & H_{22} - <E>S_{22} \end{vmatrix} = 0$$

Solving the determinant for $<E>$ requires values for each of the overlap integral and Coulomb integral; these can be obtained completely or partially from experimental data (semiempirical methods) or from full evaluation of these integrals (ab initio).

Finally, the value of $<E>$ can be substituted into the set of:

$$c_1(H_{i1} - <E>S_{i1}) + c_2(H_{i2} - <E>S_{i2}) + \cdots + c_n(H_{in} - <E>S_{in}) = 0$$

to solve for the c_{ij}.

This set of equations can be expressed in the form of a matrix equation: $HC = SCE$ where H, S, C, and E are matrices of Coulomb integrals, overlap integrals, the c_{ij}, and the $<E>$, respectively.

E.2.2 Hartree–Fock (HF) Approach

Modern molecular orbital computational methods (ab initio and semiempirical) make use of the Hartree–Fock (HF) approach to approximating the molecular wave function. Here, the Hamiltonian considers each electron, i, in the average field of all other electrons in the molecule. The Hamiltonian that describes this approximation (called the *Fock operator F*) is given by [453]:

$$F_i = -\frac{1}{2}\nabla^2 - \sum_{j=1}^{N} \frac{Z_j}{r_{ji}} + \sum_{l=1}^{k} (2J_l - K_l)$$

where J (called the *Coulomb integral*) reflects the average interaction potential of electron i due to all other electrons, and K is called the *exchange integral* (K has no formal physical interpretation). The terms J and K are functions of the one-electron molecular orbitals; it follows that F_i are also functions of the one-electron molecular orbitals. The wave equation can be written as:

$$(h_i + B_i)\psi_i = \varepsilon_i \psi_i$$

Because the ψ_i (the functions being solved for) are part of B, the solution is obtained iteratively [453]:

1. A set of ψ_i are initially guessed, from which B_i is computed.
2. The Fock operator is then used to solve for a new ψ_i, which is used to compute a new B_i.

This process is repeated until ψ_i becomes constant. The molecular orbitals are constructed from a basis set of one-electron atomic orbitals, ϕ. The electrons can be assigned to the molecular orbitals in various ways [e.g., 1s(1) 1s(2) 2s(3) 2s(4), 1s(1) 1s(2) 2s(3) 2p(4), etc.]. The LCAO method is used to approximate ψ_i, with the molecular orbital coefficients, c_{ij}, calculated via linear variation where $\langle\varepsilon_i\rangle$ is now calculated using the Fock operator. Hence, each iteration involves solving for a new set of c_{ij}. The process is repeated until the c_{ij} approach constant values. The quality of the HF result depends on the size and quality of the basis set, but, in general, the effects of electron–electron repulsion tend to be underestimated by these methods, limiting the accuracy of the resulting wave functions.

E.2.3 Configuration Interaction Method

Repulsive interactions can be minimized by allowing the electrons to exist in more orbitals. The configuration interaction (CI) method addresses this problem by allowing multiple sets of electron assignments to be used in constructing the molecular wave functions. Molecular wave functions representing different configurations [e.g. 1s(1) 1s(2) 2s(3) 2s(4), 1s(1) 1s(2) 2s(3) 2p(4), etc.] are combined in a manner analogous to the LCAO approach [453]:

$$\psi_{ci} = a_1 \psi_1 + a_2 \psi_2 + \cdots$$

The a_i terms are estimated using linear variation. CI can significantly increase the accuracy of the results, but this comes at the expense of computational complexity for solving the molecular wave equation.

E.2.4 Semiempirical Molecular Orbital Methods

Semiempirical approximations involve the neglect of a number of integrals arising in the J and K terms of the Hartree–Fock equation. For molecules with large numbers of electrons, evaluation of these integrals can be computationally impractical. Experimentally determined parameters (or in some cases, parameters determined from ab initio calculations on model systems) are used to compensate for the missing integrals [453].

E.2.5 Modeling for Nanomaterials

The semiconductor industry is constantly pushing toward ever-smaller devices, and it is expected that we will see commercial devices with gate lengths less than 10 nm within the next decade; such small devices have active regions that are smaller than relevant coherence lengths, so that full quantum modeling will be required [466]. (Computer) modeling of nanomaterials can range from ab initio (first principles) quantum theory, to molecular mechanics, to mesoscale simulation methods [92].

APPENDIX F

Basic Molecular/Nanotechnology Instrumentation

This appendix provides an overview of basic nanotechnology and molecular (imaging) instrumentation. This equipment is also known as nanoinstrumentation.

Classical methods used to study the structure and composition of solids include the light microscope, X-ray diffraction, and infrared and ultraviolet spectroscopy. In fact, spectroscopy was originally used to determine the chemical composition of materials, and new techniques were added to detect structural characteristics. Newer methods (such as atomic layer epitaxy, molecular-beam epitaxy, and the Langmuir–Blodgett technique) have been developed to fabricate thin, two-dimensional nanostructures with an accuracy of a single atomic layer. The development of new tools and techniques for nanostructure research has become a science in itself, particularly in the field of materials science and its applications to nanotechnology [22].

Conventional optical microscopy (also known as far-field optical microscopy) is a technique employed to produce magnified images of specimens. The standard setup consists of a system of lenses that focus light from the sample into a virtual, magnified image. Unfortunately, the use of lenses introduces diffraction effects and places a limit on the resolution of the technique (Abbé diffraction barrier). A lens can only focus traveling waves (this is the cause of diffraction) and as a result, point objects become smeared out into Airy functions [467]. The Rayleigh criterion states that to just resolve two structures in proximity, the maximum of one Airy function should lie on the first minimum of its neighbor. In practice, this makes the theoretical resolution limit of any lens-based microscope $\sim\lambda/2$. This is a fundamental limitation of lens-based imaging and cannot be overcome without utilizing a radically different technique.

The systems discussed in this appendix address in various ways, manners, and approaches imaging techniques aimed at obtaining resolution beyond the Abbé diffraction limit. Some of these techniques are based on (near-field) optics, and others are based on completely different mechanisms.

Nanotechnology Applications to Telecommunications and Networking, By Daniel Minoli

In the optical domain, recent technological advances in near-field optics have led to a revolution in optical microscopy. Optical microscopy and spectroscopy are now possible with resolution approaching 10 nm. This resolution is advantageous in the study of mesoscale optical properties because it allows these properties to be measured *directly* instead of being inferred from far-field spectroscopic data. Far-field spectroscopic data represents a spatial average due to the limited spot size of conventional lenses ($\lambda/2 \sim 0.25\,\mu m$ for visible light). In near-field optics, this spot size is reduced by up to an order of magnitude, thereby obviating much, if not all, of the spatial averaging. The result is clearer, more readily interpreted spectroscopic features and the ability to directly map the optical properties of the sample [108].

Over the past decade, the techniques that utilize near-field optics, near-field scanning optical microscopy (NSOM) and near-field optical spectroscopy (NFOS), have matured considerably. These techniques are useful for application in nanostructured materials such as semiconductor quantum-confined structures (quantum wells, wires, and dots), light-emitting polymers, and layered organic self-assemblies, where the optical and optoelectronic properties are defined on a length scale ideal for NSOM and NFOS. Applications of NSOM and NFOS to these materials have led to important new insights into the mesoscopic absorption, emission, and transport properties and a fuller fundamental understanding of these properties. In many cases the NSOM and NFOS data have even led directly to the discovery of new structural forms such as phase domains and novel forms of self-assembly [108]. NSOM methodologies is still relatively new, and a lot of work published in the field is directed toward understanding and modifying the technique itself. While this is important, applications of NSOM and NFOS to a variety of different materials (semiconductor heterostructures, quantum wells, quantum wires and quantum dots, polymers and molecular crystals, films and layered organic self-assemblies) are equally important and are expected to yield new technological advances in the years to come.

Other microscopy methods are based on X-ray, electron beam, ion beam, nucear magnetics, and atomic force principles, to list a few. In early 2004, researchers reportedly broken the 1-Å image resolution barrier with transmission electron microscope (TEM); 1-Å is approximately one-third the size of a carbon atom and is a key dimension for atomic level research [468, 469]. These advancements have been achieved by exploiting novel techniques to correct the aberrations in a scanning transmission electron microscope; although researchers realized half-a-century ago that it was possible to make these corrections, the technology needed to do this has only emerge of late. The aberration-correction technology fixes imperfections on the microscope's electron lenses. At press time Oak Ridge National Laboratory researchers announced that they have achieved an image resolution at 0.6 Å, breaking the previous record of 0.7 Å, which the lab set earlier this year; the electron microscope image can distinguish the individual, dumbbell-shaped atoms of a silicon crystal [470]. In 1999 the Oak Ridge Laboratory set a world record with a resolution at 1.3 Å without the aberration-correction technology. The researchers indicate the next frontier will be seeing atoms in three dimensions.

F.1 OVERVIEW OF GENERIC MICROSCOPY TOOLS

The following microscopy tools are used in physics, chemistry, and biology [471]:

- Confocal and laser microscopy
- Two-photon laser scanning confocal microscopy
- Electron beam techniques [e.g., transmission electron microscopes (TEMs)]
- Auger electron spectrometry (AES)
- Scanning auger microscopy (SAM)
- Energy dispersive X-ray spectrometry (EDS)
- Scanning electron microscopy (SEM)
- Field emission scanning electron microscopy (FE-SEM)
- X-ray photoelectron spectroscopy (XPS)
- Electron spectroscopy for chemical analysis (ESCA)
- Light microscopy
- Near-field scanning optical microscopy (NSOM) and near-field optical spectroscopy (NFOS)
- Mass spectrometry
- Time-of-flight secondary ion mass spectrometry (TOF-SIMS)
- Secondary ion mass spectrometry (SIMS)
- Nuclear magnetic microscopy
- Scanning probe microscopy
- Atomic force microscopy (AFM)
- Magnetic force microscopy (MFM)
- Scanning tunneling microscopy (STM)
- Spectrometry
- Rutherford backscattering spectrometry (RBS)
- Hydrogen forward scattering spectrometry (HFS)
- X-ray microscopy
- Particle-induced X-ray emission (PIXE)
- Total reflection X-ray fluorescence (TXRF)

Table F.1 depicts some of the parameters associated with a number of the analytical techniques used in microscopy [472]. Figure F.1 maps the analytical resolution of various techniques versus the detection limit, while Figure F.2 shows the depth capability of various analysis methods [472].

The rest of Section F.1 (courtesy of Charles Evans & Associates, except as noted) provides an overview of key microscopy techniques; the section that follows (Section F.2) focuses on nanotechnology-supportive techniques.

TABLE F.1 Analytical Techniques Used in Microscopy

Analytical Technique	Typical Applications	Signal Detected	Elements Detected	Organic Information	Detection Limits	Depth Resolution	Imaging/ Mapping	Lateral Resolution (Probe Size)
Auger	Surface analysis and high-resolution depth profiling	Auger electrons from near-surface atoms	Li–U		0.1–1 at %	<2 nm	Yes	100 nm
FE auger	Surface analysis, microanalysis, microarea depth profiling	Auger electrons from near-surface atoms	Li–U		0.01–1 at %	2–6 nm	Yes	<15 nm
AFM/SPM	Surface imaging with near atomic resolution	Atomic-scale roughness				0.01 nm	Yes	1.5 5 nm
Micro-FTIR	Identification of polymers, plastics, organic films, fibers, and liquids	Infrared absorption		Molecular groups	0.1–100 ppm		No	5 μm
TXRF	Metallic contamination on semiconductor wafers	Fluorescent X-rays	S–U		$1e^{9}$–$1e^{12}$ at/cm^{2}		Yes	10 mm
XPS/ESCA	Surface analysis of organic and inorganic molecules	Photoelectrons	Li–U	Chemical bonding	0.01–1 at %	1–10 nm	Yes	10 μm 2 mm

TABLE F.1 *(Continued)*

Analytical Technique	Typical Applications	Signal Detected	Elements Detected	Organic Information	Detection Limits	Depth Resolution	Imaging/ Mapping	Lateral Resolution (Probe Size)
HFS	Hydrogen in thin films (quantitative)	Forward scattered hydrogen atoms	H, D		0.01 at % 50 nm		No	2 mm × 10 mm
RBS	Quantitative thin-film composition and thickness	Backscattered He atoms	Li–U		1–10 at % ($Z < 20$) 0.01–1 at % ($20 < Z < 70$) 0.001–0.01 at % ($Z > 70$)	2–20 nm	Yes	2 mm
SEM/EDS	Imaging and elemental microanalysis	Secondary and backscattered electrons and X-rays	B–U		0.1–1 at %	1–5 µm (EDS)	Yes	4.5 nm (SEM)1 µm (EDS)
FE SEM	High-resolution imaging of polished precision cross sections	Secondary and backscattered electrons					Yes	1.5 nm
FE SEM (in lens)	Ultra-high-resolution imaging with unique contrast mechanism	Secondary and backscattered electrons					Yes	0.7 nm

SIMS	Dopant and impurity depth profiling, surface, and microanalysis	Secondary ions	H–U		$1e^{12}$–$1e^{16}$ at/cc (ppb-ppm)	5–30 nm	Yes	1 μm (Imaging), 30 μm (Depth profiling)
Quad SIMS	Dopant and impurity depth profiling, surface, and microanalysis, insulators	Secondary ions	H–U		$1e^{14}$–$1e^{17}$ at/cc	<5 nm	Yes	<5 μm (Imaging), 30 μm (depth profiling)
TOF SIMS	Surface microanalysis of polymers, plastics, and organics	Secondary ions, atoms, molecules	H–U	Molecular ions to mass 10,000	<1 ppma, $1e^{8}$ at/cm^2	1 monolayer	Yes	0.10 μm
MALDI	Protein, peptide, and polymer molecular weight distribution	Large molecular ions		Molecular ions to mass 150,000	low femtomole to picomole		No	10 μm

Note: "at %" means "1 part per 100" (i.e., 10,000 ppm) concentration of atoms.
Note: Li–U means from Lithium to uranium; etc.
Note: cc = cubic centemeter; ppb= parts per billion; ppm= parts per million.

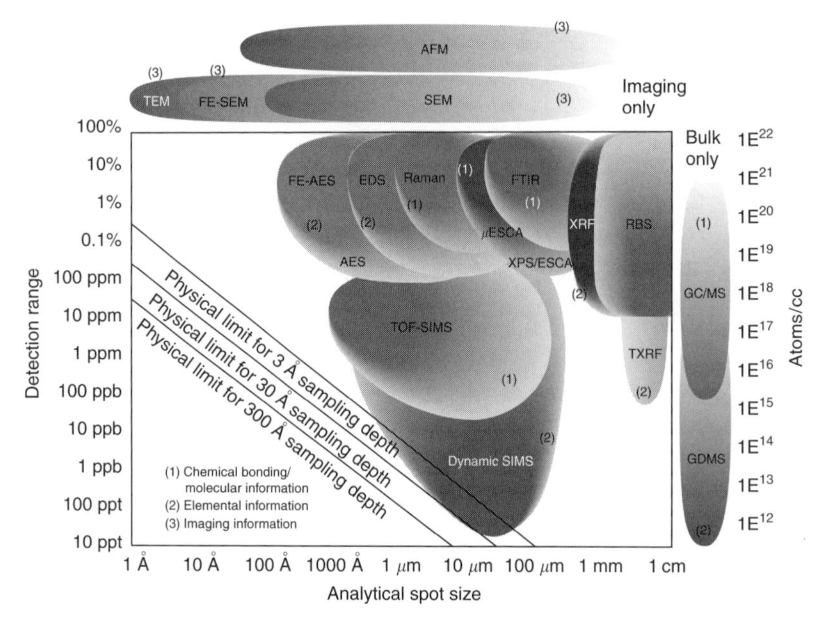

FIGURE F.1 Analytical resolution versus detection limit: AES = auger electron spectrometry, AFM = atomic force microscope, EDS = energy dispersive X-ray spectrometry, ESCA = electron spectroscopy for chemical analysis, FE-SEM = field emission scanning electron microscopy, FIB = focused ion Beam, FTIR = Fourier transform infrared spectrometry, GC/MS = gas chromatography/mass spectrometry, HFS = hydrogen forward scattering spectrometry, MFM = magnetic force microscopy, MSMS = Enhanced sensitivity for quantitation with tandem mass spectrometry, PIXE = particle-induced X-ray emission, RBS = Rutherford backscattering spectrometry, SAM = scanning auger microscopy, SEM = scanning electron microscopy, SIMS = secondary ion mass spectrometry, TEM = tunneling electron microscope (aka STM = scanning tunneling microscope), TOF-SIMS = time-of-flight secondary ion mass spectrometry, TXRF = total reflection X-ray fluorescence, XPS = X-ray photoelectron spectroscopy, XRF = X-ray fluorescence. (Courtesy: Charles Evans & Associates).

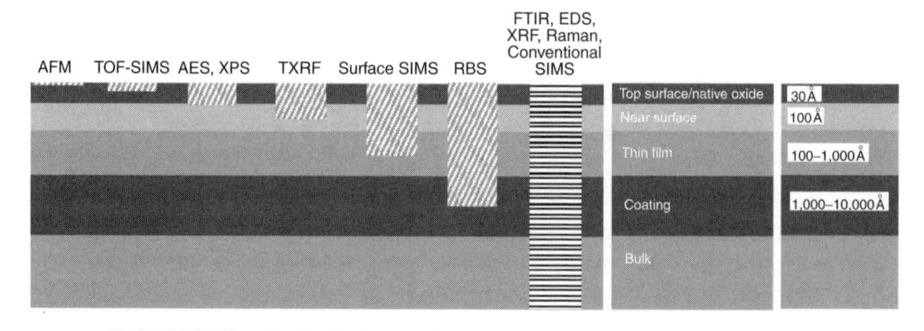

FIGURE F.2 Analysis depth. (Courtesy: Charles Evans & Associates).

F.1.1 Laser Scanning Confocal Microscopy

Laser scanning confocal microscopy (LSCM, also referred to as confocal scanning laser microscopy) provides high-resolution images and three-dimensional (3D) reconstructions of biological specimens. In LSCM, a laser light beam is expanded to make optimal use of the optics in the objective. Through an x-y deflection mechanism this beam is turned into a scanning beam focused to a small spot by an objective lens onto a fluorescent specimen. The mixture of reflected light and emitted fluorescent light is captured by the same objective and (after conversion into a static beam by the x-y scanner device) is focused onto a photodetector (photomultiplier) via a dichroic mirror (beam splitter). The reflected light is deviated by the dichroic mirror while the emitted fluorescent light passes through in the direction of the photomultiplier. A confocal aperture (*pinhole*) is placed in front of the photodetector, such that the fluorescent light from points on the specimen that are not within the focal plane (the so-called out-of-focus light) where the laser beam was focused will be largely obstructed by the pinhole. In this way, out-of-focus information (both above and below the focal plane) is greatly reduced. The spot that is focused on the center of the pinhole is often referred to as the *confocal spot* [473, 474a].

F.1.2 Secondary Ion Mass Spectrometry (SIMS)

An energetic primary ion beam sputters a sample surface; secondary ions formed in this sputtering process are extracted from the sample and analyzed in a double-focusing mass spectrometer system. The lateral distribution of the ions is maintained through the spectrometer so that the mass-resolved image of the secondary ions can be projected onto several types of image detectors. Alternatively, microfocusing the primary ion beam permits analysis in ion microprobe mode. SIMS is an analytical instrument for depth profiling, diffusion control, microanalysis, and ion imaging. It is capable of fast elemental concentration depth profiling of ultrashallow junctions and both ultrathin and micrometer-thick layers [469].

F.1.3 Time-of-Flight Secondary Ion Mass Spectrometry (TOF-SIMS)

A large area or microfocused pulsed primary ion beam sputters the top surface layer of the sample. The secondary ions produced in this sputtering process are extracted from the sample surface and injected into a specially designed time-of-flight mass spectrometer. The ions are dispersed in time according to their velocity (which is proportional to their mass-to-charge ratio m/z), and the discrete packets of different massed ions are detected on either a microchannel plate (MCP) or resistive anode encoder (RAE) detector. The TOF-SIMS technique is capable of detecting secondary ions produced over a large mass range (typically 0 to around 5000 amu) and performs this mass analysis at relatively high mass resolutions ($>6000\, m/\Delta m$). The technique also is capable of generating an image of the lateral distributions of these secondary ions at spatial resolutions of better than 0.15 μm.

F.1.4 Scanning Electron Microscopy (SEM)

The SEMs scan the surface of a sample with a finely focused electron beam to produce an image of that surface. Electrons hitting the surface are reflected and are detected by a fluorescent screen or monitor. The attainable magnification is typically around several hundreds of thousands times, enabling detection of features that are 1 nm apart [469]. A focused beam of electrons is rastered across a sample surface, the raster scan being synchronous with that of a cathode ray tube (CRT). The brightness of the CRT is modulated by the detected secondary electron current from the sample, such that the viewing CRT displays an image of the variation of secondary electron intensity with position on the sample. This variation is largely dependent on the angle of incidence of the focused beam onto the sample, thus yielding a topographical image. Different detectors can be used to provide alternative information, for example, a backscattered electron detector will provide average atomic number information. An auxiliary energy dispersive X-ray (EDS) detector provides elemental identification analyses from boron to uranium. Some high-performance instruments have enhanced abilities due to use of a special field-emission electron source (e.g., FE-SEM).

F.1.5 Field Emission Scanning Electron Microscopy (FE-SEM)

The field emission scanning electron microscope (FE-SEM) is similarly configured to a conventional SEM, except that a cold field emission electron source is used. This permits higher image resolution to be attained, increased signal-to-noise ratio, and increased depth of field.

F.1.6 Transmission Electron Microscopes (TEMs)

The TEMs support a range of applications requiring ultra-high-resolution down to sub-Ångstrom levels. TEMs utilize very thin (0.5 µm or less) samples illuminated by an electron beam. Images are recorded by detecting the electrons that pass through the sample (similar to what happens with the light in a slide projector). These electrons pass through a system of electromagnetic lenses and are focused as an enlarged image on a fluorescent screen, a photographic film or a digital camera. Magnifications of more than 1 million times are attainable. The technology uses an electron beam monochromator to improve the resolution of a Cs-corrected electron microscope; the 0.1-nm resolution is not far off the theoretical limit of about 0.07 nm [469].

F.1.7 Energy Dispersive X-Ray Spectrometry (EDS)

An EDS attachment to an SEM permits the detection and identification of the X-rays produced by the impact of the electron beam on the sample, thereby allowing qualitative and quantitative elemental analysis. The electron beam of an SEM is used to excite the atoms in the surface of a solid. These excited atoms produce characteristic

X-rays that are readily detected. By utilizing the scanning feature of the SEM, a spatial distribution of elements can be obtained. For flat, polished homogeneous samples, quantitative analysis can provide relative accuracy of 1–3% when appropriate standards are available.

F.1.8 Auger Electron Spectrometry (AES) and Scanning Auger Microscopy (SAM)

A focused electron beam irradiates a sample surface producing Auger electrons, the energies of which are characteristic of the element from which they are generated. Compositional depth profiling is accomplished by using an independent ion beam to sputter the sample surface while using AES/SAM to analyze each successive depth.

F.1.9 X-Ray Photoelectron Spectroscopy (XPS) and Electron Spectroscopy for Chemical Analysis (ESCA)

With these tools, samples are irradiated with monochromatic X-rays that cause the ejection of photoelectrons from the surface. The electron binding energies, as measured by a high-resolution electron spectrometer, are used to identify the elements present and, in many cases, provide information about the valence state(s) or chemical bonding environment(s) of the elements thus detected. The depth of the analysis, typically the outer 3 nm of the sample, is determined by the escape depth of the photoelectrons and the angle of the sample plane relative to the spectrometer.

F.1.10 Rutherford Backscattering Spectrometry (RBS)

With this instrument, a beam of He ions (with an energy of about 2.3 MeV) impinges on the target. The He ions backscatter from the near surface region of the sample and are collected by a solid-state detector. The energy of the backscattered ions provides information on both the composition and depth distribution of elements in the target. Alignment of the ion beam with sample crystallographic axes permits crystal damage to be measured quantitatively.

F.1.11 Hydrogen Forward Scattering Spectrometry (HFS)

This is the most fundamental variation on elastic recoil detection (ERD), and it is well suited for measuring hydrogen in materials. An energetic beam of helium ions impinges on the target at a glancing angle (75° from surface normal). Hydrogen atoms are scattered forward out of the sample by the He atoms, which also scatter forward after collisions with lighter atoms. The forward scattered H atoms are collected by a solid-state detector, while the He atoms are stopped by a foil placed between the sample and the detector. The number of forward scattered H atoms provides information on the concentration of H in the sample, while the energy of the H provides depth information.

F.1.12 Particle-Induced X-Ray Emission (PIXE)

With this instrument, X-ray transitions are excited simultaneously with backscattering measurement. The use of a Li-drifted silicon X-ray detector allows the collection of emitted X-ray signals. The strength and unique family of X-ray energies allows unambiguous identification of medium- and high-*Z* elements in thin films. Traditionally performed with incident protons for maximum detection limits, PIXE is used to allow quantitative analysis of films not suited for traditional RBS. PIXE measures the total presence of elements and cannot be used for depth profiling. The capability to identify and quantitate multiple transition metal elements in a single film or determine unknown film structures expands the scope of samples that can be analyzed by accelerator techniques.

F.1.13 Atomic Force Microscopy (AFM), Scanning Tunneling Microscopy (STM), and Magnetic Force Microscopy (MFM)

These are variations on a method of imaging surfaces with atomic or near-atomic resolution, collectively called scanning probe microscopy (SPM). A small tip is scanned across the surface of a sample in order to construct a 3D image of the surface. Fine control of the scan is accomplished using piezoelectrically induced motions. If the tip and the surface are both conducting, the structure of the surface can be detected by tunneling of electrons from the tip to the surface (in STM). Any type of surface can be probed by the molecular forces exerted by the surface against the tip (in AFM). The tip can be constantly in contact with the surface, it can gently tap the surface while oscillating at high frequency, or it can be scanned just minutely above the surface. By coating the tip with a magnetic material, the magnetic fields immediately above a surface can be imaged (in MFM). Image processing software allows easy extraction of useful surface parameters. These are discussed further is Section F.2.

F.1.14 Total Reflection X-Ray Fluorescence (TXRF)

X-rays from a tungsten anode or molybdenum tube impinge the sample surface at a glancing angle, within the critical angle for total external reflectance, and excite the electrons on atoms in the top few monolayers of the sample, causing them to emit photons (fluoresce). The X-ray photons emitted by the surface atoms have energies that are characteristic of the particular element. They are detected by an Si(Li) energy dispersive spectrometer. Quantification is achieved using a sample with a known areal density of impurity atoms (e.g., Ni) on the surface, and corrections with relative sensitivity factors are used for the other elements.

F.1.15 Fourier Transform Infrared Spectrometry (FTIR)

Individual chemical bonds, as well as groups of bonds, vibrate at characteristic frequencies. When exposed to infrared (IR) radiation, molecules selectively absorb radiation at frequencies that match those of their allowed vibrational modes. Measurement of the absorption of IR radiation by the sample as a function of frequency produces a spectrum that can be used to identify functional groups and consequently structure. However, vibrations that do not yield a change in dipole moment do not absorb IR radiation. For example, O_2 and N_2 do not absorb IR radiation. Consequently, IR spectra can be obtained in air. FTIR provides specific information about chemical bonding and molecular structure, making it useful for analyzing organic materials and certain inorganic materials.

F.1.16 μ-Raman Spectroscopy

When incident light strikes a sample, part of the light is scattered. Most of the scattered light has the same wavelength as the incident light; this is called Rayleigh scattering. Some of the light is scattered at a different wavelength; this is called Raman scattering. The energy difference between the incident light and the Raman scattered light is called the Raman shift. It is equal to the energy required to get the molecule to vibrate or to rotate. Several different Raman-shifted signals will often be observed in a single sample; each being associated with different vibrational or rotational motions of molecules in the sample. The particular molecule and its environment will determine what Raman signals will be observed. In practice, because the Raman effect is so slight, a laser is used as the source of the incident light. A plot of Raman intensity vs. the frequency of the Raman shift is a Raman spectrum. It usually contains sharp bands that are characteristic of the functional groups of the compounds or materials. This information can be interpreted to determine chemical structure and identify the compounds present. It is complementary to FTIR in that it uses a different method to measure molecular vibrations. Raman spectroscopy is a good technique for qualitative analysis and for discrimination of organic and/or inorganic compounds in mixed materials. A Raman spectrum can be obtained from samples that are as small as 1 μm. The intensities of bands in a Raman spectrum depend on the sensitivity of the specific vibrations to the Raman effect and are proportional to concentration. Thus, Raman spectra can be used for semiquantitative and quantitative analysis. The technique is used for identification of organic molecules, polymers, biomolecules, and inorganic compounds both in bulk and as individual particles. Raman spectroscopy is particularly useful in determining the structure of different types of carbon (diamond, graphitic, diamondlike carbon, etc.) and their relative concentrations.

F.1.17 Gas Chromatography/Mass Spectrometry (GC/MS)

Gas chromatography separates mixtures of volatile and semivolatile organic compounds into individual components using a temperature-controlled, open tubular column. The sample is flash vaporized, and the molecules are swept onto the gas chromatography column with an inert carrier gas. Separation occurs as the components partition themselves between the stationary phase on the inner wall of the column and the mobile phase (the carrier gas). The time it takes for a given molecule to traverse the entire length of the column is known as the retention time. The retention time is a function of the chemical structure of the component, the column type and the temperature profile it was subjected to during the chromatography experiment. It depends on the relative affinity of the compound for the stationary and mobile phases. The mass spectrometer then detects the components that elute from the end of the gas chromatography column. In the mass spectrometer, energetic electrons bombard the component molecules, ionizing some of them. This ionization process can also produce fragment ions that often provide structural information about the molecule. The ions are then accelerated by an electric field and enter a mass analyzer, where they are separated according to their mass-to-charge ratios. By plotting the abundance of ions as a function of mass-to-charge ratio, a mass spectrum is generated. The mass spectrum can be a unique "fingerprint," allowing identification of unknown compounds. This fingerprint is compared with a database of over 100,000 unique chemical compounds. If a reasonable match is obtained, the analyst uses this information to help identify the compound. Frequently, however, the database is of no help. This can occur for one of two reasons: (a) The mass spectrum contains unique chemical information, that is, abundant, unusual fragment ions—yet no compound in the database is even close to matching it, or (b) the mass spectrum is not very unique at all. It matches quite well with as many as a thousand other compounds. In this case, the best that can be done is to place it within a chemical class.

Liquid Analysis and Chemical Derivatization

Direct liquid analysis is the simplest form of sample introduction for GC/MS. In it, the sample is dissolved in a suitable solvent, and then injected directly into the injection port of the GC/MS. No sample preparation or pretreatment is required. Sometimes organic molecules will not volatilize—they decompose upon heating. Frequently, these compounds can still be analyzed by GC/MS using a procedure known as chemical derivatization. Chemical derivatization of a compound or mixture of compounds prior to analysis is generally done for one of three reasons: (a) To make a compound analyzable that otherwise could not be analyzed by a particular method of analysis (gas chromatography in this case), (b) to improve the analytical efficiency for the compound, or (c) to improve the detectability of the compound. Compounds often cannot be analyzed by a particular method because they are not in a form amenable to the analytical technique. Examples of this problem are nonvolatile compounds for gas chromatographic analysis and insoluble compounds for HPLC (high-performance liquid chromatography). Many compounds that are not stable under the conditions of the technique also fall into this category. Di- and tribasic acids as well as hydroxy acids are good examples of this type of molecule. The derivatization procedure modifies the chemical structure of the compound, so that they may be analyzed by gas chromatography.

Dynamic Headspace Analysis (HSA)

The HSA technique was developed primarily for the analysis of volatile compounds in matrices that could not be directly injected into a gas chromatograph. These matrices include polymers, cosmetics and toiletries, food and beverages, environmental samples, and biological specimens not suitable for direct injection. In dynamic HSA, the sample is placed in a closed chamber, heated to a specified temperature, and the atmosphere surrounding the sample is continuously swept with a stream of dry inert gas. The components that outgas from the sample are collected and analyzed by GC/MS. The temperature normally used for this test is 85°C, and the time at temperature is typically 3 h. For refractory materials, temperatures as high as 400°C can be used.

Pyrolysis

Pyrolysis is a technique normally used to analyze nonvolatile organic compounds such as wood, paper, or polymers by GC/MS. In pyrolysis the sample is heated rapidly to 750°C or higher in order to thermally decompose it. A high temperature such as this is sufficient to break the polymer backbone, forming smaller, more volatile fragments. By examining these fragments, it is sometimes possible to deduce the structure of the polymer chain. Some polymers unzip during pyrolysis to yield only the original monomer. This technique is used frequently to examine materials for the presence of additives such as plasticizers, antioxidants, flame retardants, ultraviolet (UV) stabilizers, or sizing treatments applied to cloth samples.

Solids Probe

Direct solids probe analysis is a volatilization technique that places the sample under vacuum near the mass spectrometer's ion source. The sample's temperature can either be raised to a preset maximum or it can be heated gradually, in a temperature-programmed fashion. During this time, the molecules that volatilize from the sample continuously enter the mass spectrometer's source and are ionized much in the same fashion as described above with GC/MS. The disadvantage of this technique is that there is no separation step. This is yet another way to analyze nonvolatile samples using mass spectrometry.

F.1.18 Enhanced Sensitivity for Quantitation with Tandem Mass Spectrometry

Most of the applications of GC/MS are involved with the identification of unknowns—usually trace-level contamination studies. However, frequently one needs to verify that a particular contaminant has been successfully removed from the sample submitted. The compound in question might be a lubricating oil, or a cleaning agent, for example. In cases such as these in which compound identification is not requested, the sensitivity of the mass spectrometer for a particular analyte can be increased greatly.

In single-stage mass spectrometer systems, this is accomplished using selected ion monitoring (SIM). With this technique, the most abundant (and characteristic) ion in the mass spectrum of each component to be quantitated is selected for scanning, and the instrument is programmed to scan this ion rather than the entire mass spectrum. Using this technique, the limit of detection can frequently be lowered by a factor of 10–50. The disadvantage of this method is that most of the mass spectral information for the component is lost. This is not much better than having a gas chromatograph with a flame ionization detector. For, in trace-level work, any component that has an ion with the same mass as the one being scanned and the same retention time will be mistaken for the component of interest. These small peaks are called chemical noise.

On the other hand, with a tandem (dual-stage) mass spectrometer system, the chemical noise can be reduced almost to zero, lowering the limit of detection significantly. In tandem mass spectrometry (as the name implies), a second mass spectrometry stage is added to the first. This allows the analyst to select one of the fragment ions from the first mass spectrum for further fragmentation. Now, we have a mass spectrum of an ion, whereas in single-stage mass spectrometry we have a mass spectrum of a molecule. Recalling the discussion earlier about selected ion monitoring, one can now apply this same principle to tandem mass spectrometry. With this technique, one selects the most abundant fragment ion from the first mass spectrum for second-stage fragmentation. This is called the parent ion. A mass spectrum is obtained; from this spectrum, one selects the most abundant ion to monitor—in the same manner as SIM. This is called the daughter ion. Now, when the sample is chromatographed, only those peaks will show up at the desired retention time that contain the daughter ion originating from the parent ion that was selected from the original mass spectrum. This greatly reduces the signal background or chemical noise.

F.1.19 X-Ray Fluorescence (XRF)

X-ray fluorescence is a nondestructive technique that can be used to quantify elemental constituents of solid and liquid samples. X-rays from a rhodium X-ray tube excite atoms in the sample, causing them to emit X-rays with energies characteristic for each element. The intensity and energy of the emitted X-rays are measured using a lithium drifted silicon detector and multichannel analyzer electronics. Because the efficiency of generating and detecting X-rays is greater for heavier elements, XRF is limited to detecting elements from sodium ($Z = 11$) through uranium ($Z = 92$). XRF is capable of detecting these elements from concentrations ranging from tens of parts per million (ppm) to 100%. Different voltages can also be applied to the X-ray tube to optimize the X-ray fluorescence yield in different regions of the X-ray energy spectrum. Through the use of appropriate reference standards, XRF can accurately quantify the elemental composition of both solid and liquid samples.

F.1.20 Focused Ion Beam (FIB)

Single-column FIB systems image material surfaces by bombarding them with ions (typically gallium). These ions produce images when they are reflected and recorded by a detector. Moreover, FIB systems can add (deposition or patterning) or remove (etching or milling) material. Manipulation of material in that way is popular for advanced circuit edit and for locating and repairing very small photomask defects for complex semiconductor devices [469]. In a dual FIB/SEM instrument, a finely focused ion beam mills away a precise amount of material from the sample and an SEM images either the sidewall or the underlying layer exposed during the milling process. The sidewall is essentially a precise cross section of the sample material. The focused ion beam is a Ga liquid metal ion gun (LMIG) that impacts the sample normal to the surface and can be focused to a spot as small as 70 Å. The focused ion beam can be rastered in a user-defined pattern over larger areas to selectively sputter and mill away the surface. By flooding the surface with specific gases, for example, Pt organometallic, XeF_2, or H_2O, during ion or electron bombardment, new material can be deposited (Pt metal for charge control) or reactive species can be created that selectively etch the surface and delineate specific features. This type of etching is restricted only to areas bombarded by the ion or electron beam. As an example, water vapor, in conjunction with ion bombardment, selectively etches organic materials. The combination of both unselective ion milling and selective etching using reactive species creates a very powerful sample preparation tool. The instrument's field emission SEM images the sample, which allows imaging of the edge of the etch or sputter crater (the cross section) immediately after preparation. The SEM has an ultimate resolution of 20–30 Å, allowing the measurement of coatings as thin as 100 Å.

F.1.21 Near-Field Scanning Optical Microscopy (NSOM) and Near-Field Optical Spectroscopy (NFOS)

The confinement of light to a space smaller than $\lambda/2$ (the diffraction limit) is not possible using conventional lenses, a fact that has been known for over 100 years and, until the development of NSOM, this dimension was thought to represent the resolution limit for optical microscopy. NSOM overcomes the diffraction limit by confining light via a tiny aperture (typically 10–200 nm), which can be raster scanned over a sample surface to construct an image point by point. Light emanating from this aperture rapidly diffracts, but for a small distance (about one aperture diameter) away from the aperture the spatial extent of the light is defined by the aperture. Thus, as long as the distance between the sample surface and the aperture is maintained at a small fraction of the aperture size, the resolution of the image will be determined by the size of the aperture and not the wavelength of light. Optical contrast mechanisms such as absorption, fluorescence, polarization, refractive index, and photoconductivity are all possible with NSOM, making it a completely versatile microscopy technique [107].

The NSOM imaging scheme was proposed by Synge in 1928 and was first demonstrated by Ash and Nicholls using radio waves in 1972 but has only recently become feasible for visible wavelengths. The key technological advance that has made visible light NSOM practical is the development of the tapered optical fiber probe by Betzig and Trautman in 1991. These new NSOM tips showed an improvement of 10^4 transmission efficiency over previous aperture probes such as tapered glass pipettes or quartz rods.

F.2 DETAILS ON SOME KEY SYSTEMS

The rest of this appendix discussess some of the key microscopy systems of direct interest to nanotechnologists. This appendix is provided courtesy of Molecular Imaging,[1] Tempe, Arizona.

F.2.1 Contact Mode AFM

In contact mode atomic force microscopy (AFM) (see Fig. F.3), interatomic van der Waals forces become repulsive as the AFM tip comes in close contact with the

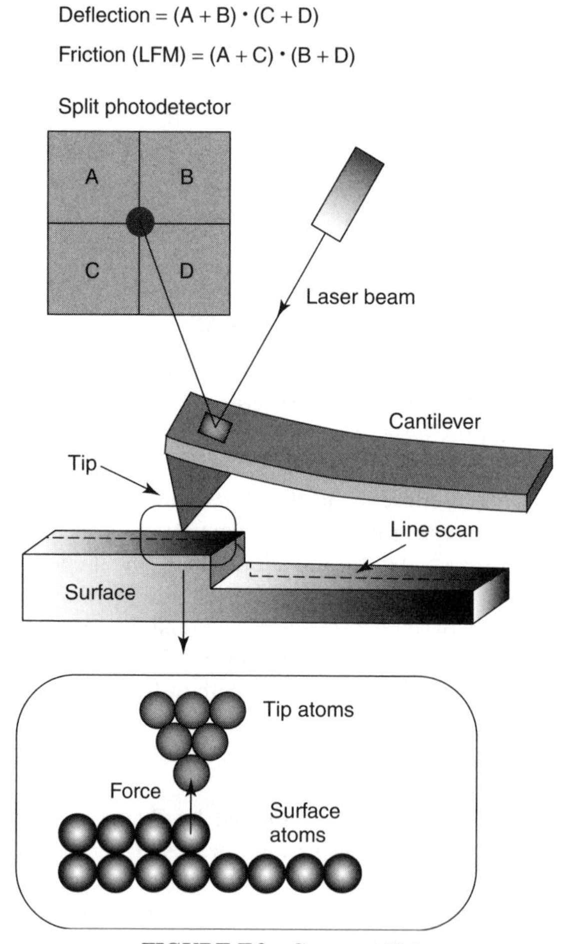

FIGURE F.3 Contact AFM.

[1]Molecular Imaging, 4666 S. Ash Avenue, Tempe, Arizona 85282 USA, Tel: (480) 753-4311; Fax: (480) 753-4312; e-mail: info@molec.com, http://www.molec.com/index_what_is_afm.html.

sample surface. The force exerted between the tip and the sample in contact mode is on the order of about 0.1–1000 nN (nanonewton). Under ambient conditions two other forces besides van der Waals interactions are also generally present. These forces include the capillary force from a thin layer of water in the atmosphere and the mechanical force from the cantilever itself. The capillary force is due to the fact that water can wick its way around the tip, causing the AFM tip to stick to the sample surface. The magnitude of the capillary force should vary with the tip–sample distance. The mechanical force resulting from the cantilever is similar to the force of a compressed spring, and its magnitude and sign (repulsive or attractive) is dependent on the cantilever deflection and the cantilever's spring constant. Consequently, in contact mode AFM, the repulsive van der Waals forces arising for the AFM tip to sample interaction must balance the sum of the forces arising from the capillary force plus the mechanical force from the cantilever.

The thin layer of water present on many surfaces in air exerts an attractive capillary force and holds the tip in contact with the surface. Thus, when scanner pulls the tip away from the surface, the cantilever bends strongly toward the surface and the scanner has to retract further so that the tip can snap off the surface. The cantilever returns to its original unbent status as the scanner moves the tip away from the surface beyond the snap-out point. In liquid, since the large capillary force is isotropic, the total force that the tip exerts on the sample can be reduced to some extent.

F.2.2 Magnetic AC Mode (MAC Mode)

Magnetic AC mode (MAC mode) is an oscillating atomic force microscopy (AFM) technique in which a magnetically coated cantilever is driven by an oscillating magnetic field. In MAC mode, a cantilever that is coated with a magnetic material is driven into oscillation by an AC magnetic field that is generated by a solenoid positioned close by the cantilever. The result is a clean cantilever response that has no artifacts or background signals when the cantilever is vibrated in air or in liquid. See Figure F.4.

Like other AC techniques, the cantilever is driven at a high frequency and the surface is monitored by changes in amplitude or phase of oscillation. Because the cantilever (and only the cantilever) is driven directly by the magnetic field, the need to shake the cantilever holder at large amplitudes is eliminated. Background resonance is absent, signal to noise is improved, and setup becomes straightforward. Better signal-to-noise means that much smaller amplitudes can be used. This decreases damage to the sample and preserves asperities on the probe, contributing to greatly improved resolution.

F.2.3 Acoustic AC Mode (AAC Mode)

Contact mode AFM often has a disadvantage for samples that are either weakly bound or soft because the tip can simply move or damage the surface feature and the resulting images are generally not high resolution. The advent of AC mode AFM, which operates in the intermittent contact regime or in the noncontact regime, provides

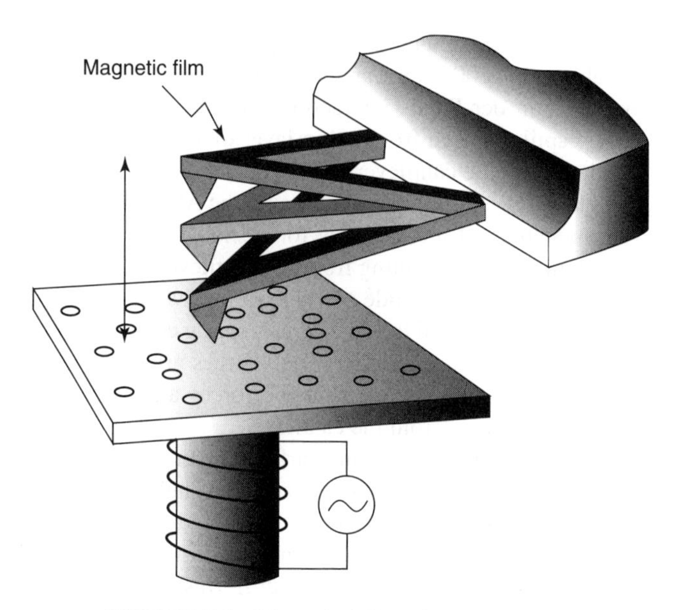

FIGURE F.4 Magnetic AC mode (MAC mode).

a solution to this problem. In AC mode, the cantilever nears its resonant frequency. AC mode AFM can be classified into two categories, intermittent contact mode and noncontact mode, depending on the force regime and the tip–sample separation distance. The interaction between the tip and the sample is predominately vertical, thus negligible lateral forces are encountered. Consequently, AC mode AFM does not suffer from the tip or sample degradation effects that are sometimes observed after many scans in contact mode AFM, and it is a technique for imaging soft samples. In AC mode, tip–sample force interactions cause changes in amplitude, phase, and the resonance frequency of the oscillating cantilever. The spatial variation of the change can be presented in height (topography) or interaction (amplitude or phase) images that can be collected simultaneously. The system monitors the resonant frequency or amplitude of the cantilever and keeps it constant by a feedback circuit that moves the scanner up and down. The motion of the scanner at each probe location is used to generate a topographic data set. The amplitude change at each probe location forms the amplitude image. The phase data is the result of the phase lag between the AC drive input and the cantilever oscillation output at each probe location. Consequently, contrast in phase images, which are due to differences in material properties, can provide very useful information. In addition, fine morphological features are easily observed in amplitude and phase images.

There are two ways to drive the cantilever into oscillation. One way is accomplished by indirect vibration, in which the cantilever is excited by high-frequency acoustic vibration from a piezoelectric transducer attached to the cantilever holder; this is called the acoustic AC mode (AAC). See Figure F.5. Another, more favored method that is much cleaner and gentler than acoustic AC mode is a direct vibration

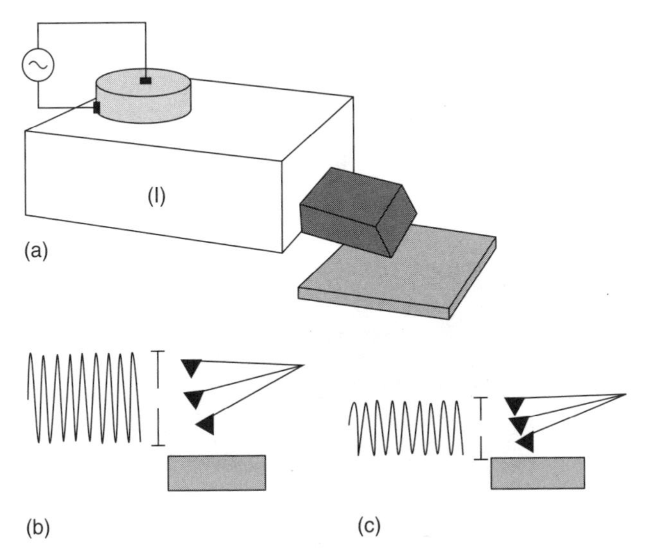

FIGURE F.5 Acoustic AC mode: (*a*) transducer attached to a cantilever housing is used to excite a cantilever into oscillation; (*b*) the amplitude of oscillation when the tip is far from the surface and; (*c*) reduced amplitude as the tip approaches the surface.

method where the cantilever is excited directly without having to vibrate the cantilever housing or other parts. This is called magnetic AC mode (MAC mode). To achieve MAC mode imaging, a cantilever coated with a magnetic material is driven into oscillation by an AC magnetic field generated by a solenoid positioned close to the cantilever housing. The result of MAC mode is a gentle, clean cantilever response that has no spurious background signals ("forest of peaks") like other AC modes can have. MAC mode has even greater advantages when the cantilever is vibrated in liquid.

F.2.4 Current-Sensing AFM

Current-sensing AFM (CSAFM) uses standard AFM contact mode along ultrasharp AFM cantilevers coated with a conducting film to simultaneously probe conductivity and topography of a sample (see Fig. F.6). By applying a voltage bias between the substrate and a conducting cantilever, a current flow is generated. This current can be used to construct a spatially resolved conductivity image, so CSAFM is often referred to as a marriage between STM and force microscopy. SCAFM has proven to be useful in joint I/V spectroscopy and contact force experiments as well as contact potential studies. SCAFM is compatible with measurements in air, under controlled environments, and measurements with temperature control. CSAFM is applicable to a diverse range of fields. CSAFM is useful in molecular recognition studies and can be used to spatially resolve electronic and ionic processes across cell membranes.

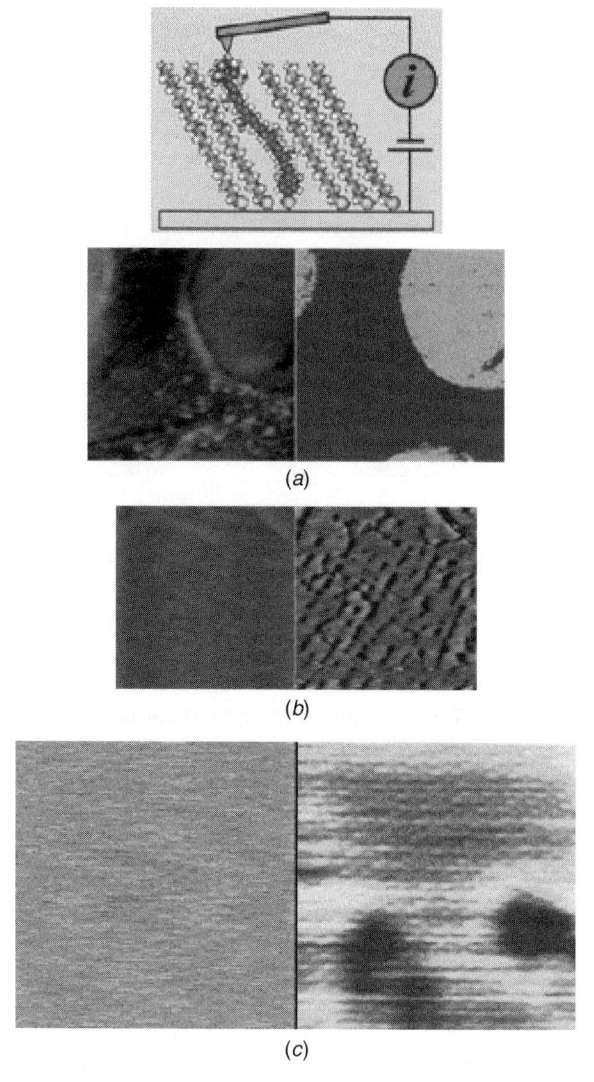

FIGURE F.6 Current sensing AFM: (*a*) 6.5-μm image of carbon epoxy composite imaged under dry N_2: topography (L) and conductivity (R); (*b*) 160-nm image of dodecanethiol thin film on Au(111) imaged in toluene: topography (L) and conductivity (R); (*c*) 12-nm image of dodecanethiol thin film on AU(111) imaged in toluene: topography (L) and conductivity (R). (Courtesy: Molecular Imaging, Tempe, Arizona).

F.2.5 Force Modulation AFM

Force modulation AFM is a fast, very sensitive imaging method that is especially useful to measure and detect variations in a surface's mechanical properties, including stiffness and elasticity (see Fig. F.7). In this technique, a modulated driving signal at

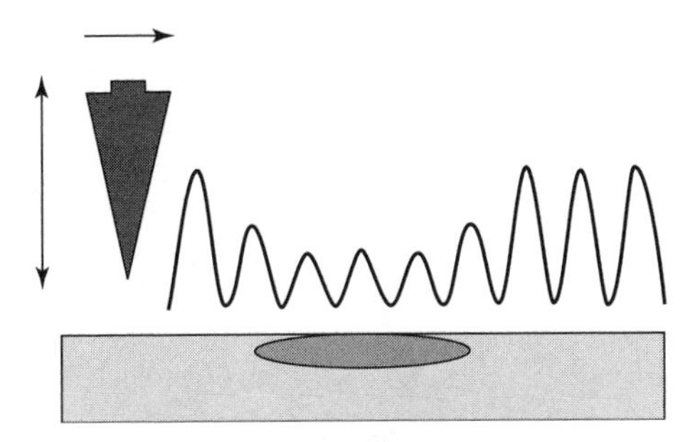

FIGURE F.7 Force modulation.

a constant frequency is applied to the AFM cantilever while the AFM tip is in contact with the sample, while amplitude variation and phase lag during the scan are measured. Force modulation provides the user with simultaneous surface topography measurements, material elasticity or stiffness (the amplitude of the modulated signal), and energy dissipation characteristics of the sample (from the phase of the cantilever response). When an AFM cantilever is modulated with the driving signal, elastic materials will result in relatively larger modulated amplitude compared to stiffer materials because the AFM tip can indent an elastic material.

Force modulation has proven its utility in life science studies, polymer studies, experiments on semiconductor materials, and the material sciences, including investigations on composite materials. Force modulation is compatible with measurements in air or liquid, under controlled environments, measurements with temperature control, MAC mode, and electrochemistry. Figure F.8 depicts an example of ATM usage.

F.2.6 Phase Imaging

Phase imaging is a powerful, dynamic force technique that can reveal many unique mechanical and chemical properties of a sample at the nanometer scale. In phase imaging (see Fig. F.9), an AFM cantilever is oscillated vertically near its mechanical resonance frequency while it is in close proximity to a sample. As the AFM tip comes in very close proximity to the sample surface, the amplitude of the cantilever's oscillation is reduced. The change in amplitude is measured and is used to track changes in the surface topography and roughness of the sample. Simultaneously, as the AFM tip encounters regions of different composition, a change in phase, relative to the phase of the drive signal, is measured and recorded. This change in phase is very sensitive to variations in material properties, including surface stiffness, elasticity, and adhesion. The phase shifts are measured and displayed in a very straightforward manner that facilitates quantitative analysis and interpretation.

Both inorganic and organic materials have been examined with phase imaging. Phase imaging has been found to be particularly useful to map the various components of composite materials, to study variations in composition and contamination in materials, and to measure adhesion, surface hardness, and elasticity. It has been applied to thin-film studies, the materials sciences, and composite characterization.

FIGURE F.8 Example of ATM usage: 9-nm lines in mica written with an AFM. (Courtesy: DFG-Center for Functional Nanostructures, Universität Karlsruhe, Germany).

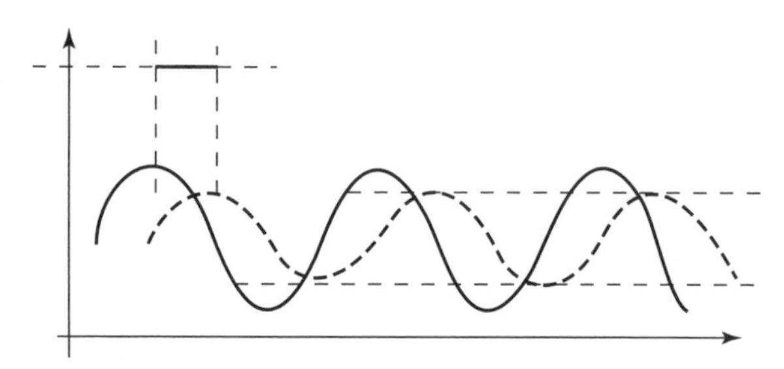

FIGURE F.9 Phase imaging.

Phase imaging studies can be conducted with or without temperature control, in air, in liquid, and even under controlled atmospheres. Phase imaging can reveal material properties that cannot be observed in surface topography, and it can identify properties that might otherwise be obscured by topography. It is a sensitive, quantitative, high lateral resolution AFM method that is often more convenient and gentler than other surface property methods that are based on contact mode operation.

F.2.7 Pulsed Force Mode (PFM)

Pulsed force mode is a technique that maps surface stiffness and adhesion in addition to topography in contact mode (see Fig. F.10). In PFM, a low-frequency (500 Hz

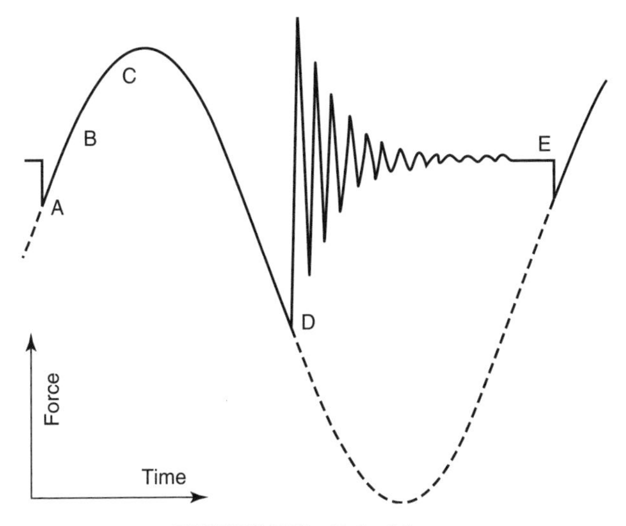

FIGURE F.10 Pulsed force.

to 1 kHz typically) sinusoidal drive is applied to the *z*-piezo to bring the tip in and out of contact. As illustrated in the left figure, tip snaps into contact at point *A*. Force increases as the piezo pushes the tip further until point *C*, where a maximum force is reached. Then the piezo pulls the tip away. At point *D*, the tip breaks off the surface. Subsequent free oscillation of the cantilever is damped and the sinusoidal drive continues. At point *E* the next cycle starts. Stiffness and adhesion can be extracted from each modulation cycle. The slope between *B* and *C* is used to map stiffness variation of the surface, while the break-off force at point *D* is measured to show the adhesion between the tip and surface. PFM is an effective imaging method on polymer blends, composite materials, and sticky surfaces.

F.2.8 Electrostatic Force Microscopy (EFM)

Electrostatic force microscopy measures local electrostatic interaction between a conductive tip and a sample through Coulomb forces (see Fig. F.11). As the tip scans across the surface, a bias is applied on the tip. Different areas of the surface may have different responses to the charged tip, depending on their local electrical properties. Such a variation in electrostatic forces can be detected in the change of oscillation amplitude and phase of the AFM probe. Because the electrostatic forces interact at greater distances than van der Waals forces, electrical force information can be separated from surface topography simply by adjusting the tip-to-sample distance. Thus, the electrical features can be resolved from topography features. Acoustic AC mode is used in EFM measurement. There are many application fields for EFM, for example, characterizing surface electrical properties, detecting defects of an integrated circuit, measuring the distribution of a particular material on a composite surface, to name a few.

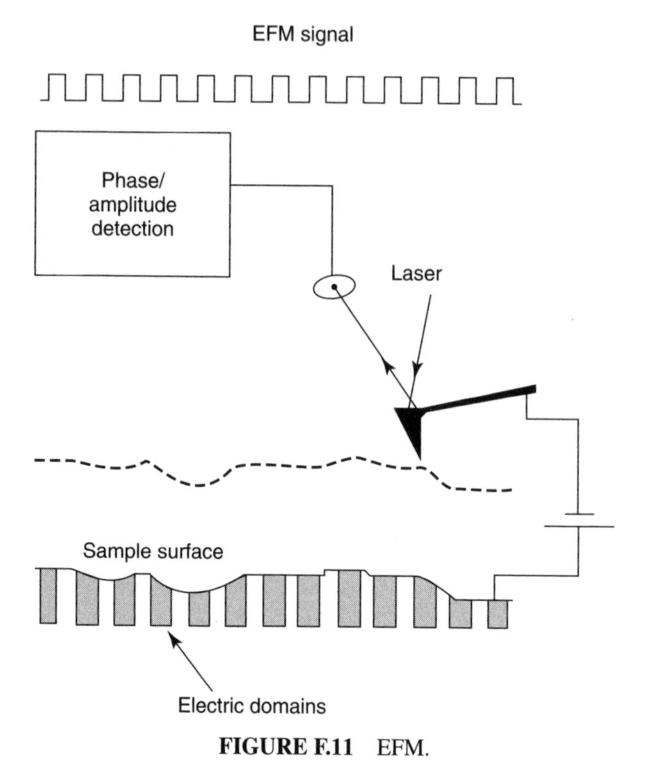

FIGURE F.11 EFM.

F.2.9 Magnetic Force Microscopy (MFM)

Magnetic force microscopy uses a magnetic cantilever to simultaneously map magnetic fields and surface topography on a sample (see Fig. F.12). By monitoring the cantilever's response to the forces that the sample's magnetic domains exert on the AFM tip as a function of x and y, we can reconstruct the magnetic features and measure the magnetic field strengths of a sample. When an MFM cantilever is scanned along a magnetic sample's surface, the tip–sample interaction includes both magnetic forces and the common van der Waals forces. However, magnetic features can be resolved from topography features because the magnetic forces interact at greater distances than van der Waals forces, so magnetic force information can be separated from surface topography simply by adjusting the tip-to-sample distance. Acoustic AC mode is a sensitive way of detecting magnetic interaction because the cantilever oscillates at resonance while changes in amplitude or phase can then be used to extract magnetic field information. The result is a high-resolution, three-dimensional map of the sample's magnetic features.

Magnetic force microscopy brings the advantages of AFM into the magnetic materials, recording, and storage media fields. It is nondestructive and requires minimal sample preparation. MFM can be used to evaluate magnetic materials and devices or to locate and map magnetic defects on a variety of materials and surfaces. MFM is compatible with imaging in fluids or in air, imaging under controlled environments,

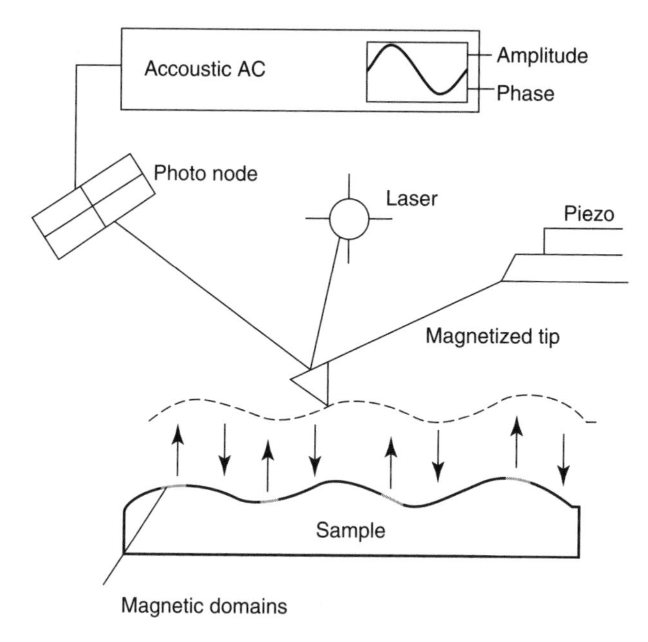

FIGURE F.12 MFM.

and with temperature control. Applications of MFM imaging include data storage media, nanoparticles, thin films, and detection of magnetic beads.

F.2.10 Lateral Force Microscopy (LFM)

During contact mode AFM scanning, as the probe is dragged over the surface, changes in surface friction and in topologic slope both can cause the cantilever to twist and give rise to forces on the cantilever that are parallel to the plane of the sample surface. Such lateral forces cause lateral deflection of the cantilever, which is sensed by the photodetector and used to form a lateral force image in a manner that is similar to a normal AFM deflection image. LFM is useful to study surfaces that have variations in friction (see Fig. F.13). However, since both surface friction and topology can contribute to the lateral deflection change, LFM and AFM images

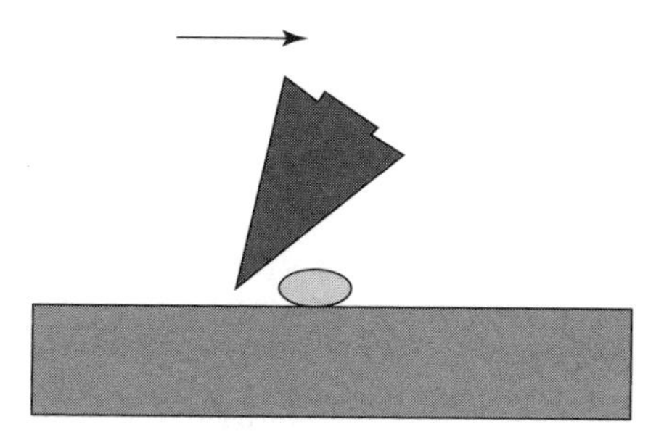

FIGURE F.13 LMF.

should be collected simultaneously in order to distinguish between the two effects, that is, to deconvolute LFM from topography effects.

F.2.11 Scanning Tunneling Microscope (STM)

The scanning tunneling microscope was invented in 1981 by G. Binnig and H. Rohrer who shared the 1986 Nobel Price in Physics for their invention. STM uses a sharp conducting tip and it applies a bias voltage between the tip and the sample (see Fig. F.14). When the tip is brought close to the sample, electrons can "tunnel" through the narrow gap either from the sample to the tip or from the tip to the sample, depending on the sign of the bias voltage. This tunneling current changes with tip-to-sample distance, it decays exponentially with the distance, which gives STM its remarkably high precision in positioning the tip (subangstrom vertically and

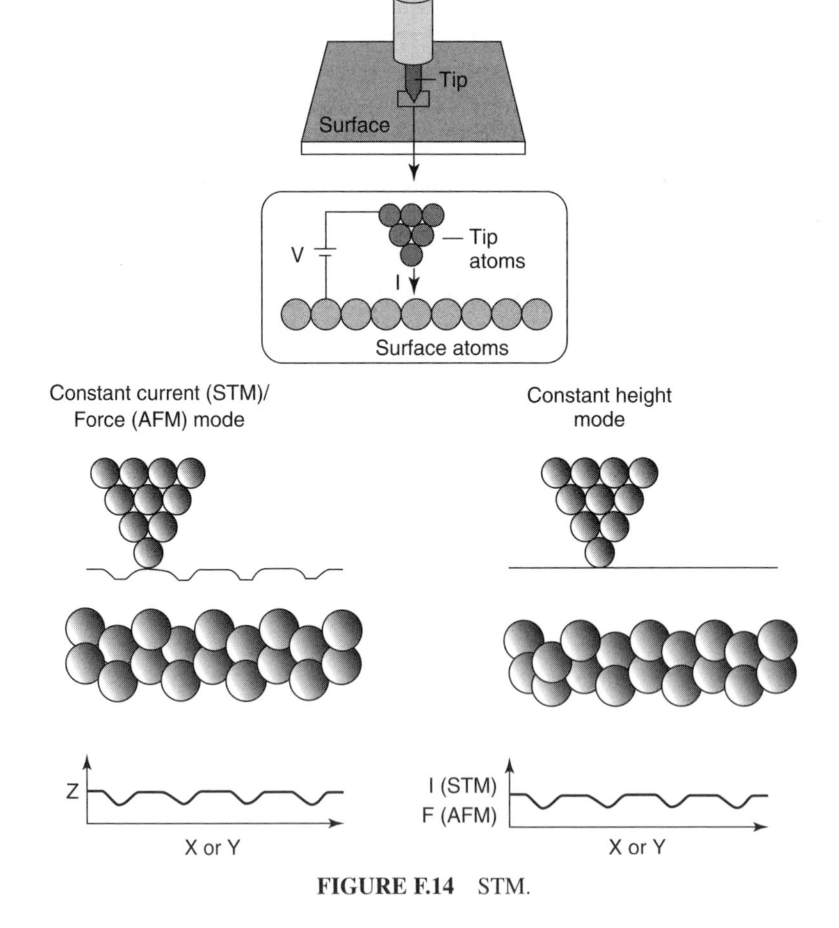

FIGURE F.14 STM.

atomic resolution laterally). For the electron tunneling to take place, both the sample and the tip must be conductive. Therefore, STM cannot be used on insulating materials.

The STM technique can image a sample surface in either constant-current mode or constant-height mode. In constant-current mode, in order to keep the tunneling current constant, STM uses feedback to adjust the height of the scanner at each measurement point, for example, when the system senses a tunneling current increase, it adjusts the voltage applied to the piezoelectric scanner so that the scanner lifts the tip and gives an increase in the tip–sample distance. The scanner height measured at each location on the sample surface constitutes the topographic image. The constant-current mode is thus generally used to acquire surface height data, its scan speed is limited by the feedback response, and thus it takes longer to image an irregular surface at a larger scan size. In constant-height mode, the tip scans at a constant height above the sample and the tunneling current changes due to the topography and the local surface electronic properties of the sample. The current image is a result of measured tunneling current at each location on the sample surface. The constant-height mode can acquire data faster because the system does not have to move the scanner in the vertical direction, so it is most often used for imaging relatively smooth surfaces.

Strictly speaking STM tunneling current is correlated to the surface electronic density of states, that is, the number of filled or unfilled electron states near the Fermi level, within an energy range determined by the bias voltage. Hence, STM measures constant tunneling probability instead of the physical topography at the surface. STM tunneling spectroscopy, looking at the current–voltage relationship at a constant tip–sample distance or the current–distance relationship at a constant bias voltage, is a useful tool to study the electronic structure and property of a sample surface at the atomic level.

APPENDIX G

Quantum Computing

G.1 INTRODUCTION

One has seen keen in interest in quantum computing (QC) [aka quantum information processing (QIP)] over the past few years; this interest originates from new and useful theoretical models for quantum turing machines and from novel (even surprising) algorithms for factoring and searching, all of which relates to complexity (computational complexity). QC is expected to become an important development during the next decade, based on nanotechnology and quantum theory principles.

The purpose of this appendix is to expose the reader to this line of research. This appendix is based on a report written by S. Lloyd, D. D. Vincenzo, and colleagues, [435], which is a U.S. government sponsored assessment of the technology and is available on a royalty-free license. The reader should first read the material on QCs in Chapter 6 before reading this appendix. Because of the range of topics discussed, this

[1] This report was prepared as an account of work sponsored by an agency of the U.S. government. Neither the U.S. government nor any agency thereof, nor any of their employees, make any warranty, express or implied, or assume any legal liability or responsibility for the accuracy, completeness, or usefulness of any information, apparatus, product, or process disclosed, or represent that its use would not infringe privately owned rights. The U.S. government strongly supports academic freedom and a researcher's right to publish; as an institution, however, the U.S. government does not endorse the viewpoint of a publication or guarantee its technical correctness. By acceptance of this appendix, the publisher recognizes that the U.S. government retains a nonexclusive, royalty-free license to publish or reproduce the published form of this contribution, or to allow others to do so, for U.S. government purposes. The U.S. government requests that the publisher identify this as work performed under the auspices of the Advanced Research and Development Activity (ARDA). ARDA is an Intelligence Community (IC) center for conducting advanced research and development related to information technology (IT) (information stored, transmitted, or manipulated by electronic means). ARDA sponsors high-risk, high-payoff research designed to produce new technology to address some of the most important and challenging IT problems faced by the intelligence community. The research is currently organized into five technology thrusts: Information Exploitation, Quantum Information Science, Global Infosystems Access, Novel Intelligence from Massive Data, and Advanced Information Assurance.

Nanotechnology Applications to Telecommunications and Networking, By Daniel Minoli

appendix is not strictly self-contained (as we attempted to do in the rest of this book), in the sense that terms and concepts are covered that are not completely defined rigorously in the text. However, the purpose of this inclusion is to expose the reader to major theoretical work being undertaken in this space, which merits tracking. Again, nanoscale mechanisms are (expected to be) needed to implement quantum computers.

G.2 FUNDAMENTAL THEORETICAL CHALLENGES

Quantum computing as a field has its roots very firmly planted in major theoretical developments in the 1980s and 1990s. The early musings of Feynman on how efficiently quantum mechanics could be simulated on a computer, Deutsch's definition of quantum Turing machines and quantum circuits, Deutsch and Jozsa's algorithm, and the study of quantum complexity theory by Bernstein and Vazirani showing that quantum Turing machines violate the modified Church–Turing thesis—all led up to Shor's remarkable polynomial (*P*) time quantum algorithms for factoring and discrete logarithm. These algorithms provided the "killer applications" that brought QC in the limelight. However, before any serious effort by experimentalists to realize quantum computers, another seemingly insurmountable hurdle had to be overcome by theoreticians. Quantum states are fragile and subject to decoherence that is continuous rather than discrete. This and the no-cloning theorem seemed to rule out the application of error correction techniques. The invention of quantum error-correcting codes by Calderbank, Shor, and Steane overturned conventional wisdom in quantum mechanics and paved the way for fault-tolerant QC and the threshold result that was independently obtained by Aharonov and Ben-Or; Knill, LaFlamme, and Zurek; and Gottesman and Preskill. Theoretical work has played a similarly central role in quantum cryptography (QCRYPT), where the 1984 protocol for quantum key distribution (QKD) due to Bennett and Brassard provided the major moving force for the field.

For the last decade, QC has brought about a remarkable collaboration between theoreticians and experimentalists. This collaboration has resulted in the elucidation of viable designs for quantum computers. The establishment by DiVincenzo and Barenco and co-workers of elementary universal families of one and two-qubit quantum gates for QC did much to simplify the quantum circuit model that the physical design needed to implement. Theorists, notably Lloyd, Cirac, and Zoller and DiVincenzo proposed the first potentially viable designs for quantum computers using ion traps and electromagnetic resonance techniques. The first prototypes of quantum computers were built by experimentalists, notably Wineland, Kimble, Cory, and Chuang—working closely with the theorists.

As the technological program of experimentally realizing quantum computers advances toward its goals, what is the future role of theory in QC? The outline below identifies some of the grand-challenge theoretical problems where progress is essential to both the success of the experimental efforts as well as the impact of QC. In addition, as the experimental effort accelerates, the collaboration between theory and experiment must continue to grow and evolve.

G.2.1 Quantum Algorithms

The search for new quantum algorithms is one of the biggest challenges in quantum computation today. Although factoring and discrete logarithms provide what some perceive to be "the killer applications" for quantum computation today, once we have quantum computers, cryptography will no longer rely on these problems—therefore greatly reducing the practical value of these algorithms.

The exploration of quantum algorithms is therefore of fundamental importance. In the years since Shor's algorithms, the framework of the hidden subgroup problem (HSP) has been developed, and the holy grail of quantum algorithms has been clearly identified as the HSP for non-Abelian groups. Two especially important cases are the dihedral group, which corresponds to the shortest lattice vector problem, and the symmetric group, which corresponds to graph isomporphism and graph automorphism, are important in their own right. The two most promising avenues are to extend the Fourier sampling approach used by Shor, and a novel approach based on adiabatic evolution as proposed by Farhi and co-workers and elaborated by Aharonov and co-workers [475, 476, 477]. Another interesting area is the use of quantum random walks to give polynomial speedups for basic problems such as element distinctness [478] and their potential for providing exponential speedups [479].

The future ability of quantum computers might be a decade or two away, their future ability to break public-key cryptography has important implications for the encryption of highly sensitive information today. For these applications, we must already design new public-key cryptosystems and one-way functions that are immune to quantum cryptanalysis. The existence of such one-way functions in an abstract setting follows from the study of Bennett et al., on exponential black-box lower bounds for inverting a random permutation [480]. Finding concrete implementations of quantum one-way functions will require a better understanding of the scope of quantum algorithms.

G.2.2 Quantum Complexity Theory

Understanding the class BQP (bounded quantum polynomial) of problems that can be solved in polynomial time on a quantum computer is the fundamental question in quantum complexity theory. Two very basic questions are the relationship between BQP and NP (nondeterministic polynomial) and between BQP and PH (the polynomial hierarchy). Although the early oracle results of Bennett et al. [480] provided evidence that BQP is not in NP, one must interpret these results carefully, especially in view of results from [481, 482]. Given the enormous payoff if NP were in BQP, this possibility remains worth exploring. Pessimists might try to prove that if BQP subset NP, then some very unlikely complexity theoretic consequence (such as the collapse of the polynomial hierarchy) would follow.

G.2.3 Fault-Tolerant Quantum Computing

The threshold result in fault-tolerant QC says that provided the decoherence rate is below a threshold η, arbitrarily long quantum computations can be faithfully carried out. (The term *decoherence* in the field of QIP refers to all manifestations of loss of unitarity in the qubit state time evolution.) Currently, the best schemes for fault-tolerant QC give a value of h between 10^{-3} and 10^{-4} [483, 484]. On the other hand, the only limit we know on η is that it is less than $\frac{1}{2}$ [485]. Narrowing this gap, and improving the achievable threshold is an essential goal for the realization of scalable, practical QC. Eventually, one would like to show that η is of the order of 1/100. Equally important is the challenge of reducing the overheads in the number of qubits and the processing time incurred in making a procedure fault-tolerant. Finally, it is important to revisit the model for fault-tolerant computation, in view of more detailed decoherence models from experimental efforts, as well as issues such as the relative delays for gate operations versus measurements.

G.2.4 Simulation of Quantum Systems

Quantum simulation is currently one of the most important applications of quantum computers. Kitaev's phase estimation method [486] provides an exponential speedup when applied to the problem of estimating eigenvalues of an operator [487], a problem of great importance in many areas of physical sciences. Grover's algorithm yields quadratic speedups when it is applied to a variety of continuous problems such as multivariate integration and path integration [488]. A recent result by Vidal [489] shows how to classically simulate 1D spin chains with logarithmically bounded entanglement length (the entanglement between a contiguous block of L spins and the rest of the spin chain; that is, the von Neumann entropy of the density matrix of the block of L spins) in polynomial time on a classical computer. Extending this classical simulation to two and three dimensions could potentially have great impact because they would be applicable to a greater range of systems.

G.3 QUANTUM COMPUTATION HISTORICAL REVIEW

G.3.1 Short Summary of Significant Breakthroughs in Quantum Information Theory

Information theory (the underpinning discipline of computing) is rooted in physics, which places limitations on how information may be processed and manipulated for computation and for communication. Before the 1980s this meant classical physics, but since that time there has been a conscious paradigm shift to the examination of benefits that may derive from basing a theory of information upon the laws of quantum physics. At least two important precursors to this paradigm shift had critical influence. The first was the demonstration of nonlocal correlations between different parts of a quantum system, correlations that possess no classical counterpart, by Bell in the early 1960s [490, 491]. The second important precursor to the new field of QIT was provided by the work of Landauer and Bennett on the thermodynamic cost of computation [492, 493]. Bennett's 1973 proof that reversible classical computation is possible [493] was the key idea in Benioff's positive response in 1980 to negative prognoses of fundamental limitations of computation provided by physics [494, 495].

In a key paradigm shift, Feynman pointed out in 1992 that simulating quantum physics on a classical computer appeared to incur an exponential slowdown [496], thus paving the way for QC. Deutch took a major step further in 1985, with the introduction of quantum circuits and universal gate sets, providing the critical leap from the restrictions of Boolean logic underlying classical computation to non-Boolean unitary operations [497]. With this critical step, the concept of QC was formalized. In 1993, Bernstein and Vazirani [498] built upon an algorithm of Deutsch and Jozsa [499], to show that quantum computers provide a superpolynomial advantage over probabilistic computers, thus showing that quantum computers violate the modified Church–Turing thesis. These algorithms as well as Simon's 1994 algorithm [500] benefited from the features of quantum superposition and entanglement, with the roots of the latter clearly identifiable with the nonclassical correlations observed by Bell in the early 1960s. This slow growth in exploration of algorithmic advantages derived from quantum circuits for computation virtually exploded in 1994 with the discovery by Shor of the polynomial time quantum algorithms for integer factorization and discrete logarithm problems [501], followed by the discovery of the quadratic speedup quantum search algorithm by Grover in 1996 [502]. Both of these theoretical results galvanized the experimental community into active consideration of possible implementations of quantum logic. Experimental interest was further stimulated by another significant result of Calderbank, Shor, and Steane, namely that error correction codes could be constructed to protect quantum states just as for classical states [503, 504, 505]. This demonstration of quantum error correction in 1995 was subsequently incorporated into a scheme by Kitaev [506], Shor [507], Aharonov and Ben-Or [508], Knill, LaFlamme, and Zurek [509], and Gottesman and Preskill [510, 511] to provide error thresholds on individual operations that show when computation can continue successfully in the presence of decoherence and errors (*fault-tolerant* computation). This result put the implementation of QC on a similar footing with classical computation using unreliable gates, and significantly altered the consciousness of the physics community with regard to experimental implementation.

Quantum complexity theory systematically studies the class of problems that can be solved efficiently using quantum resources such as entanglement. Bernstein and Vazirani's 1993 work showed that relative to an oracle the complexity class BQP, of problems that can be solved in polynomial time on a quantum computer, is not contained in MA (Merlin-Arthur), the probabilistic generalization of *NP* [498]. Thus even in the unlikely event that $P = \mathrm{NP}$, quantum computers could still provide a speedup over classical computers. The limits of quantum computers were explored by Bennett, Bernstein, Brassard, and Vazirani [480], who showed that QC cannot speed up search by more than a quadratic factor. This showed that Grover's algorithm is optimal and that, relative to a random oracle, quantum computers cannot solve NP-complete problems. They also showed a similar lower bound for inverting a random permutation by a quantum computer, thus opening up the possibility of quantum one-way functions. Recently, Aaronson showed a similar lower bound for the collision problem [512], thus showing that there is no generic quantum attack against collision-intractable hash functions. Kitaev has studied the class BQNP (bounded quantum analog of NP), the quantum analog of NP, and showed that QSAT (quantum analog of satisfiable problem), the quantum analog of the satisfiability problem, is complete for this class—thus proving that BQNP $\subseteq$ PSPACE (where PSPACE represents problems solvable with polynomial memory) [506]. Watrous considered the power of quantum communication in the context of interactive proofs and showed that the class IP (interaction proof) of problems that have interactive proofs with polynomially many rounds of communication can be simulated with only three rounds of quantum communication [513]. In the first demonstration of the power of quantum communication, Burhman, Cleve, and Wigderson showed how two parties could decide set disjointness by communicating only square root of n quantum bits, quadratically fewer than the number required classically [514]. Ambainis, Schulman, Vazirani, and Wigderson showed that for the problem of sampling disjoint subsets, quantum communication yields an exponential advantage over any protocol that communicates only classical bits [515]. Raz [516] gave a complete problem (a relation) for quantum communication complexity and showed that it had an exponential advantage over any classical protocol. Recently, Bar-jossef, Jayram, and Kerenidis, [517] showed that one-way quantum protocols are also exponentially more succinct than classical protocols.

Similar paradigm-changing advances have occurred in the theory of data transmission and communication as a result of theoretical breakthroughs in QIT. In fact the oldest branch of QIT concerns the use of quantum channels to transmit classical information, with work of Holevo dating from 1973 [518]. Since then, many significant results for the use of quantum channels to transmit both classical and quantum information have been established. It is useful to realize that these, in many cases very practical, results are derived notwithstanding the two famous results concerning inaccessibility of quantum states, namely the impossibility of distinguishing distinct quantum states (Holevo) [518] and of copying (or "cloning") an unknown quantum state (Wooters and Zurek) [519]. Notable among these quantum information theoretic results with implications for practical use in quantum communication are quantum data compression, quantum superdense coding, and teleportation. Together with quantum error correction, quantum data compression provides a quantum analog for the two most important techniques of classical information theory. The developments of quantum superdense coding in 1992 (Bennett and Wiesner) [520] and quantum transmission by teleportation (Bennett and co-workers) [521] in 1993, have no classical analog and are thus very surprising when viewed from a classical paradigm. Teleportation allows states to be transmitted faithfully from one spatial location to the other, while superdense coding allows the classical information to be transmitted with a smaller number of resources (quantum bits) via a quantum channel. A related property of quantum channels is superadditivity, namely that the amount of classical information transmitted may be increased by use of parallel channels [522, 523]. Similar to the development of theoretical techniques to deal with noise in QC mentioned above, a significant theoretical effort has also focused on the issues arising from communication with noisy channels. Several results have emerged here, but a number of open questions still remain, and this is a very active area of theoretical work. Important results arrived at in recent years include a bound on the capacity of a noisy quantum channel for transmission of classical information (Holevo–Schumacher–Westmoreland theorem) [524, 525, 526], and the development of protocols for distillation (or "purification") of entanglement [527, 528, 529].

A related area in which QIT has made remarkable advances in the last 20 years is QCRYPT. This field provides one of the most successful practical applications of quantum information to date, with the procedures for secure quantum key distribution (QKD). First developed by Bennett and Brassard in 1984 [530], several protocols now exist to make a provably secure quantum key for distribution over a public channel. These schemes rely on the uncertainty of distinguishing quantum states, with the security of the key also guaranteed as a result of the ability to detect any eavesdropping measurement by an observed increase in error rate of communication between the two parties. The remarkable security properties of QKD are a direct result of the properties of quantum information, and hence of the underlying principles of quantum physics.

These advances have demonstrated the usefulness, in many cases unexpected, of treating quantum states as information. They have also validated the field of QIT, providing a critical stimulus to experimental investigation and in some cases literally opening the path to realization of quantum processing of information for communication or computation. In fact, several of the most nonclassical or counterintuitive of the theoretical predictions have been the first to receive experimental verification (e.g., teleportation, superdense coding, and QKD). Looking back on these developments over the last 20 years, it is reasonable to expect that further investigation into the fundamentals of quantum information will continue to provide new and useful insights into issues with very practical implications. We can identify several outstanding open questions in QIT today, whose solution would impact the field as a whole. These include complete analysis of channel capacities for quantum information transmitted via quantum channels and quantification of entanglement measures for many-particle systems. Another, relatively new direction in QIT focuses on the use of measurements as an enabling tool for quantum information processing (QIP), rather than merely as a final step or source of decoherence. Measurement provides our limited access to the exponential resources intrinsic to quantum states, and recent work has shown that this access can itself be manipulated to control the processing, including some schemes to perform entire computations using only measurements in massively entangled states.

The exploration of new quantum algorithms has achieved some success over the last couple of years, following a lull of about 6 years after Shor's algorithm. These include Hallgren's 2002 quantum algorithm for Pell's equation [531] (one of the oldest problems in number theory), which breaks the Buchman–Williams cryptosystem. The framework for quantum algorithms has also been extended beyond the HSPs. van Dam, Hallgren, and Ip's 2000 quantum algorithm for shifted multiplicative characters [532, 533] breaks homomorphic cryptosystems, and the same techniques were recently extended by van Dam and Seroussi (2002) to a quantum algorithm for estimating Gauss sums [534]. The framework of adiabatic quantum algorithms introduced by Farhi, Goldman, Goldstone, and Sipser 2000 [475], and explored by van Dam, Mosca, and Vazirani 2001 [481] and by Aharonov et al. [476, 477] provides a novel paradigm for designing quantum algorithms.

G.3.2 Current Developments and Directions

This section gives more extensive and detailed descriptions of the theoretical challenges in quantum computation and places them in the context of current developments in the field.

Quantum Algorithms

The search for new quantum algorithms is undoubtedly one of the most important challenges in QC today. Following Shor's [501] discovery of quantum algorithms for factoring and discrete log in 1994 and Grover's [502] quantum search algorithm in 1995, there was a period of over 5 years with no substantially new quantum algorithms. During this period, the mathematical structure of Shor's algorithm was clarified via the formalism of the HSP—polynomial-time quantum algorithms were known for every finitely generated Abelian group. Over the last couple of years, we are starting to see some progress toward the discovery of new algorithms. In 2002, Hallgren [531] gave polynomial-time quantum algorithms for Pell's equation and the class group problem, thus breaking the Buchmann–Williams cryptosystem. This extended the framework to nonfinitely generated Abelian groups. The two most important open questions in quantum algorithms are graph isomorphism and the (gap) shortest-lattice vector problem. The first of these corresponds to the HSP in the symmetric group, and Regev [535] showed that the second can be reduced to the HSP in the dihedral group. The dihedral group is a particularly simple non-Abelian group because it has a cyclic subgroup of index 2. The standard quantum algorithm for Abelian HSP can be generalized in a natural way to non-Abelian groups. It was shown by Grigni, Schulman, Vazirani, and Vazirani [536] that for sufficiently non-Abelian groups the standard algorithm yields only an exponentially small amount of information about the hidden subgroup. On the other hand, Ettinger, Hoyer, and Knill [537] showed that the quantum query complexity of the problem is polynomial. This suggests that novel algorithmic ideas are necessary to tackle the non-Abelian HSP. Recently, Kuperberg [538] gave a $O(2\sqrt{n})$ algorithm for the dihedral HSP. The algorithm was an interesting modification of the standard algorithm. Other computational problems that are potential targets for quantum algorithms are the nonsolvable group membership, the McElise cryptosystem, and the learning AC0 circuits.

A different approach to designing quantum optimization algorithms via adiabatic evolution was proposed by Farhi et al. [539]. Initial efforts in this direction concentrated on the question about whether adiabatic optimization could solve NP-complete problems such as variants on SAT in polynomial time. Surprisingly, query lower bounds do not rule out this possibility [481]. However, van Dam and Vazirani [540] and more recently Reichardt [541] gave classes of SAT instances for which the spectral gap is exponentially small. Nevertheless, Farhi et al. [542] showed that adiabatic quantum optimization algorithms can tunnel through local optima and give an exponential speedup over local search. Aharonov and Ta-Shma [476] suggested that rather than optimization problems, adiabatic algorithms might be better suited for quantum-state generation. They also showed that every problem in the complexity class SZK can be reduced to the problem of generating an appropriate quantum state. Aharonov et al. [477] showed that a slightly more general formulation of adiabatic algorithms, when used for quantum-state generation, is in fact universal for QC. Designing quantum algorithms via quantum-state generation is a novel and potentially important direction, because it ties into classical algorithm design techniques using Markov chains and techniques such as bounds on conductance and spectral gaps. As a first step, it would be interesting to even give such an algorithm for solved problems such as quadratic residuosity or discrete logarithms.

Quantum random walks have held out the promise, over the last few years, as another interesting approach to the design of quantum algorithms. In the computational context, quantum walks were introduced by Farhi and Goldstone [543] in 1997 in their continuous-time incarnation, and in 1998 by Watrous [544] as discrete- time walks. Aharonov et al. [545] studied such walks and showed that their mixing time is polynomially related to that of the corresponding classical Markov chain. Cleve et al. [478] recently showed that in an oracle setting a quantum-walk-based algorithm gives an exponential speedup over any classical randomized algorithm. This is based on an exponential speedup by quantum walk for the hitting time between two specified vertices in a graph. The promise of quantum walks in the design of algorithms for concrete problems was recently realized by Ambainis [479] by combining it with Grover's search. He gave an optimal algorithm for element distinctness. The approach was further extended by Magniez, Santha, and Szegedy [539] to finding triangles in graphs, and by others to checking matrix multiplication. In each case, the speedup obtained is by a polynomial factor. This approach appears to be very promising. Challenges for the future include applying these new techniques to solve classical computational problems such as matrix multiplication, determinant computations, bipartite matching, or linear programming.

Quantum Error Correction and Fault-Tolerant QC

The discovery of the threshold result in fault-tolerant QC provided the theoretical basis for considering truly scalable physical implementations of QC. The original threshold result showed that as long as the decoherence rate is below $\eta = 10^{-6}$, arbitrarily long quantum computations may be carried out. The error model here is that each gate is subject to decoherence independently with probability η. More recent improvements by Aharonov and Gottesman [483] put the threshold at 10^{-4}, and Steane [484] shows that under mild assumptions the threshold is 10^{-3}. These improvements make use of quantum teleportation to prepare ancilla states [546] as well as improved use of quantum error-correcting codes. On the other hand, the best upper bound on the threshold was recently established by Razborov [485], who showed that if the threshold is below $\frac{1}{2}$, unless BQP = BQNC. For scalable QC to be practical, it is essential to improve the threshold by at least another order of magnitude.

There is clearly great room for improvement, although this will likely require new techniques. Equally important are the penalty in the number of qubits and total number of gate operations incurred to make a quantum circuit fault tolerant. These currently scale as 7^k and 343^k, respectively, for k levels of error correction. Progress in this area will likely require the study of new techniques, including the design of efficiently encodable and decodable quantum error-correcting codes, using expander-graph-based techniques, and list decoding. Another approach is to search for equivalent quantum models that are resilient to certain types of noise in the physical system under consideration for implementation. An example of this approach is the development of encodings based on recognition of symmetries in the physical interactions underlying the noise sources, referred to as *decoherence-free subspace* and *decoherence-free subsystem* encodings [547]. These provide passive error correction, in contrast to the active error correction approach of standard quantum error correction. Additional protection can be gained by engineering extra interactions to obtain supercoherent codes that provide thermal suppression of some physical noise sources in addition to complete protection against specific errors [548]. More generally, the approach of topological QC provides a powerful framework to rigorously suppress all effects of noise by encoding into topologically invariant subspaces [549, 550]. This passive approach to error correction has led to the emergence of alternative realizations of universal QC, including "encoded universality" [551] (see Section G.5) and the topological QC paradigm.

Quantum Complexity Theory

Clarifying the limitations of QC is a question of fundamental importance. One important issue is clarifying the relationship between BQP and the classical complexity classes—Is NP a subset of BQP? Does BQP lie in the polynomial hierarchy? Progress toward answering the first question was made via the oracle results of Bennett et al., who showed that relative to a random oracle NP is not a subset of BQP. This may be interpreted as saying that it is unlikely that quantum computers can efficiently solve NP-complete problems, or at least that nonrelativizing techniques are essential to resolving this question. This does not completely rule out the possibility of tackling this question, in light of the results of Arora et al. [482] showing that the principle of local checkability is nonrelativizing, and the demonstration by Mosca et al. that exponential query lower bounds do not apply to queries that examine the number of clauses left unsatisfied by the given truth assignment. Another important issue is understanding whether the limits on QC provide an opportunity to reconstitute modern cryptography despite Shor's assault on the two most important one-way functions—factoring and discrete log. Are there one-way functions that cannot be efficiently inverted even by a quantum algorithm? The complexity theoretic basis for an affirmative answer was given by Bennett et al., by showing that quantum computers require exponential time to invert a random permutation in the query model. More recently, it was shown by Aaronson that quantum computers require exponential time to solve the collision problem in the query model, thus opening the possibility of collision-intractable hash functions that are secure against quantum cryptanalysis. Interactive-proof systems have had important and unexpected applications in classical complexity theory. Kitaev and Watrous [552, 553] proved that quantum interactive-proof systems have interesting properties and are fundamentally different from classical proof systems. They showed the following:

1. Any polynomial-message quantum interactive proof can be parallelized to three messages (which does not happen classically unless AM = PSPACE).
2. Quantum interactive-proof systems can be simulated in deterministic exponential time.

The first result is interesting because it is unexpected and represents a way of taking advantage of quantum information that seems to be quite different from other applications. The second result represents one of the first applications of semidefinite programming to QC. In the classical case, the study of interactive-proof systems led to surprising and important applications, in particular with respect to the hardness of approximation problems. Are there interesting applications of quantum interactive-proof systems? For instance, can quantum interactive-proof systems give us insight into designing new quantum algorithms? Presently, we have no such applications.

The nature of quantum information is such that there is a great potential for zero-knowledge quantum interactive-proof systems. However, it turns out that perplexing mathematical difficulties are also associated with quantum variants of zero knowledge. Watrous [553] proves some fundamental limitations on one particular type of quantum zero knowledge, but this is (hopefully) just a beginning. That study also defines quantum zero knowledge in a very restrictive setting, but even the first step of giving a cryptographically satisfying general definition of quantum zero knowledge is a challenging problem.

The simplest variant of the interactive-proof-system model consists of two interacting parties, one prover and one verifier. A more complicated variant of the model allows multiple provers. In the quantum setting, fascinating connections exist between this model and the fundamental notion of a Bell inequality from quantum physics. Kobayashi and Matsumoto [554] studied this model in a very restricted setting where entanglement between the provers is not permitted.

However, it seems that entanglement is at the heart of the difficulty in understanding this model in the general case. Two-prover quantum interactive-proof systems could be more powerful, less powerful, or incomparable with classical two-prover interactive proofs—we presently know almost nothing about the power of this model, even in the case where the verifier is classical.

Quantum Simulation

Quantum simulation represents, along with the Shor and Grover algorithms, one of the three main experimental applications of quantum computers. Of the three, quantum simulation is in fact the application of quantum computers that has actually been used to solve problems that are apparently too difficult for classical computers to solve. As larger-scale quantum computers are developed over the next 5–10 years, quantum simulation is likely to continue to be the application for which quantum computers can give substantial improvements over classical computation.

Quantum simulation was in fact the first proposed application for which quantum computers might give an exponential enhancement over classical computation. In 1982, Feynman noted that simulating quantum dynamics on a classical computer was apparently intrinsically hard. Merely to write down the state of a quantum system made up of N two-state systems such as spins took up exponential amounts of space in the memory of a classical computer; and determining the dynamical evolution of such a state required the multiplication of exponentially large matrices. Suppose, Feynman continued, that it were possible to construct a "universal quantum simulator," an intrinsically quantum device whose state and dynamical evolution could be programmed to mimic the behavior of the quantum system of interest. Such a device, he concluded, could function as a quantum "analog" computer, capable of reproducing the behavior of any desired quantum system. Feynman merely noted the potential existence of such universal quantum simulators: he did not supply any prescription for how such a universal quantum analog computer might be realized in practice. In 1996, however, Lloyd, Wiesner, and Zalka showed that conventional "digital" quantum computers could be programmed to perform universal quantum simulation. Since then, Cory et al. have used room temperature nuclear magnetic resonance (NMR) QIPs to perform coherent quantum simulations of harmonic oscillators [555, 556, 557] and chaotic quantum dynamics such as the quantum Baker's map [558, 559]. Note that for the purpose of quantum simulation, the apparent lack of scalability of a room temperature NMR QIP does not prevent such a processor from supplying an apparently exponential speedup over a classical computer: Simulating high-temperature quantum systems is still apparently exponentially hard [560].

An example of a large-scale experimental realization of quantum simulation is the use of solid-state NMR QIPs to study the diffusive limit of transport of dipolar coupled spins in dielectric single crystals. The multibody dynamics were studied over times of tens of seconds, corresponding to an order of 10^8 times the spin–spin correlation time and spin transport over a distance of 1 μm. One result of these studies was to reveal that the diffusion constant for the two-spin dipolar ordered state is roughly 4 times faster than that of the single-spin, Zeeman ordered state. This speedup was not predicted by theoretical models and has been attributed to constructive interference in the transport of the two-spin state. Today solid-state NMR permits selected multibody problems to be addressed; the field does not yet have sufficient control to enable universal quantum simulation [561, 562].

Another potentially interesting source of problems relevant to the sciences are continuous, numerical problems such as integration and Feynman integrals. Because Grover's algorithm gives a quadratic speedup for not just search but also counting, it can be applied to get a quadratic speedup for integration in a natural way [488]. It remains an interesting open question whether some of the more sophisticated quantum walk techniques or other quantum algorithm techniques can be used in this context.

At the other end of the spectrum, QIT has provided novel algorithms for classically simulating quantum systems with limited entanglement. Vidal et al. [563] characterized the scaling properties of the ground-state entanglement in several 1D spin-chain models both near and at the quantum-critical regimes. They showed that the entanglement length scales logarithmically in the number of spins [it scales like $\log(L)$]. Vidal

[489] recently gave an efficient classical algorithm for simulating the dynamics of 1D spin chains that runs in time exponential in the entanglement length. Experimental results suggest that this method may be very effective in simulating a variety of systems. Extension of these results to 2D and 3D would be very interesting.

Novel Models

What are the primitives necessary to carry out QC? The answer in the quantum circuit model is clear—an implementation of qubits, a universal set of quantum gates, and the ability to measure the output. In recent years, there has been an exploration of novel models for QC that look fundamentally different from the quantum circuit model. One of the first such attempts, the topological QC, provides a different paradigm in which the qubits are no longer identified with specific atomic degrees of freedom but with collective excitations that must then be manipulated. Another approach was motivated by an attempt to prove that linear optics cannot be used to implement scalable quantum computers. In the attempt, Knill, Laflamme, and Milburn [550, 564] discovered a technique, using teleportation-based [546] use of ancillas, of implementing scalable QC using linear optics. In a different direction, Nielsen [565] showed that projective measurements can be used in the place of quantum gates as the fundamental primitive for QC. This was followed by the results of Raussendorf and Briegel [566] showing how to perform QC by preparing certain highly entangled cluster states, followed by a sequence of measurements. Adiabatic QC, first proposed by Farhi et al. [542] and then generalized by Aharonov et al. [476, 477], starts with an initial state that is the ground state of a sum of local Hamiltonians, and then gradually transforms to a different sum of local Hamiltonians whose ground state is closely related to the desired output of the QC. Aharonov et al. showed that this model is exactly as powerful as the quantum circuit model, thus providing another potential implementation of QC. The nontrivial spectral gap gives this model some natural fault-tolerant properties.

The role of entanglement in the power of QC is a fundamental theme. Two questions about this issue have arisen in the context of liquid NMR QC. The first question asks about the computational power of a mixed-state quantum computer whose state is required to be separable at every time step of the computation. Caves and Schack [567] pointed out that even though at first glance this model appears to be classical (because there is no entanglement), we do not know how to simulate it classically; nor do we know how to perform nontrivial QC with it. Another model, proposed by Knill and Laflamme [568] consists of 1 clean qubit with $n-1$ qubits in the maximally mixed state. Pulin et al. [569] give a quantum algorithm in this model to measure the average fidelity decay of a quantum map under perturbation.

G.4 QUANTUM INFORMATION THEORY

This section is a survey of the current and prospective future development of QIT. Continuing progress in QIT is crucial to the ultimate success of the laboratory implementation of QC. QIT addresses itself to performing useful processing tasks with noisy resources, and doing so optimally. The laboratory work in quantum information is and will be plagued by noise, and knowing the strategies for dealing with these (e.g., using a well-chosen quantum error-correcting code) will be very important for making progress. In addition, QIT invents fundamentally new applications for distributed quantum processing. These are in the form of uniquely quantum mechanical cryptographic primitives such as quantum key distribution, quantum data hiding, and private remote database access. For the purposes of this write-up, "QIT" should be understood as the information-theoretic analysis of quantum mechanical systems. Information theory quantifies the correlations between separated systems and the amount by which these correlations can be enhanced using the communications resources at hand. This subject sits at a more abstract level than the analysis of particular information-processing systems; that is, it does not address itself to the particularities of optical or electrical systems, but attempts to give a general framework within which the analysis of any such particular system can be performed. QIT is also distinct from algorithm theory, which seeks efficient procedures for solving mathematical problems; it does interface with it on the point of distributed algorithms, in which procedures using both local computation and communication are employed. The manifold uses of quantum teleportation are a prime example here.

We have chosen to discuss QIT below in terms of three big organizing themes: capacities (i.e., carrying capabilities of different communication resources), entanglement (i.e., quantification of the correlations, quantum and otherwise, between different subsystems), and cryptography (i.e., what we do with these capacities and correlations when we have them).

Very close to this subject, but distinct enough that they will not be discussed here, include the studies of communication and sampling complexity in the quantum setting [570], distributed quantum algorithm design, and quantum Kolmogorov complexity [571, 572].

G.4.1 Capacities

One of the two important quantifications of information theory is the calculation of capacities. Capacities measure the rate at which correlations (e.g., knowledge of a message text, shared randomness, quantum entanglement) grow per use of the given communications resource, in an "asymptotic" setting where arbitrarily many uses of the communication resource are available. More than one type of capacity is definable in the classical setting, and the number of different capacities grows substantially in a quantum setting because there are more distinct types of channel resources available, as well as more distinct types of correlations. Historically, one can consider Holevo's investigations in the 1970s [573] as the starting point of this subject, when he considered the classical capacity of a quantum state; this work remains seminal in that it established that, in general, a two-level quantum state is not capable of carrying more than one bit of information, despite the large amount of information needed to describe such a quantum state. One can say that it is the evasions of this theorem of Holevo, in the various special circumstances where one qubit can amount to more than one bit of information, that have been one of the important general themes of QIT.

In current language, Holevo's result pertains to the transmission of classical correlations (i.e., a classical message text) from sender to receiver (frequently "Alice" and "Bob" below) using a particular kind of quantum channel, which conveys a certain ensemble of quantum states ψ_i perfectly. This kind of channel is now known as a "cq" channel [574], in which a classical instruction, i, indicates that the quantum state ψ_1 should be synthesized and then conveyed undisturbed to the receiver. This is now considered as a special case of a more general resource, the quantum channel, which is described by some general completely positive trace preserving linear map between a quantum input state and a quantum output state [575]. The general question of the text-carrying capacity of such a general channel has been partly solved in that there is a formal expression (the Holevo capacity) for this quantity [525, 526]. A big open question remains, however, about the evaluation of this expression, which is one of several "additivity" questions that remain open in QIT [576]. The Holevo capacity expression involves an optimization over some number N of uses of the quantum channel, where N could be unboundedly large. The capacity is "additive" if the optimal is achieved for $N = 1$. For $N = 1$ the optimization is quite easy, and an explicit form (the Holevo c function) is known. But this and other additivity questions remain high on the priority list for solution in this area.

Perhaps the simplest quantum capacity is what has been called Q [529], the capacity of a noisy quantum channel to faithfully convey quantum states. Q is important from various points of view; achieving it requires the use of quantum error correction codes, and the optimization of Q can and will drive the optimization of these codes. Q also provides a bound on D, an important measure of the entanglement of mixed quantum states, the distillable entanglement [529] (see the next subsection). An entropic expression is now known for Q, the so-called coherent information [577]. It is known not to be additive, and its evaluation even for most qubit channels remains open.

Of the multitude of mixed capacities that can be considered, the first one to be studied was the one involving the same task as Q, that is, faithfully conveying quantum states from sender to receiver; but a dual resource was considered, namely a noisy quantum channel plus a classical side channel. It was shown that a forward side channel cannot increase Q, but that a two-way classical channel does, introducing a new capacity, Q_2 [529] (referring to the case of unlimited two-way use of the side channel). Bounds can be given for Q_2, and there are known to be quantum channels for which $Q_2 > 0$ but $Q = 0$; but there is no known entropic expression for Q_2, and there are no obvious strategies for making the present bounds on Q_2 tighter.

The other dual-resource capacity that has received a lot of attention is one for which both a channel and shared entanglement are available. The prototypes of these problems are quite famous: If the channel is a noiseless quantum channel, and the task is the conveyance of classical data, then this is the "superdense coding" [520] problem, in which one use of the channel, and the consumption of one entangled EPR (Einstein, Podolsky, Rosen) pair, results in two bits sent. The generalization of this to a noisy quantum channel gives a capacity that has been called C_E [578, 579]; useful entropic expressions for C_E have been derived, and it is known to be additive.

The dual problem, in which the channel resource is classical, but quantum states are to be transmitted, is teleportation [521]. The fully quantum version of this, in which the channel is quantum and the data to be transmitted is quantum, gives a capacity known as Q_E. For all channels, $Q_E = \frac{1}{2} C_E$ [521], showing that added resources can sometimes simplify the quantification of capacities.

Several other tasks that have no analog in the classical world have been considered in recent work. One is *remote state preparation* [580]—given a sender who has complete knowledge of a quantum state, the objective is for the recipient to come into possession of a faithful specimen of that quantum state. If the resources to be used are shared EPR pairs and a classical channel, the scenario resembles teleportation; but unlike in teleportation, the "capacity," that is, the minimal resources needed to perform the task, are highly nontrivial [581, 582]. (More use of the bit channel can reduce the number of EPR pairs needed.) Another uniquely quantum task is the "remote POVM" in which the sender has a set of quantum states, and the recipient is to obtain a bit string that represents a fair draw from the output of the POVMs performed on these states. This is to be done using a classical bit channel between sender and receiver, plus preshared randomness. The optimal capacity for this problem is also highly nontrivial, and has introduced new methods for the analysis of a host of other capacity problems [583].

To summarize this work, capacities are defined with respect to the following tasks:

- Bit transmission
- Qubit transmission
- Remote state preparation
- Remote POVM
- Private key transmission
- Sharing of entanglement
- Intersimulation (e.g., simulating a noisy channel by a noiseless one)

Employing the following means:

- Classical channel (noisy or noiseless, one-way or two-way)
- Quantum channel (noisy or noiseless)
- Shared correlations
    - Quantum (noisy or noiseless entanglement)
    - Classical (shared randomness)
- Quantum interaction (i.e., two-body Hamiltonian acting over time t)

Matching all possible tasks with all possible means, and including multiple parties, leads to the observation that the amount of work to be done in this area is practically infinite. It appears that the community will continue to tackle various cases among these infinite possibilities as the interest arises.

G.4.2 Entanglement and Correlations

Because, from some point of view, entanglement is simply one of the correlation resources available in quantum communication, it would seem that it might not deserve a heading of its own in a survey such as this. But this would be unfair to the unique role that it plays in the quantum setting; it is the feature of the quantum world that distinguishes it from the classical world [584, 585], saying that for a single pair of systems, a description of each system's state is not sufficient to describe the entire state of the system; it is the property that permits the violation of Bell's inequalities [586]. It is also the feature of quantum systems that makes exponential speedup of computations possible [587]. Thus, entanglement is of special interest, both from the foundational and the practical point of view. And thus, not surprisingly, it has received a large amount of special attention within the quantum-information community and will doubtless continue to do so.

A great deal of work has been done and continues to be done on the problem of measuring entanglement. For pure states of two parties, there is a single measure that, for most information-theoretic purposes, is satisfactory for quantifying entanglement: the von Neumann entropy of the reduced density matrix [527]. (By "information-theoretic," one means that, as above, one considers an asymptotic situation in which many copies of the states of interest are available.) For almost any other circumstance, it seems impossible to devise a single measure that will quantify entanglement in physically meaningful ways. The prototype example of this is the mixed state of two parties. If the state is separable (can be written as a convex mixture of product projectors), then for almost all purposes the state may be considered to be unentangled [529]; the state has correlations, but for most purposes (some exceptions occur in the next section on cryptography) these correlations behave as in the classical world. So, if a mixed state is inseparable, it is entangled. But how entangled is it? Here is a list of some of the measures that have been described:

- *Distillable entanglement* (D) [529]. This measure is an answer to the question: How good is my entangled mixed state for doing quantum teleportation? Thus, it has an operational significance in quantum capacities. The distillable entanglement is also the number of EPR singlets that can be obtained from a set of copies of the given mixed state, assuming that the parties can do only "local" operations, where "locality" includes the possibility of classical communication. This was the first setting in which the class of quantum operations denoted by "local quantum operations and classical communication" [usually LOCC (local operations and classical communication)] was introduced—although in some sense it was already implicit in discussions of Bell inequalities. This class of operationally local quantum dynamics has now been considered in many other contexts. The effort to calculate D explicitly has been difficult. It turns out to have none of the convexity or additivity properties that one would desire for an information-theoretic measure to apply to D [588]. Also, D is not nonzero for all inseparable states [589]; but this relates to the PPT story discussed below.

- *Entanglement of formation* (E_F) [529]. This is defined as the minimum average entanglement of a pure state ensemble making up the mixed state. Thus, it is not an operational measure of entanglement, but it is one that is amenable to exact calculation, and it is an upper bound on D. It is nonzero for all inseparable states.
- *Entanglement cost* (E_C) [590], which is the smallest number of EPR pairs needed to create a given number of copies of a mixed state, ρ, by LOCC operations. This may equal the entanglement of formation, but it turns out that this is one of the "additivity" questions that has not been settled and is equivalent to the additivity conjecture for the Holevo capacity [576].

This by no means exhausts the list of entanglement measures of mixed states:

- *Relative entropy of entanglement* [591, 592]. This measure is based on the idea that entanglement should be measured by "how far" ρ is from the set of unentangled (separable) states. One way of measuring how far for quantum states is by their relative entropy. This measure is upper bounded by the entanglement cost and lower bounded by the distillable entanglement. It is relatively easy to compute. It also has the property that it cannot be increased under "separable" quantum operations [593]. This class is not the same as LOCC, but it does include it. This result is illustrative of a more general principle in the quantification of entanglement: Because it is supposed to represent uniquely quantum correlations, it should not be possible to increase it using only classical communication between the parties.
- *Entanglement monotones* [594]. This is a kind of metameasure—in that it potentially includes an infinity of specific measures. It simply states that any functional of the quantum state that is nonincreasing under LOCC should be considered a measure of entanglement. It is known that there is a whole continuum of such measures, which (under sensible restrictions) lie between the entanglement cost and the distillable entanglement.
- *Negativity* [595]. This quantification arises from a different idea about the characterization of entanglement, that arising from the "partial transpose." Peres [596] noted that if the partial transposition, that is, matrix transposition applied only to the indices of one of the parties, is performed on the matrix describing a separable mixed state, the result is always another mixed state (i.e., it is another matrix with nonnegative eigenvalues). On the other hand, if it is applied to the density matrix of an EPR pair, the result is a matrix with some negative eigenvalues. The "Peres criterion" for entanglement states that r is entangled if its partial transpose is negative. For small Hilbert spaces this is a necessary and sufficient condition for entanglement [597]; but in higher Hilbert space there are entangled r's that are positive under partial transpose [589]. Recognizing this flaw, it is still possible to give another quantification of entanglement that is the sum of the negative eigenvalues of the partial transpose. This measure is easy to compute and has been used to develop bounds in some calculations pertaining to entanglement.

So, this relatively innocent exercise of trying to associate a number with a degree of entanglement has led to a very complex discussion that raises questions on various fundamental aspects of quantum theory. First, one can ask, can entanglement be reversibly converted from one form to another? For pure bipartite states the answer is yes [527]; this is related to the fact that there is considered to be only one information-theoretic measure of pure state entanglement. Thus, a large supply of partially entangled mixed states can be converted, by purely local operations, to a smaller supply of EPR pairs (*entanglement concentration*), and converted back again to the same number of partially entangled states (*entanglement distillation*). But for mixed states the answer is the reverse [527, 590], thanks to the known gap between the entanglement cost and the distillable entanglement. This is connected with another basic question: Why does the partial transpose criterion sometimes fail to detect the entanglement of a state? One answer to this is that states exist for which the entanglement cost is finite but the distillable entanglement is zero, so the irreversibility is complete. States for which this happens are said to have *bound entanglement* [589], meaning that it cannot be freed up by LOCC operations.

A great deal is known about bound-entangled states now, for example, how to construct instances of such states [589, 598, 599, 600], but there remain many unanswered questions about them. Also, this is related to a final question that is only partially answered: What is a good notion of locality for joint operations involving two parties? It was once thought that the LOCC class captured everything of interest; that is, all LOCC operations resulted in only classical correlations (they do not produce or increase entanglement), and that all operations outside the LOCC class could produce quantum correlations.

This is no longer so clear. We mentioned that in the context of the relative entropy of entanglement, the "right" characterization of local quantum operations is the "separable" class, in which each Krauss operator of a superoperator can be written in a product form. It is somewhat surprising that this class is strictly larger than the LOCC class [593]. Yet, from most points of view, such an operator seems incapable of generating any entanglement. There is yet a larger class, which is called the "ppt preserving" class [601], which by definition includes all bipartite quantum operations such that if the input state is positive under partial transposition, so is the output. These operations can definitely produce entanglement, but only of the bound variety. (So, e.g., it cannot produce the kind of entanglement that would be useful for teleportation.) Thus, many entanglement measures of interest are well behaved even within this large class. This class has been very useful because its mathematical characterization turns out to be much simpler than either the LOCC or the separable class. But it remains unclear whether this class of quantum operations has any real physical significance. The experimental detection of entanglement has been a subject of more recent theoretical interest. The simplest way to approach this, which requires no new ideas, is that a state can be characterized by quantum tomography; then, if the tomography is sufficiently precise, any of the measures of entanglement discussed above can be calculated for the state. But there are potentially more direct ways in which this determination can be made. Terhal's "entanglement witness" [602] is a Hermitian operator, W, that has the property that its expectation value $Tr\ (W\rho)$ is positive for all unentangled states, but is negative for some entangled states. (Unfortunately, it is impossible for it to be negative for all entangled states.) Thus, determination of the expectation value of W by repeated measurement can detect entanglement (a negative answer means entangled), and the value of this expectation value becomes another quantification of entanglement. Nonlinear functionals can also detect entanglement: One can find quantum operators for which the variance is only zero for entangled states, being nonzero for all unentangled states [603]. Finally, there are modifications of tomography such that, with only a subset of the measurements performed for full tomography, it can be determined whether a state is entangled or not [603]. It is expected that future work in this area will connect these means of detecting entanglement more directly with the applications of entanglement in cryptography, communications, and computing.

All of the characterizations of entanglement that we have discussed so far are information theoretic, that is, apply to a setting where there is a large supply of identical copies of the state r of interest; many of the measures of entanglement we discussed, for instance, involve taking the limit of the number of copies of the state to infinity. But there is another, potentially more practical, area of investigation in which the number of copies of the state is considered to be limited. For example, one can ask: If only one specimen of the bipartite state r is held by two parties, is it possible for them to convert this state, by LOCC operations, to a single specimen of the state ρ'? If ρ and ρ' are pure, then there is a very beautiful answer to this question involving the statistical concept of majorization [565]. But almost all other problems in this area are open.

Finally, it should be mentioned that the theory of entanglement has a direct bearing on QC itself. The theory of quantum error-correcting codes, and their application to fault-tolerant QC, is from some point of view a theory of the properties of special kinds of entangled states. It is a paradoxical truth that has emerged from quantum information research that sometimes highly entangled states can be more robust against decoherence than apparently more classical unentangled states [511]. This robustness has also had application in areas of QCRYPT (see secret sharing, below). Entanglement can also be used in the implementation of quantum logic gates; teleporting through the right kind of entangled quantum state can result in two-bit gate operations applied to a pair of qubits [604]. Generalizations of these ideas have resulted in the discovery that linear optics is sufficient for QC [564]. Also, it is now known that with the right kind of entanglement (the *cluster state*), QC can be reduced completely to a sequence of local quantum measurements, with all information flows in the computer being classical [566]. There is likely to be considerably more work to be done in this area, to connect these remarkable features of entanglement to other workable approaches to QC in the laboratory.

G.4.3 Cryptographic Primitives

Broadly defined, cryptography considers distributed information-processing tasks constrained by requirements of privacy, secrecy, and security. Quantum mechanics has offered a new toolkit for the construction (and demolition) of cryptographic tasks, and this remains an extremely active area of research.

In many people's minds, cryptography is defined as the sending of secret messages from one party to another. While cryptography actually encompasses much more than this task alone, the *key distribution* problem is still central to QCRYPT, and it is the only one for which there is active laboratory work. The theory of secure key distribution using quantum channels has been undergoing a continuing rapid evolution in recent years. The basic idea of using the unclonability and unmeasurability of single unknown quantum states to make secret messages intrinsically unreadable to an eavesdropper (without disturbance) dates back to Wiesner's work in the 1970s [605], and the explicit protocols for doing this style of cryptography were all established more than 10 years ago, independently by Bennett and Brassard [530, 606] and by Ekert [607]. This work was enough to stimulate serious experimental work [608], which continues to this day. But the security of these protocols remained unproved in the general setting for many years, although proofs for restricted "Eves" were known some time ago. In addition, there was an early insight that entanglement distillation would be a crucial ingredient in this proof [609, 610], although the details were a long time in coming. But the real revolution in this area theoretically was initiated by Mayers in the late 1990s [611]. He found a proof that BB84 is absolutely secure for sufficiently low detected bit error rate for quantum transmission. His proof was difficult and was not understood by much of the community for some years; but the revolution was made general by Shor, who, with Preskill [612], redid Mayers proof in much more transparent language.

Shor's starting point was a different proof by Lo and Chau [546] that a different key-distribution protocol involving the distillation of perfect entanglement is secure. This proof was much easier than Mayers' and established that the Ekert [607] style of "quantum Vernam cypher" QCRYPT was actually valid, but assumed that Alice and Bob have the full power of QC. Shor and Preskill showed that using a particular style of quantum error correction code in the Lo-Chau purification permitted a reduction of this proof to BB84. Their approach to this proof has been workable enough that more results are now flowing out; one result involves the strengthening of the BB84 by use of two-way classical (insecure) communication; it is now known that this resource permits secure key distribution in a more noisy environment (i.e., a more aggressive eavesdropper). Also, B92 has been proved secure now by an ingenious variant of the Shor reduction [613].

It appears that this activity in security proofs for key distribution still has a long way to go. Very important fundamental and practical questions involving imperfect sources persist. Fundamental questions also remain open about the relation of security to the violation of Bell inequalities. Also, because experiments are underway, there are a host of technical questions (e.g., involving the use of weak coherent sources) that deserve theoretical attention.

As stated above, cryptography is not just secret-message transmission. We give a brief survey here of the other areas of cryptography that have been reconsidered in the light of quantum theory:

- *Bit commitment.* Bit commitment, a primitive for many other forms of cryptography (e.g., secure function evaluation) involves

- Choosing of a bit value by Alice
- Commitment by Alice of this bit value to Bob in an unreadable form
- Unveiling of this bit value to Bob at a later time

Mayers [614] showed that bit commitment is impossible in the standard quantum model of the world, by showing that Alice can always cheat by using quantum entanglement. Partially secure bit commitment is possible and has been analyzed [615]. An interesting recent development here is to consider the effect of various additional fundamental and practical physical effects on the security of bit commitment. For example, special relativity makes a limited form of secure bit commitment possible. Recent work has focused on the role of selection rules. It is now believed that fundamental selection rules (e.g., charge superselection) do not modify the no-go theorem for bit commitment, although the proof is considerably more technical. Perhaps more interesting is the fact that nonfundamental, technological restrictions (e.g., the inability to change spin angular momentum in the lab) may enable a new kind of conditionally secure bit commitment. Current theoretical work in this area is very active.

- *Remote coin tossing*. As with bit commitment, there are quantum no-go results [616]. However a closely related primitive, weak coin tossing, in which Alice would prefer a "heads" and Bob would prefer a "tails" is sufficient for most of the applications of coin tossing. Ambainis and Kerenidis and Nayak gave protocols for weak coin tossing that beat Kitaev's bound, thus showing that his no-go theorem does not apply in this case. Whether protocols that achieve arbitrarily small bias exist is an open question.
- *Quantum secret sharing*. Secret sharing is a concept in classical cryptography in which many parties receive "shares" of a secret that are unintelligible to the individual parties, or to small groups, but can be faithfully reconstructed if any "quorum" of these parties is brought together or can communicate among themselves. There are protocols that perform similar functions in which a quantum state is the secret [617]. That is, parties receive shares of a quantum state, whose identity is unintelligible to single parties (i.e., the reduced density matrix is proportional to the identity operator). Classical or quantum communication among a subquorum of parties also is incapable of revealing anything about the identity of the secret.
- *Quantum data hiding*. This is dual to the previous: The idea is that the parties receive "shares" representing ordinary classical data, but the idea is to enforce security in the presence of arbitrary classical communication. Thus, reconstruction of the secret is only possible with quantum communication. The existence of states that perform this task is known [618], and, surprisingly, it is known that they can be separable mixtures (i.e., they need not involve any entanglement) [619]. Also recently, it has been shown that a variant of quantum data hiding can be used in conjunction with quantum secret sharing to strengthen the security of the latter [620].

- *Quantum fingerprinting*. Fingerprinting is a classical technique for associating with each large data set a small bit string such that the bit string for each data set is distinct. It has been shown that using quantum techniques, more efficient construction of fingerprints for distributed data sets is possible [621].
- *Secure remote computation*. In this protocol, the premise is that Alice has a computation she wants to do on a quantum computer; she has only a very small computer, but she has a quantum channel connecting her to Bob, who has a large quantum computer. She wants to have a computation performed by Bob, but she does want him to know the nature of the computation or for him to be able to obtain any information about the answers without her detecting it. A quantum protocol exists that meets all these requirements [622, 623].
- *Private quantum channels*. Quantum channels can be made private, that is, containing only transmissions that are completely unintelligible to an interceptor, with the use of shared classical randomness between sender and receiver. For exact privacy, two bits per sent qubit are necessary and sufficient. For asymptotically perfect privacy, it is now known that one bit per qubit is sufficient [624, 625, 626, 627]. If this shared resource is quantum, then there are scenarios in which the shared resource can be recycled [628] (if a negligible amount of eavesdropping is detected).
- *Quantum digital signatures*. With this scheme, a sender (Alice) can sign a message in such a way that the signature can be validated by a number of different people, and all will agree either that the message came from Alice or that it has been tampered with. To accomplish this task, each recipient of the message must have a copy of Alice's "public key," which is a set of quantum states whose exact identity is known only to Alice. Quantum public keys are more difficult to deal with than classical public keys: For instance, only a limited number of copies can be in circulation, or the scheme becomes insecure. However, in exchange for this price, unconditionally secure digital signatures are claimed. Sending an m-bit message uses up $O(m)$ quantum bits for each recipient of the public key (adapted from [629]).
- *Privacy in remote database access*. Private-information-retrieval (PIR) systems allow a user to extract an item from a database that is replicated over $k \geq 1$ servers, while satisfying various privacy constraints. Quantum k-server symmetrically private information-retrieval (QSPIR) systems have been found that:
  - Use sublinear communication.
  - Do not use shared randomness among the servers.
  - Preserve privacy against honest users and dishonest servers. Classically, SPIRs without shared randomness do not exist at all (adapted from [630]).
- *Quantum interactive proofs*. Certain computational problems (e.g., graph nonisomorphism) are defined as requiring the participation of two parties; of interest is the case where one knowledgeable party is trying to prove something to an ignorant but intelligent party. It is know that these "interactive proofs" may

require arbitrarily many rounds of communication between the two parties. It is now known that in a quantum settings, just three rounds of quantum communication are sufficient [552].

- *Authentication of quantum messages*. Authentication is a well-studied area of classical cryptography: a sender, S, and a receiver, R, sharing a classical private key want to exchange a classical message with the guarantee that the message has not been modified by any third party with control of the communication line. Authentication of messages composed of quantum states is possible. Assuming S and R have access to an insecure quantum channel and share a private, classical random key, a noninteractive scheme exists that enables S both to encrypt and to authenticate (with unconditional security) an m qubit message by encoding it into $m + s$ qubits, where the failure probability decreases exponentially in the security parameter, s. The classical private key has $2m + O(s)$ bits. Any scheme to authenticate quantum messages must also encrypt them. (In contrast, one can authenticate a classical message while leaving it publicly readable.) This gives a lower bound of $2m$ key bits for authenticating m qubits, and it shows that digitally signing quantum states is impossible, even with only computational security (adapted from [622, 623]).
- *Secure multiparty QC*. Secure multiparty computing, also called *secure function evaluation*, has been extensively studied in classical cryptography. This task can be extended to computation with quantum inputs and circuits. The protocols are information-theoretic secure, that is, no assumptions are made on the computational power of the adversary. For the weaker task of verifiable quantum secret sharing, there is a protocol that tolerates any $t < n/4$ cheating parties (out of n). This is optimal. This tool can perform any multiparty QC as long as the number of dishonest players is $<n/6$ (adapted from [622, 623]).

G.5 QUANTUM COMPUTER ARCHITECTURES

Large-scale quantum computers, if they can be built, will be complex quantum systems with many parts, all of which must work together coherently to perform large-scale quantum computations. To construct a large-scale quantum computer, it is not enough to exhibit components (qubits, quantum logic gates, input–output devices, etc.) sufficient for attaining the DiVincenzo criteria, and each of which on its own attains the limits required for fault-tolerant QC. The components of a large-scale quantum computer must be designed to fit together and to work together. That is, a large-scale quantum computer must have an architecture—a unified overall design in which each component plays an integral role. In addition, each of these components must be designed for optimal efficiency. For example, the quantum Fourier transform is a fundamental building block in all quantum algorithms, and recent work has shown that we can significantly enhance the performance of this component by implementing quantum circuits for the quantum Fourier transform with only logarithmic depth [631, 632].

Note that theory, coupled strongly to experiment, is a necessary part of developing a viable quantum computer architecture. Designing an architecture for a quantum computer is fundamentally a theoretical task: One is creating specifications and solving problems for a device that does not yet exist. Of course, because a viable architecture must marry theoretical concept with experimental reality, the design of such an architecture is a theoretical task at which experimentalists can excel as well as theorists. As will be seen below, in the specification of the stages and development of quantum computing architectures, designing and building quantum computers is a task that must be performed by experimentalists and theoreticians working together. For example, approach of *encoded universality*, which emerged from theoretical work in decoherence-free subspaces, has potential for simplifying spin-based computation in solid-state QC since it relies exclusively on tuning the exchange interaction and does not require local magnetic fields [633].

A quantum computer architecture specifies not only the components of a quantum computer [qubits, quantum logic gates, input–output (I/O) devices, etc.], but provides protocols and mechanisms for how those components are to work together. Even at the early stages of development of a quantum computing technology, as in the case of semiconductor quantum computers, considerable effort must be made to design architectures that allow the different pieces of the quantum computer to function together.

Quantum computer architectures have played a key role in the development of quantum computers. The Cirac–Zoller proposal for ion-trap QC provides an architecture for medium-scale quantum computers with $O(10^1)$ qubits. Cirac and Zoller specified explicit designs for qubits (hyperfine levels of ions), quantum logic gates (optical resonance), quantum "wires" (the use of a shared vibrational mode as a quantum "bus" to transfer information from one qubit to another), as well as readout (fluorescence via cycling transitions). Most important, they showed how all of these different components for a small- to medium-scale quantum computer could, in principle, be put together to perform simple QC coherently. Their proposal was based on quantum technologies that had been pioneered by experimentalists in atomic and optical physics (Wineland, Monroe, and Blatt). Because it supplied a well-thought-out design together with explicit proposals for implementing the pieces of that design in an integrated fashion, the Cirac–Zoller proposal was swiftly implemented by Wineland and Monroe. The Cirac–Zoller proposal met with swift success exactly because it specified an architecture for QC.

A detailed quantum computing architecture is a necessary proof of principle that a particular method for performing QC has a chance of succeeding. The initial work on QC of Benioff, Feynman, and Deutsch in the 1980s took place in the absence of any specific ideas on how a quantum computer might, in fact, be built. It was not until the explicit demonstration of a universal architecture for QC using electromagnetic resonance [634] that it became clear that quantum computers might actually be built. The techniques for using electromagnetic resonance to perform universal QC subsequently matured in simple NMR QIPs, which were then used to demonstrate the first quantum algorithms.

In short, a well-thought-out architecture is the key to successful quantum computer design. Given the importance of QC architectures, it should be no surprise that the development of such architectures has played and continues to play a key role in ARDA's Quantum Computing Roadmap. We can identify a set of stages in the development of QC architectures. Each stage is associated with advances in the quantum technologies required to realize that architecture. Each stage represents, in essence, a test that a QC architecture must pass if it is to form the basis for constructing a viable quantum computer.

G.5.1 Initial Conceptual Development

In this stage, the basic concepts for meeting the DiVincenzo criteria for constructing a viable quantum computer are developed. Potential answers are supplied to the questions of how quantum information is to be registered (qubits), how it is to be processed (quantum logic gates), how it is to be moved from one place to another (quantum "wires" and quantum "buses"), and how it is to be programmed in and read out (I/O devices). The initial conceptual development can be purely theoretical but must be fully informed by existing quantum technologies or quantum technologies under development. Care must be taken to ensure that the quantum computer architecture is integrated (i.e., that the various components of the quantum computer can act coherently and in concert together).

G.5.2 Testing the Components

In this stage, the different components of the architecture are subjected to experimental tests and to more detailed theoretical investigations to determine whether or not they "meet spec."

- Qubits are prepared, manipulated, and read out.
- Relaxation and decoherence times are measured.
- Quantum operation and state tomography are performed.

The testing stage for the components of a quantum computer architecture forms the basis for an extended experimental program. As tests reveal the strengths and weaknesses of a particular approach, the architecture is revised and refined to emphasize those strengths and to minimize the effects of the weaknesses. (An example of such revision and refinement is Wineland's development of techniques for moving ions coherently from one ion trap to another, to get around the problem of the finite size of ion traps.)

G.5.3 Assembling the Components into a Working Device

In this stage, the various components of the quantum computing architecture are assembled to construct a working QIP capable of performing QC. The ability to perform sequences of coherent quantum manipulations and to put them together in a quantum algorithm is a strong test of the viability of a quantum computing architecture. To date, only a few architectures have succeeded in performing extended sequences of coherent logic manipulations. Room temperature NMR QIPs, despite their intrinsic lack of scalability, have been strikingly successful at performing demonstrations of quantum algorithms such as the Deutsch–Jozsa algorithm, Grover's algorithm, and Shor's algorithm, as well as quantum error correction, decoherence-free subspaces (DFSs), and so forth. The success of such demonstrations bodes well for the ability of lower-temperature (e.g., optically pumpable) scalable NMR devices to perform larger-scale quantum computations. Similarly, ion-trap quantum computers have been used to exhibit a wide variety of techniques for coherently manipulating quantum information, including the recent performance of a quantum algorithm on an ion-trap quantum computer. The recent demonstration of coherent one- and two-qubit quantum logic operations on superconducting quantum bits suggests that superconducting quantum computers may soon be capable of performing quantum algorithms. Actually operating a quantum computer with a particular architecture is, of course, the proof in practice that the architecture can indeed function at a particular scale (i.e., number of qubits and number of quantum logic operations).

G.5.4 Scaling up the Architecture

Once a quantum computing architecture has been developed, tested, and put into practice, it can then be scaled up to more qubits and to more coherent quantum logic operations. As the architecture is scaled up, stages one, two, and three above must be revisited again and again. Often, the testing of the components and their assembly into a coherently functioning whole will reveal a weakness of the initial conceptual scheme, which must be readdressed at the fundamental conceptual level if the architecture is to be scaled to the next level. (Once again, Wineland's movable ions are an example of the recognition of a weakness and the development of a fundamental quantum technology to correct that weakness. Similarly, the development of methods for performing optical pumping for NMR-based systems addresses and corrects the problem of state preparation for liquid-state NMR.)

Each increase in the number of qubits and the number of coherent operations supplies a strong test of the scalability of a QC architecture. Each doubling of the number of qubits and number of quantum logic operations typically brings with it a host of new quantum technological problems, which must be addressed and solved in detail before the quantum computing architecture can be brought to the next level.

To optimize and test scalable quantum computers requires theoretical software that can simulate the dynamics of algorithms involving a large number of qubits. Some pioneering work has been done to create perturbation theories and software that enable one to calculate the dynamics of a restricted set of logic involving a large number of qubits [635, 636]. These perturbation theories are essential for minimizing the error rates of quantum computers involving more than 30 qubits. Related theoretical progress has been in resolving dynamical issues for single-qubit measurement technologies based on magnetic resonance force microscopy, scanning tunneling microscopy, optical magnetic resonance and resolving dynamical problems for utilizing and measuring charge-based qubits using single-electron transistors and other nanodevices based on semiconductor and superconductor materials [637]. In order to meet the 5- and 10-year goals of the ARDA's Quantum Computing Roadmap, all four stages of the development of QC architectures must be accomplished at least once for each viable QC technology. In order to construct a quantum computer with eight or more qubits, a QC architecture must undergo at least three doublings from its initial demonstration of a viable quantum bit. Theory plays a key role in the development of QC architectures. The initial conceptual development of such an architecture is a purely theoretical task. As the architecture is tested, assembled, and scaled up, the development of theoretical concepts and solutions is married ever more closely with the experimental development of specific quantum technologies.

G.5.5 Type-II Quantum Computing

Type-II QC is a particular application of quantum simulation in which quantum "microprocessors" are connected via classical links. Type-II QC is useful for simulating systems in which coherent quantum behavior is important at small scales. Systems that could potentially benefit from the application of Type-II QC include nanofluids, quantum gases, Bose–Einstein condensates, and plasmas at high temperatures and pressures. Unlike quantum simulation in general, Type-II QC does not afford an exponential speedup over classical computation. However, there are specific and important problems for which the exponential power of QC can be brought to bear to simulate using a few tens of qubits an intrinsically quantum piece of a larger system that would require a supercomputer to simulate classically. Such few-qubit quantum microprocessors might then be hooked up using classical communication links to perform mixed quantum/classical simulation of extended quantum systems.

G.6 DECOHERENCE ROADBLOCKS FOR QUANTUM INFORMATION PROCESSING

G.6.1 Theoretical Terminology

Quantum information processing relies to a large extent upon the ability to ensure and control unitary evolution of an array of coupled qubits for long periods of time. There are a number of physical effects that act against this coherent evolution. These include interaction of the qubits with a larger environment, unwanted or uncontrolled interactions between qubits, and imperfections in applied unitary transformations. The latter can be either systematic or random, and can also give rise to additional unitary errors. The term *decoherence* referred originally explicitly to errors that arise in the wave function phase, that is, to decay of off-diagonal terms in the density matrix. This decay of phase is basis-set dependent. It also does not constitute the only source of loss of unitarity. Today, the term is therefore more generally understood in the field of QIP to refer to all manifestations of loss of unitarity in the qubit state time evolution. It thereby includes

1. Explicit loss of coherence
2. Dissipative or energy relaxation effects
3. Leakage out of the qubit state space

There are many theoretical languages in which decoherence may be framed and usefully understood. Nonunitary evolution of qubit states and density matrices may be generally regarded as resulting from entanglement of the qubit states with those of a larger quantum system whose quantum evolution is of no intrinsic interest, such as the environment or a measuring device [638]. This entanglement with the environment converts pure qubit states into mixed states and results in a loss of information from the qubit system that can be quantified by an associated increase in entropy. The resulting qubit density matrix is referred to as the *reduced density matrix.*

- The density matrix allows analysis of decoherence resulting from physical interactions via formulation and solution of many different levels of master equations that have been developed to study the dynamics of reduced density matrices [639] (see below). These constitute one set of languages for analysis, systematization, and quantification of decoherence.
- Another type of decoherence language deriving from the reduced density matrix is that of superoperators. So named because they act on the density matrix, which is itself an operator, superoperators provide a very useful formalism for general analysis of the evolution of pure states into mixed states. An important distinction between unitary evolution operators and superoperators is that the former always constitute a group while the latter may sometimes define a dynamical semigroup that lacks an inverse. The language of superoperators is naturally related to that of generalized measurements, allowing useful connections between decoherence and measurements to be established. The operator sum representation provides a compact way to obtain the superoperators that result from any specific Hamiltonian describing the qubit system and its interaction with the environment [640].
- Nonunitary time evolution can also be expressed as the action of quantum noise operations [641]. These are maps that describe the introduction of errors onto qubit states. They are written in a digitized form (error occurs with probability p) analogous to the noise channels employed in classical information theory.

G.6.2 Studies of Decoherence and Ways to Overcome It

Over 2000 publications have appeared between 2000 and 2004 discussing decoherence. Theoretical studies of decoherence and its mitigation to date have tended to fall into four broad categories.

1. Physical studies of origin and magnitude of decoherence for specific candidate qubit states in specific physical systems. Such studies generally seek to predict values of the decay times T_1 (energy dissipation or population relaxation) and T_2

(dephasing) for qubit states, starting from specific models of coupling mechanisms and of the spectral distribution of the environment (bath), and assumptions as to Markovian or non-Markovian dynamics of the environment on the intrinsic time scale of the qubit states. For a recent review of these approaches, see, for example, [642].

2. Mitigation of decoherence by either encoding to allow subsequent quantum error correction (active error correction) or encoding to eliminate or suppress decoherence (passive error correction). The former includes quantum error-correcting codes that have been developed to correct a wide variety of errors [551, 643]. Construction of fault-tolerant protocols using these codes has been demonstrated. The passive error correction approach includes use of decoherence-free subspaces and subsystems and in its most ambitious form is represented by topological QC (below).

3. Work on topological QC that seeks to develop naturally fault-tolerant codes may be viewed as an ambitious alternative paradigm that would provide a powerful set of self-correcting codes immune to many of the usual sources of decoherence if the required Hamiltonians could be physically realized [644].

4. Suppression of decoherence by dynamical decoupling techniques. These employ external pulse fields in a controlled manner that is specifically designed to cancel or minimize errors by averaging them out. These methods are related to coherent averaging methods in pulsed magnetic resonance spectroscopy and have recently been extended from the original techniques requiring arbitrarily strong, instantaneous control pulses ("bang-bang control") to realistic bounded-strength Hamiltonians ("Eulerian decoupling") [645].

Some work has been done on combining several of the above approaches to obtain combined error correction techniques for QC architectures that have the capability of correcting errors deriving from very different physical sources [646]. There are a number of further directions beyond these characterization and mitigation studies that would be valuable to pursue in the next period of research into control of decoherence. These include:

- Relatively few studies have addressed the effect of decoherence on short time qubit dynamics, that is, within T_2, and possibly over the time period during which control pulses would be applied. Studies of electron spin decoherence due to hyperfine interactions with nuclear spin are a first step in this direction, analyzing the effect of very short time nonexponential electron spin dynamics. Measures of decoherence times based on the density matrix norm rather than on exponential time scales for decay of matrix elements have been proposed to quantify such short time dynamics [647]. Weakly coupled situations where decoherence can produce nonexponential behavior that can give rise to "prompt" loss of coherence amplitude [648] or, under appropriate conditions, manifest itself solely as a reduction in the norm of an effective system wave function [649] may provide a useful new avenue to explore coherent control of intrinsically noisy qubit systems. This is particularly relevant to qubit implementations displaying "reduced visibility" or "reduced contrast" Rabi oscillations [650, 651].

- There have also been few studies of decoherence that might arise specifically during gate switching of control fields. Some studies of pulse shaping and of compensation techniques to stabilize control pulses against imperfections have been made [652]. We expect such studies to become routine and to benefit from interaction between theory and experiment.
- Complete simulations of controlled manipulations of coupled qubits with realistic decoherence effects are rare. A few such simulations of small-scale algorithms on coupled qubits have been made [653].
- Despite much theoretical work on fault-tolerant protocols, complete analysis of the error threshold for fault-tolerant QC applicable to a specific set of errors for a given physical implementation is lacking. This represents a highly desirable direction of theoretical and simulation research and would usefully be combined with the algorithmic simulations described above.
- Develop realistic microscopic description of the parameters for quantum noise operators, to enable a unification of microscopic physical studies of decoherence with information-theoretic description of noise channels.

G.6.3 Physical Sources of Decoherence

The following is a summary of physical sources of decoherence that have been identified and/or discussed for the physical implementations.

1. NMR
   1.1 Liquid state
      1.1.1 External random fields due primarily to dipoles of spins in other molecules going past the molecule in question
      1.1.2 Modulation of through-space dipolar interactions between spins in the same molecule through rotational diffusion of the molecule changing the direction of the tensor with respect to the external field
      1.1.3 Modulation of the chemical shift of a spin through its dependence on the orientation of the molecule with respect to the external field and rotational diffusion
      1.1.4 Quadrupole/electric field gradient coupling modulation (spin $> \frac{1}{2}$)
   1.2 Solid state
      1.2.1 Chemical shift/dipole coupling dispersion in inhomogeneous samples
      1.2.2 Entanglement of spins through dipole coupling with their neighbors
      1.2.3 Spontaneous phonon emission and Raman spin/phonon interactions (the latter dominates at high temperatures)
      1.2.4 Spectral diffusion due to other nuclear species and magnetic impurities

2. Trapped Ions
   - 2.1 Spontaneous emission from ions
   - 2.2 Cross talk in ion addressing due to imperfect laser focusing
   - 2.3 Mode–mode couplings due to anharmonicities of the trap
   - 2.4 "Heating" of ion motion due to stray RF fields, patch potentials, and the like
   - 2.5 Coupling of thermal vibrations into internal ion states
   - 2.6 Leakage losses into other atomic levels (i.e., breakdown of the two-level qubit approximation)
   - 2.7 Ionization
   - 2.8 Inefficiencies in readout
3. Neutral atoms
   - 3.1 Photon scattering from trapping laser fields
   - 3.2 Photon scattering from Raman laser fields during single qubit transitions
   - 3.3 Spontaneous emission from Rydberg states during a Rydberg gate operation (including effects of black-body radiation)
   - 3.4 Background gas collision (includes qubit loss and leakage and also standard qubit errors)
   - 3.5 Fluctuating trap potentials
   - 3.6 Background magnetic fields
   - 3.7 Heating of atoms (i.e., vibrational excitation in the optical lattice potential)
   - 3.8 Scattering to atomic states outside the computational basis during collisional gates
4. Cavity QED
   - 4.1 Motional decoherence from trap fluctuations and environmental noise
   - 4.2 Motional decoherence from gate operations, noise in driving fields
   - 4.3 Photon qubit decoherence when strong coupling regime not achieved or exited during operations
   - 4.4 Differential Stark shifts from optical trapping fields
   - 4.5 Spontaneous emission, background gas collisions, photon scattering, and other sources of decoherence for ions and neutral atoms (see items 2 and 3 above)
5. Optical
   - 5.1 Scattering from the electromagnetic vacuum, leading to possible photon loss
     - 5.1.1 Loss at the source (failure of single-photon source)
     - 5.1.2 Loss in processing/transit
     - 5.1.3 Loss in detection
   - 5.2 Photon addition (from failure of the source or a detector, mistaking one photon for two photons, e.g., as a result of detector noise)
   - 5.3 Failure of a teleportation gate (corresponding to a detected qubit measurement error)

5.4 Phase errors deriving from failure to carefully tune interferometers or from timing errors in teleportation protocols

6. Solid state
   6.1 Spin based
      6.1.1 Spontaneous phonon emission mediated by spin–orbit coupling
      6.1.2 Dipolar couplings with magnetic impurities and other trapped electrons
      6.1.3 Hyperfine interaction with nuclear spins, giving rise to
         6.1.3.1 Direct electron–nuclear spin flip (may or may not include phonon emission)
         6.1.3.2 Spectral diffusion whereby dipolar coupling induced fluctuation of nuclear spins leads to a fluctuating hyperfine field acting on electron spin
      6.1.4 Inhomogeneous qubit environments (magnetic fields, impurities, quantum-dot sizes, interface strains, defects, frozen hyperfine fields)
      6.1.5 Gate errors due to inhomogeneities
      6.1.6 Current and voltage fluctuations
      6.1.7 Switching errors due to imperfect gate operations on qubits
      6.1.8 Measurement process
   6.2 Charge based
      6.2.1 Spontaneous photon emission
      6.2.2 Spontaneous phonon emission
      6.2.3 Gate voltage fluctuations (due to thermal noise, trapped charges, electromagnetic environment)
      6.2.4 Electron tunneling and co-tunneling in the dots/donors
7. Superconducting
   7.1 Electromagnetic environment
   7.2 Phonons
   7.3 (Hot) quasi-particles
   7.4 Background charges
   7.5 Critical current noise
   7.6 Spurious resonances (and critical current noise)
   7.7 Gate voltage fluctuations
   7.8 Nuclear spins
   7.9 Paramagnetic impurities

G.6.4 Decoherence Analyses

The following decoherence models and theoretical approaches have been used to analyze decoherence in the above QC implementations.

1. NMR
   1.1 Liquid state
      1.1.1 Hadamard product formalism
      1.1.2 Redfield theory and Redfield kite structure of NMR relaxation superoperators
      1.1.3 Spherical harmonic tensor expansions of dipole–dipole and other interactions, combined with Langevin analysis
      1.1.4 Stochastic Liouville method
      1.1.5 Quantum noise channels
   1.2 Solid state
      1.2.1 The method of moments
      1.2.2 Spin-boson models parameterized by experiment
      1.2.3 A wide variety of semiclassical models
2. Trapped Ion
   2.1 Standard first-order perturbation theory
   2.2 Weisskopf–Wigner/Markov approximation techniques for spontaneous emission modeling
   2.3 Quantum Monte Carlo numerical modeling
   2.4 Analytic non-Markovian stochastic models for some effects (e.g., heating)
3. Neutral Atoms
   3.1 Wave-packet simulations
   3.2 Stochastic Schrödinger equation (Monte Carlo wave function approach); applicable for large-scale simulations
   3.3 Master equation approach (including Redfield or Lindblad model of dissipative superoperator)
   3.5 Analysis of heating/decoherence rates due to trap fluctuations and collisions
   3.6 Analysis of gate leakage due to collisions
4. Cavity QED
   4.1 Perturbation theory
   4.2 Weisskopf–Wigner/Markov approximation techniques for spontaneous emission modeling
   4.3 Monte Carlo wave function (stochastic trajectory) approach
   4.4 Master equations
   4.5 Analysis of heating rates due to trap fluctuations and gas collisions
5. Optical
   5.1 Gate fidelity calculations in presence of photon loss, modeled by beam-splitter that mixes mode with vacuum state
   5.2 Quantum error encodings and protocols to correct for photon loss

6. Solid State
   - 6.1 Spin based
     - 6.1.1 Master equation in extended Bloch–Redfield description for single-spin decoherence
     - 6.1.2 Single-spin decay due to phonon emission by perturbative and basis-set calculations within effective mass theory
     - 6.1.3 Many-spin decay due to inhomogeneities within tight-binding description
     - 6.1.4 Method of moments for dipolar coupling to impurities
     - 6.1.5 Spin-bath theory
     - 6.1.6 Gate fidelity calculations for effects of inhomogeneities, switching errors, spin–orbit coupling
     - 6.1.7 Master equation in Born–Markov limit for analysis of measurement efficiency and n-shot read out
     - 6.1.8 Exact solution with Laplace transforms for effect of hyperfine coupling in fully polarized nuclear spin field
     - 6.1.9 Perturbative analyses of hyperfine coupling effects for general (partially polarized) nuclear spin field, evidence for nonexponential decay
     - 6.1.10 Stochastic noise theory combined with method of moments for analysis of indirect effects of hyperfine coupling via nuclear spectral diffusion
   - 6.2 Charge based
     - 6.2.1 Perturbation theory for photon/phonon emission
     - 6.2.2 Bloch–Redfield theory for photon/phonon emission and for electron tunneling/co-tunneling
     - 6.2.3 Stochastic noise theory to describe charge noise and gate fluctuations
7. Superconducting
   - 7.1 Generalized spin-Boson theory
   - 7.2 Spin-bath model
   - 7.3 Fano–Anderson/Dutta–Horne model (for $1/f$ noise)
   - 7.4 Mesoscopic transport models (for readout)
   - 7.5 Bloch–Redfield theory
   - 7.6 Real-time path integrals
   - 7.7 Diagrammatic Keldysh technique and exact solutions for simplified Hamiltonians
   - 7.8 Quantum Monte Carlo
   - 7.9 Renormalization group
   - 7.10 Bloch vector diffusion (stochastic differential equation)
   - 7.11 Analysis of qubit depolarization in readout

Glossary

μ-Raman spectroscopy Microscopy tool. When incident light strikes a sample, most of the scattered light has the same wavelength as the incident light (Rayleigh scattering); some of the light is scattered at a different wavelength; this is called Raman scattering. The energy difference between the incident light and the Raman scattered light is called the Raman shift; it is equal to the energy required to get the molecule to vibrate or to rotate. Several different Raman-shifted signals will often be observed in a single sample, each being associated with different vibrational or rotational motions of molecules in the sample. The particular molecule and its environment will determine what Raman signals will be observed; this information can be interpreted to determine chemical structure and identify the compounds present [472].

3R Term used in optical transmission systems: reshape, reamplify, re-time.

Ab initio calculations Ab initio calculations do not use semiempirical models, preconceived ideas, or experimental interpretations to achieve results. These calculations are used in combination with experimental observations. In calculating the total energy of a solid, they employ a combination of density functional theory, pseudopotential theory, and iterative minimization techniques. Ab initio techniques are applied to various phenomena such as surface growth and reconstruction, structural phase transitions, and chemisorption. Technique can be used to secure theoretical predictions of surface geometries and behavior [406].

Absolute temperature scale Thermodynamic temperature scale, named for Lord Kelvin (1848), in which temperatures are given in kelvins (K). (In the SI system, the degree sign and the word *degree* are not used for kelvin temperatures.) The absolute zero of temperature is zero (0) K, or −273.16°C (Celsius or centigrade), or −459.7°F (Fahrenheit). The size of the kelvin unit (degree) is the same as that of the Celsius degree [255].

Accelerator Machine used to accelerate particles to high speeds (and, thus, high energy compared to their rest mass–energy) [50].

Acid Compound that releases hydrogen ions (H^+) in solution and/or that can accept a pair of electrons from a base.

Acidic Describes a solution with a high concentration of H^+ ions.

Addition compound Compound that contains two or more simpler compounds that can be packed in a definite ratio into a crystal. Hydrates are a common type of addition compound.

AFM Atomic force microscope.

Alkaline earth Oxide of an alkaline earth metal that produces an alkaline solution in reaction with water.

Alkane Series of organic compounds with general formula C_nH_{2n+2}. A carbon–carbon double bond, or a molecule containing a carbon–carbon double bond. Alkane names end with -ane. Examples include propane (with $n = 3$) and octane (with $n = 8$).

Allotrope Different forms of matter made from the same set of atoms but bonded differently. Atoms can be bonded in a number of ways, even if they are all atoms of the same element. Different forms of matter can be made from a single type of atom. An example includes ozone and dioxygen. Another pertinent example is the diamond, buckyball, and graphite: They all are allotropes of carbon.

Alloy Metal that contains more than one element.

Alpha (α) particle Nucleus of a helium (^{4}He) atom, carrying a positive charge of 2e. Its proton number and neutron number are both 2, resulting in a particularly stable particle. It has a relative atomic mass of 4.00260 [50].

Amplitude For a wave or vibration, the maximum displacement on either side of the equilibrium (midpoint) position [654].

Angstrom (Å) Unit of measurement for atomic distances equal to 1×10^{-10} m or 0.10 nm. Most atoms are between 1 and 5 Å in diameter.

Anhydrous Compound where all water has been removed.

Anion Negatively charged ion. Nonmetals typically form anions.

Anisotropy (optical anisotropy) Condition refers to the fact that the refractive index of a crystal depends on the direction of the electric field in the propagating light beam. Hence, the velocity of light in a crystal depends on the direction of propagation and on the state of its polarization (i.e., the direction of the electric field). The outcome is that (except along certain special directions), any unpolarized light ray entering an optically anisotropic crystal breaks into two different rays with different phase velocities and polarizations. Contrast with *optically isotropic* where the refractive index is the same in all directions.

Examples of optically isotropic materials are most noncrystalline materials (e.g., glasses, liquids) and all cubic crystals. For all other classes of crystals (excluding cubic structures), the refractive index depends on the propagation direction and the state of polarization. When one views an image through an optically anisotropic crystal, one sees two images, each constituted by light of different polarization passing through the crystal.

Annihilation Process where a particle meets its corresponding antiparticle and both disappear. The energy appears in some other form, perhaps as a different particle and its antiparticle (and their energy), perhaps as many mesons, perhaps as a single neutral boson such as a Z^0 boson. The produced particles may be any combination allowed by conservation of energy and momentum and of all the charge types and other rules [50].

Antibonding orbital Molecular orbital, produced by the overlap of atomic orbitals, that lies at higher energy than the atomic orbitals that composed it. Occupation of such an orbital destabilizes two atoms close to each other, causing them to move apart [655].

Antimatter Material made from antifermions. One defines the fermions that are common in the universe as matter and their antiparticles as antimatter. In the particle theory there is no a priori distinction between matter and antimatter. The asymmetry of the universe between these two classes of particles has not yet been completely explained [50].

Antiparticle For each kind of atomic particle, there is an associated antiparticle with the same mass but opposite electromagnetic, weak, and strong charges, as well as spin. Each quantum number of a real neutral particle is identical with its antiparticle's one. Some particles, e.g., photons, have no distinct antiparticle (stated alternatively, these particles are identical to their antiparticle).

For every fermion type there is another fermion type that has exactly the same mass but the opposite value of all other charges (quantum numbers). For example, the antiparticle of an electron is a particle of positive electric charge called the positron. Bosons also have antiparticles, except for those that have zero value for all charges, e.g., a photon or a composite boson made from a quark and its corresponding antiquark. In this case there is no distinction between the particle and the antiparticle, they are the same object [50].

Antiquark Antiparticle of a quark. By nomenclature, an antiquark is denoted by putting a bar over the corresponding quark ($\bar{d}$, $\bar{u}$, $\bar{s}$, etc.) [50].

Arrayed waveguide grating (AWG) Optical multiplexing device that uses interference effects between different waveguides of progressively longer optical path length on a planar substrate (typically silicon). The interference effect directs each wavelength (typically arriving on different ports on the AWG, onto an output port that is coupled into a fiber output.

Aspect ratio Ratio of depth to width of a via or contact structure [656].

Asperity Roughness of surface; unevenness; contrasted to smoothness.

Assembler General-purpose machine for molecular manufacturing capable of guiding chemical reactions by positioning molecules. A molecular machine that can be programmed to build any molecular structure or device from simpler chemical building blocks.

Atmospheric pressure CVD (APCVD) Refers to systems whose deposition environments operate at or near atmospheric pressure. Typically, wafers are placed horizontally on belt-driven flat susceptors that move through the deposition zone. Belt speed and gas flow determine the film thickness [656].

Atom Smallest particle of an element that can take part in a chemical reaction.

Atomic force microscope (AFM) Instrument able to image surfaces to the nanoscale by mechanically probing the surface contours of molecules. Under favorable conditions it provides an atomic-level-resolution topographic map of the surface. Aka, scanning force microscope (SFM).

AFM is a tool for analyzing the surface of a rigid material all the way down to the level of the atom. AFM uses a mechanical probe to magnify surface features up to 100,000,000 times, and it produces 3D images of the surface [657].

Atomic mass unit (amu) Units of mass used to describe atomic particles. An atomic mass unit is equal to 1.66054×10^{-24} g.

Atomic number Number of positively charged protons in the nucleus of an atom. Atomic number, Z, is a characteristic property of an element, equal to the number of protons present in the nucleus of an atom. In neutral species, it is also equal to the number of electrons present in the atom [61, 62, 63, 64, 65, 66, 67].

Atomic orbital Wave function that defines the properties (location, etc.) of an electron in an atom. One speaks of an electron that obeys such a wave function as occupying the relevant orbital [655].

Atomic radius Atomic radius is usually referred to as one half of equilibrium internuclear distance between two adjacent atoms (which may either bonded covalently or present in a closely packed crystal lattice) of an element.

Atomic volume Atomic (or molecular) volume V_m is the average volume per $10^3 N_0$ of atoms in the structure, where N_0 is Avogadro's number (6.022×10^{23}/mol) [61, 62, 63, 64, 65, 66, 67].

Atomic weight Average relative weight of the atoms of an element referred to an arbitrary standard of 16.0000 for the atomic weight of oxygen. The atomic weight scale used by chemists takes 16.0000 as the average atomic weight of oxygen atoms as they occur in nature. The scale used by physicists takes 16.00435 as the atomic weight of the most abundant oxygen isotope. Division by the factor 1.000272 converts an atomic weight on the physicists' scale to the corresponding atomic weight on the chemists' scale. *See also* Atomic number [61, 62, 63, 64, 65, 66, 67].

Atomic layer deposition Semicondutor manufacturing method that employs a gas comprised of small molecules that naturally stick to the surface but do not bond to one another. With this technique, a molecule-thick film can be laid down by exposing the underlying wafer to the gas until every spot becomes covered. Treatment with a second gas (one that reacts with the first to form the material in the coating), creates a thin veneer: Repeated sequential applications of these two gases deposit layer over layer of this substance until the desired thickness is achieved [19]. *See also* Chemical vapor deposition.

Atomistic simultations Atomic motion computer simulations of macromolecular systems are increasingly becoming both possible and essential for materials science and nanotechnology. Recent advances in supercomputer simulation techniques support computations on nanoscale objects containing as many as 0.25 million atoms and on materials simulated with 1 million atoms.

Attenuation Measure of how much of the light injected into the fiber actually reaches the other end. Attenuation determines either how much fiber one can use in an application or how much light an optical source must produce. It is normally measured in decibels per kilometer (dB/km) and is highly wavelength dependent. In general, attenuation decreases with increasing wavelength for wavelengths below 2 μm (2000 nm). A number of independent loss mechanisms combine to determine the overall attenuation of an optical fiber (e.g., Rayleigh scattering, infrared absorption) [658].

Auger electron Electron ejected from the solid as a result of two-stage ionization of atoms bombarded with high-energy ions (Auger process); the Auger electron carries energy that is specific to the atom from which it was ejected.

Auger electron spectroscopy (AES)/scanning Auger microscopy (SAM) Microscopy tools. A surface characterization and depth profiling method (tool) based on the determination of the energy of Auger electrons ejected from the solid surface bombarded with high-energy ions (only elements with atomic number >2 can be detected with this method). A focused electron beam irradiates a sample surface producing Auger electrons, the energies of which are characteristic of the element from which they are generated. Compositional depth profiling is accomplished by using an independent ion beam to sputter the sample surface while using AES/SAM to analyze each successive depth [472].

Automatic defect classification (ADC) Defects found by wafer inspection systems are classified by the system into several categories based on their physical and optical properties [656].

Avogadro's number Number representing the number of molecules in 1 mol: 6.023×10^{23}.

Ballistic magnetoresistance (BMR) Manner by which spin orientation can modify electrical resistance in a nearby circuit, so as to accomplish the sensing of that orientation. Spin orientation refers to the encoding information on a storage medium such as a disk drive.

Ballistic/ballistic transport Motion of electrons in ultrasmall (highly confined) regions in semiconductor structures at very high electric field with velocities much higher than their equilibrium thermal velocity; ballistic electrons are not subjected to scattering, and, hence, they can move with ultrahigh velocity; ballistic transport is determined by electronic structure of semiconductor and is different for different semiconductors; allows ultrafast devices [39]. Ballistic (as opposed to the term *diffusive*) refers to the passage of electrons through a conductor whose length is less than the mean free path of electrons in the conductor, with the result that most of the electrons pass through the conductor without scattering [31].

Carbon nanotubes, molecular wires, and crystals, among other systems, provide examples of ballistic electron transport. In optimal environments, the electrical conductance of the quantum conduction channel has very low electrical resistivity. The conductance, however, is impacted by the (i) kind of imperfections in the quantum conductor, (ii) the density of the imperfections in the quantum conductor, and (iii) by the geometry of the contacts (i.e., how the conductor is attached to a source or sink of the electrons). To achieve optimal signal transmission, the conductor must exhibit an atomic or molecular structure with long-range periodicity and a significant electronic overlap between adjacent atomic sites in order to form extended electron states. Also, to be useful in computer chip device and interconnect applications, the conductor must possess quantum mechanical electron energy levels that reasonably match those of the metal leads and contacts [229].

Band-engineered transistor Transistor with enhanced mobility of holes or electrons in the channel enabled by, e.g., use of SiGe layer or a strained Si on a relaxed SiGe layer. Transistor structure is essentially unchanged. Can be bulk Si or SOI [401].

Bandgap A range of frequencies where propagating modes of a signal or wave are absent. The difference in energy between bands formed from the overlap of orbitals of different symmetry, though it is typically taken to be the energy difference between the highest fully occupied and the lowest fully unoccupied bands [655].

Bandgap (electronics) In a semiconductor material, the minimum energy necessary for an electron to transfer from the valence band into the conduction band, where it moves more freely [255].

Barrier Physical layer designed to prevent intermixing of the layers above and below the barrier layer [656].

Baryon Hadron made from three quarks. They may also contain additional quark–antiquark pairs. For example, the proton (uud) and the neutron (udd) are both baryons [50, 51].

Baryon–antibaryon asymmetry Observation that the universe contains many baryons but few antibaryons; a fact that needs explanation [50].

Base Compound that reacts with an acid to form a salt and/or produces hydroxide ions in aqueous solution. It can also be seen as a molecule or ion that captures hydrogen ions or that donates an electron pair to form a chemical bond. Substance that gives off hydroxide ions (OH^-) in solution.

Beam Particle stream produced by an accelerator; usually clustered in bunches.

Beam pipes Particle bunches in the accelerator travel in a vacuum in metal structures called beam pipes. All the air must be removed or the particles would collide with the air molecules and would lose their energy and direction very quickly [50].

Benzene ring C_6H_6. A hexagonal aromatic molecule.

BioMEMS (NEMS) MEMSs (NEMSs) used in the field of medicine. A term used when referring to MEMS devices for the biotechnology field. A BioMEMS is a specific subset of MEMSs.

Biomimetic "Copying," "emulating," "imitating," or "learning" from nature. The study and process of optimizing functions in nature to look for design solutions in biology. Nanodevices already exists in nature; hence, researchers have a plethora of components and techniques already available. Biomimetics is the study of the structure and function of biological substances with the goal of making artificial products that mimic nature. Designing molecules, molecular assemblies, and macromolecules having biomimetic functions. Set of techniques used to develop novel synthetic materials; processes and sensors through advanced understanding and exploitation of design principles found in nature. Biomimetic chemistry is the field of experimental simulation or modeling of biological simulations where chemists attempt to mimic the complex systems found in nature using less complicated artificial systems. An example includes genetically engineered high-strength silk fibers. These fibers, which are about 5 μm in diameter, are 100% tougher than Kevlar fibers.

Biomimetic chemistry Confluence of biochemistry, analytical chemistry, polymer science, and chemistry as applied to research in designing new molecules, molecular assemblies, and macromolecules having biomimetic functions [695]. Biomimetic materials are materials that "copy," "emulate," "imitate," or "learn" from nature.

Biopolymeroptoelectromechanical systems (BioPOEMS) Discipline combining optics and microelectromechanical systems and used in biological applications.

Birefringence Material is described as birefringent when a light wave traveling through it can have two distinct velocities. In other words, the material has two distinct values of refractive index. Birefringence has units of 1/mm [658].

Body-centered cubic (bcc) atomic structure This structure is comprised of a unit cube with atoms at the corners and center of the cube [90]. The bcc structure is less closely packed than fcc or hcp. Often bcc is the high-temperature form of metals (these being close packed at lower temperatures).

Bohr radius Quiescent (natural, preferred) distance of separation between the positive and negative charges in the excited state of matter. Radius defined in terms of the hydrogen atom, in which a single negative charge (the electron) is attracted to a single positive charge (the proton in the nucleus) via the electrostatic potential $V(r) = -e^2/r$. The solution of Schrödinger's equation for the electron wave function is then [659]

$$\psi = C \exp(-r/a_0)$$

where a_0 is the Bohr radius; $a_0 = h^2/4\pi^2 me^2 = 0.053$ nm.

Boiling point Temperature at which the vapor pressure of a liquid is equal to the atmospheric pressure. The normal boiling point is the boiling point at normal atmospheric pressure (101.325 kPa).

Bonding orbital Molecular orbital, produced by the overlap of atomic orbitals, which lies at lower energy than the atomic orbitals that composed it. Occupation of such an orbital stabilizes two atoms close to each other, forming a bond between them [655].

Born–Oppenheimer approximation Approximation that permits the use of classical mechanics in modeling molecular and atomic motions.

Boron nitride Compound that is a structural equivalent of carbon; it is mostly found in the same phases and produces similar nanostructures. Boron nitride is intensively used as a carbon substitute for its much higher chemical inertness, especially at high temperature. Hexagonal boron nitride is nonreactive to molten metals (Al, Fe, Cu, Zn) or to hot Si and stable against air oxidation up to 1000°C. Hexagonal boron nitride also offers an electrically resistant counterpart to the semimetallic graphite. Substitution of boron nitride in carbon nanotubes has been studied by a number of researchers [660].

Bose–Einstein statistics (aka B–E statistics) The quantum statistics obeyed by integer-spin particles (bosons) where any number of particles may occupy a given state; specifically, the Bose–Einstein statistics describes physical systems in which the particle number is nonconservative, particles are indistinguishable, and cells of phase space are statistically independent. B–E systems allow two polarization states, and exhibit totally symmetric wavefunctions. It has been shown that B–E statistics emerge from quantum field theory.

Boson Particle that has integer intrinsic angular momentum (spin) measured in units of $\hbar$ (spin $= 0, 1, 2, \ldots$). All particles are either fermions or bosons. The particles associated with all the fundamental interactions (forces) are bosons. Composite particles with even numbers of fermion constituents (quarks) are also bosons [50].

Bottom quark (B) Fifth flavor of quark (in order of increasing mass), with electric charge $-\frac{1}{3}$ [50].

Bottom-up Construction of larger objects from smaller building blocks. Technique is used by chemists in attempting to create structure by connecting molecules. Nanotechnology looks to utilize atoms and molecules as basic building blocks. Note: the technique is properly called "bottom-up" and not the (more English-friendly) "bottoms-up."

Method for building larger objects from smaller building blocks. Nanotechnology seeks to use atoms and molecules as those building blocks. The advantage of bottom-up design is that the covalent bonds holding together a single molecule are far stronger than the weak interactions that hold more than one molecule together [661].

Bragg angle/Bragg's law The Bragg angle (θ) is defined by the expression $2d \sin \theta = n\lambda$, where θ is the angle between a crystal plane and the diffracted X-ray beam; λ *is* the wavelength of the X-rays, d is the crystal plane spacing, and n is the diffraction order n (any integer). The Bragg law is the foundation of X-ray diffraction analysis, it allows one to make quantification of the results of experiments carried out to determine crystal structure. Bragg's law was formulated in 1912 by W. L. Bragg in order to explain the observed phenomenon that crystals only reflected X-rays at certain angles of incidence [256]. An X-ray diffraction occurs from a crystal structure only when the Bragg condition is satisfied; this condition depends on the angle of the incident X-ray beam as it enters the crystal structure and the direction at which the diffracted beam exits the structure.

An X-ray incident upon a sample will either be transmitted, in which case it will continue along its original direction, or it will be scattered by the electrons of the atoms in the material. All the atoms in the path of the X-ray beam scatter X-rays. In general, the scattered waves destructively interfere with each other, with the exception of special orientations at which Bragg's law is satisfied. A Bragg condition is such that the scattered rays from two parallel planes interact with each other in such a way as to create constructive interference. The angle between the transmitted and Bragg diffracted beams is always equal to 2θ as a consequence of the geometry of the Bragg condition. This angle is readily obtainable in experimental situations, and hence the results of X-ray diffraction are frequently given in terms of 2θ. However, it is important to note that the angle used in the Bragg equation *must* always be that corresponding to the angle between the incident radiation and the diffracting plane, i.e., θ [256].

Bragg grating Filter that separates light into many colors under the principles of Bragg's law. Specifically, a fiber-based Bragg grating used in optical communications to separate wavelengths.

Breakdown Change in or failing of the physical principles governing materials and machines as they approach the nanoscale [36].

Brillouin zone Property of a crystal. This geometrical shape can be considered to contain the valence electrons of the crystal. Its planes define, in momentum space (k space), the location of the bandgap. Wave vectors lying within the zone are in the same energy band: There is a jump in energy before the state given by the shortest vector in the next Brillouin zone. The Fermi sphere is usually considered to have the same origin as the zones and is thought of as filling the zone [659].

Buckminsterfullerene (aka Buckyball, C60) Most famous of the fullerenes, consisting of 60 carbon atoms. It is a nanostructure composed of 60 atoms of carbon arranged in a perfectly symmetric closed cage. Discovered in 1985 by Richard Smalley, Harold Kroto, and Robert Curl for which they won the 1996 Nobel Prize in Chemistry [36].

Bulk micromachining Fabrication process of creating structures by etching into (and through) silicon wafers.

Calorimeter Device that can measure the energy deposited in it. (Originally, devices to measure heat energy deposited, using change of temperature; particle physicists use the word for any energy-measuring device) [50].

Capacitance Ability to store electrical charge.

Carbon Element. Carbon has six protons, six neutrons, and six electrons. The electronic configuration is two electrons in the K shell, and four electrons in the L shell. In theory, while forming compounds carbon should either give up four electrons or borrow four electrons. However, carbon does not form ionic bonds; the mechanism, instead, is to share its four electrons with other atoms and form covalent bonds. When a carbon atom forms a compound, it always forms covalent bonds.

In the periodic table of the elements, carbon is listed as crystallizing in the hexagonal structure. This structure consists of planar layers of carbon atoms arranged in honeycomb lattices called graphene sheets. Within a sheet, three of the four valence electrons of a carbon atom form three strong trigonal bonds to three equidistant neighbors 0.14 nm away. The fourth valence electrons from different carbon atoms interact to form weak π bonds perpendicular to successive sheets that are loosely piled up on top of each other every 0.34 nm in an alternating ABAB ... sequence producing a 3D hexagonal unit cell. Although there are various other stacking arrangements, this allotrope, known as "graphite," is the most stable and most abundant solid form of pure carbon found in nature. A slightly less stable and vastly less abundant crystallographic form is diamond, which has a cubic structure in which each atom is covalently bound to four neighbors at the apexes of a regular tetrahedron. Until 20 years ago, these were the only known crystalline forms of solid carbon. In 1985, the science of carbon was unexpectedly enlarged by the discovery of an entirely new class of structures: fullerenes. The fullerenes first discovered are spheroidal molecules, and such molecular clusters are sometimes called "curved graphite" because of their obvious appearance as curved sheets of graphene, with the typical threefold coordination of each atom in a honeycomb lattice. However, occasional pentagonal rings occur in the hexagonal network, and these cause the curvature and eventual closure of the graphene sheets. Fullerene molecules are in turn able to crystallize in a variety of 3D structures [662].

Carbon nanotube Cylinder-shaped structure resembling a rolled-up sheet of graphite that can be a conductor or semiconductor depending on the alignment of its carbon atoms. It is about 100 times stronger than steel of the same weight, although due to high fabrication costs, widespread commercial use is still distant [36].

Casimir effect In the late 1940s Hendrik Casimir predicted that two uncharged parallel metal plates have an attractive force pressing them together. This attraction is now called the Casimir effect. This force is only measurable when the distance between the two plates is ultrasmall, on the order of several atomic diameters.

Catalysis Phenomenon supported by the action of a substance that either promotes a particular reaction or accelerates that reaction. In catalysis, the catalyst itself remains unchanged through the process.

Catalyst Compound that accelerates the rate of a chemical reaction and is not itself consumed in the reaction. The catalyst lowers the total amount of energy required to execute the reaction, by reacting with the starting materials in a series of sequential steps. The amount of catalyst is generally small with respect to the quantities of starting material.

A species that increases the rate of a reaction by providing an alternative (lower energy) pathway for the reaction to follow. It is only needed in very small concentration because it is regenerated after use (this is part of the definition of a catalyst) [655].

Catenation Formation of chains of bonded atoms.

Cathode Electrode where electrons are gained (reduction) in redox reactions.

Cation Positively charged ion. Metals typically form cations.

CD-SEM Type of scanning electron microscope used to measure critical dimension [656].

Central atom In a Lewis structure, usually the atom that is the most electronegative.

Central tracker Charged particles leaving the interaction region travel first through a set of detectors that measure their position very precisely. This central tracker has a solenoidal magnetic field that causes the charged particles to follow curving paths; the amount of curvature is a measure of the momentum of the charged particle [50].

Charge Quantum number carried by a particle. Determines whether the particle can participate in an interaction process. A particle with electric charge has electrical interactions; one with strong charge has strong interactions, etc. [50].

Describes an object's ability to repel or attract other objects. Protons have positive charges while electrons have negative charges. Like charges repel each other while opposite charges, such as protons and electrons, attract one another [663].

Charge conservation Observation that charge is conserved in any process of transformation of one group of particle into another [50].

Charm quark (C) Fourth flavor of quark (in order of increasing mass), with electric charge $+\frac{2}{3}$ [50].

Chemical changes Processes or events that have altered the fundamental structure of something.

Chemical equation Expression of a fundamental change in the chemical substances.

Chemical mechanical polishing (CMP) Process that uses an abrasive and corrosive slurry to physically grind flat the microscopic topographic features on a partly processed wafer (planarization) so that subsequent processes can begin from a flat surface [656].

Chemical vapor deposition (CVD) Chemical growth process where a reaction between gaseous reactants creates products that are deposited as solid. It is a fabrication technique used to deposit thin layers of material on a substrate. A process for depositing thin films from a chemical reaction of a vapor or gas [656]. A technique used to deposit coatings; chemicals are first vaporized and then applied using an inert carrier gas such as nitrogen.

Often, a crystalline substrate is employed leading to epitaxial growth of the deposited material. Deposition occurs when a heated substrate is placed in a stream of vapor containing materials that react on the hot substrate surface, leading to growth. CVD is the general term used to describe MCVD, OVD, VAD, and other preform manufacturing processes where chemical vapors (presence of heat) are deposited on the surface of a substrate.

Chip Small piece of a silicon wafer that contains a complete integrated circuit [656].

Chiral Chiral molecule is one without a plane of symmetry. A chiral molecule is one that cannot be superimposed on its mirror image. A chiral molecule and its mirror image are enantiomers of each other [655].

Chirality Term describing the geometric property of a rigid object (or spatial arrangement of points or atoms) that is nonsuperimposable on its mirror image; such an object has no symmetry elements of the second kind (a mirror plane, a center of inversion, a rotation reflection axis). If the object is superimposable on its mirror image the object is described as being achiral [657]. Possessing "handedness," i.e., existing in right- and left-handed forms. Chiral molecules are not superimposable on their mirror images.

Chromatic dispersion Determines both the data-carrying capacity of a single-mode optical fiber and the optimum distance between repeaters in a communications link. Different wavelengths of light will travel at different velocities within the core of an optical fiber because the core material (germania-doped silica) is "dispersive." This phenomenon is important because the optical power produced by the laser, LED, or LD (laser diode) sources used in communications and sensors is usually distributed throughout a range of wavelengths (or spectrum) and not within a single, discreet "line." This means that as the pulse of light travels within the core of the optical fiber, it spreads out because the different wavelengths of the pulse travel at different velocities. If these pulses form part of a communication signal, they begin to overlap and corrupt the data [658].

Dispersion increases both with the length of fiber involved and the spectral width of the optical source used and is measured in "picoseconds per nanometer kilometer." That is, if a communications link has a dispersion of 2 ps/nm × km, is 2 km long, and is used with a source that has a linewidth of 5 nm, then the duration of the pulse will have increased by 20 ps by the time it reached the detector. Dispersion also increases with wavelength and changes signs from negative to positive between the 1300- and 1550-nm transmission windows. The wavelength at which the value of dispersion passes through zero is known as the "lambda zero" (λ_0); dispersion-shifted fibers are specially designed so that their lambda zero occurs within the 1550-nm transmission window [658].

Circuit Combination of a number of connected electrical elements to accomplish a desired function [656].

Cladding Material that surrounds the core of an optical fiber. The core has a higher index of refraction than the cladding. The lower index of refraction in the cladding causes the transmitted light to travel downstream through the core.

Close packing Structure of many compounds is based on the stacking of spheres in arrangements where the occupied volume of the structure is maximized. This is known as close packing of spheres [655].

CMOS Complementary metal–oxide semiconductor structure; consists of N-channel and P-channel MOS transistors; due to very low power consumption and dissipation as well minimization of the current in the "off" state, CMOS is a very effective device configuration for implementation of digital functions; CMOS is a key device in state-of-the-art silicon microelectronics [39].

Coarse wavelength division multiplexing (CWDM) A WDM system where only a few channels are needed or supported. Here wider wavelength spacing is possible compared to a dense WDM system. CWDMs do not need to be optically amplified; this typically reduces cost by allowing uncooled lasers and simpler termination equipment.

Coherence From quantum physics, coherence is a property exhibited by matter and energy when it remains unmeasured and is thus able to exist in more than one state simultaneously (superposition). The "wave function" is a math formula that describes all possible futures and possible pasts for a system. Until a measurement is made, the system is coherent and all possibilites are both true and false. Once a measurement is made, the wave function collapses and the state becomes determined. Any interaction can be considered a measurement, which is why coherence is such a delicate property to maintain [661].

Collider Accelerator in which two beams traveling in opposite directions are steered together to provide high-energy collisions between the particles in one beam and those in the other [50].

Colliding-beam experiments Experiments done at colliders.

Colloid (aka, colloidal dispersion) Intermediate form of matter between a solution and a mixture, where microscopic particles of one substance are distributed throughout another. The former is said to be in the dispersed or solute phase; the latter is said to be in the dispersing, continuous, or solvent phase. Particle size can vary from 1 to 1000 nm. Stable dispersions of microscopic solid particles (ca. 1 μm) in fluids. Examples include colloidal aerosols, colloidal emulsions, colloidal foams, or colloidal suspensions.

An ensemble of particles 1–1000 nm in diameter dispersed in another phase. Term means "gluelike." Most paints, foods, cosmetics, and inks consist of colloidal structures. Colloids comprise soaps, emulsions, suspensions of solids (with sizes from microns to nanometers), and latex particles whose dimensions can be tuned in composition and size. A common feature of colloids is that the particles are floating in a solvent; at low enough concentrations, a colloid can flow freely, which is of considerable practical importance. Colloidal systems, particularly systems of electrically charged colloids, have important practical significance and are also of considerable theoretical interest [31].

Color charge Quantum number that determines participation in strong interactions; quarks and gluons carry nonzero color charges [50].

Color neutral (quark terminology) Object with no net color charge. For composites made of color-charged particles the rules of neutralization are complex. Three quarks (baryon) or a quark plus an antiquark (meson) can both form color-neutral combinations [50].

Complex numbers Extension of the real numbers, where all nonconstant polynomials have roots. The complex numbers contain a number i, the imaginary unit, with $i^2 = -1$. Every complex number can be represented in the form $x + iy$, where x and y are real numbers; x is called the real part and y the imaginary part of the complex number.

Complimentary metal–oxide semiconductor (CMOS) Common type of integrated circuit/fabrication process based on insulated gate field-effect transistors, similar to MOS, but uses complimentary MOS transistors to provide the circuits basic logic functions. The process layers from bottom to top are: semiconductor, insulating oxide, metal.

Compound Two or more atoms joined together chemically, with covalent or ionic bonds.

Compton effect Change in wavelength and direction of a photon when it collides with a particle, usually an electron; also known as Compton scattering. Some of the photon's energy is transferred to the particle, and the photon is reradiated at a longer wavelength in the electron's initial rest frame. Compton wavelength is defined as $\lambda_c = h/mc$, where h is Planck's constant, m is the electron rest mass, and c is the speed of light. It is the length scale below which a particle's quantum mechanical properties become evident in relativistic quantum mechanics [664].

Compton wavelength $\lambda_c = h/mc$, where h is Planck's constant, m is the electron rest mass, and c is the speed of light, is the length scale below which a particle's quantum mechanical properties become evident in relativistic quantum mechanics [664]. Direct calculations shows that $\lambda_c = 2.4263 \times 10^{-12}$ m. Hence, relativistic quantum mechanics applies at distances 1000 times smaller than the nanoscale [50].

Computing: amorphous computing Concept based on molecular biology and microfabrication. Each of these is the basis of a kernel technology that makes it possible to build or grow huge numbers of almost identical information-processing units, with integral actuators and sensors (e.g., MEMS) [665]. Microelectronic components could become so inexpensive that one can imagine mixing them into materials that are produced in bulk, such as paints, gels, and concrete; these "smart materials" could be used in structural elements and in surface coatings, such as skins or paints [666].

Computing: autonomic computing Self-managed computing environments with a minimum of required human interference (autonomic originates from an analogy with the human body's autonomic nervous system, which controls key functions without conscious awareness or involvement) [667, 668].

Computing: grid computing (aka utility computing) Environment that can be built at the local (data center), regional, or global level, where individual users can access computers, databases, and scientific tools in a transparent manner, without having to directly take into account as where the underlying facilities are located [668].

Computing: massively parallel computing Mechanisms (e.g., quantum computing, among others) where large-scale problems can be tackled in an independently parallelized fashion. For example, sequencing and assembling the genome required teraflop clusters; proteomics could require 10–100 times as much computing power; optimal target identification with design of intervention may require petascale computing [665]. Simulation and optimization are dependent on floating-point operations, involve less input–output, and require an effective communications fabric among processors.

Computing: molecular computers Computers whose input, output, and state transitions are carried out by biochemical interactions and reactions. At this time, researchers are exploring physical processes that can be put to use as computing substrates: chemical, biomolecular, optical computing via photonics, and quantum systems. Applications for non-silicon-based computing, including cryptography, pharmaceutical development, protein folding, data storage, and data mining [669].

Computing: pervasive computing Emerging trend in which computing devices are increasingly ubiquitous, numerous, and mobile [670].

Computing: quantum computing Mechanism of computing where the underlying computer takes advantage of quantum theory properties (e.g., entanglement) present at the nanoscale. The idea of a computational device based on quantum theory was first explored in the 1970s and early 1980s by physicists and computer scientists such as Charles H. Bennett of the IBM Thomas J. Watson Research Center, Paul A. Benioff of Argonne National Laboratory in Illinois, David Deutsch of the University of Oxford, and Richard P. Feynman of the California Institute of Technology (Caltech) [665, 671]. The concept of quantum computing emerged when scientists were pondering the fundamental limits of computation: Scientists understood that if technology continued to abide by Moore's law, then the continually shrinking size of circuitry packed onto silicon chips would eventually reach a point where individual elements would be no larger than a few atoms. Here a problem arose because at the atomic scale the physical laws that govern the behavior and properties of the circuit are inherently quantum mechanical in nature, not classical. This then raised the question of whether a new kind of computer could be devised based on the principles of quantum physics. Feynman was among the first to attempt to provide an answer to this question by producing an abstract model in 1982 that showed how a quantum system could be used to do computations. In 1985, Deutsch realized one could conceive a general-purpose quantum computer; Deutsch published a seminal paper showing that *any* physical process, in principle, could be modeled perfectly by a quantum computer; hence a quantum computer would have capabilities far beyond those of any traditional classical computer. Work has continued in earnest since then [671, 672, 673, 674]. Refer to text for complete discussion.

Computing: ubiquitous computing Term coined by Mark Weiser in 1988; concept envisions computers embedded in walls, in tabletops, and in common objects. In this environment, an individual could interact with hundreds of computers at a time, each invisibly embedded in the environment and wirelessly communicating with each other [675]. In this environment, the technology recedes into the background. Some call this "third paradigm" computing (the first paradigm being mainframes and the second paradigm being the personal computer). Related to the ubiquitous computing paradigm is the idea of smart rooms, where a room might contain multiple sensors that keep track of the comings and goings of the people around [665, 676].

Computing: utility computing (aka grid computing) Computing power on-demand (similar to an electric grid providing ready access to electricity). IBM, Sun, and HP, among others, provide utility computing systems/services [668].

Concentration The amount of substance in a specified space.

Condensed phase State of matter in which the constituent particles (atoms or molecules) are densely packed in space and in constant interaction with neighbors; usually refers to liquids, liquid crystals, glasses, crystalline solids, or quasi-crystals [677].

Condensed-matter physics In condensed-matter physics studies, quantitative measurements are used to probe the electrical, optical, thermal, and magnetic properties of a material; these measurements also focus on a material's response to externally applied electrical fields and temperature gradients [406].

Conductor Material that conducts current.

Confinement Property of the strong interactions that quarks or gluons are never found separately but only inside color-neutral composite objects [50].

Confocal microscopy Approach that makes use of spatial filtering to eliminate out-of-focus light or flare in specimens that are thicker than the plane of focus. Confocal microscopy offers several advantages over conventional optical microscopy, including controllable depth of field, the elimination of image degrading out-of-focus information, and the ability to collect serial optical sections from thick specimens. There has been a lot of interest in confocal microscopy in recent years, due to the relative ease with which extremely high quality images can be obtained from specimens prepared for conventional optical microscopy, and due to the large number of applications in many areas of current research interest [678].

Current instruments are highly evolved from the earliest versions, but the principle of confocal imaging advanced by Marvin Minsky, and patented in 1957, is employed in all modern confocal microscopes. In a conventional widefield microscope, the entire specimen is bathed in light from a mercury or xenon source, and the image can be viewed directly by eye or projected onto an image capture device or photographic film. In contrast, the method of image formation in a confocal microscope is fundamentally different. Illumination is achieved by scanning one or more focused beams of light, usually from a laser or arc-discharge source, across the specimen. This point of illumination is brought to focus in the specimen by the objective lens, and laterally scanned using some form of scanning device under computer control. The sequences of points of light from the specimen are detected by a photomultiplier tube (PMT) through a pinhole (or in some cases, a slit), and the output from the PMT is built into an image and displayed by the computer. Although unstained specimens can be viewed using light reflected back from the specimen, they usually are labeled with one or more fluorescent probes [678, 679, 680, 681, 682].

Conjugate acid Substance that can lose a H^+ ion to form a base.

Conjugate base Substance that can gain a H^+ ion to form an acid.

Conservation When a quantity (e.g., electric charge, energy, or momentum) is conserved, it is the same after a reaction between particles as it was before [50].

Contact diameter Diameter of the metal structure used to connect the doped contact area formed in the silicon base material to the metal interconnect [656].

Continuous wave (CW) Refers to the constant optical output from an optical source when it is turned on but not (yet) modulated with a signal. An electromagnetic wave of constant amplitude and frequency, e.g., in a laser application. CW is an unmodulated carrier wave: It carries no information embedded within the signal itself (information is transfered in the rhythm and spacing with which the signal is sent).

Continuous wave (CW) laser Continuously on laser. The laser is used as a light source for external test equipment or modulators; the laser is not modulated by drive voltage or current.

Copper seed layer A thin copper layer deposited by physical vapor deposition over the barrier layer. It acts as a wetting and nucleation layer for growth of the subsequent copper film deposited by electroplating [656].

Coulomb blockade Blockade of quantum mechanical tunneling produced by specific energy barrier constraints.

Coulombic force Electrical force between two charged particles.

Covalent bond Very strong attraction between two or more atoms that are sharing their electrons. In structural formulas, covalent bonds are represented by a line drawn between the symbols of the bonded atoms. The sturdy bond between two atoms that share a pair of electrons [36].

Covalent compound Compound made of molecules (not ions.) The atoms in the compound are bound together by shared electrons. Also called a molecular compound.

Covalent radius Half the distance between the nuclei of two identical atoms when they are joined by a single covalent bond.

Critical dimension (CD) Width of a patterned line or the distance between two lines of the submicron-sized circuits in a chip [656].

Crystal Identical structural units, consisting of one or more atoms, that are regularly arranged with respect to each other in space. A three-dimensional periodic arrangement of atoms with translational periodicity along its three principal axes. Thus, it is possible to obtain an infinitely extended crystal structure by aligning building blocks called unit cells until the space is filled up. Crystal structures can be described by one of the 230 space groups, which describe the rotational and translational symmetry elements present in the structure. Diffraction patterns of these normal crystals, therefore, show crystallographic point symmetries (belonging to one of the 11 Laue groups). One can assign three integer values (Miller indices) to label the observable reflections; this is due to the three-dimensional translational periodicity of the structure [683, 684, 685, 686].

Crystal field theory Description of the effect on the metal ion energy levels when in a complex, based on the electrostatic interaction of point charges representing the ligands with the metal center [655].

Cubic close-packed (CCP) atomic structure This structure is comprised of atoms that reside on the corners of a cube, with additional atoms residing at the centers of each cube face [also called face-centered cubic (fcc)]. The symmetry is described by nomenclature as Fm-3m where F means face centered, m signifies a mirror plane (there are two), and 3 indicates that there is a threefold symmetry axis (along the body diagonal) as well as inversion symmetry. Many metals have this fcc structure, e.g., gold) [90].

Cutoff wavelength Propagation of light (in a fiber) is a function of the wavelength. The cutoff wavelength is the wavelength at which an optical fiber operates as single mode: At wavelengths shorter than cutoff, several optical modes may propagate and the fiber operates as multimode. As the cutoff wavelength is approached (usually around 100 nm), progressively fewer modes may propagate until, at "cutoff," only the fundamental mode may propagate. At wavelengths longer then cutoff the guidance of the fundamental mode becomes progressively weaker until, eventually (usually at a wavelength several hundred nanometres above cut-off) the fiber ceases to guide—the fiber loses all optical functionality.

Damascene Integrated circuit process where a metal conductor pattern is embedded in a dielectric film on the silicon substrate, resulting in a planar interconnection layer. The creation of a damascene structure most often involves chemical mechanical polishing of a nonplanar surface resulting from multiple process steps [656].

Decay Process in which a particle disappears and in its place two or more different particles appear. The sum of the masses of the produced particles is always less than the mass of the original particle [50].

Decoherence Originally the term referred explicitly to errors that arise in the wave function phase, i.e., to decay of off-diagonal terms in the density matrix. This decay of phase is basis set dependent. Today, the term decoherence is therefore more generally understood in the field of QIP to refer to all manifestations of loss of unitarity in the qubit state time evolution. It includes explicit loss of coherence, dissipative or energy relaxation effects, and leakage out of the qubit state space [435]. Also, the interaction of the qubits with a larger environment, unwanted or uncontrolled interactions between qubits, and imperfections in applied unitary transformations.

Degrees of freedom Number of independent parameters required to specify the configuration of a system.

Delocalization In Lewis representation it is the ability to put the formal charge in different locations. An electron is delocalized if it can occupy more than one molecular orbital and, hence, is free to move throughout a compound [655].

Design rule Rules that outline the allowable dimensions of features used in the design and layout of integrated circuits, such as limits for feature size, and layer-to-layer overlap [656].

Detector Any device used to sense the passage of a particle. Also a collection of such devices designed so that each serves a particular purpose in allowing physicists to reconstruct particle events [50].

Diatomic molecule Molecule that contains only two atoms. All of the noninert gases occur as diatomic molecules; e.g., hydrogen, oxygen, nitrogen, fluorine, and chlorine are H_2, O_2, N_2, F_2, and Cl_2, respectively.

Dielectric Material that conducts no current when it has a voltage across it; an insulator. Two dielectrics commonly used in semiconductor processing are silicon dioxide (SiO_2) and silicon nitride (SiN) [656].

Differential equation Describes the relationship between an unknown function and its derivatives. The order of a differential equation identifies the most times any function in it has been differentiated.

Diffraction grating Optical device that behaves similarly to a prism. An array of fine, parallel, equally spaced reflecting or transmitting channels that mutually enhance the effects of difraction to concentrate the diffracted light in a few directions determined by the spacing of the channels and by the wavelength of the light.

Diode laser (aka semiconductor lasers or laser diode) Laser where the active element is a p-n semiconductor junction. When current flows across the junction, light is emitted from the edge of the chip in the plane of the junction.

Dipole Pair of electric charges or magnetic poles of equal magnitude but opposite polarity, separated by some (usually small) distance.

Dipole–dipole forces Intermolecular forces that exist between polar molecules. Active only when the molecules are close together. The strengths of intermolecular attractions increase when polarity increases [663].

Dipole–dipole interaction This is a dipole–dipole force and represents electrostatic attraction between oppositely charged poles of two or more dipoles.

Dispersion-shifted fiber (DSF) Fibers optimized for operation at 1550 nm. Regular single-mode fibers exhibit lowest attenuation performance at 1550 nm and optimum bandwidth at 1310 nm. DSFs are made so that both attenuation and bandwidth are optimal at 1550 nm.

Dissociation Breaking down of a compound into its components.

Distributed feedback laser diode (DFB-LD) Injection laser diode that has a Bragg reflection grating in the active region; the grating is used to suppress multiple longitudinal modes and enhance the properties a single longitudinal mode.

DNA Deoxyribonucleic acid; usually 2′-deoxy-5′-ribonucleic acid. DNA is a code used within cells to form proteins. Typically discussed in the context of genetic engineering (genetic engineering refers to directed modification of the gene complement of a living organism by such techniques as altering the DNA, substituting genetic material by means of a virus, transplanting whole nuclei, transplanting cell hybrids, etc.). In the context of this book the term is of interest for the study/development of molecular computers.

Dopants Impurity added in a controlled amount to a material in order to modify some intrinsic characteristic, such as resistivity/conductivity or melting point [656].

Dopeyballs Superconducting Buckyballs. Exhibit the highest critical temperature of any known organic compound.

Doping Adding a controlled amount of impurities to a material in order to modify some intrinsic characteristic, e.g., resistivity/conductivity, melting point. The substitution of atoms in a semiconductor by species of a different valence to the host species, such that extra charge carriers, electrons, or positive holes are introduced, and the conductivity of the semiconductor is increased [655].

Double-clad doped fiber amplifiers Fibers where the single-mode doped core is surrounded by a large multimode core, which guides the pump light. Double-clad doped fibers are a special structure used in high-power devices. Such fiber amplifiers are very efficient because they can be pumped with high-power laser diodes. Due to the large outer core area, the system is not very sensitive to pump misalignment; however, a large outer core reduces the absorption of the pump power, resulting in an increase of the required amplifier length [687].

Double-gate transistor One having surface conduction channels on two opposite horizontal surfaces and with current flow in the horizontal direction. Channel length is given by the horizontal separation between source and drain and is defined by lithography and etch [401].

Down quark (D) Second flavor of quark (in order of increasing mass), with electric charge $-\frac{1}{3}$ [50].

Dry etching Gas-phase-based technique for material removal. Typically, this process is conducted under vacuum and involving beams of ions or electrons to affect material removal. There are two types: physical etching (sometimes called "sputtering") and reactive etching. In reactive etching, the etching typically results from a chemical reaction occurring between incident species and surface atoms. Processes commonly in use for semiconductor dry etching include *reactive ion etching* and *chemically assisted ion beam etching*.

Dual-window fiber For single-mode fibers, the term implies that the fiber supports operation at 1310 and at 1500 nm. For multimode fibers, the term means that the fiber is optimized for 850- and 1310-nm transmission.

EDFA Amplifier based on optical fibers doped with erbium; commercially successful technology being used in the context of WDM. An amplifier is a device that boosts the strength of an electronic signal. An EDFA acts as an all-optical amplifier. A device that uses doped fiber and a secondary pump laser to optically amplify a signal. EDFAs operate as basic transmission network elements that eliminate the need for intermediate regeneration and retransmission functions. The doped fiber can amplify light in the 1550-nm region when pumped by an external light source. In a cable system (including fiber-optic-based systems), amplifiers are spaced at regular intervals throughout the system to maintain signal strength. EDFAs are optical fibers doped with the rare-earth element erbium, which can amplify light in the 1550-nm region when pumped by an external light source.

Efflorescent Substances that lose water of crystallization to the air. The loss of water changes the crystal structure, often producing a powdery crust.

Effusion Movement of gas molecules through a small opening.

Eigenfunction If the application of an operator to a function returns the original function multiplied by some constant, then the function is termed an eigenfunction of that operator [655].

Eigenvalue If the application of an operator to a function gives back the original function multiplied by some constant, then this constant is known as the eigenvalue of that function. In quantum theory, eigenvalues often correspond to observables [655]. A scalar value λ that permits nonzero solutions y of equations of the form

$$Ly = \lambda y$$

where L is a linear operator and where y can represent a vector or a function that is subject to certain boundary conditions. When $\mathbf{y}$ is a vector, $\mathbf{L}$ represents a matrix and $\mathbf{y}$ is termed an eigenvector.

In linear algebra, a scalar λ is called an eigenvalue of a linear mapping $\mathbf{L}$ if there exists a nonzero vector $\mathbf{y}$ such that $\mathbf{Ly} = \lambda\mathbf{y}$. The vector $\mathbf{y}$ is called an eigenvector. When y is a function, L can represent a differential or integral form, in which case y is called an eigenfunction.

Electric charge Quantum quantity that determines participation in electromagnetic interactions.

Electric dipole moment Measure of the degree of polarity of a polar molecule. Dipole moment is a vector with magnitude equal to charge separation times the distance between the centers of positive and negative charges. Chemists point the vector from the positive to the negative pole; physicists point it the opposite way. Dipole moments are often expressed in units called debyes.

Electrical conductivity Characterizes the conduction capacity (electrical and thermal) of a substance. Electrical conductivity is expressed in siemens per unit of length. Electronic or ionic conduction is the phenomenon by which an electron or an ion moves in a material {sideline: thermal conduction is the phenomenon by which, in a given medium, heat flows from a high-temperature region to a lower temperature region or between two media in contact with each other. Ionic or protonic conductivity quantifies the easiness with which an ion or a proton moves in a material} [61, 62, 63, 64, 65, 67].

Electrical resistivity (aka specific electrical resistance) Measure indicating how strongly a material opposes the flow of electric current; low resistivity indicates a material that readily allows the movement of electrons.

Electroabsorption (EA) modulator Electroabsorptive optical device (e.g., LiNbO3) used to electrically attenuate the laser light at microwave rates. Applicable at digital data rates to over 10 GHz [257].

Electroabsorption modulated laser (EML) Has integrated a continuous wave (CW) laser and an electroabsorption (EA) modulator on the same semiconductor chip.

Electrodes Device that moves electrons into or out of a solution by conduction.

Electrolysis Changing the chemical structure of a compound using electrical energy.

Electromagnetic calorimeter This is a dense and finely instrumented metal structure that measures the position and energy of electrons and photons [50].

Electromagnetic fluctuating near field A (less conventional) class of surface near fields has a considerable impact in local probe-based experiments. It concerns the fluctuating electromagnetic field existing spontaneously near the surface of any material [375].

Electromagnetic interaction Interaction due to electric charge; this includes magnetic effects that have to do with moving electric charges [50].

Electron (e) Elementary particle having a negative charge $e = 1.602192 \times 10^{-19}$ coulombs and a mass $m = 9.109381 \times 10^{-28}$ g ($= 5.486 \times 10^{-4}$ amu). It is the most common lepton, with electric charge -1 [50, 51].

Electron transport Electron transport in conductors is usually described by so-called Fermi-liquid theory, which assumes that the energy states of the electrons near the Fermi level are not qualitatively altered by Coulomb interactions. In 1D systems, however, even weak Coulomb interactions cause strong perturbations (the resulting system, known as a "Luttinger liquid" named after Joaquim Luttinger, who worked on such systems in the 1950s) [31].

Electronegativity Parameter that describes, on a relative basis, the power of an atom or group of atoms to attract electrons from the same molecular entity [61, 62, 63, 64, 65, 66, 67].

Electronic bandgap Band of forbidden energies between the top of the valence band and the bottom of the conduction band in a solid. Electrons may not adopt states that have energies in the bandgap. Doping a semiconductor to make it extrinsic produces extra states that may lie in the bandgap. Metals generally have no apparent bandgap, although there may be gaps between the bands in any one direction in the crystal. Semiconductors have bandgaps between 0 and 2 eV. Insulators have bandgaps greater than about 2 eV [659].

Electronic configuration (structure) Arrangement of electrons in an atom when it is in its ground state. All the properties of elements depend on their electronic configuration.

Electronic wave function at a metal surface Metal–vacuum interface (the surface charge density near metal) can be merely described with the free-electron Sommerfeld approximation in which the ground-state properties of the electron gas are obtained by filling up the conduction band with *N* free electrons obeying a Fermi–Dirac distribution. This free-electron scheme can be completed by applying the density functional method inside a "jellium" environment in which the ion cores are smeared out into a uniform positive background truncated by the surface [375].

Electroplating Deposition process in which metals are removed from a chemical solution and deposited on a charged surface [656].

Electropositivity Opposite of electronegativity. The more electropositive an atom is, the less tightly it holds onto electrons [655].

Electrostatic force microscopy (EFM) Instrument that measures local electrostatic interaction between a conductive tip and a sample through Coulomb forces.

Electrostatic surface fields Example of permanent electric near field can be found close to the surface of ionic or metal–oxide crystals [375].

Electroweak interaction In the standard model, electromagnetic and weak interactions are related (unified); physicists use the term electroweak to encompass both of them.

Electroweak theory Presents a unified description of two of the four fundamental forces of nature: electromagnetism and the weak nuclear force. Although these two forces appear very different at low energies, the theory models them as two different aspects of the same force.

Element Simple substance that cannot be resolved into simpler substances by normal chemical means.

Elementary particles Basic particles of matter; elements of a model to isolate irreducible constituents of matter. A particle by which other, larger particles are composed. For example, atoms are made up of smaller particles known as electrons, protons, and neutrons; the proton and neutron are, in turn, composed of more elementary particles known as quarks. One of the outstanding problems of particle physics is to identify the most fundamental elementary particles (fundamental particles) that make up all the other particles found in nature and are not themselves made up of smaller particles.

In the second half of the 20th century, experiments based on the bombardment of matter by protons or by electrons have revealed a large variety of subnuclear objects. A relatively large numbe/variety of particles was identified through scattering experiments. To systemitize these composite particles, the standard model was developed during the 1970s. This model posits that large numbers of particles can be explained as combinations of a (relatively) small number of fundamental particles.

Embossing Nonlithographic patterning technique that has found widespread industrial use in the manufacture of diffraction gratings, compact disks, and security features such as holograms, but that is also capable of imprinting nanoscale patterns into single, sacrificial polymer layers that can be transferred subsequently into a functional layer by conventional etching [76].

Empirical formula Formula showing the simplist ratio of elements in a compound. A formula that shows which elements are present in a compound, with their mole ratios indicated as subscripts. For example, the empirical formula of glucose is CH_2O, which means that for every mole of carbon in the compound, there are 2 moles of hydrogen and one mole of oxygen [83].

Energies of electrons Measured and expressed in terms of a unit called an *electron volt* (eV), the most commonly used unit of energy, is defined as the energy acquired by an electron when it is accelerated through a potential difference of 1V [61, 62, 63, 64, 65, 66, 67].

Energy dispersive X-ray spectrometry (EDS) Microscopy tool. An attachment to an SEM that permits the detection and identification of the X-rays produced by the impact of the electron beam on the sample, thereby allowing qualitative and quantitative elemental analysis. The electron beam of an SEM is used to excite the atoms in the surface of a solid. These excited atoms produce characteristic X-rays that are readily detected [472].

Energy level Quantified stable energy that a physical system can have. In the context of this book, the term is used in reference to the electron configuration of electrons, in atoms or molecules. Quantum theory dictates that only certain energy levels are possible; the higher an electron's level, the greater its potential energy. The set of all energy levels of a system is known as its energy spectrum.

Enhanced sensitivity for quantitation with tandem mass spectrometry (MSMS) Microscopy tool. Most of the applications of GCMS are involved with the identification of unknowns—usually trace-level contamination studies. However, frequently, one needs to verify that a particular contaminant has been successfully removed from the sample submitted. In single-stage mass spectrometer systems, this is accomplished using selected ion monitoring, or SIM. With this technique, the most abundant (and characteristic) ion in the mass spectrum of each component to be quantitated is selected for scanning, and the instrument is programmed to scan this ion rather than the entire mass spectrum. On the other hand, with a tandem (dual-stage) mass spectrometer system, the chemical noise can be reduced almost to zero, lowering the limit of detection significantly [472].

Entanglement (quantum mechanics) Relationship between two objects in which they both exhibit superposition, but, once the state of one object is measured, the state of the other is also known.

Enthalpy Change in heat.

Epitaxy (aka epi) Process technology used in some semiconductor designs where a pure silicon crystalline structure is deposited or "grown" on a bare wafer, enabling a high-purity starting point for building the semiconductor device [656].

Equilibrium When the reactants and products are in a constant ratio. The forward reaction and the reverse reactions occur at the same rate when a system is in equilibrium [663].

Equilibrium constant Value that expresses how far the reaction proceeds before reaching equilibrium. A small number means that the equilibrium is toward the reactants side while a large number means that the equilibrium is toward the products side [663].

Equilibrium expressions Expression giving the ratio between the products and reactants. The equilibrium expression is equal to the concentration of each product raised to its coefficient in a balanced chemical equation and multiplied together, divided by the concentration of the product of reactants to the power of their coefficients [663].

Equivalence point Occurs when the moles of acid equal the moles of base in a solution [663].

Erbium-doped fiber amplifier *See* EDFA.

Etching Process for removing material in a specified area through a chemical reaction [656]. Removal of material from a surface. Wet etching of silicon uses a chemical bath (usually potassium hydroxide). Dry etching uses gas, plasma, or the blasting of particles.

Event What occurs when two particles collide or a single particle decays. Particle theories predict the probabilities of various possible events occurring when many similar collisions or decays are studied. They cannot predict the outcome for any single event [50].

Excited state Atom with an electron in a higher energy level than it normally occupies.

Exciton Quasi-particle comprised of a negatively charged electron bound together with a positively charged "hole."

Exothermic Reaction that gives off heat to the environment.

Extinction ratio Ratio of the optical output in the "on" state (rated output power) to the optical power in the "off" state (threshold power).

Extreme ultraviolet (EUV) EUV-based aligners are fabrication tools that could replace optical lithography for photolithography at $<$70-nm linewidth (extreme ultraviolet typically refers to a wavelength of 13 nm used for exposure, although the EUV band itself covers the wavelength of 10–100 nm).

Fab Facility for manufacturing semiconductors.

Fabry–Perot laser diode (FP–LD) Semiconductor laser diode that uses a "Fabry Perot" filter. The filter selects wavelengths utilizing a light interference pattern produced by precisely spaced parallel surfaces.

Face-centered cubic (fcc) lattice Lattice that has one lattice point at each corner of a cube and one at the center of each face. If there is an atom or a molecule at each lattice point, then the structure is an fcc crystal.

Face-centered cubic array Close-packed array of spheres that give a structure with spheres at each of the corners and in the center of each of the faces [655].

Fermi energy Energy of the highest occupied state at zero temperature. It is depicted by the symbol E_F. The average energy per particle when adding particles to a distribution but without changing the entropy or the volume. Chemists refer to this quantity as being the electrochemical potential [710]. Energy at which the probability of finding an electron is 0.5; below the Fermi energy, orbitals are largely filled with electrons. Above the Fermi energy, the orbitals are largely unfilled with electrons [688].

Fermi level Highest occupied energy level in a band at $T = 0$ K.

Fermi sphere Shape in k space that encloses the wave vectors of the lowest energy electrons in the crystal. Any electron in a state outside the Fermi sphere can lower its energy by adopting a wave vector at the surface of the sphere [659]

Fermi–Dirac (F–D) statistics Determine the statistical distribution of fermions over the energy states for a system in thermal equilibrium (fermions are particles that obey the Pauli exclusion principle, i.e., no two particles may occupy the same state at the same time). Statistical thermodynamics is used to describe the behavior of large numbers of particles. F–D statistics are related to Maxwell–Boltzmann statistics: F–D statistics holds for fermions, B–E (Bose–Einstein) statistics plays the same role for bosons—the other type of particle found in nature. F–D statistics were introduced in 1926 by Enrico Fermi and Paul Dirac and were applied in 1927 by Arnold Sommerfeld to electrons in metals.

The Fermi–Dirac function $f(E)$ gives the probability that an electron, e.g., has energy E at temperature T given by [659]:

$$f(E) = \frac{1}{\exp(E - E_f)/kT + 1}$$

where E_f is the Fermi energy.

Fermi–dirac distribution Describes the population of levels above the Fermi level, which become occupied at temperatures above absolute zero ($T > 0\,\text{K}$) [655].

Fermion Any particle that has odd half-integer ($\frac{1}{2}$, $\frac{3}{2}$, ...) intrinsic angular momen-tum (spin), measured in units of $\hbar$ (Planck's constant). Fermions, named after Enrico Fermi, are subatomic particles that form totally antisymmetric composite quantum states. These particles are subject to the Pauli exclusion principle and obey Fermi–Dirac statistics (see below). Fermions have half-integer spin. One way of "visualizing" spin is that particles with a $\frac{1}{2}$ spin (such as fermions) have to be rotated by two full rotations to return them to their initial state. Many of the properties of ordinary matter arise because of this rule. Electrons, protons, and neutrons are all fermions, as are all the fundamental matter particles, both quarks and leptons. As a consequence of the peculiar angular momentum, fermions obey a rule called the Pauli exclusion principle, which states that no two fermions can exist in the same state at the same place and time [50, 51].

Fiber amplifiers (aka Raman amplifiers) Optical amplifiers that employ a doped optical fiber (a fiber with an impurity added to alter some of the properties) that carries the communication signal and is optically pumped by a laser. The signal is intensified by Raman amplification (hence, the name). The laser has a high-powered continuous output at an optical frequency slightly higher than that of the communication signal.

Fiber Bragg grating (FBG) Grating that consists of fiber segment whose index of refraction varies along its length; the variations of the refractive index constitute discontinuities that emulate a Bragg structure.

Fiber laser Laser where the lasing medium is an optical fiber doped with rare-earth atoms to make it capable of amplifying light. Because of the fiber laser's low threshold power, laser diodes can be used for pumping.

Field Domain or region throughout which a force may be exerted. Examples of fields include (but are not limited to) the electric and magnetic fields that envelop electric charges and magnets. Fields are used to describe the situation where two bodies separated in space exert a force on each other. The alternative to using a field-based view is to postulate that physical influences can be transmitted through empty space without any material or physical mechanism; this action-at-a-distance model, particularly if it occurs instantaneously, violates relativity theory (since it states that nothing can travel faster than light) and also violates common sense. (The concept of the field was originally developed by M. Faraday based on work on magnetism.)

Field-effect transistor (FET) Transistor where control of the current flow is based on controlling the conductance of a channel between two electrodes. The control is accomplished by the application of an external field. The current is generated by *majority carrier* drift from the source to the drain and is controlled by the voltage applied to the gate.

Field emission scanning electron microscopy (FE-SEM) Device configured as a conventional SEM, except that a cold field emission electron source is used, which permits higher image resolution to be attained, increased signal to noise ratio, and increased depth of field [472].

FinFET Type of double-gate transistor with horizontal current flow and opposing conduction channels on horizontal surfaces, but channel length is defined by lithography and sidewall spacer etch [401].

First ionization potential Ionization energy is the minimum energy required to remove an electron from an isolated atom or molecule (in its vibrational ground state) in the gaseous phase.

Fixed-target experiment Experiment in which the beam of particles from an accelerator is directed at a stationary (or nearly stationary) target. The target may be a solid, a tank containing liquid or gas, or a gas jet [50].

Flavor (quark terminology) Name used for the different quark types (*up, down, strange, charm, bottom, top*) and for the different lepton types (*electron, muon, tau*). For each charged lepton flavor there is a corresponding *neutrino* flavor. In other words, flavor is the quantum number that distinguishes the different quark/lepton types. Each flavor of quark and charged lepton has a different mass. For neutrinos one does not yet know if they have a mass or what the masses are [50].

Fluorine-doped silicate glass (FSG) Reduced dielectric constant (k = approximately 3.5) material made by doping SiO_2 with fluorine [656].

Focused ion beam (FIB) Microscopy tool. In a dual FIB/SEM instrument, a finely focused ion beam mills away a precise amount of material from the sample, and an SEM images either the sidewall or the underlying layer exposed during the milling process. The sidewall is essentially a precise cross section of the sample material. The focused ion beam is a Ga liquid metal ion gun (LMIG) that impacts the sample normal to the surface and can be focused to a spot as small as 70 Å. The focused ion beam can be rastered in a user-defined pattern over larger areas to selectively sputter and mill away the surface [472].

Formula weight Sum of the atomic weights of the atoms in an empirical formula. Formula weights are usually written in atomic mass units (u).

Fourier transform infrared spectrometry (FTIR) Microscopy tool. Individual chemical bonds, as well as groups of bonds, vibrate at characteristic frequencies. When exposed to infrared (IR) radiation, molecules selectively absorb radiation at frequencies that match those of their allowed vibrational modes. Measurement of the absorption of IR radiation by the sample as a function of frequency produces a spectrum that can be used to identify functional groups and consequently structure [472].

Free electron Electron that is not attached to a nucleus.

Friedel oscillations Charge density of particles in space oscillates in a marked way in the proximity of objects and less strongly away from said objects. These oscillations in the charge density are called Friedel oscillations (this is in contrast particles in unbounded space where the probability of finding a particle at a given point is the same everywhere—if the particles had charge, the observed charge density is the same everywhere). For example, an impurity atom placed inside a metal produces electron density oscillations near by the atom; these oscillations decay as one moves away from the atom. Friedel worked on the impurity problem in a metal; the condensed matter researchers have generalized the term beyond its original meaning. (Also related to the Fermi degeneracy.)

Fullerenes Molecular form of pure carbon that has a cagelike structure. A third form of carbon, after diamond and graphite. Fullerenes are large molecules composed entirely of carbon, with the chemical formula $C(n)$, where n is any even number from 20 to over 100. In other words, the carbon fullerene family is a group of molecules, with the chemical formula C_{2n}, $(20 < n < 50)$. Can be spherical or tubular in shape. The most common is buckminsterfullerene (buckyball) with 60 carbon atoms arranged in a spherical structure. Much larger fullerenes also exist. Fullerenes have the structure of a hollow spheroidal cage with a surface network of carbon atoms connected in hexagonal and pentagonal rings. Carbon nanotubes are similar to fullerenes, except their shape is tubular. Fullerenes were discovered by Iijima in 1991. Fullerenes come in both multiwall and single-wall versions, with single-wall nanotubes having a diameter of approximately 1 nm and multiwall versions having diameters of 10–30 nm [31].

Fullerenes were discovered serendipitously in the soot formed when a hot carbon vapor (several thousand degrees) cools off and condenses into clusters in an inert gas atmosphere. The most abundant and most well-known such molecule, C_{60}, comprises 60 carbon atoms, regularly arranged in 12 pentagonal and 20 hexagonal rings, in a soccer ball arrangement. The 1996 Nobel Prize in Chemistry was awarded to R. Curl, H. Kroto, and R. Smalley, the discoverers of this molecule. Beyond C_{60}, the next most abundant fullerene in the condensed carbon vapor is C_{70}; this molecule can be conceived of being constructed by addition of a ring of 10 atoms at one of the fivefold equators of C_{60}. By adding successively (n) such parallel rings, while maintaining the graphite threefold coordination, one can theoretically produce a series of cigar-shaped molecules C_{60+10n}. In the limit of large (n), these are particular members of a subfamily of fullerenes called "single-wall nanotubes." In general, single-wall nanotubes can be conceived as any such long strip of graphene rolled up into a seamless cylinder. The latter can be left open or can be capped by hemifullerenes [662].

Function (mapping, map, transformation and operator are usually used synonymously) Relation such that each element of a set is associated with a unique element of another (possibly the same) set. Concept of a function is fundamental to nearly every branch of mathematics and every quantitative science.

Function domain Description of the possible input values to a function. Given a function, the set A is called the domain or domain of definition of f. The set of all values in the codomain that f maps to is called the range of f, or $f(A)$.

Fundamental interaction In the standard model the fundamental interactions are the strong, electromagnetic, weak, and gravitational interactions. There is at least one more fundamental interaction in the theory that is responsible for fundamental particle masses. Five interaction types are all that are needed to explain all observed physical phenomena [50].

Fundamental particle Particle with no internal substructure. In the standard model the quarks, leptons, photons, gluons, $W^{\pm}$ bosons, and Z^0 bosons are fundamental, and all other objects are made from these [50, 51].

Gas Form of matter without fixed shape or volume. Shape: this form of matter conforms to the shape of its container. Volume: can be *compressed* or *expanded* (up to a certain limiting point) to encompass different volumes.

Gas chromatography/mass spectrometry (GCMS) Microscopy tool. Gas chromatography separates mixtures of volatile and semivolatile organic compounds into individual components using a temperature-controlled, open tubular column. The sample is flash vaporized, and the molecules are swept onto the gas chromatography column with an inert carrier gas. Separation occurs as the components partition themselves between the stationary phase on the inner wall of the column and the mobile phase (the carrier gas). The time it takes for a given molecule to traverse the entire length of the column is known as the retention time. The retention time is a function of the chemical structure of the component, the column type, and the temperature profile it was subjected to during the chromatography experiment [472].

Gate Electrode that adjusts the flow of current in a metal–oxide semiconductor transistor [656].

Generation Set of one of each charge type of quark and lepton, grouped by mass. The first generation contains the up and down quarks, the electron, and the electron neutrino [50].

Geometry Circuit line or etched feature on a chip [656].

Giant magnetoresistance (GMR) Phenomenon (200 times stronger than ordinary magnetoresistance) that results from electron-spin effects in ultrathin "multilayers" of magnetic materials, causing large changes in their electrical resistance. When a magnetic field is applied, GMR enables sensing of significantly smaller magnetic fields. This technology allows the development of hard disk that have storage capacity 20× larger than previous technology.

Gibbs energy of reaction (reaction Gibbs energy) Partial derivative of the Gibbs energy with respect to the extent of reaction [655].

Gibbs' free energy Thermodynamic state function, represented by *G*. The free energy is related to enthalpy and entropy by the equation: $\Delta G = \Delta H - T\,\Delta S$ [655].

Gluons (g) Carrier particle of strong interactions.

Grand unified theory Any of a class of theories that contain the standard model but go beyond it to predict further types of interactions mediated by particles with masses of order 10^{15} GeV/c^2. At large energies compared to this mass (times c^2) the strong, electromagnetic, and weak interactions are seen as different aspects of one unified interaction [50].

Gravitational interaction Interaction of particles due to their mass energy.

Graviton Carrier particle of the gravitational interactions; not yet directly observed.

Ground state System's lowest energy quantum mechanical state (by contrast, an excited state is any state with energy greater than the ground state). The lowest energy configuration of a system.

GSI (gigascale integration) Processor or chip has over 1 billion transistors, but less than 1 trillion.

Hadron Particle made of strongly interacting constituents such as quarks and/or gluons. (Hadrons include the mesons and baryons.) Such particles participate in residual strong interactions [50, 51].

Hadronic calorimeter This measures the energy and position of strongly interacting particles like pions, kaons, and protons. The hadronic calorimeter must be very large and very dense to collect all the energy of particles that interact in it [50].

Halide Compound or ion containing fluorine, chlorine, bromine, iodine, or astatine.

Hamiltonian (H) (1) In classical physics, H is a function that describes the state of a mechanical system in terms of position and momentum variables (it, in fact, is the framework for a reformulation of classical mechanics known as Hamiltonian mechanics.) (2) In quantum mechanics, H is the observable corresponding to the total energy of a system.

Heat of fusion Heat of fusion is energy absorbed during the change of a mole of a solid to liquid without a change in temperature. (The heat of atomization is energy needed to decompose one mole of a certain substance into atoms.)

Heat of vaporization Heat of vaporization is energy absorbed during the change of a mole of liquid to a vapor without a change in temperature.

Heisenberg uncertainty principle This principle states that it is not possible to know a particle's location and momentum *precisely* at any time. Principle that states that the product of the uncertainty in measurement of one variable, say momentum p, multiplied by the uncertainty of measurement of another variable, say position x, can never be smaller than Planck's constant h:

$$\Delta p \, \Delta x = h$$

Thus, if one knows the position of a particle very accurately, one cannot determine its momentum with great precision. The same relationship occurs between other pairs of variables such as energy and time [659].

Hexagonal close-packing (hcp) atomic structure This structure is comprised of layers with stacking (e.g., the structure of sodium at low temperatures). The formation dynamic between ccp and hcp is determined by longer-range forces between the atoms [90].

Higgs boson Carrier particle or quantum excitation of the additional force needed to introduce particle masses in the standard model. Not yet observed [50].

High *k* High dielectric constant. Materials with high k (e.g., hafnium oxide and strontium titanate) are being used in state-of-the-art ULSI/GSI/TSI computer chips.

Highest occupied molecular orbital (HOMO) Highest energy molecular orbital (of an atom or molecule) that contains an electron. If the atom or molecule were to loose an electron, it would most likely loose it from this orbital [661].

Hilbert space Inner product space that is complete with respect to the norm defined by the inner product. Hilbert space is an infinite-dimensional vector space. Hilbert spaces serve to clarify and generalize the concept of Fourier expansion, and/or certain linear transformations such as the Fourier transform. In Fourier analysis one can express a given function as an infinite sum of multiples of specified base functions (e.g., sine and cosine terms); this can be studied more abstractly in the context of Hilbert spaces.

Histograms Histograms are a way of presenting information about the relative frequency of different values of a particular variable. The horizontal axis shows the range of that variable; it is divided into a number of bins—successive intervals of the value, so a particular observation will fall in one of the bins like letters into a bin in the post office. The vertical axis represents how many times a particular value of the variable has been observed to fall in a particular bin. Cases where the value of the variable is below the low edge of the lowest bin are "underflows," and when the value is higher than the high edge of the highest bin it is an "overflow." These and the number of cases that fall in the range of the plot are registered in the under/in/over counters [50].

Holey fibers New kind of optical fiber that has microscopic holes running along the length of the fiber and/or a hollow core in the center of the light guide, that give the fiber advantages for transmitting information. By controlling the size a placement of the holes, it is possible to control the physical properties of the light as it propagates through the fiber and/or the hollow core (light travels through air or a vacuum more efficiently than through glass or plastic.) Photonic crystals can be employed in the manufacturing of holey fibers.

HOMO *See* Highest occupied molecular orbital.

Hybridization Carbon exhibits tetra-valency: since the 2s and the 2p orbitals are very close in energy, one electron from the 2s orbital jumps to the $2p_z$ orbital. The one 2s and three 2p orbitals mix together and give rise to four new altogether different types of orbitals. This arrangement, seen only in carbon atom, is called hybridization.

Hydrate Addition compound that contains water in weak chemical combination with another compound.

Hydrocarbon Organic compounds that contain only hydrogen and carbon.

Hydrogen bond Relatively strong dipole–dipole force between molecules X-H ··· Y, where X and Y are small electronegative atoms (usually F, N, or O) and "..." denotes the hydrogen bond. Hydrogen bonds are responsible for the unique properties of water and they loosely pin biological polymers like proteins and DNA into their characteristic shapes.

Hydrogen forward scattering spectrometry (HFS) Microscopy tool. A variation on elastic recoil detection (ERD); well suited for measuring hydrogen in materials. An energetic beam of helium ions impinges on the target at a glancing angle. Hydrogen atoms are scattered forward out of the sample by the He atoms, which also scatter forward, after collisions with lighter atoms. The forward scattered H atoms are collected by a solid-state detector, while the He atoms are stopped by a foil placed between the sample and the detector [472].

Hydrolysis When water reacts with another substance and as a result the oxygen in water makes a bond with the substance.

Hydrophobic effect From the Greek for "water fearing," refers to the force that causes oil and water to separate, as the attraction between water molecules is stronger than oil–water attraction. Its opposite is the hydrophilic effect [36].

Hygroscopic Able to absorb moisture from air.

Ideal gas law $PV = nRT$. Describes the relationship between pressure (P), temperature (T), volume (V), and moles of gas (n). This law is not completely accurate and becomes less accurate as conditions become less ideal.

Infrared absorption Absorption is caused by the vibration of molecular bonds of both constituent elements and impurities in the glass. This absorption occurs in distinct bands that become progressively closer together, as wavelength increases, rendering the region beyond 2 μm (2000 nm) unusable for transmission in a silica-based glasses [658].

Injection laser Another term for a semiconductor or laser diode.

Insulator Nonconductive dielectric films used to isolate electrically active areas of the device or chip from one another. Some commonly used insulators are silicon dioxide, silicon nitride, boro-phospho-silicate glass (BPSG), and phospho-silicate glass (PSG) [656].

Integrated circuit (IC) Fabrication technology that combines components of a circuit on a wafer [656]. An interconnected array of active and passive elements integrated with a single semiconductor substrate or deposited on the substrate by a continuous series of compatible processes and capable of performing at least one complete electronic circuit function. Also known as integrated semiconductor [689].

Integrated photonic circuits (microphotonic integrated circuits) Circuits where light is guided in optical waveguides that are rendered on a planar substrate. For example, passive devices such as optical splitters and multiplexers can be made by proper materials design and engineering. Next-generation devices that could replace the CMOS technology below the 10- to 5-nm frontier will likely be based on new technologies: optoelectronic and quantum properties of nanometer-scale devices will probably be the basis of alternative technologies [293].

Intensity Measure of the time-averaged energy flux. Defined as the energy density (i.e., the energy per unit volume) multiplied by the velocity at which the energy is moving. The resulting vector has the units of W/m^2.

Interaction Process in which a particle decays or responds to a force due to the presence of another particle (as in a collision). Also used to mean the underlying property of the theory that causes such effects [50].

Interconnect Wiring in an integrated circuit that connects the transistors to one another [656].

Interlayer dielectric (ILD) Films used between metal layers of an IC for insulation [656].

Intermetal dielectric (IMD) Insulating films used between adjacent metal lines; typically silicon dioxide [656].

Intermolecular force Attraction or repulsion between molecules. Intermolecular forces are much weaker than chemical bonds. Hydrogen bonds, dipole–dipole interactions, and London forces are examples of intermolecular forces.

Internal reflection Reflection of an electromagnetic wave traveling in a medium 1 with high refractive index n_1, when it hits the boundary with a medium 2 of lower refractive index n_2 ($n_2 < n_1$); however, some of the energy may be transmitted (escape) into medium 2.

Intramolecular forces Forces within molecules. Forces caused by the attraction and repulsion of charged particles.

Ion Electrically charged atom, molecule, or group of atoms or molecules. Removing or adding electrons to an atom creates an ion (a charged object very similar to an atom).

Ion implantation Process technology in which ions of dopant chemicals (boron, arsenic, etc.) are accelerated in intense electrical fields to penetrate the surface of a wafer, thus changing the electrical characteristics of the material [656].

Ion–dipole forces Intermolecular force that exists between charged particles and partially charged molecules.

Ionic bond (also used in the forms of "ionically bound" and/or "ionic bonding"). An attraction between ions of opposite charge. Potassium bromide consists of potassium ions (K^+) ionically bound to bromide ions (Br^-). Unlike covalent bonds, ionic bond formation involves transfer of electrons, and ionic bonding is not directional.

Ionic bonding Electrostatic attraction between oppositely charged ions; when two oppositely charged atoms share at least one pair of electrons, but the electrons spend more time near one of the atoms than the other.

Ionic compound Compound made of distinguishable cations and anions, held together by electrostatic forces.

Ionization When a substance breaks into its ionic components.

Ionization energy Energy required to remove an electron from a specific atom.

Isotopes Two or more nuclides having an identical nuclear charge (i.e., same atomic number) but differing atomic mass. Elements with the same number of protons but have different numbers of neutrons, and thus different masses.

Jet Depending on their energy, the quarks and gluons emerging from a collision will materialize into 5–30 particles (mostly mesons and baryons). At high momentum, these particles will appear in clusters called "jets," that is, in groups of particles moving in roughly the same direction, centered about the original quark or gluon [50].

Josephson effect Effect observed in two superconductors that are separated by a thin dielectric when a steady potential difference V is applied. An oscillatory current is set up with a frequency proportional to V [51].

Josephson junction Electronic circuit (named for Brian David Josephson, the physicist who designed it) operating at temperatures approaching absolute zero (0 K) and capable of switching at very high speeds. The device makes use of the phenomenon of superconductivity. A Josephson junction is comprised of two superconductors, separated by a nonsuperconducting layer; the nonsuperconducting layer is so thin that electrons can cross (tunnel) through the barrier under certain conditions. The movement of electrons across the barrier is known as Josephson tunneling (when a voltage is applied, the current stops flowing through the barrier). Superconductivity is the ability of certain materials to conduct electric current with near-zero resistance. A Josephson interferometer is made up of two or more junctions joined by superconducting paths. Josephson junctions are utilized in highly sensitive microwave detectors magnetometers (the Josephson effect is influenced by magnetic fields in the proximity, a capability that allows the Josephson junction to be utilized in devices that measure extremely weak magnetic fields—e.g., subtle changes in the human body's electromagnetic energy field), and superconducting quantum interference devices (SQUIDs).

Kelvin SI unit of temperature. It equates to the degrees celsius plus 273.

Kinetic energy Energy of motion, described by the relationship: Kinetic energy $= (\frac{1}{2}) \times (\text{mass}) \times (\text{speed})^2$ [654].

Kondo effect Large (counterintuitive) increase in the resistance of certain alloys of magnetic materials in nonmagnetic hosts as the temperature is lowered. The Kondo effect, named after Jun Kondo, is observed in nonmagnetic crystals when magnetic ions occur as impurities. The effect originates from the interactions between the localized spins of the magnetic impurities and the free electrons that support conduction. Typically, the Kondo effect occurs when an impurity atom with an unpaired electron is placed in a metal, producing an interaction of localized electrons with delocalized electrons. This phenomenon is encountered in the context of quantum dots and semiconductors. There is a temperature, (the "Kondo temperature," it being related to the impurity and the nature of host crystal), at which the resistivity starts to increase. If the crystal is a metal, the magnetic impurities contribute to the electrical resistivity; it follows that the conduction electrons scattered by the magnetic impurity. The scattering is conspicuous at low temperatures and increases as temperature decreases; this is a counterintuitive resistivity–temperature relation [31].

Kondo resonance (also known as Abrikosov–Suhl resonance) For materials that display the Kondo effect, a long-lived scattering resonance of electronic states near the Fermi level that forms at temperatures less than the Kondo temperature; accounts for the qualitative behavior of resistivity and magnetic susceptibility as functions of temperature [690].

Kondo temperature Temperature at which the Kondo effect predominates. For an iron-in-copper system, Kondo temperature is 24 K; in general, the Kondo temperature varies from a fraction of 1–1000 K. In specific cases, the Kondo temperature can be tuned by means of a gate voltage as a single-particle energy state nears the Fermi energy [691].

Langmuir–Blodgett (LB) film Set of monolayers or layers of organic material one molecule thick, deposited on a solid substrate. Langmuir and Blodgett discovered unique properties of thin films in the early 1900s. Langmuir's work involved the transfer of monolayers from liquid to solid substrates; Blodgett later expanded on Langmuir's work to include the deposition of multilayer films on solid substrates. An LB film can consist of a single layer or several layers, up to a depth of several wavelengths of visible light. LB films exhibit various electrochemical and photochemical properties. By transferring monolayers of organic material from a liquid to a solid substrate, one can control the structure of the film at the molecular level. This has led researchers to pursue LB films as a possible structure for integrated circuits. Some believe it might be possible to construct an LB film memory chip in which each data bit is represented by a single molecule. Complex switching networks could also be fabricated onto multilayer LB films chips.

Laser diode (LD) Semiconductor diode lasers are the standard light sources in fiber-optic systems.

Laser diode wavelengths Typical wavelengths for laser diodes are 1550 and 1310 nm. For WDM applications, laser diodes may be specified at different subwavelengths.

Laser scanning confocal microscopy (LSCM), aka confocal scanning laser microscopy High-resolution imaging system for biological specimens. Through an $x - y$ deflection mechanism a laser light beam is turned into a scanning beam, focused to a small spot by an objective lens onto a fluorescent specimen. The mixture of reflected light and emitted fluorescent light is captured by the same objective and is focused onto a photodetector via a beam splitter. The reflected light is deviated while the emitted fluorescent light passes through in the direction of the photomultiplier. A confocal aperture *(pinhole)* is placed in front of the photodetector, such that the fluorescent light from points on the specimen that are not within the focal plane where the laser beam was focused will be largely obstructed by the pinhole [473, 474].

LASER/Laser Originally an acronym for light amplification by stimulated emission of radiation. The laser is a device that uses a quantum mechanical effect (specifically stimulated emission) to generate a coherent beam of light. Laser-generated light is directional, spans a narrow range of wavelengths, and is more coherent than ordinary light. A laser beam enjoys the following features: (i) is intense (high power), (ii) is coherent, (iii) has a very low divergence, (iv) can be compressed in time up to few femtoseconds. Basic laser references include: (i) V. A. Fabrikant, "A method for the application of electromagnetic radiation (ultraviolet, visible, infrared, and radio waves)" patented in the Soviet Union (1951); (ii) Townes and Arthur L. Schawlow, "Infrared and optical masers," *Physical Review* (1958); (iii) Gordon Gould definition of "Laser" as *light amplification by stimulated emission of radiation* (1958); (iv) Schawlow and Townes, U.S. Patent No. 2,929,922 (1960).

Law of conservation of mass This law states that there is no change in total mass during a chemical change [83]. The demonstration of conservation of mass by Antoine Lavoisier in the late 18th century was a milestone in the development of modern chemistry.

Law of definite proportions This law states that when two pure substances react to form a compound, they do so in a definite proportion by mass [83]. For example, when water is formed from the reaction between hydrogen and oxygen, the "definite proportion" is 1 g of H for every 8 g of O.

Law of multiple proportions This law states that when one element can combine with another to form more than one compound, the mass ratios of the elements in the compounds are simple whole-number ratios of each other [83]. For example, in CO and in CO_2, the oxygen-to-carbon ratios are 16:12 and 32:12, respectively. Note that the second ratio is exactly twice the first because there are exactly twice as many oxygen atoms in CO_2 per carbon as there are in CO.

LCD (liquid crystal display) Represents the prevalent technology currently utilized in flat-panel displays. A crystal's alignment can be altered with an electric current: When the crystal is lined up one way, it allows the light waves to pass through a polarized filter, but as the electric current alters the crystal's alignment, it guides light in such a manner that the polarized filter blocks the light. By packing red, blue, and green light-emitting crystals next to each other on a sheet ("substrate"), one can create a full-color display.

Le Chatlier's principle States that a system at equilibrum will oppose any change in the equilibrium conditions.

LEDs (light-emitting diodes) Semiconductor diode that emits chromatically pure but incoherent light (spontaneous emission.) Light is emitted at the junction between p- and n-doped materials. Device (manufactured with two semiconductors) such that by running current in one direction across the semiconductor, the LED emits light of a particular frequency (hence a particular color). These semiconductors are durable and do not require much power. By packing red, blue, and green LEDs next to each other on a sheet ("substrate"), one can create a full-color display.

Lepton (K) Fundamental fermion that does not participate in strong interactions. A meson containing a strange quark and an antiup (or an anti-down) quark, or an antistrange quark and an up (or down) quark. The electrically charged leptons are the *electron* (e), the *muon* (μ), the *tau* (τ), and their antiparticles. Electrically neutral leptons are called *neutrinos* (ν) [50, 51].

Lewis structures/Lewis representation Way of representing molecular structures based on valence electrons.

LIGA Micromachining technique used to create very tall, straight-walled structures for microsystems (a German acronym from the words for lithography, electroplating, and molding).

Light guide Optical fiber or light conducting material.

Light localization State where light of a given frequency is totally confined to a small and finate region of space and cannot propagate except through a nonlinear interaction.

Limiting reagent Reactant that will be exhausted first.

Linac Abbreviation for linear accelerator; an accelerator that is has no bends in it.

Line spectra Spectra generated by excited substances. Consists of radiation with only specific wavelengths.

Linewidth Width of laser beam frequency.

Linewidth (IC fabrication) Width of a metal interconnect [656].

Liquid Form of matter where there is a fairly definite volume, but there is no specific *shape*. Shape: This form of matter conforms to the shape of its container. Volume: Liquids can be compressed but only to a limited degree.

Liquid crystals (LC) Fourth phase of matter: They flow like a liquid, but there is order in at least one dimension in the arrangement of the molecules. "Nematic crystals" are liquid crystals with long molecules all aligned in the same direction. "Cholesteric" and "smectic" liquid crystals have molecules arranged in distinct layers: In cholesteric crystals, the axes of the molecules are parallel to the plane of the layers; in smectic crystals, the axes of the molecules are perpendicular to the plane of the layers. A liquid-crystal polymer is a polymer with a self-organized liquid-crystal structure that combines strength with lightness [31].

Lithography Process of copying a pattern onto a surface using light, electron beams, or X-rays (etymology: "writing on small rocks"). Semiconductor manufacturing technique where, after a wafer insulator is put in place, parts of the insulator are selectively removed to obtain the appropriate pattern on the wafer. Lithography employs a photographic mask to generate a pattern of light and shadows, which is in turn projected onto the wafer after it is coated with a light-sensitive substance called photoresist. Chemical processing and baking harden the unexposed photoresist, which protects those places in shadow from follow-on stages of chemical etching [19].

Load locks Isolation chamber that allows a process chamber to be protected from ambient conditions [656].

Local field theory In this context, "locality" is the condition that two events at spatially separated locations are entirely independent of each other, provided that the time interval between the events is less than that required for a light signal to travel from one location to the other. For example, the quantum mechanical wave function is a "local" field [31, 53].

Local fields Name given to the mathematical function that defines the probability that particles of given quantum numbers (charge, mass, spin, etc.) will be created or destroyed at given points of space and time. Quantum field theory postulates that at every point of space and time one can define such a probability [692].

London force Intermolecular attractive force that arises from a cooperative oscillation of electron clouds on a collection of molecules at close range. These forces (also known as transitory forces) arise when electron clouds oscillate in step on two molecules at close range. Bond vibrations in molecules may produce the oscillations or they may be triggered by random, instantaneous pileups of electrons in atoms.

Low ***k*** Dielectric material having relatively greater insulating ability than silicon dioxide (SiO_2), usually with a $k < 3.5$ [656].

Low-pressure CVD (LPCVD) Refers to systems that process wafers in an environment with less than atmospheric pressure. LPCVD systems may be furnaces that process wafers in batches, or single-wafer systems [656].

Lowest unocupied molecular orbital (LUMO) Lowest energy molecular orbital of an atom or molecule that does not contain an electron. If the atom or molecule were to accept an electron, it would be most likely to do it with this orbital [661].

LSI (large-scale integration) Chip contains thousands of transistors—but less than a million.

LUMO *See* Lowest unoccupied molecular orbital.

Mach–Zehnder (MZ) modulator Intensity modulation (IM) approach that relies on an interference effect between two waveguides. By modulating the refractive index in one portion of the device, IM is achieved at the output.

Magnetic force microscopy (MFM) MFM uses a magnetic cantilever to simultaneously map magnetic fields and surface topography on a sample by monitoring the cantilever's response to the forces that the sample's magnetic domains exert on the AFM.

Mask Pattern used in lithography that determines which areas are exposed and which are not. A flat, transparent plate that contains the photographic image of wafer patterns to define one process layer [656].

Mass number Number of protons and neutrons in an atom.

Megascopic Observable over very large scales.

Meissner effect When a superconducting loop (*see* superconductivity) or hollow tube, in a weak magnetic field, is cooled through its transition temperature (T_c) the magnetic flux is trapped in the loop, this is the Meissner effect. The flux is constant, being unchanged by variations in the external field. It is sustained by supercurrents circulating around the loop, any field variation is countered by the induction of an appropriate supercurrent [51].

Melting point Temperature at which the solid and liquid phases of a substance are in equilibrium at a specified pressure (normally taken to be atmospheric unless stated otherwise).

MEMS—microelectromechanical systems Generic term to describe micron-scale electrical/ mechanical devices. *See also* MOEMS.

Meson Hadron made from an even number of quark constituents. The basic structure of most mesons is one quark and one antiquark [50, 51].

Mesoscopic/mesoscale Scale of between 10 and 10000 nm in diameter; refers to objects larger than an atom but smaller than what can be manipulated with human hands. "Middle observation," observable within a scale in between macroscopic and megascopic. The physical properties of mesoscopic systems, which are not as small as a single atom, but small enough so that properties are dramatically different from those in a larger piece of a material. When electrons become confined in the mesoscopic regime, they display quantum mechanical behavior [22].

Metallization Deposition of a layer of high-conductivity metal such as aluminum used to interconnect devices on a chip by CVD or PVD. Metals typically used include aluminum, tungsten, copper, and so on [656].

Metalorganic chemical vapor deposition (MOCVD) Type of chemical vapor deposition (CVD) process in which metalorganic materials (group III elements such as Ga and Al) are employed as source gases for the deposition process. One of the two main techniques for fabricating epitaxial layers of compound semiconductors [the second technique being molecular beam epitaxy (MBE)].

Metamaterials Artificial (new) types of materials with electromagnetic properties not found in nature. Metamaterials are electromagnetic and multifunctional artificial materials, created in order to comply with certain specifications. New "designer" materials [298]. Composites offering a range of magnetic properties that cannot be secured using known naturally occurring materials.

Metric tensor Mathematical statement (involving a set of quantities) that describes the deviation of the Pythagoras theorem in a curved space.

Microresonator Miniature component that enable frequency-selective coupling between waveguides. Integrated optical devices operating on a fixed wavelength, such as add–drop filters, switches, and demultiplexers can be built. Optical microresonators are a promising basic building block for filtering, amplification, modulation, and switching.

Microscopic Very small; indistinguishable without a microscope.

Microtechnology Technology dealing with matter on the size scale of microns (1 millionth of a meter). Microtechnology is a broad term and can refer to microelectronics, MEMS, or any technology that manipulates matter on a micron scale.

Microwave Electromagnetic wave with wavelength in the micron range.

Mitochondrial DNA (mtDNA) DNA (deoxyribonucleic acid) contained within mitochondria. Mitochondria: Organelles in eukaryotic cells involved in energy metabolism. Organelle: Body with specialized function, found in eukaryotic cells. Eukaryote: An organism having cells containing a nucleus, e.g., fungi, plants, and animals. All mitochondrial DNA comes from the mother. Of the over 3 billion base pairs ("DNA letters") in all of human DNA, only less than 17,000 are in the mitochondrial DNA; the balance (>99.999%) of the human DNA is nuclear DNA (DNA found in the chromosomes).

Mixture Composed of two or more substances, but each keeps its original properties.

Modal dispersion Temporal dispersion arising from differences in the travel time that different modes take to travel through multimode fibers.

Modified chemical vapor deposition (MCVD) Chemical vapor deposition (CVD) process for manufacturing preforms where glass layers are deposited on the inside surface of a starting tube.

MOEMSs (microoptoelectromechanical systems) (aka MEMSs) Optical switching is possible with the aid of MEMS-based micromirrors, which mechanically deflect the input optical signal into desired output port directly with micromirrors mounted on tiltable cantilevers. MOEMS is already used in components for telecom switches; it is also being used in projection display systems; optical displays, scanners, maskless lithography, and optical spectroscopy [250].

Molar Term expressing molarity, the number of moles of solute/liters of solution.

Molarity Number of moles of solute (the material dissolved) per liter of solution. Used to express the concentration of a solution.

Mole Collection of 6.023×10^{23} number of objects. Usually used to mean molecules.

Mole fraction Number of moles of a particular substance expressed as a fraction of the total number of moles.

Molecular beam epitaxy (MBE) Process used to make compound (multilayer) semiconductors. The process consists of depositing alternating layers of materials, layer by layer, one type after another (such as the semiconductors gallium arsenide and aluminum gallium arsenide). A form of vacuum evaporation where the vacuum levels (referred to as *ultra high vacuum* conditions) are around 10^{-11} torr, which permit molecular flow (i.e., molecules from the source arrive at the substrate without suffering collisions with other molecules). The sources provide beams of material used for deposition. MBE is one of the most sophisticated epitaxial techniques available, offering high flexibility during growth and highest quality material. Considering the slow growth rates, this technique permits atomically engineered device structures to be fabricated (specifically, *nanostructures*).

Molecular electronics (ME) (aka, moletronics) Any system with atomically precise electronic devices of nanometer dimensions, especially if made of discrete molecular parts rather than the continuous materials found in today's semiconductor devices [693]. Also: Using molecule-based materials for electronics, sensing, and optoelectronics. ME is the set of electronic behaviors in molecule-containing structures that are dependent upon the characteristic molecular organization of space; ME behavior is fixed at the scale of the individual molecule (the nanoscale) [694, 695].

Molecular formula Notation that indicates the type and number of atoms in a molecule. The molecular formula of glucose is $C_6H_{12}O_6$, which indicates that a molecule of glucose contains 6 atoms of carbon, 12 atoms of hydrogen, and 6 atoms of oxygen. Shows the number of atoms of each element present in a molecule [83].

Molecular geometry Shape of a molecule, based on the relative positions of the atoms.

Molecular integrated microsystems (MIMS) Microsystems in which functions found in biological and nanoscale systems are combined with manufacturable materials [695].

Molecular manipulator Device for atomically precise positioning of molecules.

Molecular model (aka stick model, ball-and-stick model, space-filling model) A representation of a molecule. The model can be purely computational or it can be an actual physical object. Stick models show bonds, ball-and-stick models show bonds and atoms, and space-filling models show relative atomic sizes [83].

Molecular nanoelectronics Nanoelectronics based on the nanometer-scale building blocks such as organic molecules, nanoparticles, nanocrystals, nanotubes, and nanowires. Issues relate to transport theory through nanostructures, processing, self-assembly, device fabrication, and architecture for the nanoelectronic device applications [405].

Molecular nanotechnology (MNT) Techniques for the control of the structure of matter based on molecule-by-molecule engineering; the products and processes of molecular manufacturing. Thorough but inexpensive control of the structure of matter based on molecule-by-molecule control of products and byproducts; the products and processes of molecular manufacturing, including molecular machinery [693].

Molecular self-assembly (MSA) Efficient mechanism for the self-regulated creation and/or fabrication of nanoscale elements and machinery. A chemistry-based method for assembling atomically precise materials. Similar, in some ways, to biological-based mechnism

Molecular systems engineering Design, analysis, and manufacturing of systems of molecular parts

Molecular weight Average mass of a molecule, calculated by summing the atomic weights of atoms in the molecular formula. The combined weight (as given on the periodic table) of all the elements in a compound.

Molecular wire (aka nanowire) Quasi-one-dimensional molecule that can transport charge between its endpoints.

Molecular-scale manufacturing Manufacturing using molecular/nanoscale machinery, providing molecule-by-molecule control of products through positional chemical synthesis.

Molecule Electrically neutral entity consisting of more than one atom ($n > 1$). Rigorously, a molecule must correspond to a depression on the potential energy surface that is deep enough to confine at least one vibrational state [696]. Unlike ions, molecules carry no electrical charge. A chemical comprised of two or more atoms; a group of atoms arranged to interact in a particular way; smallest particle of a compound that has all the chemical properties of that compound (a single atom is usually not referred to as a molecule, and ionic compounds are not comprised of molecules). A substance that is made up of molecules is called a molecular substance.

Molecules can be either polyatomic (composed of several atoms) or monoatomic (as in noble gases, which are composed of single-atom molecules). Polyatomic molecules are electrically neutral clusters of two or more atoms joined by shared pairs of electrons (covalent bonds) that behave as a single particle.

Momentum Product of an object's mass and its velocity.

Monomolecular computing Embedding inside a single molecule of the functional capabilities or "circuits" to realize a calculation.

Moore's law In 1965 Gordon Moore made his now well-known observation that an exponential growth in the number of transistors per integrated circuit could be observed and predicted that this trend would continue. Through technology advances, Moore's law, the doubling of transistors every 18 months has been maintained, and still holds true today. Observers (such as Intel) expect that it will continue at least through the end of the current decade [25, 389].

Morphology Shape and size of the particles making up the object; direct relation between these structures and materials properties (ductility, strength, reactivity, etc.) [697].

MOSFET Metal–oxide–semiconductor field-effect transistor; FET with MOS structure as a gate; current flows in the channel between source and drain; channel is created by applying adequate potential to the gate contact and inverting semiconductor surface underneath the gate; MOSFET structure is implemented almost uniquely with Si and SiO_2 gate oxide; efficient switching device that dominates logic and memory applications; PMOSFET (p-channel, n-type Si substrate) and NMOSFET (n-channel, p-type Si substrate) combined form basic CMOS cell [39]. A MOSFET is thick-film transistor devices. A more general device can be envisioned, called MISFET, when the insulator is not silicon oxide.

MSI (medium-scale integration) Chip contains hundreds of transistors but not thousands.

Multimode fiber Fiber used in localized telecommunication applications (e.g., data centers, central offices, etc.). Multiple optical modes of light propagate. Transmission typically occurs at 850–1300 nm.

Multiwall carbon nanotubes Tubular form of carbon, were discovered in 1991 by S. Iijima. Carbon nanotubes are comprised of a rolled graphite sheet ("graphene") and closed by fullerene-like caps. Depending on the way the graphene is rolled, different chiralities are possible and are commonly distinguished by their chiral vector (n, m). The (n, n) tubes are called "armchair" and the (n, 0) tubes are called "zigzag" nanotubes. A simple analysis imposing appropriate boundary conditions on the graphene band structure predicts that the armchair tubes are metallic (i.e., the bandgap is zero due to band crossing), whereas the zigzag tubes are either semimetals or semiconductors, depending on the value of (n) [698].

Muon (μ) Second flavor of charged lepton (in order of increasing mass), with electric charge -1.

Muon chamber Outer layers of a particle detector capable of registering tracks of charged particles. Except for the chargeless neutrinos, only muons reach this layer from the collision point.

Nanoarray Nanoscale (approximately 1/10,000th of the surface area occupied by a conventional microarray) array for biomolecular analysis.

Nanobarcode Technology that utilizes cylindrically shaped colloidal metal nanoparticles, where the metal composition is alternated along the length and the size of the individual metal segment is controllable.

Nanobiotechnology Molecular nanotechnology tools and processes to develop devices for studying biosystems. An emerging area of scientific study that applies the tools and processes of nano/microfabrication to build devices (e.g., nanobioprocessors, implantable nanoscale processors that can integrate with biological pathways and modify biological processes), for studying and analyzing biosystems.

Nanobot (aka nanoagent) Robot of microscopic dimensions built through nanotechnology means; as yet, only a speculative concept.

Nanobubbles Tiny air bubbles on colloid surfaces intended to reduce drag.

Nanochemistry The (evolving) science of confining single molecules in solution to perform chemistry at the lowest level possible and/or delivering microvolumes of chemicals to single subcellular components. To effectively analyze nanoscale systems (e.g., single molecules or single cells) one needs ultrasmall tools: The precision and sensitivity of the manipulation and detection schemes used are critical and are addressed by the science.

Nanochips "Next-gen" computer chips making use of nanotechnology principles and structures. They enjoy higher density, greater speed, and lower cost. A nanocomputer is a computer made from nanochip components.

Nanochondria Nanoscale machines symbiotically existing inside living cells and participating in their biochemistry.

Nanocomputer Computer whose fundamental components measure only a few nanometers in size. State-of-the-art current computer components are no smaller than about 60–100 nm (at press time). Researchers believe that nanoscale devices may lead to computer chips with billions of transistors, instead of millions—which is the range in current semiconductor technology.

Nanocones Carbon-based nanostructures with fivefold symmetry; nanotube caps (but also available as freestanding structures.) Nanocones form due to defects in graphene sheets.

Nanocrystals Aggregates ~100 to ~10000 atoms that combine into a crystalline form of matter measuring ~10 nm in diameter. Nanocrystals (aka nanoscale semiconductor crystals or nanoclusters) are larger than molecules but smaller than solids; they exhibit chemical and physical properties that are a hybrid between those of molecules and solids.

Nanocubic coating Very thin layer coating to support improved digital memory storage (e.g., TB range for tape and GB range for floppy diskettes).

Nanoelectronics Nanoscale electronic systems; includes molecular electronics and nanoscale devices semiconductor devices.

Nanofabrication (nanomanufacturing) Construction of items using nanoscale engineering. Chemical processes and techniques to construct nanostructures and composites (some also use terms such as nanoengineering, femtoengineering, picoengineering, and microengineering). Nanofabrication methods can be divided into two categories: *top-down* methods, which carve out or add aggregates of molecules to a surface, and *bottom-up* methods, which assemble atoms or molecules into nanostructures [699].

Nanofluidics Approaches for controlling nanoscale amounts of fluids. Uses nanogates that are devices that precisely outputs the flow of tiny amounts of fluids.

Nanoimprinting Set of techniques comparable to traditional molds (masters)- or form-based printing technology, but that uses masters with nanoscale dimensions. There are two techniques: (i) employs pressure to make indentations in the form of the mold on a surface; (ii) employs the application of "transfer materials" applied to the mold to stamp a pattern on a surface.

Nanoindentation Nanoscale indentation processes (either for hardness testing or for some atomic-level modification of a material).

Nanolithography Imprinting at the nanoscale. Writing on the nanoscale. From the Greek words *nanos* dwarf, *lithos* rock, and *grapho* to write, this word literally means "small writing on rocks" [661].

Nanomachining Process similar to traditional machining where the goal is to remove or modify portions of the structure, but done on a nonoscale; the goal is changing the structure of nanoscale materials or molecules.

Nanomanipulation Nanoscale process of manipulating items at an atomic or molecular level in order to produce precise structures.

Nanomanipulator Virtual reality (VR) methods to provide a way to study/interact with the atomic world.

Nanomaterials Bottom-up (quantum theory) designed materials where one engineers structures and functional capabilities from the ground up; materials are designed and assembled in controlled molecular fashion. Materials include nanoparticles, nanofilms, and nanocomposites.

Nanomedicine Medical applications of nanotechnology. Specifically, the monitoring, repairing, construction, and control of (human) biological systems at the molecular level, utilizing engineered nanodevices, nanostructures, and nanopharmaceuticals.

Nanooptics Nanoscale-level phenomena originated by the interaction of light and matter. *See also* Nanophotonics.

Nanopharmaceuticals Nanoscale particles used for drug delivery applications.

Nanophase carbon materials Form of matter where small clusters of atoms form the building blocks of a larger structure. Examples include carbon nanotubes, nanodiamond, and nanocomposites.

Nanophotonics Technology to fabricate and operate nanoscale photonic devices, that utilize local electromagnetic interactions between a small nanoscale element and an optical near field. Since an optical near field is free from the diffraction of light due to its size-dependent localization and size-dependent resonance features, nanophotonics enables the fabrication, operation, and integration of nanoscale devices. Atom-photonics manipulates atoms by using an optical near field, which enables the fabrication of novel matter on the atomic scale [250].

Nanoprobe Nanoscale machines used to image, manimulate, and treat biological functions (typically in a living body).

Nanoreplicators Set of nanomachines capable of self-replication.

Nanorods Multiwall carbon nanotubes.

Nanoropes Nanotubes connected and strung together.

Nanosensors Nanoscale-size sensors.

Nanosources Sources that emit light from nanoscale components.

Nanostructure Particle of nanometer size. It includes items such as clusters, big molecules, nanocrystals embedded in a matrix, etc. In the specific case of layered materials, nanometer-scale structures are topologically closed and hollow, forming tubular or spherical morphologies. It is especially the case for carbon, which presents a remarkable diversity of structures on the nanometer scale. At the nanoscale structures are typically formed of concentric hexagonal layers, folded as tubes or as spheres [660].

Nanotechnology Creation and utilization of materials, devices, and systems through the control of matter on the nanometer-length scale, i.e., at the level of atoms, molecules, and supramolecular structures [37].

Nanoterrorism Use of nanotechnology products to carry out terrorist acts.

Nanotube A 1D fullerene (a convex cage of atoms with only hexagonal and/or pentagonal faces) with a cylindrical shape. Sheets of graphite rolled up to make a tube. More generally, any tube with nanoscale dimensions, e.g., a boron-nitride-based tube.

Nanotweezers Chip-scale devices for the optical sorting of submicron particles.

Nanowires Electrical conductors that function like wires but are at the nanoscale; can be used to manufacture faster computer chips, higher-density memory, and smaller lasers. Wires have been manufactured in the 40- to 80-nm diameter range.

Near field An "interaction" field existing spontaneously or induced artificially in immediate proximity to the surface of materials or at the interface of two materials. A surface limiting a solid body locally modifies the physical properties of many materials (dielectric, metal, or semiconductor). The concept is not restricted to specific areas in physics but actually covers numerous domains of contemporary physics (electronics, photonics, interatomic forces, phononics, etc.) Addresses phenomena involving evanescent fields (electronic density surface wave, evanescent light, local electrostatic and magnetic fields, etc.) or localized interatomic or molecular interactions [375]. Do not confuse with local fields or local-field theory.

Near-field optical spectroscopy (NFOS) Tool to study semiconductor nanostructures such as single quantum dots or nanocrystals and molecular nanoobjects. Also allows optical and/or electrical manipulations on the latter.

Near-field scanning optical microscopy (NSOM) Tool for imaging at the nanoscale level capable of recording simultaneously the topography and optical response of a surface.

NSOM allows spatial resolution with more than an order of magnitude improvement over the best conventional optical methods, including laser scanning confocal microscopy. Although optical characterization is the most widespread method to analyze materials from biology to the semiconductor industry, it suffers from one inherent problem: The diffraction limit provides a spatial resolution limit of about half of the wavelength of light. Thus, features smaller than 250 nm cannot be imaged or spectrally characterized with visible light. NSOM combines scanning probe microscopy instrumentation with optical microscopy and spectroscopy to provide optical characterization with, in some cases, 15-nm resolution using visible light. The technique employs a sharpend optical fiber that is coated with metal such that a small aperture (approximately 25 nm diameter) is formed at the tip of the fiber. This aperture serves to illuminate a small spot on the sample that is much smaller than the conventional diffraction limit. The sample is then scanned beneath the tip and the image is formed in the same fashion that a dot matrix printer prints a picture [700].

NEMS (nanoelectromechanical systems) Nanoscale electromechanical devices; comparable to MEMs but at the reduced physical dimensions.

Neutral Having a net charge equal to zero. Unless specified otherwise, it usually refers to electric charge.

Neutrino (ν) Lepton with no electric charge. Neutrinos participate only in weak and gravitational interactions and therefore are very difficult to detect. There are three known types of neutrino all of which are very light and could possibly even have zero mass [50].

Neutron (n) Particle found in the nucleus of an atom. Particle is almost identical in mass to a proton but carries no electric charge. An elementary particle, having zero charge and a rest mass of $1.6749286 \times 10^{-24}$ g (939.6 MeV/c^2), that is a constituent of the atomic nucleus. The mass equates to 1.0087 amu. It is a baryon with electric charge zero; it is a fermion with a basic structure of two down quarks and one up quark (held together by gluons). The neutral component of an atomic nucleus is made from neutrons. Different isotopes of the same element are distinguished by having different numbers of neutrons in their nucleus [50, 51].

Node Point of zero amplitude in a standing wave. Antinodes are points of maximum amplitude.

Nonreturn to zero (NRZ) Optical line coding where a "1" or "0" is designated by a constant levels of opposite polarity. Used by SONET (synchronous network) transmission systems.

Nonzero dispersion shifted fiber (NZDSF) Fiber that introduces a small amount of dispersion without the zero-point crossing being in the C-band (1528–1565 nm).

Nonlinear optics Branch of optics that describes the behavior of light in *nonlinear media*, i.e., media in which the polarization responds nonlinearly to the electric field of the light. This nonlinearity is typically only observed at very high light intensities such as provided by pulsed lasers.

When the intensity of the incident light to a material system increases to a large value the response of medium is no longer linear. The response of an optical medium to the incident electromagnetic field is the induced dipole moments inside the medium [378]. Typically, it takes fields greater than 10^5 V/m to observe most nonlinear optical phenomena. These optical fields are easily generated by lasers; in fact, while nonlinear optical phenomena can be formulated by Maxwell's and Schrödinger's equations, it was not until the advent of the laser that most nonlinear optical phenomena could be tested; since the invention of the laser researchers have indeed confirmed a large number of nonlinear optical effects and have applied them to practical uses [377].

Non-local-field theory History of classical theoretical physics, from Maxwell's electrodynamics to Einstein's general relativity, can be seen as a trend toward *local-field theories* and away from action-at-a-distance theories (non-local-field theory). Some, however, have argued that the key property of quantum mechanics is precisely its *nonlocality*, as expressed in Bell's theorem and its generalizations. Therefore, a non-local-field theory, although counterintuitive, may not only be natural but, in fact, *preferred* in the quantum context: while classical general relativity is a local-field theory, quantum gravity is inherently nonlocal) [701].

NP "Nondeterministic polynomial (time)"; these are very computationally complex problems [307].

Nuclear magnetic resonance (NMR) Phenomenon that occurs when the nuclei of certain atoms are immersed in a static magnetic field and exposed to a second oscillating magnetic field. Some nuclei experience this phenomenon, and others do not, dependent upon whether they possess a spin. Spectroscopy is the study of the interaction of electromagnetic radiation with matter. Nuclear magnetic resonance spectroscopy is the use of the NMR phenomenon to study physical, chemical, and biological properties of matter. NMR spectroscopy is routinely used by chemists to study chemical structure using simple 1D techniques. 2D techniques are used to determine the structure of more complicated molecules. These techniques are replacing X-ray crystallography for the determination of protein structure. Time-domain NMR spectroscopic techniques are used to probe molecular dynamics in solutions. Solid-state NMR spectroscopy is used to determine the molecular structure of solids [702].

Nucleation layer Thin layer of film that promotes the growth of the subsequently deposited film [656].

Nucleon Proton or a neutron; that is, one of the particles that makes up a nucleus.

Nucleus Collection of neutrons and protons that forms the core of an atom (plural: *nuclei*). The central (intrinsic) part of an atom that contains the protons and neutrons.

Numerical aperture Factor that determines how strongly a fiber guides light and so how resistant it is to bend-induced losses. Numerical aperture may be defined in a number of ways [658]:

$$\mathrm{NA} = \sin \theta/2$$

where θ is the angle of the cone of light emitted from the fiber.

$$\mathrm{NA} = \sqrt{n_1^2 - n_2^2}$$

where n_1 and n_2 are the core and cladding refractive indices, respectively. It may also be approximated as $\sqrt{2n\ \delta n}$ where δn is the index difference between the core and cladding.

An optical fiber with "high" numerical aperture will guide a single optical mode over a greater range of wavelengths than is possible with a fiber with a "low" numerical aperture fiber. An optical fiber with "high" numerical aperture will guide a single optical mode when coiled or bent to a smaller diameter than is possible with a "low" numerical aperture fiber.

Observables In quantum physics a "system observable" is a property of the system state that can be determined by some sequence of physical operations. In systems governed by classical mechanics any experimentally observable value is given by a real-valued function on the set of all possible system states. In quantum theory the relation between system state and the value of an observable requires linear algebra to describe (e.g., states are given by vectors in a Hilbert space)

OEO Electical–optical–electrical conversions, e.g., in a transmission system.

OLED (organic LED) Device constructed made from carbon-based molecules. The carbon-based molecules are much smaller; this could be the driver to potentially replacing LEDs. OLEDs are also brighter, thinner, lighter, less power demanding, and faster than the LCD display in use today.

Onions Particles (typically 10–300 nm) roughly rounded, constituted of atomic layers piled as an onion. Often, an onion is closed by facets and angles, rather than by a continuous curvature, forming a "nanopolyhedron." This is especially true for diatomic materials such as boron nitride, with no "pentagon flexibility": Pentagons allow a hexagonal plane to curve. Except for boron nitride, the formation of a pentagon is thought to be energetically costly [660].

Opalescence Effect seen in some minerals, principally opal (hence, its name), that cause it to exhibit a glimmer of different colors when rotated or seen from different angles.

Operator Function or operand. A common usage is a mapping between vector spaces; this kind of operator is realized by taking one vector and returning another. In many cases, operators transform functions into other functions (an operator maps a function to another). The operator itself is a function but has an attached type, indicating the correct operand, and the kind of function returned (this data can be defined formally, using type theory).

If the operator name is called T and operates on a function f, one writes Tf and not usually $T(f)$; however, this notation may be used for clarity if there is a product for instance, e.g., $T(fg)$. Operators are described usually by the number of operands: (i) *monodic*, or *unary*; one argument, (ii) *dyadic*, or *binary*; two arguments, (iii) *triadic*, or *ternary*; three arguments, etc. There are a number of ways of writing operators and their arguments:

- *Prefix*: where the operator name comes *first* and the arguments follow, e.g., $T(x_1, x_2, \ldots, x_n)$.
- *Postfix*: where the operator name comes *last* and the arguments precede, e.g., $(x_1, x_2, \ldots, x_n)\ T$

The most common types of operators are *linear operators* (aka linear transformations) (e.g., the differential operator and Laplacian operator.) Linear operators are those that satisfy the following conditions; take the general operator T, the function acted on under the operator T, written as $f(x)$, and the constant a:

$$T(f(x) + g(x)) = T(f(x)) + T(g(x))$$
$$T(af(x)) = aT(f(x))$$

The convolution of two functions is a mapping from two functions to one other, defined by an integral: If $x_1 = f(t)$ and $x_2 = g(t)$, define the operator Q such that

$$Q(x_1, x_2) = \int f(t)g(\tau - t)\, dt$$

The Fourier transform is used in many areas, not only in mathematics, but also in physics and in signal processing. Any continuous periodic function can be represented as the sum of a series of sine waves and cosine waves:

$$f(t) = \frac{a_0}{2} + \sum_{n=1}^{\infty} a_n \cos(wnt) + b_n \sin(wnt)$$

The *Laplace transform* is another integral operator and is involved in simplifying the process of solving differential equations. Given $f = f(s)$, it is defined by:

$$F(s) = (Lf)(s) = \int_0^\infty e^{-st} f(t)\, dt$$

There are three main operators in vector calculus:

- The operator ∇, known as gradient, where at a certain point in a scalar field forms a vector that points in the direction of greatest change of that scalar field.
- The divergence, which is an operator that measures a vector field's tendency to originate from or converge upon a given point.
- The curl, in a vector field, is a vector operator that shows a vector field's tendency to rotate about a point.

In physics, an operator often takes on a more specialized meaning than in mathematics. Operators as observables are a key part of the theory of quantum mechanics; here *operator* often means a linear transformation from one Hilbert space to another

Optical amplifier Device that amplifies the input optical signal without converting it to electrical form.

Optical mode "Ray" of light. Light entering a waveguide can be regarded as confined and is referred to as an optical mode. The properties of the optical mode are determined from the characteristics of the propagating light and the refractive indices of the absorbing cladding and/or substrate regions. Propagation of the confined mode can be defined unambiguously by a property of the mode called its *effective index*. Propagation is a function of the wavelength.

Observations show that the optical mode is not totally confined in an ideal fashion within the waveguide region but also extends into the neighboring regions. The shape of the optical modes in these regions is referred to as evanescent.

Single-mode fibers transmit a single (one) mode of light. Multimode fibers transmit multiple modes. Telecom systems are based on single-mode fibers (multimode fibers find some applications in short-run applications in data centers, central offices, and interrack and/or collocated interrack cabling).

Propagation is a function of the wavelength. The cutoff wavelength is the wavelength at which an optical fiber operates as single mode: At wavelengths shorter than cutoff several optical modes may propagate and the fiber operates as multimode. As the cutoff wavelength is approached (usually around 100 nm), progressively fewer modes may propagate until, at "cutoff," only the fundamental mode may propagate. At wavelengths longer then cutoff the guidance of the fundamental mode becomes progressively weaker until, eventually (usually at a wavelength several hundred nanometers above cutoff) the fiber ceases to guide—the fiber loses all optical functionality.

Optical multiplexer Device that combines two or more optical wavelengths into a single output or fiber.

Optical nanotweezers Chip-scale devices for the optical sorting of submicron particles.

Optical near fields Optical nonfluctuating near fields are not permanent and consequently must be generated by an external light source. The simplest method consists of illuminating the surface of a sample by external reflection. In this case, the structure of the electromagnetic field above the sample critically depends on the incident angle. This effect is particularly important outside the Brewster angle, where the field intensity tends to be modulated by the interferences between incident and reflected waves. The physics of optical evanescent waves (which is the central concept used in near-field optics instrumentation) has been familiar in traditional optics for a long time: the analysis of the skin depth effect at metallic surfaces about a century ago was probably the first recognition of the existence of evanescent electromagnetic waves [375].

Optical properties of metallic nanostructures Nanoscale objects can amplify and focus light via a mechanism based on plasmons; plasmons are ripples of waves in the plasma (ocean) of electrons flowing across the surface of metallic nanostructures. The type of plasmon that exists on a surface is related to its geometric structure (e.g., curvature of a nanoscale gold sphere or a nano-sized pore in metallic foil). When light of a specific frequency strikes a plasmon that oscillates at a compatible frequency, the energy from the light is absorbed by the plasmon, converted into electrical energy that propagates through the nanostructure, and eventually converted back to light [265, 294].

Optical pumping Exciting the lasing medium by the application of light.

Optical tunneling Quantum mechanical phenomena resulting from photon delocalization that allows light to cross propagation barriers such as an interface [661].

Optical waveguide Any structure that can guide light, e.g., optical fiber, planar light waveguides, etc.

Oracle access In the context of a Turing machine, two machines to have oracle access, if they are allowed to make membership queries to a language A, called the oracle, and to receive a correct response in constant time. Such machines have a separate query tape and three extra states: a query state, a yes state, and a no state. Is of interest in the context of quantum computing.

Oracles are meta-mathematical results delineating the limitations of proof techniques and indicating what results might be possible to achieve and which are likely beyond our current reach. Oracle results concern relativized computations. One states that a computation is carried out "relative to an oracle set O" if the computation has access to the answers to membership queries of O. That is, the computation can query the oracle O about whether or not a string x is in O. The computation obtains the answer (in one step) and proceeds with the computation, which may depend on the answer to the oracle query [703].

Orbitals Energy state in the atomic model that describes where an electron will likely be. A mathematically probabilistic expression of a 3D topology where the electron is most likely to be located

Osmosis Transfer of a solvent through a semipermeable membrane separating two solutions of different concentrations. A semipermeable membrane is one through which the molecules of a solvent can pass, but the molecules of most solutes cannot. There is a thermodynamic tendency for solutions separated by such a semipermeable membrane to become equal in concentration, the water (or other solvent) flowing from the weaker to the stronger solution. Osmosis will cease when the two solutions reach equal concentrations. Osmosis can also be stopped by applying a pressure to the liquid on the stronger solution side of the membrane; the pressure required to stop the flow from a pure solvent into a solution is a characteristic of the solution and is called the "osmotic pressure" (it depends only on the concentration of particles in the solution, not on their nature) [31].

Outside vapor deposition (OVD) Chemical vapor deposition (CVD) process for manufacturing preforms where layers of glass particles are deposited on the outside surface of a target rod.

Oxidation number Number assigned to each atom to help keep track of the electrons during a redox reaction.

Oxidation reaction Reaction where a substance loses electrons.

P "Polynomial (time)"; these are problems that can be tackled more easily [307].

Particle Subatomic object with a definite mass and charge.

Particle in a box In quantum mechanics, a common model in which a particle is confined to a region by potential energy curves that reach infinity. It is commonly, though not entirely accurately, used as a synonym for an infinite square well [655].

Particle-induced X-ray emission (PIXE) Microscopy tool. X-ray transitions are excited simultaneously with backscattering measurement. The use of a Li-drifted silicon X-ray detector allows the collection of emitted X-ray signals. The strength and unique family of X-ray energies allows unambiguous identification of medium- and high-Z elements in thin films [472].

Particle physics (aka high-energy physics) Branch of physics that studies the elementary constituents of matter and radiations and the interactions between them. "Elementary particle" refers to a particle of which other, larger particles are composed. For example, atoms are made up of smaller particles such as electrons, protons, and neutrons; the proton and neutron are, in turn, composed of more elementary particles known as quarks [50, 51].

Particle-wave duality In quantum theories, energy and momentum have a definite relationship to wavelength. All particles have properties that are wavelike (such as interference) and other properties that are particlelike (such as localization). Whether the properties are primarily those of particles or those of waves, depends on how one observes them. For example, photons are the quantum particles associated with electromagnetic waves. For any frequency f the photons each carry a definite amount of energy. Only by assuming a particle nature for light with this relationship between frequency and particle energy could Einstein explain the photoelectric effect. Conversely, electrons can behave like waves and develop interference patterns [704].

Passivation Final layer in a semiconductor device that forms a hermetic seal over the circuit elements. Plasma nitride and silicon dioxide are the materials primarily used for passivation [656].

Pauli exclusion principle Principle asserts that no two fermions can exist in the same state at the same place and time [50, 51]. The Pauli exclusion principle is a quantum mechanical principle that states that no two identical fermions may occupy the same quantum state. This principle was formulated by Wolfgang Pauli (an Austrian-Swiss physicist known for his work on the theory of spin). It is also referred to as the *exclusion principle* or *Pauli principle*. The Pauli principle only applies to fermions, particles that form antisymmetric quantum states and have half-integer spin: protons, neutrons, and electrons, the three types of elementary particles that constitute matter. The Pauli exclusion principle establishes many of the distinctive characteristics of matter. Particles such as the photon and graviton do not follow the Pauli exclusion principle because these particles are bosons. (All elementary particles are either bosons or fermions.)

The principle states that in any atom no two electrons may have the same four quantum numbers. Another way of stating this is to say that electrons in the same orbital must have opposite spins [655].

The Pauli exclusion principle plays a fundamental role in a number of physical phenomena, including the electron shell structure of atoms. An electrically neutral atom contains bound electrons equal in number to the protons in the nucleus. Because electrons are fermions, the Pauli exclusion principle forbids them from occupying the same quantum state. E.g., consider a neutral helium atom, which has two bound electrons: Both of these electrons can occupy the lowest energy (1s) states; they do so by acquiring opposite spin. This predicament does not violate the Pauli principle because spin is part of the quantum state of the electron; hence, the two electrons are occupying different quantum states. However, the spin can take only two different values; therefore in a lithium atom, which contains three bound electrons, the third electron cannot exist in a 1s state, and must, therefore, occupy one of the higher-energy 2s states.

PECVD Plasma-enhanced chemical vapor deposition is a process where plasma is used to lower the temperature required to deposit film onto a wafer [656].

Periodic table Grouping of the known elements by their number of protons. There are many other trends such as size of elements and electronegativity that are easily expressed in terms of the periodic table [663].

Pervasive computing Environment/state where computers (and sensors and actuators) become practically invisible and are used in almost every aspect of human commerce, interaction, and life.

pH Measures the acidity of a solution. It is the negative log of the concentration of the hydrogen ions in a substance.

Phase In quantum mechanics, the phase angle of a complex wave function or order parameter; in statistical mechanics, the state of aggregation of matter (e.g., crystal, liquid, vapor, quasi-crystal, liquid crystal); in dynamics, the position–momentum specification [677].

Phonon In the lattice vibrations of a crystal, the phonon is a quantum of thermal energy (given by hf, where h is the Planck constant and f the vibrational frequency, and $h = 6.6260755 \times 10^{-34}$ J s) [50, 51].

Photo diode Diode that can produce an electrical signal proportional to the light falling upon it.

Photolithography Process by which a mask pattern is transferred to a wafer, usually using a "stepper" [656]. Carving through use of light. Often, a photosensitive surface (a photoresist) is selectively exposed to light using a template. The exposed areas are subsequently etched (carved by chemical means) [661].

Photon (γ) Quantum of electromagnetic radiation. The carrier particle of electromagnetic interactions. Photons move at the speed of light: 299,792,460 m/s [50, 51]. (From Greek φοτοζ, meaning light.)

Photonic bandgap (PBG) Periodic dielectric structures that forbid propagation of electromagnetic waves in a certain frequency range. An optical effect of nanochannel structured optical materials (usable in miniaturized optoelectronic devices), that relates to spectral regions inhibiting photons from traveling through the structured materials.

Photonic bandgap material Non-light-absorbing material that contains a bandgap for electromagnetic waves propagating in any and all directions.

Photonic crystal fibers Microstructured fibers; one of the first commercial products based on two dimensionally periodic photonic crystals. Fibers that use a nanoscale structure to confine light with radically different characteristics compared to conventional optical fiber for applications in nonlinear devices, guiding exotic wavelengths, among others [292].

Photonic crystals Non-light-absorbing material with a refractive index that exhibits periodic modulation in two or three orthogonal (vector) spatial directions.

Structure that provides means to manipulate, confine, and control light in one, two, or three dimensions of space: 1D, 2D, or 3D devices with ordered variations in refractive index. Device constructed of ultrathin layers of nonconducting material that reflect various wavelengths of light; a highly engineered material with superior optical properties; may be used to develop optical circuits. Device designed to create a bandgap structure with forbidden regions and allowed energies that can select or confine electromagnetic waves. Photonic crystals are periodic dielectric or metallodielectric (nano)structures that are designed to affect the propagation of electromagnetic waves in the same way as the periodic potential in a semiconductor crystal affects the electron motion by defining allowed and forbidden electronic energy bands. The absence of allowed propagating electromagnetic modes inside the structures, in a range of wavelengths called a photonic bandgap, gives rise to distinct optical phenomena such as inhibition of spontaneous emission, high-reflecting omnidirectional mirrors, and low-loss waveguiding among others [292]. Photonic crystals have periodically varying indexes of refraction. They can be thought of as "optical analogs" to electronic semiconductors. The periodically varying index of refraction permits the control of the propagation of photons inside the crystals, similar to the manner by which electrons are excited in a semiconductor crystal [291]. Photonic crystals are expected to enter commercial applications in the immediate future.

Photonic switching Use of photonic devices to make or break connections within integrated circuits, rather than electronic devices.

Photonics Technology of generating and harnessing light and other forms of radiant energy whose quantum unit is the photon. The science includes light emission, transmission, deflection, amplification, and detection by optical components and instruments, lasers and other light sources, fiber optics, electrooptical instrumentation, related hardware and electronics, and sophisticated systems. The range of applications of photonics extends from energy generation to detection to communications and information processing [257].

Photons Massless packet of energy, which behaves like both a wave and a particle.

Photoresist Light-sensitive organic polymer that is exposed by the photolithography process, then developed to produce a pattern that identifies some areas of the film to be etched [656].

Photoresist Resist that is sensitive to light and/or UV light.

Physical vapor deposition (PVD) Also called sputtering, it is a process technology in which molecules of conducting material (aluminum, titanium nitride, etc.) are "sputtered" from a target of pure material, then deposited on the wafer to create the conducting circuitry within the chip [656].

Pi (π) **bonds** Type of covalent bond in which the electron density is concentrated around the line bonding the atoms. A bond between two atoms that consists of two electrons occupying a bonding molecular orbital that has a nodal plane in its wave function [655].

Piezoelectric Having the ability to generate a voltage when mechanical force is applied, or to produce a mechanical force when a voltage is applied, as in a piezoelectric crystal [689].

Pion (π) Least massive type of meson, pions can have electric charges ± 1 or 0.

Planar devices Semiconductor devices having planar electrodes in parallel planes, made by alternate diffusion of p- and n-type impurities into a substrate. P-type semiconductors have electron hole densities exceeding the conduction electron densities. N-type semiconductors have conduction electron densities exceeding the hole densities [689].

Planar lightguide circuit (PLC) (aka planar lightwave circuits) Device (typically manufactured in wafer form, say over a silicon substrate) that is used to guide light, such as planar light waveguide. A circuit (waveguide) that is fabricated on flat material, such as a thin film. A device that incorporates a planar waveguide. This is a waveguide fabricated in a flat material such as thin film.

In a channel waveguide the light propagates a ribbonlike channel that is embedded in a planar substrate. To confine light within the channel, it is necessary for the channel to have a refractive index greater than that of the substrate. This type of waveguide is a good choice for fabricating integrated photonic devices. Given that the substrate is planar, the technology associated with integrated optical circuits is also called *planar lightwave circuit.*

Planck Planck contributed to the understanding of the electromagnetic spectrum by realizing that the relationship between the change in energy and frequency is quantized according to the equation delta $E = hv$ where h is Planck's constant [663].

Plasma Gas of charged particles. Ionized gases that have been highly energized—e.g., by a radio frequency energy field [656].

Plasmodic devices Ultrasmall metal structures of various shapes capture and manipulate light. The goal of the field of plasmonics is to develop new optical components and systems that are the same size as today's smallest integrated circuits and that could ultimately be integrated with electronics on the same chip [265, 294].

Plasmon (surface plasmon) Quantum associated with a plasma oscillation in the electron gas of a solid; the quantum associated with longitudinal waves propagating in matter through the collective motion of large numbers of electrons. Surface plasmons are a subset of these "eigenmodes" of the electrons, which are bound to regions in the material where the optical properties reverse, i.e., the interface between a dielectric and conducting medium. However in the low k-vector limit these surface modes couple with the free EM field to yield a polariton-type excitation [657].

Plasmonics Emerging field of optics aimed at the study of light at the nanometer scale.

Polariton Quantum associated with the coupled modes of photons and optical phonons in an ionic crystal.

Polarization maintaining fiber (PMF) Fiber where light is able to propagate in one mode and maintain a fixed polarization.

Polarization mode dispersion (PMD) Light transmitted on a single mode fiber is decomposable into two perpendicular polarization components. Polarization mode dispersion is distortion that results due to each polarization propagating at different velocity.

Polyatomic ion Charged particle that contains more than two covalently bound atoms.

Polyatomic molecule Uncharged particle that contains more than two atoms.

Polycide Material formed by reaction of polysilicon with a metal [656].

Polymer Large molecule (molecular weight ~10,000 or greater) composed of many smaller molecules (monomer) covalently bonded together. The term *polymer* derives from the Greek *polymeros* and means "many parts." Natural polymer molecules dominate the field of biology, whereas artificial polymers are used as plastics or emulsifiers in a gamut of modern industrial and consumer products.

A substance consisting of molecules characterized by the repetition (neglecting ends, branch junctions, and other minor irregularities) of one or more types of monomeric units. The individual units of polymers are called *monomers*, and most common polymers are composed of regular repetitions of one or more monomers. There is no formal restriction on the composition of a polymer. A simple linear polymer is a chain molecule composed of monomers with two reactive sites (bifunctional monomers), with monofunctional terminal units. If more than one bifunctional monomer is present, the chain is known as a *copolymer*. A copolymer in which a number of units of the same monomer are located adjacent to one another (in "blocks" of monomers) is called a *block copolymer*. A *diblock copolymer* is composed of two types of monomers (e.g., A and B), and may be depicted as: AAAAAABBBBBAAAAAABBBBBAAAAAAA [31].

Polysilicon (poly) Polycrystalline silicon; extensively used as conductor/gate materials in a highly doped state. Poly films are typically deposited using high-temperature CVD technology [656].

Positron (E^+) Antiparticle of the electron.

Potential energy Stored energy that an object possesses by virtue of its position with respect to other objects. For example, gravitational potential energy by virtue of the position of one mass relative to other(s) [654].

Principal quantum number Number related to the amount of energy an electron has and therefore describing which shell the electron is in.

Probability amplitude Formal definition of probability amplitude is: the squared norm of an amplitude with respect to a chosen orthonormal basis.

Process (IC fabrication) Group of sequential operations in the manufacture of an integrated circuit [656].

Process chamber Enclosed area in which a process-specific function occurs during wafer manufacturing [656].

Proton (p) Positively charged elementary particle that forms the nucleus of the hydrogen atom and is a constituent particle of all nuclei. Rest mass $m = 1.6726231 \times 10^{-24}$ g (=1.0073 amu). The proton has a charge of $+1$ electron charge (or $+1.602 \times 10^{-19}$ C). The proton is the most common hadron, a baryon with electric charge $(+1)$ equal and opposite to that of the electron (-1). Protons have a basic structure of two *up* quarks and one *down* quark (bound together by gluons). The nucleus of a hydrogen atom is a proton. A nucleus with electric charge Z contains Z protons; therefore, the number of protons is what distinguishes the different chemical elements [50, 51].

Pulsed laser Laser that emits light in a series of pulses rather than continuously. Used for testing fiber systems.

Pulsing (laser) Allows the production of light pulses with extremely high peak intensity, much higher than would be produced by the same laser if it were operating in a continuous wave (constant output) mode. A laser can be made to produce a pulsed output beam using "Q-switching" (*See Q*).

Pump laser Power laser used to drive optical amplifiers by exciting the rare-earth doped fiber.

Pumping Addition of energy (thermal, electrical, or optical) into the atomic population of the laser medium, necessary to produce a state of population inversion.

Q (High Q) Measure of the quality of a resonator; in optics Q is a measure of how much light from the gain medium of the laser is fed back into itself by the resonator. Resonant systems respond to frequencies close to the natural frequency f_0 much more strongly than they respond to other frequencies. The Q factor is defined as the resonant frequency (center frequency) f_0 divided by the bandwidth (BW):

$$Q = \frac{f_0}{\mathrm{BW}}$$

with $\mathrm{BW} = f_2 - f_1$, where f_2 is the upper and f_1 the lower cutoff frequency.

QIP Quantum information processing; quantum computing.

Quanta (a) Fundamental units of energy [38]. (b) Light can carry energy only in specific amounts, proportional to the frequency, as though it came in packets. The term *quanta* was given to these discrete packets of electromagnetic energy by Max Planck [38]. (c) The smallest physical units into which something can be partitioned, according to the laws of quantum mechanics. For example, photons are the quanta of the electromagnetic field [38]. (d) Each particle is surrounded by a field for each of the kinds of charges it carries, such as an electromagnetic field, if it has electric charge. In the quantum theory, the field is described as made up of particles that are the quanta of the field. More loosely, the smallest amount of something that can exist [38].

Quantization Restriction of various quantities to certain discrete values; or, more generally, to deriving the quantum mechanical laws of a system from its corresponding classical laws [38].

Quantizing In experimental physics, a quantized variable is a variable taking only discrete multiple values of a quantum mechanical constant. In theoretical physics, quantizing means the consistent application of certain rules that lead from classical to quantum mechanics. In general, quantization is a transition from a classical theory or a classical quantity to a quantum theory or the corresponding quantity in quantum mechanics [31, 53].

Quantum When used as a noun (plural quanta): a discrete quantity of energy, momentum, or angular momentum, given in units involving Planck's constant h. For example, electromagnetic radiation of a given frequency f is composed of quanta (also called photons) with energy hf [705]. It is the smallest amount of energy that can be absorbed or radiated by matter at that frequency [38].

When used as an adjective (as in quantum theory, quantum mechanics, quantum field theory): defines the theory as involving quantities that depend on Planck's constant h. In such theories radiation comes in discrete quanta as described above; angular momenta must be integer units of h, except that the intrinsic angular momenta of fundamental particles are integer multiples of $\frac{1}{2}h$; and solutions for the possible states of a particle in a potential (such as the states of an electron due to an atomic nucleus) occur only for certain discrete energies [705].

Quantum chemistry Application of quantum mechanics to the study of chemical phenomena.

Quantum chromodynamics (QCD) Physical theory describing the strong interaction (one of the fundamental forces.) Uses quantum field theory to describe the interaction of quarks and gluons. Proposed in the early 1970s by Politzer, Wilczek, and Gross [711].

(a) The quantum field theory describing the interactions of quarks through the strong "color" field (whose quanta are gluons) [38]. (b) QCD is the accepted theory of the forces that bind quarks together to form protons, neutrons, and other strongly interacting particles [38]. (c) The quantum theory of the strong nuclear force, which it envisions as being conveyed by quanta called gluons. The name derives from the assignment of a quantum number called color to designate how quarks function in response to the strong force [38]. (d) QCD: relativistic quantum field theory of the strong force and quarks, incorporating special relativity [38]. (e) The modern theory of the strong forces between quarks and, hence, of the forces between hadrons. It is a generalization of quantum electrodynamics, with color charge replacing electric charge and gluons replacing photons [38].

Quantum computer Computer that takes advantage of quantum theory properties (e.g., entanglement) present at the nanoscale.

Quantum confinement Light emission from bulk (macroscopic) semiconductors such as LEDs results from exciting the semiconductor either electrically or by shining light on it, creating electron–hole pairs that, when they recombine, emit light. The energy, and therefore the wavelength, of the emitted light is governed by the composition of the semiconductor material. If, however, the physical size of the semiconductor is considerably reduced to be much smaller than the natural radius of the electron–hole pair (Bohr radius), additional energy is required to "confine" this excitation within the nanoscopic semiconductor structure leading to a shift in the emission to shorter wavelengths [198].

Quantum cryptography System based on quantum mechanical principles. Eavesdroppers alter the quantum state of the system and so are detected. Developed by Brassard and Bennett, only small laboratory demonstrations have been made so far [706].

Quantum defect Principal quantum number responsible for a spectral series, minus the Rydberg denominator for any actual spectral term of the series (also called Rydberg correction) [38].

Quantum determinism Property of quantum mechanics that knowledge of the quantum state of a system at one moment completely determines its quantum state at future and past moments. Knowledge of the quantum state, however, determines only the probability that one or another future will actually ensue [38].

Quantum device Semiconductor device whose operation is based on quantum effects [36, 39].

Quantum dot (QD) Nanometer-scale "boxes" for selectively holding or releasing electrons; the size of the box can be from 30 to 1000 nm [40, 41]. Something (usually a semiconductor island) capable of confining a single electron, or a few, and in which the electrons occupy discrete energy states just as they would in an atom [42]. Grouping of atoms so small that the addition or removal of an electron will change its properties in a significant way [36]. The term emphasizes the quantum confinement effect; the term typically refers to the subclass of nanocrystals that are small enough to exist in the quantum confinement regime, and more typically refers to fluorescent nanocrystals in the quantum confined size range [198]. (In this book the term QD refers to the former, and the term quantum dot nanocrystals (QDNs) refers to the latter.)

Quantum dots are small metal or semiconductor boxes that electrostatically confine/hold a specified number of electrons (the number can be adjusted from zero to several hundred by changing the dot's electrostatic status). During the late 1990s and early 2000s QDs evolved from laboratory constructs to the building blocks for a future computer applications. QD elements can be used for next-generation telecom devices and can be incorporated into planar lightwave circuits for high-speed signal-processing applications in optically routed networks.

An artificial atom. As realized in the laboratory, quantum dots are small electrically conducting regions, typically less than 1 μm in diameter that contain from one to a few thousand electrons. Because of the small volume, the electron energies within the dot are quantized, and the behavior of the quantum dot is intermediate between that of an atom and that of a classical macroscopic object [31].

Nanometer-sized semiconductor crystals or electrostatically confined electrons. Something (usually a semiconductor island) capable of confining a single electron, or a few, and in which the electrons occupy discrete energy states just as they would in an atom (quantum dots have been called "artificial atoms") [695]. Other terminology reflects the focus of different areas of research: microelectronics researchers may refer to a *single-electron transistor* or *controlled potential barrier*; quantum physicists may speak of a *Coulomb island* or *zero-dimensional gas*; chemists speaks of a *colloidal nanoparticle* or *semiconductor nanocrystal*. All of these terms are, at various times, used interchangeably with "quantum dot," and they refer more or less to the same thing: a trap that confines electrons in all three dimensions [707].

Quantum dot nanocrystals (QDNs) "Tiny" semiconductor nanocrystals that glow in various colors when excited by laser light. QDNs are used to tag biological molecules [quadot]. QDNs have sizes of 5–10 nm. QDNs' cores contain paired clusters of atoms that combine to create a semiconductor; the clusters are surrounded by a shell made of an inorganic substance to protect the clusters. The cluster releases light of a specific color when stimulated by ultraviolet light. Since the term emphasizes the quantum confinement effect, the term typically refers to the subclass of nanocrystals that are small enough to exist in the quantum confinement regime, and more typically refers to fluorescent nanocrystals in the quantum-confined size range [198]. QDNs in solid matrix (composite) materials allow product developers to control the form factor of nanocrystals. QDN matrix materials allow creation of films, beads, fibers, and micron-sized particles for numerous applications. QDNs composites accelerate engineering and development of nanocrystal applications including photonics, LEDs, ink, and paints [199].

The QDNs are used to tag biological molecules; by varying the number of atoms in the QDN, they can be made to emit light of different colors. As noted, QDNs are comprised of three components [695]: (i) cores that contain paired clusters of atoms such as cadmium and selenium that combine to create a semiconductor; the release of light of a specific color when stimulated by ultraviolet of a wide range of frequencies; (ii) clusters are surrounded by a protective shell constituted form an inorganic substance; and, (iii) a coating with an organic surface, to allow the attachment of proteins or DNA molecules.

Quantum effect Properties of transistors and wires become altered at the nanoscale level, so that they can no longer be characterized by classical electronic circuit theory. Quantum effects, such as the quantization of electronic charge and the interfering wave properties of electrons as they propagate through transistors and wires, need to be take into account [36].

Quantum efficiency (QE) (a) Efficiency of a counter in detecting photons; the probability that a photon will liberate an electron and thus be detected [38]. (b) QE: The ratio of the number of photoelectrons released for each incident photon of light absorbed by a detector. This ratio cannot exceed unity.

Quantum electrodynamics (QED) Quantum field theory of electromagnetism. It describes phenomena involving electrically charged particles interacting through electromagnetic force. QED allows one to make accurate predictions of quantities such as the magnetic moment of the muon and energy levels of hydrogen. QED is one of the most well-tested and successful theories in physics [711].

(a) The quantum field theory describing the interactions between electrically charged particles through the electromagnetic field (whose quantum is the photon) [38]. (b) The theory of photons and electrons (or other electrically charged particles) and their interactions. It is called "quantum" when the electromagnetic radiation (i.e., light, etc.) is being treated by quantum theory, so that its discrete photon nature is important [38]. (c) The quantum theory of the electromagnetic force, which it envisions as being carried by quanta called photons [38]. (d) This is the accepted theory of electromagnetic interactions, including all the effects of relativity and quantum theory. The photon acts as the carrier of the electromagnetic force [38]. (e) Relativistic quantum field theory of the electromagnetic force and electrons, incorporating special relativity [38].

Quantum electronics Name used for those parts of quantum optics that have practical device applications [36, 38].

Quantum entanglement State of a two-particle system is said to be entangled when its quantum mechanical wave function cannot be factorized into two single-particle wave functions [306]. A starting photon can spontaneously split into a pair of entangled photons inside a nonlinear crystal.

Quantum field Distribution of energy that is constantly creating and destroying particles, according to the probabilities of quantum mechanics, and transmitting the forces of nature [38].

Quantum field theory (QFT) Study of the quantum mechanical interaction of elementary particles and fields. The application of quantum mechanics to fields. It provides a theoretical framework used in particle physics and condensed-matter physics [711].

When applied to the understanding of electromagnetism, it is called quantum electrodynamics (QED) When applied to the understanding of the strong interactions between quarks and between protons, neutrons, and other baryons and mesons, it is called quantum chromodynamics (QCD). The fundamentals of QFT were developed between the late 1920s and the 1950s, by researchers such as Dirac, Fock, Pauli, Tomonaga, Schwinger, Feynman, and Dyson.

A quantum mechanical theory applied to systems having an infinite number of degrees of freedom: (a) The relativistically invariant version of quantum mechanics [38]. (b) The theory used to describe the physics of elementary particles. According to this theory, quantum fields are the ultimate reality and particles are merely the localized quanta of these fields [38]. (c) The theory that describes the quantum effects of a classical system of fields defined on space–time and satisfying various partial differential equations [38]. (d) When interactions among particles are described as transmitted via the exchange of bosons, the methods of quantum field theory are used [38].

Quantum fluctuations (a) The spontaneous fluctuation of energy in a volume of space. A consequence of the Heisenberg uncertainty principle [38]. (b) Turbulent behavior of a system on microscopic scales due to the uncertainty principle [38]. (c) Continuous variations in the properties of a physical system caused by the probabilistic character of nature as dictated by quantum mechanics. For example, the number of photons in a box with perfectly reflecting walls is constantly varying because of quantum fluctuations. Quantum fluctuations can cause particles to appear and disappear [38].

Quantum geometry Modification of Riemannian geometry required to describe accurately the physics of space on ultramicroscopic scales, where quantum effects become important [38].

Quantum gravity (a) Theory of gravity that would properly include quantum mechanics. To date, there is no complete and self-consistent theory of quantum gravity, although successful quantum theories have been found for all the forces of nature except gravity [38]. (b) General term used to describe attempts to quantize gravity. The elementary particle of the gravitational field is the graviton. (c) Theory that successfully mergers quantum mechanics and general relativity, possibly involving modifications of one or both. String theory is an example of a theory of quantum gravity.

Quantum Hall effect Effect in which Hall resistivity changes by steps so that it is a fraction of h/e^2, where h is Planck's constant and e is the electronic charge [51]. In a two-dimensional electron system at sufficiently low temperature and in sufficiently high magnetic field, the ratio of the current to the voltage applied in a direction perpendicular to the current is very accurately a multiple (integer or fraction with small odd denominator) of e^2/h [38].

Quantum leap Disappearance of a subatomic particle, e.g., an electron, at one location and its simultaneous reappearance at another. The counterintuitive weirdness of the concept results in part from the limitations of the particle metaphor in describing a phenomenon that is also in many respects a wave [38].

Quantum liquid System of particles that are sufficiently mobile and operate at sufficiently low temperature to display the effects of quantum mechanical indistinguishability. Examples include the electrons in superconducting metals and the atoms in liquid helium [38].

Quantum mechanical amplitude Mathematical quantity in quantum mechanics whose absolute square determines the probability of a particular process occurring [38].

Quantum mechanics Laws of physics that apply on very small scales. The essential feature is that energy, momentum, and angular momentum as well as charge come in discrete amounts called quanta [705]. Quantum mechanics is a physical theory derived from a small set of basic principles, which at very small distances generates results that are very different and much more accurate than the results of classical mechanics. Quantum mechanics is the underlying framework of many fields of physics and chemistry, including quantum chemistry, condensed-matter physics, and particle physics [711].

(a) Theory that explains the dual wavelike and particlelike behavior of matter and the probabilistic character of nature. According to quantum mechanics, it is impossible to have complete and certain information about the state of a physical system, just as a wave cannot be localized to a single point in space but spreads out over many points. This uncertainty is an intrinsic aspect of the system or particle, not a reflection of our inaccuracy of measurement. Consequently, physical systems must be described in terms of probabilities. For example, in a large collection of uranium atoms, it is possible to accurately predict what fraction of the atoms will radioactively disintegrate over the next hour, but it is impossible to predict which atoms will do so. As another example, an electron with a well-known speed cannot be localized to a small region of space but behaves as if it occupied many different places at the same time. Any physical system, such as an atom, may be viewed as existing as a combination of its possible states, each of which has a certain probability. Quantum theory has been successful at explaining the behavior of nature at the subatomic level, although many of its results violate some commonsense intuition [38]. (b) Theory of the interaction of matter and radiation, that rests on the original idea of Planck that radiating bodies emit energy in discrete units or quanta of radiation whose energy is proportional to the frequency of the light [38]. (c) Framework of laws governing the universe whose unfamiliar features such as uncertainty, quantum fluctuations, and wave–particle duality become most apparent on the microscopic scales of atoms and subnuclear particles [38].

Quantum mirage Phenomenon at the nanoscale by which it is possible to transmit information using the wave properties of electrons, possibly enabling future nanoscale machines to operate without wires [36].

Quantum number Number that labels a quantum state; it denotes the number of quanta of a particular type that the state contains (electric charge given as an integer multiple of the electron's charge is an example of a quantum number) [705]. The number determines various properties of that system. There are principal, orbital angular momentum, magnetic angular momentum, and spin angular momentum quantum numbers [655].

Quantum optics Science concerned with the applications of the quantum theory of optics; i.e., optics defined in terms of the quanta of radiant energy, or photons [34].

Quantum physics Physics based upon the quantum principle that energy is emitted not as a continuum but in discrete units [38].

Quantum solid Degenerate gas in which the densities are so great that the nuclei are fixed with respect to each other so that they resemble a crystalline lattice [38].

Quantum space Vacuum with the potential to produce virtual particles [38].

Quantum state Describes the state of a quantum system. In quantum theory "state" is described using a mathematical representation such as a state vector (also called a wave function) or a density operator. Dirac developed a powerful and intuitive mathematical notation to talk about states, known as the *bra and ket* notation.

Quantum theory (a) Theory that seeks to explain that the action of forces is a result of the exchange of subatomic particles [38]. (b) Theory used to describe physical systems that are very small, of atomic dimensions or less. A feature of the theory is that certain quantities (e.g., energy, angular momentum, light) can only exist in certain discrete amounts, called quanta. (c) Initially, the theory developed by Planck that radiating bodies emit energy not in a continuous stream but in discrete units called quanta, the energy of which is directly proportional to the frequency. Now, all aspects of quantum mechanics. (d) Quantum theory provides the rules with which to calculate how matter behaves. Once scientists specify what system they want to describe and what the interactions among the particles of the system are, then the equations of the quantum theory are solved to learn the properties of the system.

Quantum tunneling (a) Quantum leap through a barrier. The mechanism where one uses the wave properties (the wave function) of an electorn (electron tunneling) through a thin potential barrier [26]. (b) Process by which a quantum system can suddenly and discontinuously make a transition from an initial configuration to a final one, even if the system does not have enough energy to classically attain the configurations between the two [38]. (c) Feature of quantum mechanics showing that objects can pass through barriers that should be impenetrable according to Newton's classical laws of physics [38]. When electrons pass through a barrier, without overcoming it or breaking it down [695]. The tunneling of electrons below the tops of energy barriers [406].

A quantum mechanical phenomenon involving an effective penetration of an energy barrier by a particle resulting from the width of the barrier being less than the wavelength of the particle [31]. ("Wavelength" refers to the de Broglie wavelength of the particle, which is given by $\lambda = h/mv$, with λ the wavelength of the moving particle, h the Planck constant, m the mass of the particle, and v the velocity of the particle.)

Quantum well (QW) A p-n-p junction with an n layer of ~10 nm. With this arrangement an *electron trap* is created. As the layers become thin (say ~10 nm), quantum effects effects begin to dominate the behavior of the electrons. The quantum well consists of a thin layer of a narrower-gap semiconductor between thicker layers of a wider-gap material. In a diode laser, a region between layers of gallium arsenide and aluminum gallium arsenide, where the density of electrons is very high, resulting in increased lasing efficiency and reduced generation of heat [34, 257]. Semiconductor heterostructure fabricated to implement quantum effects in electronic and photonic applications; typically an ultrathin layer of narrower bandgap semiconductor is sandwiched between two layers of larger bandgap semiconductor; electrons and holes are free to move in the direction perpendicular to the crystal growth direction but not in the direction of crystal growth, hence, are "confined" [36, 39, 695].

Quantum wells are real-world implementation of the "particle in the box" problem; QWs act as potential wells for charge carriers and are typically experimentally realized by epitaxial growth of a sequence of ultrathin layers consisting of semiconducting materials of varying composition [33].

Quantum well intermixing (QWI) Manufacturing a technique that permits the properties of a semiconductor material to be modified, allowing multiple optical communications functions to be integrated on a monolithic chip. Specifically, the energy bandgap is controlled making the semiconductor opaque or transparent to light [386].

Quantum wire Narrow channel created by cleaving a crystal made of alternating layers of gallium arsenide and aluminum gallium arsenide, and adding additional layers on the cleaved end face, at right angles to the first, resulting in an efficient diode laser [34]. A wire that is so thin that quantum properties of electrons are needed to understand how the current passes through them.

Can be viewed as a form of quantum dot, but the movement of electrons in the quantum wire is restricted (confined) to only two dimensions; this allows the electrons to propagate in a "particlelike" fashion. Typically built onto a semiconductor base.

Another form of quantum dot, but unlike the single-dimension "dot," a quantum wire is confined only in two dimensions—i.e., it has "length" and allows the electrons to propagate in a particlelike fashion. Constructed typically on a semiconductor base, and (among other things) used to produce very intense laser beams, switchable up to multigigahertz per second [695].

Quarks (q) Fundamental fermion that has strong interactions. Names up, charm, top, down, strange, and bottom are used to characterize different types of quarks. Quarks have electric charge of either $\frac{2}{3}$ (up, charm, top) or $-\frac{1}{3}$ (down, strange, bottom) in units where the proton charge is 1.

Quasi-crystal Solids exhibiting long-range orientational coherence of local atomic coordination geometry but no crystallographic periodicity or unit cell [677]. Material that shows diffraction patterns with rotational symmetries, that are "forbidden" by classical crystallography; the structure of quasi-crystals comprises of atoms that are arranged in a nonperiodic fashion. Quasi-crystal structures show long-range order, but no translational periodicity. Quasi-crystal structures can be approximated by filling an appropriate quasi-periodic tiling with atoms. Materials with perfect long-range order, but with no three-dimensional translational periodicity. The former is manifested in the occurrence of sharp diffraction spots and the latter in the presence of a noncrystallographic rotational symmetry. In 1984, however, Shechtman, Blech, Gratias, and Cahn published a study that marked the discovery of quasi-crystals. The authors showed electron diffraction patterns of an Al–Mn alloy with sharp reflections and 10-fold symmetry; the whole set of diffraction patterns revealed an icosahedral symmetry of the reciprocal space. Since then, many stable and metastable quasi-crystals were found. These are often binary or ternary intermetallic alloys with aluminium as one of the constituents. The icosahedral quasi-crystals form one group and the polygonal quasi-crystals another (8-,10-,12-fold symmetry.) Since quasi-crystals lost periodicity in at least one dimension, it is not possible to describe them in 3D space as easily as normal crystal structures; hence, it becomes more difficult to find mathematical formalisms for the interpretation and analysis of diffraction data [683, 684, 685, 686].

Qubit Quantum computing bit. Qubits admit superposition, namely, a qubit can be both 1 and 0 at the same time. Bits of information physically implemented by quantum systems [307].

Radical (or free radical) Molecular entity possessing an unpaired electron.

Raman effect Effect where part of the energy in a photon is transferred to (or from) the vibration/rotational energy of a molecule.

Rayleigh scattering Scattering of radiation as it passes through a medium containing particles, when the size of which is small compared with the wavelength of the radiation.

Scattering is caused by the presence of submicroscopic variations in the density and composition of the glass. This scattering decreases according to a forthpower law as wavelength increases [658].

Reactants Substances initially present in a chemical reaction.

Reactive Tending to participate readily in chemical reactions (also, reactivity). More formally, per reference [696]: As applied to a *chemical species*, the term expresses a kinetic property. A species is said to be more reactive or to have a higher reactivity in some given context than some other (reference) species if it has a larger rate constant for a specified *elementary reaction*. The term has meaning only by reference to some explicitly stated or implicitly assumed set of conditions. The term is not to be used for reactions or reaction patterns of compounds in general. The term is also more loosely used as a phenomenological description not restricted to elementary reactions; when applied in this sense the property under consideration may reflect not only rate but also equilibrium constants

Reactive ion etch (RIE) Combination of chemical and physical etch processes carried out in a plasma [656].

Reagents Chemicals used to carry out a reaction, usually excluding the starting molecule itself.

Reflectance Ratio of the amount of reflected-light to light-falling on the object.

Reflection Return of radiant energy (incident light) by a surface, with no change in wavelength.

Refraction Change of direction of propagation of any wave when it passes from one medium to another in which the wave velocity is different. Also, the bending of incident rays as they pass from one medium to another.

Refractive index Linear refractive index of a material, n, is defined as the ratio of light speed in vacuum, c, to the speed of light in the material, v.

Refractive index gradient Change in refractive index with respect to the distance from the axis of an optical fiber.

Regenerator Receiver–transmitter pair that receives a weak signal, reshapes it, then retransmits it.

Residual interaction Interaction between objects that do not carry a charge but do contain constituents that have charge. Although some chemical substances involve electrically charged ions, much of chemistry is due to residual electromagnetic interactions between electrically neutral atoms. The residual strong interaction between protons and neutrons, due to the strong charges of their quark constituents, is responsible for the binding of the nucleus [50].

Residual strong interactions Interaction between objects that do not carry a charge but do contain constituents that have charge. Residual strong interactions provide the nuclear binding force.

Resist Irradiation-sensitive chemical emulsion that is coated as a thin layer onto the surface of a wafer. The resist's chemical solubility (or etchability) depends on the mechanism of irradiation (UV light, X-rays, or electron beam). *See also* Photoresists.

Resonant tunneling diode (RTD) based devices Nanoelectronic quantum mechanics-based tunnel devices that have potential for a variety of high-speed electronic applications including terrahertz oscillators and logic circuits with switching speeds as low as 2 ps at room temperature.

The RTD structure is the electrical analog of the Fabry Perot resonator: Negative differential resistance regions arise due to the electrical confinement and coupling in one dimension. Resonant tunneling occurs when one of the quantum well bound states is monoenergetic with the input electrode Fermi level. Peaks in the electrical current as a function of bias voltage are thus observed in the current–voltage characteristics [404]. RTDs involve a device with two electrodes with two tunnel barriers between the electrodes. The barrier thickness is 10 nm or less, hence, the nanostructure nature of this device. The quantum well that is created by the confinement of the electron wave function between the two barriers produces a discrete set of allowed electron energy states in the QW; only when an electron from the electrode has an energy that corresponds to the allowed state in the QW can it tunnel through the two barriers and the QW (based on quantum mechanics principles) and reach the other electrode [26].

Resonator Cavity resonator utilizes resonance to amplify a wave. The cavity has interior surfaces that reflect one type of wave (specifically, a center frequency f_0); when a wave that is resonant with the cavity enters, it bounces back and forth within the cavity, with low loss. As more wave energy enters the cavity, it combines with and reinforces the standing wave, increasing its intensity (an example of cavity resonators include the klystron tube in a microwave oven, a laser cavity, and a tube of an organ).

Rest mass The rest mass (m) of a particle is the mass defined by the energy of the isolated (free) particle at rest, divided by c^2. When particle physicists use the word *mass*, they always mean the *rest mass* (m) of the object in question. The total energy of a free particle is given by $E=\sqrt{p^2c^2+m^2c^4}$ where p is the momentum of the particle. Note that for $p=0$ this simplifies to Einstein's equation $E=mc^2$. For a general particle with mass and momentum, it can also be written as $E=\gamma mc^2$ where $r=1/\sqrt{1-v^2/c^2}$. The quantity E includes both the mass energy and the kinetic energy [50].

RIE (reactive ion etching) Form of dry etching where ions are blasted at a wafer's surface.

Rutherford backscattering spectrometry (RBS) Microscopy tool. A beam of He^2 ions impinges on the target. The He ions backscatter from the near surface region of the sample and are collected by a solid-state detector. The energy of the backscattered ions provides information on both the composition and depth distribution of elements in the target. Alignment of the ion beam with sample crystallographic axes permits crystal damage to be measured quantitatively [472].

Rydbergs Unit of energy used in atomic physics, value = 13.605698 eV.

s Orbital Electron with no angular momentum is contained in a (spherical) s orbital.

Salts Ionic compounds that can be formed by replacing one or more of the hydrogen ions of an acid with another positive ion.

SAMs Self-assemble monolayers. An organic film is the self-assembled. In SAMs, long alkane thiols self-organize on gold surfaces and form highly ordered hydrophobic surfaces [712].

Scanning electron microscopy (SEM) Microscopy tool. A focused beam of electrons is rastered across a sample surface, the raster scan being synchronous with that of a cathode ray tube (CRT). The brightness of the CRT is modulated by the detected secondary electron current from the sample, such that the viewing CRT displays an image of the variation of secondary electron intensity with position on the sample. Different detectors can be used to provide alternative information, e.g., a backscattered electron detector will provide average atomic number information [472].

Scanning near-field optical microscopy Approach for observing optical characteristics of a surface that can be smaller than the wavelength of the light used.

Scanning near-field optical microscopy (SNOM) (or NSOM: near-field scanning optical microscopes) Method for observing local optical properties of a surface that can be smaller than the wavelength of the light used [661]. Scanning near-field optical microscopy (SNOM) is an imaging technique used to obtain resolution beyond the Abbé diffraction limit. The operational principle behind near-field optical imaging involves illuminating a specimen through a subwavelength sized aperture while keeping the specimen within the near-field regime of the source. If the aperture-specimen separation is kept roughly less than half the diameter of the aperture, the source does not have the opportunity to diffract before it interacts with the sample and the resolution of the system is determined by the aperture diameter as oppose to the wavelength of light used. An image is built up by raster scanning the aperture across the sample and recording the optical response of the specimen through a conventional far-field microscope objective [467].

Scanning probe microscope (SPM) General term used to describe scanning instruments that brings a probe (tip) very close to a surface of the sample (near 1 nm). Tool allows researcher to obtain information at the nanometric level. SPMs appeared in the 1980s. The first SPM was the scanning tunneling microscope (STM), invented by G. Binnig and H. Rohrer in 1981 (awarded Nobel prize in 1986). G. Binnig, C. F. Quate, and C. Gerber introduced another SPM in 1986: the atomic force microscope (AFM).

Scanning probe microscopy (SPM) Atomic force microscopy (AFM), scanning tunneling microscopy (STM), and magnetic force microscopy (MFM) are variations on a method of imaging surfaces with atomic or near-atomic resolution, collectively called SPM. A small tip is scanned across the surface of a sample in order to construct a 3D image of the surface. Fine control of the scan is accomplished using piezoelectrically induced motions. If the tip and the surface are both conducting, the structure of the surface can be detected by tunneling of electrons from the tip to the surface (STM). Any type of surface can be probed by the molecular forces exerted by the surface against the tip (AFM). The tip can be constantly in contact with the surface, it can gently tap the surface while oscillating at high frequency, or it can be scanned just minutely above the surface. By coating the tip with a magnetic material, the magnetic fields immediately above a surface can be imaged (MFM). Image processing software allows easy extraction of useful surface parameters [472].

Scanning thermal microscopy Approach method for observing temperatures and temperature gradients on a surface.

Scanning tunneling microscope (STM) Instrument able to image conducting surfaces to atomic accuracy. Instrument that uses a sharp conducting tip, and it applies a bias voltage between the tip and the sample. When the tip is brought close to the sample, electrons can "tunnel" through the gap between the sample to the tip (or from the tip to the sample).

Scattering Number of related phenomena by which particles are deflected by collisions with other particles.

Schrödinger equation Fundamental postulate of quantum mechanics, which states $H\Psi = E\Psi$, i.e., that the application of the Hamiltonian operator to the wave function of a system will give the energy of that system as the eigenvalue of the function. Thus, it relates the wave function to the allowed energies of the wave function [655].

Second law of thermodynamics $\Delta G = \Delta H - T\Delta S$ where ΔG is the change in Gibbs' free energy and ΔS is the change in entropy. The second law of thermodynamics states that for a reaction to be "spontaneous," ΔG must be negative overall. Another way of stating this is that for a reaction to be spontaneous, the overall entropy increase must be positive [655].

Secondary ion mass spectrometry (SIMS) Microscopy tool. A tool used in nanotechnology employed to characterize the surface and in-depth chemical composition of organic and semiconductor materials. An energetic primary ion beam sputters a sample surface; secondary ions formed in this sputtering process are extracted from the sample and analyzed in a double-focusing mass spectrometer system [472].

Self-assembled monolayer field-effect transistor (SAMFET) Transistor where a few molecules act as FETs, exhibiting very strong gain and rapid response.

Self-assembly Manufacturing method that uses chemical solutions and where the molecular joining of complementary surfaces results from the random motion of the molecules and the affinity of their binding sites. The process of self-assembly is a coordinated action of independent entities under distributed control that produces a larger structure or achieves a group effect [406]. Examples of self-assembly can be observed in biology (e.g., embryology and morphogenesis) and in chemistry, where groups of molecules form loosely bound supramolecular structures.

Self-replication (aka exponential replication) Process of growth involving doubling of the base entity within a given period.

SEM Scanning electron microscopy.

Semiconductor Material whose electrical conductivity is intermediate between that of metals (conductors) and insulators (nonconductors) and can be modified physically or chemically to increase or decrease its conductivity from a "normal" state by "dopants" [656]. A compound that has a conductivity value that is intermediate between that of a metal and an insulator and whose conductivity increases with temperature [655].

Semiconductor laser Laser where the injection of current into a semiconductor diode produces light by recombination of holes and electrons at the junction between p- and n-doped regions.

Semimetal Compound with a zero bandgap between the filled valence band and the empty conduction band [655].

Shells Where the electrons generally stay. There are four types of electron shells: s, p, d, and f shells.

Short-channel effects Undesired physical phenomena occurring in the scaled-down channel of a MOSFET; special measures must be taken to prevent short-channel effects; in other words certain "scaling rules" must be followed while scaling down MOSFETs geometry [39].

SI system/SI unit Systeme International d'Unites, an international system that established a uniform set of measurement units.

Sigma bond Type of covalent bond in which most of the electrons are located in between the nuclei. A bond between two atoms that consists of two electrons occupying a bonding molecular orbital with cylindrical symmetry about the internuclear axis. (Note this means that there are no nodes in the wave function as one rotates about the internuclear axis.) [655]

Signal-to-noise (S/N) ratio Ratio of the integrated energy within the passband envelope to the energy outside this envelope and within the free spectral range.

Silane (SiH_4) Gas that readily decomposes into silicon and hydrogen, silane is often used to deposit silicon-containing compounds. It also reacts with ammonia to form silicon nitride, or with oxygen to form silicon dioxide [656].

Silicide Film compound of silicon with a refractory metal. Common silicide semiconductor films (used as interconnects) include tantalum, tungsten, titanium, and molybdenum [656].

Silicon Brownish crystalline semimetal used to make most semiconductor wafers [656].

Silicon dioxide (SiO_2) Most common insulator in semiconductor device technology, particularly in silicon MOS/CMOS where it is used as a gate oxide [39].

Single mode Containing only one mode.

Single-electron transistors (SET) Quantum/nanoscale devices that have switching properties controlled by the injection or removal of one electron or through which only one electron can be transported at any one time [26].

Single-mode fiber Most common fiber used in telecommunication applications. Only one optical mode of light propagates. Transmission typically occurs at 1300–1550 nm.

Smart materials Materials capable of complex behavior due to the incorporation of nanomachines and nanocomputers.

Snell's law Principle that relates the angles of incidence and refraction to the refractive indices of the media. For example, if light is traveling in a medium, with index n_1, incident on a medium of index n_2, and the angles of incidence and refraction (transmission) are θ_i and θ_t, then $n_1 \sin \theta_i = n_2 \sin \theta_t$.

SOI (silicon on insulator) Silicon substrate of choice in future generation CMOS ICs; basically a silicon wafer with a thin layer of oxide (SiO_2) buried in it; devices are built into a layer of silicon on top of the buried oxide; SOI substrates provide superior isolation between adjacent devices in an integrated circuit as compared to devices built into bulk wafers (elimination of "latch-up" in CMOS devices); also, improved performance of SOI devices due to reduced parasitic capacitances [39].

Solid Form of matter where there is definite shape and volume. This form of matter is rigid. Volume: Solids can be compressed but only to an extremely limited degree.

Solitons Solitary solutions of the wave equation. Hence, a soliton is a wave packet that propagates without changing its shape. The first soliton ever observed was a running water wave in a river, maintaining its shape over a long distance. An optical soliton requires a nonlinear optical medium to provide the effect of *self-focusing* that compensates for the effects of *dispersion* and *diffraction*, which would alter the shape of the wave packet in the regular case. In photorefractive optics, spatial solitons have been discovered only recently [713]. An optical pulse that regenerates to its original shape at certain points as it travels along an optical fiber. Solitons can be combined with optical amplifiers to carry signals very long distances [257].

Solubility Amount of a substance that will dissolve per unit volume of a particular solvent.

Solute What is dissolved in a solution, e.g., the salt in saltwater.

Solution Mixture of a solid and a liquid where the solid never settles out, e.g., saltwater.

Solvent Liquid in which something is dissolved, e.g., the water in saltwater.

Specific heat Amount of heat it takes for a substance to be raised one degree Celsius.

Specific heat capacity Heat capacity of a system divided by its mass, (or also specific heat). It is a property solely of the substance of which the system is composed. As with heat capacities, specific heats are commonly defined for processes occurring at either constant volume (c_v) or constant pressure (c_p). Heat capacity (also called thermal capacity) is the ratio of the energy or enthalpy absorbed (or released) by a system to the corresponding temperature rise (or fall) [61, 62, 63, 64, 65, 66, 67].

Spectroscopy Study of the interaction of electromagnetic radiation with matter [655]. The study of spectra, that is, the dependence of a physical measure to frequency. A device for recording a spectrum is a spectrometer. Spectrometers are often used in physical and analytical chemistry for the identification of substances, through the spectrum emitted or absorbed. Spectroscopy was originally used to determine the chemical composition of materials; now new techniques have been added to detect structural characteristics [406].

Spin In the quantum mechanics context, it is the intrinsic angular momentum of a particle, given in units of $\hbar$, the quantum unit of angular momentum, where $\hbar = h/2\pi$ and $h = 6.62 \times 10^{-34}$ Js [50]. Spin is an intrinsic angular momentum associated with particles. Elementary particles, such as the electron, possess spin angular momentum, even though they are considered point particles; other subatomic particles, such as neutrons, which have no electrical charge, also have spin.

Spintronics Devices that rely on an electron's spin to perform their functions (short for spin-based electronics).

Spintronics (aka quantum spintronics, magnetoelectronics, spin electronics) Electronic science that exploit the spin of electrons as well as the electrons' charge (in conventional electronics is based on number of charges and their energy; spintronics is based on the direction of electron spin and spin coupling).

Spontaneous reaction Reaction that will proceed without any outside energy.

SQUID Superconducting quantum interference device.

SSI (small-scale integration) Chip contains transistors, but not hundreds.

Stable Does not decay. A particle is stable if there exist no processes in which a particle disappears and in its place two or more different particles appear [50].

Standard model Model for the theory of fundamental particles and their interactions. The standard model is used to describe the realm of subnuclear particles up to the current experimental limit energy, with probing distances as small as 10^{-18} m. The model contains 24 fundamental particles that are the constituents of matter: 12 species of elementary fermions (matter particles) and 12 species of elementary bosons (radiation particles), plus their corresponding antiparticles. It describes the strong, weak, and electromagnetic fundamental forces, using mediating bosons known as "gauge bosons." The species of gauge bosons are the photon, W− and W+ and Z bosons, and the gluons. The model predicts the existence of a type of boson known as the Higgs boson, but these are yet to be discovered. While it is widely tested and is currently accepted as correct by particle physicists, the standard model is currently perceived to be a provisional theory (until a more comprehensive theory is developed), also because it appears that there may be some elementary particles that are not properly described by the model (such as graviton—the hypothetical particle that carries gravitational force) [50, 51].

Standing wave (aka stationary wave) Wave that remains in a constant position. This occurs when the medium is moving in the opposite direction to the wave. A standing wave can also arise in a stationary medium as a result of interference between two waves traveling in opposite directions.

State property Quantity that is independent of how the substance was prepared. Examples of state properties are altitude, pressure, volume, temperature, and internal energy.

States of matter Solid, liquid, gas, and plasma. Plasma is a "soup" of diassociated nuclei and electrons, normally found only in stellar objects.

Steric hindrance Physical hindrance to something.

Steric relief Energetic stabilization afforded when steric strain is relieved during the course of a reaction [655].

Steric strain Molecule will be subject to steric strain if there are large bulky pieces of molecule forced together; this is an unfavorable state to be in [655].

Sterics General physical properties of molecules.

Stoichiometry Study of the relationships between amounts of products and reactants. A branch of chemistry that quantitatively relates amounts of elements and compounds involved in chemical reactions, based on the law of conservation of mass and the law of definite proportions. (Also it can refer to the ratios of atoms in a compound or to the ratios of moles of compounds in a reaction) [83]. The number of species of each type of element in the formula for a compound or reaction [655].

Strange quark (*S*) Third flavor of quark (in order of increasing mass), with electric charge $-\frac{1}{3}$ [50].

String theory In particle physics, string theory is a theory of elementary particles based on the idea that the fundamental entities are not pointlike particles but finite lines (strings), or closed loops formed by strings, the strings one-dimensional curves with zero thickness and lengths (or loop diameters) of the order of the Planck length of 10^{-35} m [31, 53]. String theory is a model with fundamental building blocks consisting of one-dimensional extended objects (strings); this is in contrast with the zero-dimensional points (particles) that were the basis of most physics at earlier times. It follows that string theories are able to avoid problems associated with the presence of pointlike particles in a physical theory. String theories do not just describe strings but other objects as well, including points, membranes, and higher-dimensional objects. String theory has not yet made the kind of predictions that would allow it to be experimentally tested.

Strong interaction Interaction responsible for binding quarks, antiquarks, and gluons to make hadrons. Residual strong interactions provide the nuclear binding force [50].

Strong, electroweak, and gravitational processes Fundamental forces comprise the gravitational force, the electromagnetic force, the nuclear strong force, and the nuclear weak force. The electroweak interactions are a unification of the electromagnetic and nuclear weak interactions and are described by the Weinberg–Salam theory (sometimes called "quantum flavor dynamics"; also called the Glashow–Weinberg–Salam theory) [31, 53].

Structural formula Diagram that shows how the atoms in a molecule are bonded together. Atoms are represented by their element symbols and covalent bonds are represented by lines. The symbol for carbon is often not drawn. Most structural formulas do not show the actual shape of the molecule (they are like floor plans that show the layout but not the 3D shape of a house) [83].

Subatomic particle (aka as elementary particle or subnuclear particle) Any particle that is small compared to the size of the atom.

Subnuclear particles Another name for elementary particles.

Substrate Wafer that is the basis for subsequent processing operations in the fabrication of semiconductor devices [656].

Subwavelength phenomena and plasmonic excitations There is research interest in surface electromagnetic waves and the extraordinary light transmittance (EOT) through an optically thick metal film that is perforated with subwavelength-size holes. EOT was first discussed in the late 1990s and then was intensively investigated [296]. These phenomena and properties in nanoengineered structures can be used as integrated elements in various optoelectronic and photonic devices, including quite sophisticated ones, such as optical computers [297]. In the optical and infrared spectral ranges, the excitation of the electron density coupled to the electromagnetic field results in a surface plasmon polariton (SPP) traveling on the metal surface. At the metal–air interface, the SPP is a wave, with the direction of the magnetic field parallel to the metal surface. In the direction perpendicular to the interface, SPPs exponentially decay in both media. The SPP can propagate not only on the metal surface but also on the surface of artificial electromagnetic crystals, for example, on wire-mesh crystals. Since the SPP propagation includes rearrangement of the electron density, its speed is less than the speed of light; as a result, the SPP cannot be excited by an electromagnetic wave impinging on a perfectly flat metal surface. The situation, however, changes when the film is modulated: when one of the spatial periods of the modulation coincides with the wavelength of the SPP, the latter can be excited by a normally incident electromagnetic wave [297].

Superconducting quantum interference device (SQUID) One of a family of devices capable of measuring extremely small currents, voltages, and magnetic fields. Based on two quantum effects in superconductors: (1) flux quantization and (2) the Josephson effect [51]. The device is utilized to measure extremely weak signals.

The SQUIDs are employed for a variety of engineering, medical (neuroscience, imaging), and geological testing applications that require very high sensitivity. Because SQUIDs measure changes in a magnetic field with such high sensitivity, they do not have to come in contact with the system under testing. A SQUID consists of miniature loops of superconductors employing Josephson junctions to achieve superposition. With superposition, each electron moves simultaneously in both directions. Because the current is moving in two opposite directions, the electrons have the ability to act as qubits in a quantum computer environment.

Superconductivity Phenomenon occurring in many metals and alloys. If these substances are cooled below a transition temperature, T_c, close to absolute zero, the electrical resistance becomes vanishingly small [51].

Superconductor Substance that conducts electricity with near-zero impedance.

Supercritical fluid By controlling temperature and pressure every substance can be set into a *supercritical* state. Supercritical fluid, as its called, is heavy like liquid but with the penetration power of gas. These qualities make supercritical fluids effective and selective solvents. Some see this as yet another state, in addition to the "better-known" three physical states: solid, liquid, and gas. Fluids become supercritical when their temperature and pressure are above the critical temperature and pressure. The critical temperature (TC) is unique for every fluid and is defined as the temperature above which a gas cannot be liquefied by simply increasing its pressure. The critical pressure (PC), also unique for every fluid, is the gas–liquid equilibrium pressure that corresponds to the critical temperature. The triple point is the point at which the gas, liquid, and solid phases all exist in equilibrium. Carbon dioxide (CO_2) has been selected as the solvent of choice for extrusion processing and may be used as the solvent for supercritical deposition techniques [708].

Superlattice nanowire Interwoven bundles of nanowires created from substances with different properties.

Superposition Quantum mechanical phenomenon where an entity exists in more than one state simultaneously.

Superprism effects Phenomenon is the extremely large angular dispersion experienced by a light beam when entering a photonic crystal. This arises from the anisotropy of the photonic band structure that can be present even in systems without a complete photonic bandgap [295]. These effects can be exploited for sensing and filtering applications.

Supersymmetry In particle physics supersymmetry is a symmetry that relates bosons and fermions. In supersymmetric theories, every fundamental fermion has a superpartner that is a boson of equal mass and vice versa. While supersymmetry has yet to be observed in the real world, it remains a key element of many proposed theories of physics, including various extensions to the standard model as well as modern superstring theories.

Supramolecular Composed of more than one molecule, more complex than one molecule. A system of two or more molecular entities held together and organized by means of intermolecular (noncovalent) binding interactions [696].

Surface micromachining Fabrication process for MEMS based on standard CMOS microelectronic processes. MEMS structures are photolithographically patterned in alternating layers of deposited polysilicon and silicon dioxide, and then are "released" by dissolving away the silicon dioxide layers.

Surface plasmon polariton (SPP) Wave with the direction of the magnetic field parallel to the metal surface.

Surface plasmon resonance (SPR) Phenomenon that occurs when light is reflected off thin metal films; this can be used in a plethora of applications in bioscientific industries.

Symmetry If a theory or process does not change when certain operations are performed on it, the theory or process is said to possess a symmetry with respect to those operations (e.g., a circle remains unchanged under rotation or reflection, and a circle therefore has rotational and reflection symmetry) [31, 53].

Symmetry breaking Term refers to the deviation from exact symmetry exhibited by many physical systems, and, in general, symmetry breaking encompasses both "explicit" symmetry breaking and "spontaneous" symmetry breaking. Explicit symmetry breaking is a phenomenon in which a system is not quite, but almost, the same for two configurations related by exact symmetry. Spontaneous symmetry breaking refers to a situation in which the solution of a set of physical equations fails to exhibit a symmetry possessed by the equations themselves [31, 53].

Synchrotron Type of circular accelerator in which the particles travel in synchronized bunches at fixed radius.

Synergic bond Formed by a ligand to a metal ion, which has a σ-bond from the ligand to the metal ion, and the back donation of electron density from the metal to the π^*orbitals on the ligand [655].

Tau lepton Third flavor of charged lepton (in order of increasing mass), with electric charge -1.

Temperature Thermal state of matter with respect to its ability to transfer heat to other matter. Heat is the energy that is transferred between matter by means of radiation, conduction, and/or convection. The common scales for measuring temperature are Celsius (centigrade), Fahrenheit and Kelvin.

Thermal conductivity Rate of heat flow divided by area and by temperature gradient.

Thermodynamics Study of heat and energy flow in chemical reactions.

Thin film Thin layer of substance on a substrate. Thickness of a film is generally in the range of 10–10 000 Å. X-ray diffraction, specular X-ray reflectivity, and nonspecular X-ray scattering are three techniques that provide information about a thin film. A thin layer of a substance deposited on an insulating base in a vacuum by a microelectronic process. Thin films are most commonly used for antireflection, achromatic beamsplitters, color filters, narrow passband filters, semitransparent mirrors, heat control filters, high-reflectivity mirrors, polarizers, and reflection filters [709]. Term used to describe a process of working with films of thicknesses less than a few microns.

Throughput Number of wafers per hour through a machine, assuming 100% equipment uptime and a fully loaded machine [656].

Time-of-flight secondary ion mass spectrometry (TOF-SIMS) Microscopy tool. A large area or microfocused pulsed primary ion beam sputters the top surface layer of the sample. The secondary ions produced in this sputtering process are extracted from the sample surface and injected into a specially designed time-of-flight mass spectrometer. The ions are dispersed in time according to their velocity, and the discrete packets of different massed ions are detected on either a microchannel plate (MCP) or resistive anode encoder (RAE) detector [472].

Titration Reacting a solution of unknown concentration with a solution of a known concentration for the purpose of finding out more about the unknown solution.

Top-down molding/manufacturing (aka mechanical nanotechnology) Fabricating and/or carving small materials and components by using larger objects such as tools and lasers, etc.

Top quark Sixth flavor of quark (in order of increasing mass), with electric charge $\frac{2}{3}$. Its mass is much greater than any other quark or lepton [50].

Top-down Molding, carving, and fabricating small materials and components by using larger objects such as our hands, tools and lasers, respectively [661].

Total reflection X-ray fluorescence (TXRF) Microscopy tool. X-rays from a tungsten anode or molybdenum tube impinge the sample surface at a glancing angle, within the critical angle for total external reflectance, and excite the electrons on atoms in the top few monolayers of the sample, causing them to emit photons (fluoresce). The X-ray photons emitted by the surface atoms have energies that are characteristic of the particular element [472].

Track Record of the path of a particle traversing a detector.

Tracking Reconstruction of a "track" left in a detector by the passage of a particle through the detector.

Transistor On/off electronic switch of microscopic size (micron and submicron range) and high switching speed (mega- or gigahertz range). A transistor consists of three layers of a semiconductor material: of a source (where electrons originate from), a drain (where electrons go), and a gate that controls the flow of electrons through a channel that connects the source and the drain. Two types of transistors exist: the junction transistor (aka, bipolar transistor) and the field-effect transistor (FET).

An active component of an electronic circuit consisting of a small block of semiconducting material to which at least three electrical contacts are made, usually two closely spaced rectifying [i.e., converting an alternating current (AC) to a unidirectional one] contacts and one ohmic (nonrectifying) contact; it may be used as an amplifier, detector, or switch [689]. The transistor was invented at Bell Telephone Laboratories in 1947. Texas Instruments built the first integrated circuit (IC) in 1958 using germanium (Ge) devices. Later that same year Fairchild Semiconductor announced the development of a planar double-diffused Si IC. The complete transition from the original Ge transistors with grown and alloyed junctions to silicon (Si) planar double-diffused devices took about a decade. The success of Si as an electronic material was due partly to its wide availability from silicon dioxide (SiO_2) (sand), resulting in potentially lower material costs relative to other semiconductors [258].

Transistor gate Portion of the transistors that switch between blocking electric current and allowing it to pass [19].

TSI (tera-scale integration) Processor or chip has over one trillion (10^{12}) transistors, but less than 1 quadrillion (10^{15}).

Tunneling Penetration of particles into areas of potential that classically they would be forbidden to enter.

Type theory Branch of mathematics and logic that concerns itself with classifying entities into sets called types. Modern type theory was invented partly in response to Russell's paradox, and features prominently in Russell and Whitehead's *Principia Mathematica*.

ULSI (ultra-large-scale integration) Microchip has over one million transistors.

Ultrathin-body SOI Silicon-on-insulator transistor that is built in an extremely thin silicon body (<20 nm) with underlying oxide layer. The fully depleted condition means that the thin silicon body or entire substrate is depleted of mobile carriers under all operating bias conditions. In contrast, a partially depleted SOI CMOS transistor has a quasi-neutral body or substrate region under all operating bias conditions. This quasi-neutral body is left floating with no external electrical connections (floating-body effect) [401].

Uncertainty principle Quantum principle, first formulated by Heisenberg, that states that it is not possible to know exactly both the position x and the momentum p of an object at the same time. $\Delta x \Delta p \geq h$. It can be written as $\Delta E\ \Delta t \geq h$ where ΔE means the uncertainty in energy and Δt the uncertainty in lifetime of a state [50].

Up quark Least massive flavor of quark, with electric charge $\frac{2}{3}$.

Valence Number of electrons needed to fill out the outermost shell of an atom. Example: a carbon atom has 6 electrons, with an electron shell configuration of $1s^2 2s^2 2p^2$. Hence, carbon has a valence of 4, since 4 electrons can be accepted to fill the 2p orbital. Some exceptions exist; therefore, the more general definition of valence is the number of electrons with which a given atom generally bonds or number of bonds an atom forms (e.g., iron may have a valence of 2 or a valence of 3).

Valence electrons Electrons in the outermost shell of an atom.

van Der Waals force Force acting between nonbonded atoms or molecules. Includes the following forces: dipole–dipole, dipole–induced dipole, and London forces.

Vapor deposition (VAD) Chemical vapor deposition (CVD) process utilized for making preforms where a chemical reaction of silicon vapors doped with other chemicals forms submicroscopic particles that deposit on the surface of a starting object.

VCSEL Vertical-cavity surface-emitting laser; a type of laser that emits light vertically out of the element, not out the edge.

Vector spaces Generalization of the set of all geometrical vectors; concept employed throughout mathematics. Fundamental concept in linear algebra. A vector is informally described as an object with a "magnitude" and "direction"; more formally defined by its relationship to the spatial coordinate system under rotations. Can also be defined in a coordinate-free manner via a tangent space of a three-dimensional manifold in the language of differential geometry. Vector fields are utilized in physics to model, e.g., the strength and direction of some force, such as the magnetic or gravitational force, as it changes from point to point.

Vertical transistor Transistor has surface conduction channels on two or more vertical surfaces. Current flows in the vertical direction so channel length, given by the vertical separation between source and drain, is determined by the thickness of an epi layer, not lithography [401].

Via Holes through dielectric layers, opened by etching. Metal will be deposited in the via to form a plug and create an interconnect between two metal lines [656].

Virtual particle Particle that exists only for an extremely brief instant as an intermediary in a process. The intermediate or virtual particle stages of a process cannot be directly observed. If they were observed, one might think that conservation of energy was violated; however, the Heisenberg uncertainty principle allows an apparent violation of the conservation of energy. If one sees only the initial decaying particle (such as a meson with the c quark) and the final decay products, one observes that energy is conserved. The *virtual* particle (such as the $W^{\pm}$) exists for such a short time that it can never be observed [50].

Virtual photons Particles that do not have a permanent existence are called virtual particles (aka, vacuum fluctuations.) Virtual particles always come in pairs: a particle and antiparticle. Virtual particles exist for an extremely short time: virtual particles are always created as a particle–antiparticle pair, and they mutually annihilate each other in short order (in some instances, however, it is possible to boost the pair apart using external energy so that they avoid annihilation and become real particles).

Antiparticle pairs (and virtual photons) form out of nothing and then vanish back into nothing an instant later. In the description of the interaction between elementary particles in quantum field theory, a virtual particle is a temporary elementary particle, used to describe an intermediate stage in the interaction.

The QED method studies the interactions of charged particles with the electromagnetic field, describing mathematically not only all interactions of light with matter but also those of charged particles with one another. (QED is a relativistic theory in that Albert Einstein's theory of special relativity is built into each of its equations.) QED rests on the concept that charged particles (e.g., electrons and positrons) interact by emitting and absorbing photons. These photons are virtual; i.e., they cannot be seen or detected in any way because their existence violates the conservation of energy and momentum.

VLSI (very-large-scale integration) Processor has on the order of 100,000 or more transistors, but not over a million.

$W^{\pm}$ **bosons** Carrier particle of the weak interactions; it is involved in all electric-charge–changing weak processes.

Wafer Thin (thickness depends on wafer diameter, but is less than 1 mm), circular slice of pure silicon on which semiconductors are built [656]. Round, thin slices of silicon that form the base substrate for semiconductor processing. Typical diameter sizes include 4-, 5-, 6-, 8-, and 12-inch (300 mm).

Circular slice of single-crystal semiconductor material used in manufacturing of semiconductor devices and integrated circuits; depending on material wafer diameter may range from about 25 to 300 mm; cut from the ingot of single-crystal semiconductor [39].

Wave vector Wave vector of an electron is proportional to its momentum. Its direction indicates the direction of motion of the electron, while its magnitude is $2\pi/\lambda$, where λ is the electron wavelength. Wave vectors have units of 1/length and are therefore plotted in reciprocal space or *k*-space [659].

Wave function Function of position and time that is a solution to a differential equation. The square of the wave function that is the solution to the Schrödinger equation predicts a probability density [654].

Waveguide Structure that guides electromagnetic waves along its length. An optical fiber is an optical waveguide.

Waveguide dispersion Portion of chromatic dispersion that arises from the different speeds light travels in the core and cladding of a single-mode fiber.

Wavelength Distance between successive crests, troughs, or identical parts of a periodic wave. Also, in quantum terms, the electron wavelength, λ, is related to electron energy via the de Broglie relationship.

Wavelength division multiplexing (WDM) Multiplexing of optical signals by transmitting them at different wavelengths through the same fiber.

Wavelength–momentum relation Wavelength = Planck's constant/momentum, namely, $\lambda = h$/momentum.

Weak interaction Interaction responsible for all processes where "flavor" changes, hence, for the instability of heavy quarks and leptons and particles that contain them. Flavor is the name used for the different quark types (up, down, strange, charm, bottom, top) and for the different lepton types (electron, muon, tau). Hence, flavor is the quantum number that distinguishes the different quark/lepton types. Each flavor of quark and charged lepton has a different mass [50].

Window Some of the bands are: the S band is defined in the range 1280–1350 nm. The C band is defined in the range 1528–1565 nm. The L band is defined in the range 1561–1620 nm (inspired by the optical attenuation of the fiber).

X-ray diffraction Atomic planes of a crystal cause an incident beam of X-rays (if wavelength is approximately the magnitude of the interatomic distance) to interfere with one another as they leave the crystal. The phenomonen is called X-ray diffraction [659].

X-ray fluorescence (XRF) Nondestructive technique that can be used to quantify elemental constituents of solid and liquid samples. X-rays from a rhodium X-ray tube excite atoms in the sample, causing them to emit X-rays with energies characteristic for each element. The intensity and energy of the emitted X-rays are measured using a lithium-drifted silicon detector and multichannel analyzer electronics [472].

X-ray photoelectron spectroscopy (XPS), electron spectroscopy for chemical analysis (ESCA) Microscopy tool. Approach where samples are irradiated with monochromatic X-rays that cause the ejection of photoelectrons from the surface. The electron binding energies, as measured by a high-resolution electron spectrometer, are used to identify the elements present and, in many cases, provide information about the valence state(s) or chemical bonding environment(s) of the elements thus detected [472].

Yield (IC fabrication) Percentage of wafers or die produced in a process that conforms to specifications [656].

Z^0 bosons Carrier particle of weak interactions; it is involved in all weak processes that do not change flavor.

Zero dispersion wavelength Wavelength at which the net chromatic dispersion of an optical fiber is nominally zero. This arises when waveguide dispersion cancels out material dispersion.

References

[1] C. M. Vest, Societal Implications of Nanoscience and Nanotechnology, National Science Foundation Workshop, Sept. 28–29, 2000. National Science Foundation. http://www.wtec.org/loyola/nano/NSET.Societal.Implications/nanosci-appendices.pdf.

[2] National Nanotechnology Initiative, federal R&D program established to coordinate the multiagency efforts in nanoscale science, engineering, and technology, National Nanotechnology Initiative; Research and Development Supporting the Next Industrial Revolution, Supplement to President's FY 2004 Budget, Oct. 2003, http://www.nano.gov.

[3] Forbes Nanotech Report, www.Forbes-Nanotech.net. Online Journal. 2003–2005 material.

[4] S. Lenhert, Introduction to Nanotechnology, Nanoword.net Newsletter, Quanteq, Interface Physics Department of the University of Münster, Germany.

[5] J. Walker, Nanotechnology in Manufacturing, Fourmilab, Switzerland, Jan. 15, 2004.

[6] Asian Technology Information Program (ATIP), ATIP99.061, Commercial & Industrial Applications, for Micro Engineering & Nanotechnology, London, Apr. 26–27, 1999.

[7] R. Merke, Nanotechnology, http://www.zyvex.com/nano/.

[8] Staff, Israelis Bring Nanoscale Motors Move Closer to Reality, *Nanotechnology News*, Mar. 31, 2004.

[9] Z. H. Zhou, Biologically Inspired Nanotechnology, Institute of Biomaterials and Biomedical Engineering, E. S. Rogers, Sr., Department of Electrical and Computer Engineering, University of Toronto, Toronto, Ontario, Canada, http://www.utoronto.ca.

[10] P. Falstad and T. Talbert, A Sense of Scale, A Visual Comparison of Various Distances, http://www.falstad.com/scale/.

[11] N. C. Seeman and A. M. Belcher, On Building Nanostructures from the Bottom Up, *Proc. Natl. Acad. Sci.* **99**, 6451 (2002).

[12] D. Leff, The Truth About Nanotechnology, ZDNet Australia, Dec. 6, 2002, info@zdnet.com.au.

[13] M. Kanellos, HP to Unveil Nanotech Breakthrough, ZDNet, Sept. 9, 2002, ZDNet Australia, Pyrmont, Australia, info@zdnet.com.au.

[14] C. Warris, *Nanotechnology Benchmarking Project*, Australian Academy of Science, Feb. 2004.

[15] S. Olson, Nanotechnology Conference of March 31–April 2, 2004 Emphasizes Molecular Electronics, NanoApex Corp./RAWNET Co., Cleveland, OH, News.nanoapex.com.

Nanotechnology Applications to Telecommunications and Networking, By Daniel Minoli

[16] Promotional Material, Darmstadt University of Technology, Institute of Materials Science, Thin Films Division, Darmstadt, Germany.

[17] E. Giannelis, Discovering Materials Science and Engineering, Department of Materials Science and Engineering, Cornell University, Ithaca, NY, http://www.mse.-cornell.edu.

[18] DFG-Center for Functional Nanostructures(CFN), Universität Karlsruhe (TH), Karlsruhe, Germany, http://www.cfn.uni-karlsruhe.de/projects.html.

[19] G. Dan Hutcheson, The First Nanochips, *Sci. Am.*, Apr. 2004, pp. 76–83.

[20] University of California at Berkeley, MEMS and Nano, Promotional Material.

[21] L. Gasman, What Will the Future Nanotech Industry Look Like? NanoMarkets White Paper, Mar. 2004, Glen Allen, VA, lawrence@nanomarkets.net.

[22] Staff, The Foundations of Next-Generation Electronics: Condensed-Matter Physics at RLE, *RLE Currents* **10**(2) (1998).

[23] V. Ho, D. Scansen, and E. Keyes, The 1 Billion Transistor Processor: Who Will Be First? Semiconductor International, Mar. 1, 2003.

[24] P. Clarke, Breaching the Great Divorce, *EE Times*, Nov. 24, 2003.

[24a] Staff, Intel: First 65nm Chips. Problems and Prospects, International Nano Business Directory, Dec. 9, 2003, NanoVIP.com.

[25] Promotional Materials, Intel, Santa Clara, CA.

[26] D. J. Paul, Nanoelectronics, Cavendish Laboratory Paper, University of Cambridge, Madingley Road, Cambridge, U.K.

[27] Electronics and Electrical Engineering, University of Glasgow, Glasgow, Enquiries@elec.gla.ac.uk.

[28] C. Lieber, *Science*, Nov. 8, 2001.

[29] Advances in Nanoelectronics by Nanosys Scientific Founders Honored as the Breakthrough of the Year by Science Magazine; Nanosys Founder and Harvard University Professor, Charles Lieber, Wins the Feynman Prize for Nanotechnology, *PR Newswire*, Jan. 12, 2002.

[30] N. Mokhoff, Nanotech Forum Reflects on Technology's Mission, *EE Times*, Sept. 11, 2003.

[31] *Science Week*, ScienceWeek/Spectrum Press, Chicago, IL, editors@scienceweek.com.

[32] G. D. Skidmore, E. Parker, M. Ellis, N. Sarkar, and R. Merkle, Exponential Assembly, Zyvex Corporation, Richardson, TX.

[33] Y. Tzeng, ELEC 7970: Special Topics in Electrical Engineering on Nanoscale Science and Technology, Outline, Summer 2003, Electrical and Computer Engineering, Auburn University, Auburn, AL, tzengy@eng.auburn.edu.

[34] *The Photonics Dictionary*, Photonics, Laurin Publishing Co., Pittsfield, MA, photonics@laurin.com, www.Photonics.com.

[35] *Sci. Mag*. **302** (2003).

[36] Nanotechnology Glossary, Smalltimes, Small Time Media, Ann Arbor, MI.

[37] Nanotechnology Research Directions, IWGN Workshop Report, Sept. 1999.

[38] Quantum Glossary, NASA/IPAC Extragalactic Database (NED), Jet Propulsion Laboratory, California Institute of Technology.

[39] Semiconductor Glossary, Semiconductor OneSource, Prosto/J. Ruzyllo, 2002.

[40] M. A. Reed, Quantum Dots, *Sci. Am.*, Jan. 1993, pp.118–123.

[41] M. A. Kastner, Artificial Atoms, *Phys. Today*, Jan. 1993, pp. 24–31.

[42] Glossary, *Nanotechnology Now*, 7th Wave Inc., http://www.nanotech-now.com/about.htm.

[43] J. M. Elzerman, R. Hanson, J. S. Greidanus, L. H. Willems van Beveren, S. De Franceschi, L. M. K. Vandersypen, S. Tarucha, and L. P. Kouwenhoven, Mesoscopic Systems and Quantum Hall Effect, *Phys. Rev. B* **67**, 161308(R) (2003).

[44] Staff, Quantum Dots, The Delft Spin Qubit Project, Kavli Institute of Nanoscience Delft, The Faculty of Applied Sciences at Delft University of Technology, Delft, http://qt.tn.tudelft.nl/research/qdots/.

[45] A. Gibson, NSF Grant Funds Ohio University Nanotechnology Research, Outreach Programs, Press Release, Ohio University, Nov. 20, 2003.

[46] N. Mokhoff, U.S. Official Calls for Closer Cooperation on Nanotechnology, *EE Times*, Dec. 9, 2003.

[47] P. Di Justo, "Big Bucks From Little Science," *Wired Magazine*, May 23, 2002.

[48] R. C. Johnson, President Signs Nanotechnology R&D Act Into Law, *EE Times*, Dec. 8, 2003.

[49] Quantum Physics, Definitional Material. Academy of Science of Saint Louis, Saint Louis, MO.

[50] Physics Department, *Glossary of High-Energy Physics Terms*, Boston University, Boston, MA.

[51] Quantum Glossary, National Physical Laboratory, Teddington, Middlesex, U.K., enquiry @npl.co.uk.

[52] D. Gammon and D. G. Steel, On Quantum Dots, *Phys. Today*, Oct. 2002.

[53] F. Wilczek, Theoretical Physics: On Limitations, *Phys.* Today, Jan. 2004.

[54] U.S. Department of Energy, Office of High-Energy Physics, SC-20/Germantown Building, U.S. Department of Energy, Washington, DC.

[55] Particle Physics, Wikipedia, http://en.wikipedia.org/wiki/Particle_physics.

[56] F. Halzen and A. Martin, *Quark and Leptons*, Wiley, New York, 1984.

[57] D. Griffiths, *Introduction to Elementary Particles*, Wiley, New York, 1987.

[58] D. Perkins, *Introduction to High Energy Physics*, Addison-Wesley, Reading, MA, 2000.

[59] R. F. W. Bader, *An Introduction to the Electronic Structure of Atoms and Molecules*, Clarke Irwin, 1970.

[60] R. F. W. Bader, *Theory of Atoms in Molecules*, Oxford University Press, 1995.

[61] A. Vardhan and E. Generalic, Ed., *Glossary of Chemical Concepts*, Faculty of Chemical Technology, KTF, Split, Croatia.

[62] Granta Design Limited, Glossary of Materials Attributes, Granta Design Ltd., Cambridge, U.K.

[63] K. Kalpakis, K. Markowitz, D. Zaelke, S. Jamar, S. Hoban, R. Medina, and N. Kozura, Eds., *Environmental Legal Information Systems Glossary.*

[64] A-M. Birac, Chief Editor, Clefs CEA no 44, Commissariat a L'Energie Atomique, France.

[65] T. S. Glickman, Ed., *Glossary of Meteorology*, 2nd ed., American Meteorological Society, Boston, MA, 2000.

[66] J. A. Dutton, *Dynamics of Atmospheric Motion*, Dover Press, 1995.

[67] A. Sommerfeld, *Thermodynamics and Statistical Mechanics*, Academic Press, New York, 1964, p. 45.

[68] D. Watson, *A Visual Interpretation of the Table of Elements*, Strathclyde University, Glasgow, Scotland, maintained by the Royal Society of Chemistry (RSC).

[69] M. Winter, WebElements Periodic Table: Group Numbers, The University of Sheffield and WebElements Ltd., U.K.

[70] International Union of Pure and Applied Chemistry, Commission on the Nomenclature of Inorganic Chemistry, *Nomenclature of Inorganic Chemistry Recommendations 1990*, G. J. Leigh, Ed., Blackwell Science, Oxford, 1990.

[71] W. C. Fernelius and W. H. Powell, Confusion in the Periodic Table of the Elements, *J. Chem. Ed.* **59**, 504–508 (1982).

[72] Periodic Table, any volume of *Inorganic Oxford Chemistry Primers*, Oxford University Press, Oxford, U.K.

[73] Promotional Material, Quantum Dot Corporation, Hayward, CA, http://www.qdots.com.

[74] G. M. Whitesides and J. C. Love, On Nanofabrication and Nanotechnology, *Sci. Am.* Sept. 2001.

[75] Staff, Duke University Chemists Describe New Kind of "Nanotube" Transistor, *AScribe Newswire*, Mar. 29, 2004.

[76] N. Stutzmann et al., Self-Aligned, Vertical-Channel, Polymer Field-Effect Transistors, *Science* **299**, 1881 (2003).

[77] R. Mezzenga et al., Templating Organic Semiconductors via Self-Assembly of Polymer Colloids, *Science* **299**, 1872 (2003).

[78] M. G. Peters, S. G. den Hartoga, and J. I. Dijkhuisb, Single Electron Tunneling and Suppression of Short-Channel Effects in Submicron Silicon Transistors, *J. Appl. Phys.* **84** (91) (1998).

[79] D. Foty and M. Bucher, New Techniques for Modern Analog Design: CMOS Design Methodologies and the EKV MOSFET Model, 9th IEEE International Conference on Electronics, Circuits and Systems—ICECS 2002, Sept. 15–18, 2002, Dubrovnik, Croatia.

[80] Y. J. Choi, CMOSFET Below 0.1 μm: Suppression of Short Channel Effect, SMDL Annual Report, School of Electrical Engineering, Seoul National University, 1998.

[81] M. Blaber, *Chemistry 1—A Virtual Textbook*, Florida State University, Tallahassee, FL, 1998.

[82] D. Ebbing and S. Gammon, *General Chemistry*, 6th ed., Houghton Mifflin Company, Boston/New York, 1999.

[83] F. Senese, Glossary, Frostburg State University, Frostburg, MD.

[84] http://home.att.net/%7Ecat6a/carb_bonds-I.htm.

[85] F. Senese, What Are van der Waals forces? Frostburg State University, Frostburg, MD.

[86] V. Tohver et al., A New Colloid Stabilization Mechanism, *Proc. Natl. Acad. Sci.* **98**, 8950 (2001).

[87] Staff, Catalysis, *ScienceWeek* **7**(24A) (2003).

[88] Carbon Nanotechnologies, Houston, TX, BusDev@cnanotech.com, http://www.cnanotech.com.

[89] D. Colberts, Single-Wall Nanotubes: A New Option for Conductive Plastics and Engineering Polymers, in *Plastics Additives and Compounding*, Elsevier Science, 2003.

[90] ILL's ICSD-for-WWW database. Originally featured in *Science*, Mar. 12, 1999, and *La Recherche*, Oct. 1999. Produced by Marcus Hewat's 3D VRML structure drawing application xtal-3d, written during 1994 summer student work experience. The database is used by scientists at the European High Flux Reactor, Synchrotron and elsewhere for studying the atomic structure of materials. The latest version of these pages is on theILL's 3D structure gallery, and more 3D computer graphics are on Marcus's WWW pages and on those of his employer, SERA-CD designers of concept cars.

[91] P. J. Herer, Ferromagnetic Nanostructured Materials, Soft Magnetic Nanocrystalline Alloys, World Technology Evaluation Center (WTEC), managed by Loyola College through a cooperative agreement with NSF, 1998.

[92] A. Maiti, Multiscale Modeling in Nanotechnology, 2004 NSTI Nanotechnology Conference and Trade Show, Nanotech 2004, Mar. 7–11, 2004, Boston, MA.

[93] S. Heinze, J. Tersoff, and Ph. Avouris, Electrostatic Engineering of Nanotube Transistors for Improved Performance, *Appl. Phys. Lett.* **83**(24), 5038 (2003).

[94] T. Dürkop, E. Cobas, and M. S. Fuhrer, High-Mobility Semiconducting Nanotubes, *AIP Conf. Proc.* **685**(1), 524 (2003).

[95] A. P. Graham, G. S. Duesberg, F. Kreupl, R. Seidel, M. Liebau, E. Unger, and W. Hönlein, Towards the Integration of Carbon Nanotubes in Microelectronics, *AIP Conf. Proc.* **685**(1), 587 (2003).

[96] Intel enlists Nanosys to Probe Nanomemory Research, *Silicon Strategies*, Jan. 15, 2004.

[97] Center for Nanoscale Science and Technology at Rice University.

[98] S. Pratsinis, Nanoparticles: Synthesis and Applications, 2004 NSTI Nanotechnology Conference and Trade Show, Nanotech 2004, Mar. 7–11, 2004, Boston, MA.

[99] S. Iijima, Helical Microtubules of Graphitic Carbon, *Nature* **354**, 56 (1991).

[100] S. Iijima and T. Ichihashi, Single-Shell Carbon Nanotubes of 1-nm Diameter, *Nature* **363**, 603 (1993).

[101] D. S. Bethune, C. H. Kiang, M.S. DeVries, G. Gorman, R. Savoy, and R. Beyers, Cobalt-Catalysed Growth of Carbon Nanotubes with Single-Atomic-Layer Walls, *Nature* **363**, 605 (1993).

[102] NanoteC'03: Nanotechnology in Carbon and Related Materials, University of Sussex, Brighton, U.K., Aug. 27–30, 2003.

[103] M. Radosavljevi, J. Appenzeller, V. Derycke, R. Martel, Ph. Avouris, A. Loiseau, J. L. Cochon, and D. Pigache, Electrical Properties and Transport in Boron Nitride Nanotubes, *Appl. Phys. Lett.* **82**(23), 4131 (2003).

[104] S. Talbot, Plant to Produce Kilograms of EviDots Semiconductor Quantum Dot Crystals per Week, Evident Technologies, Troy, NY, Press Release, Mar. 18, 2003, stalbot@ evidenttech.com.

[105] C. C. Koch, A Top-Down Synthesis of Nanostractured Materials: Mechanical and Thermal Processing Methods, *Rev. Adv. Mater. Sci.* **5**, 91–99 (2003).

[106] H. Engelkamp, S. Middelbeek, and R. J. M. Nolte, *Science* **284**, 785 (1999).

[107] J. A. DeAro, K. D. Weston, and S. K. Buratto, Near-Field Scanning Optical Microscopy of Nanostructures, Department of Chemistry, University of California, Santa Barbara, CA.

[108] T. W. Ebbesen, H. J. Lezec, H. Hiura, J. W. Bennett, H. F. Ghaemi, and T. Thio, Electrical-Conductivity of Individual Carbon Nanotubes, *Nature* **382**, 54 (1996).

[109] J. Qiu, Z. Ying, Z. Fan, W. Linna, S. C. E. Tsang, and P. J. F. Harris, Carbon Nanomaterials and Fullerenes from Chinese Coals, NanoteC'99 Conference, Sept. 8–10, 1999, University of Sussex at Brighton, U.K.

[110] D. Tomanek, Morphology, Growth and Destruction of Carbon Nanotubes, American Physical Society (APS) March Meeting, Kansas City, MO. 1997.

[111] H. R. Sadeghpour, B. E. Granger, and P. Král, Interaction of Laser Light and Electrons with Nanotubes, 2004 NSTI Nanotechnology Conference and Trade Show, Nanotech 2004, March 7–11, 2004, Boston, MA.

[112] J. W. Mintmire, B. I. Dunlap, and C. T. White, Are Fullerene Tubules Metallic? *Phys. Rev. Lett.* **68**, 631 (1992).

[113] N. Hamada, S. Sawada, and A. Oshiyama, New One-Dimensional Conductors—Graphitic Microtubules, *Phys. Rev. Lett.* **68**, 1579 (1992).

[114] R. Saito, M. Fujita, G. Dresselhaus, and M. S. Dresselhaus, Electronic Structure of Graphene Tubules Based on C60, *Phys. Rev. B* **46**, 1804 (1992).

[115] Y. Ohno, S. Kishimoto, T. Mizutani, T. Okazaki, and H. Shinohara, Chirality Assignment of Individual Single-Walled Carbon Nanotubes in Carbon Nanotube Field-Effect Transistors by Micro-Photocurrent Spectroscopy, *Appl. Phys. Lett.* **84**(8), 1368 (2004).

[116] P. McEuen, McEuen Group Research, Carbon Nanotubes—A New Class of 1D Conductors, talk at ITP, UCSB, Sept. 17, 1998.

[117] M. P. Anantram, Which Nanowire Couples Better Electrically to a Metal Contact: Armchair or Zigzag Nanotube? *Appl. Phys. Lett.* **78**(14), 2055 (2001).

[118] J. Appenzeller and D. J. Frank, Frequency Dependent Characterization of Transport Properties in Carbon Nanotube Transistors, *Appl. Phys. Lett.* **84**(10), 1771 (2004).

[119] G. Overney, W. Zhong, and D. Tománek, Structural Rigidity and Low Frequency Vibrational Modes of Long Carbon Tubules, *Z. Phys. D* **27**, 93 (1993).

[120] NanoteC'99, British Carbon Group, Nanotechnology in Carbon and Related Materials, Conference Sept. 8–10, 1999, University of Sussex at Brighton, U.K.

[121] Y. Umeno and T. Kitamura, Ab Initio Simulation on Mechanical and Electronic Properties of Nanostructures Under Deformation, 2004 NSTI Nanotechnology Conference and Trade Show, Nanotech 2004, Mar. 7–11, 2004, Boston, MA.

[122] R. Saito, G. Dresslhaus, and M. S. Dresslhaus, *Physical Properties of Carbon Nanotubes*, Imperial College Press, London, 1998.

[123] S. Roche and R. Saito, On Magnetoresistance of Carbon Nanotubes, *Phys. Rev. Lett.* **87**, 246803 (2001).

[124] P. Singer, Nanotube Alchemy: Metal to Semiconductor and Back, *Semiconductor International*, July 1, 2004.

[125] R. H. Baughman et al., Carbon Nanotubes—The Route Toward Applications, *Science* **297**, 787 (2002).

[126] A. Thess, R. Lee, P. Nikolaev, H. Dai, P. Petit, J. Robert, C. Xu, Y. H. Lee, S. G. Kim, A. G. Rinzler, D. T. Colbert, G. Scuseria, D. Tománek, J. E. Fischer, and R. E. Smalley, Crystalline Ropes of Metallic Carbon Nanotubes, *Science* **273**, 483 (1996).

[127] A. Hassanien, M. Holzinger, A. Hirsch, P. Venturini, A. Prodan, M. Tokumoto, and P. (2002).

[128] XVIIIth International Winterschool on Electronic Properties of Novel Materials, Euroconference Molecular Nanostructures, Kirchberg, Tirol, Austria.

[129] C. Li and T-W. Chou, Vibrational Behaviors of Multiwalled-Carbon-Nanotube-Based Nanomechanical Resonators, *Appl. Phys. Lett.* **84**(1), 121 (2004).

[130] M. Yudasaka, H. Kataura, T. Ichihashi, L. C. Qin, S. Kar, and S. Iijima, Diameter Enlargement of HiPco Single-Wall Carbon Nanotubes by Heat Treatment, *Nano Letters* **1**(9), 487–489 (2001).

[131] B. H. Hong et al., Fabrication of Silver Nanowires, *Science* **294**, 348 (2001).

[132] S. J. Tans, M. H. Devoret, H. Dai, A. Thess, R. E. Smalley, L. J. Geerligs, and C. Dekker, Individual Single-Wall Carbon Nanotubes as Quantum Wires, *Nature* **386**, 474 (1997).

[133] A. Javey, M. Shim, and H. Dai, Electrical Properties and Devices of Large-Diameter Single-Walled Carbon Nanotubes, *Appl. Phys. Lett.* **80**(6), 1064 (2002).

[134] P. C. Collins, M. S. Arnold, and P. Avouris, Engineering Carbon Nanotubes and Nanotube Circuits Using Electrical Breakdown, *Science* **292**, 706 (2001).

[135] A. Bachtold, P. Hadley, T. Nakanishi, and C. Dekker, Logic Circuits with Carbon Nanotubes, *AIP Conf. Proc.* **633**(1), 502 (2002).

[136] Y. Cui and C. M. Lieber, Functional Nanoscale Electronic Devices Assembled Using Silicon Nanowire Building Blocks, *Science* **291**, 891–893 (2001).

[137] E. Smalley, Coax Goes Nano, *Technology Research News (TRN)*, Nov. 13/20, 2002.

[138] B. Lamprecht, J. R. Krenn, G. Schider, H. Ditlbacher, M. Salerno, N. Felidj, A. Leitner, F. R. Aussenegg, and J. C. Weeber, Surface Plasmon Propagation in Microscale Metal Stripes, *Appl. Phys. Lett.* **79**(1), 51–53 (2001).

[139] J. C. Weeber, A. Dereux, C. Girard, J. R. Krenn, and J. P. Goudonnet, Plasmon Polaritons of Metallic Nanowires for Controlling Submicron Propagation of Light, *Phys. Rev. B* **69**(12), 9061–9068 (1999).

[140] R. M. Dickson and L. A. Lyon, Unidirectional Plasmon Propagation in Metallic Nanowires, *J. Phys. Chem. B* **104**, 6095–6098 (2000).

[141] J. Takahara, S. Yamagishi, H. Taki, A. Morimoto, and T. Kobayashi, Guiding of a One-dimensional Optical Beam with Nanometer Diameter, *Opt. Lett.* **22**(7), 475–477 (1997).

[142] S. A. Maier, P. G. Kik, L. A. Sweatlock, H. A. Atwater, J. J. Penninkhof, A. Polman, S. Meltzer, E. Harel, A. A. G. Requicha, and B. E. Koel, Energy Transport In Metal Nanoparticle Plasmon Waveguides, *Mat. Res. Soc. Symp. Proc.* **777** (2003).

[143] J. P. Kottmann, and O. J. F. Martin, Plasmon Resonant Coupling in Metallic Nanowires, *Opt. Express* **8**(12), 655–663 (2001).

[144] J. P. Kottmann, O. J. F. Martin, D. R. Smith, and S. Schultz, Plasmon Resonances of Silver Nanowires with a Nonregular Cross Section, *Phys. Rev. B* **64**, 235402 (2001).

[145] J. Wang et al., Photoluminescent Nanowires, *Science* **293**, 1455 (2001).

[146] N. P. Padture, Nanotechnology, Dept. of Metallurgy and Materials Engineering, Institute of Materials Science, University of Connecticut, Storrs, CT.

[147] P. J. F. Harris, S. C. Tsang, et al., High-Resolution Electron Microscope Studies of a Microporous Carbon Produced by Arc-Evaporation, *J. Chem. Soc. Faraday Trans.* **90**, 2799 (1994).

[148] IWFAC'2003, Fullerenes and Atomic Clusters, 6th Biennial International Workshop, St. Petersburg, Russia, June 30–July 4, 2003.

[149] S. Chopra, K. McGuire, N. Gothard, A. M. Rao, and A. Pham, Selective Gas Detection Using a Carbon Nanotube Sensor, *Appl. Phys. Lett.* **83**(11), 2280 (2003).

[150] H. Kajiura, S. Tsutsui, K. Kadono, M. Kakuta, M. Ata, and Y. Murakami, Hydrogen Storage Capacity of Commercially Available Carbon Materials at Room Temperature, *Appl. Phys. Lett.* **82**(7), 1105 (2003).

[151] P. G. Collins, M. S. Arnold, and P. Avouris, Engineering Carbon Nanotubes and Nanotube Circuits Using Electrical Breakdown, *Science* **292**, 706 (2001).

[152] S. J. Oh, Y. Cheng, J. Zhang, H. Shimoda, and O. Zhou, Room-Temperature Fabrication of High-Resolution Carbon Nanotube Field-Emission Cathodes by Self-Assembly, *Appl. Phys. Lett.* **82**(15), 2521 (2003).

[153] J. Robertson, W. I. Milne, K. B. K. Teo, and M. Chhowalla, Field Emission Applications of Carbon Nanotubes, *AIP Conf. Proc.* **633**(1), 537 (2002).

[154] M. Milas, R. Foschia, A. Kulik, R. Gaál, E. Ljubovi, and L. Forró, Carbon Nanotubes as Scanning Probe Tips, *AIP Conf. Proc.* **633**(1), 614 (2002).

[155] M. Freitag and A. T. Johnson, Nanoscale Characterization of Carbon Nanotube Field-Effect Transistors, *AIP Conf. Proc.* **633**(1), 513 (2002).

[156] J. Guo, M. Lundstrom, and S. Datta, Performance Projections for Ballistic Carbon Nanotube Field-Effect Transistors, *Appl. Phys. Lett.* **80**(17), 3192 (2002).

[157] I. Alexandrou, E. Kymakis, and G. A. J. Amaratunga, Polymer–Nanotube Composites: Burying Nanotubes Improves Their Field Emission Properties, *Appl. Phys. Lett.* **80**(8), 1435 (2002).

[158] J-M. Bonard, T. Stöckli, O. Noury, and A. Châtelain, Field Emission from Cylindrical Carbon Nanotube Cathodes: Possibilities for Luminescent Tubes, *Appl. Phys. Lett.* **78**(18), 2775 (2001).

[159] Q. H. Wang, M. Yan, and R. P. H. Chang, Flat Panel Display Prototype Using Gated Carbon Nanotube Field Emitters, *Appl. Phys. Lett.* **78**(9), 1294 (2001).

[160] E. S. Snow, J. P. Novak, P. M. Campbell, and D. Park, Random Networks of Carbon Nanotubes as an Electronic Material, *Appl. Phys. Lett.* **82**(13), 2145 (2003).

[161] M. A. Lantz, B. Gotsmann, U. T. Dürig, P. Vettiger, Y. Nakayama, T. Shimizu, and H. Tokumoto, Carbon Nanotube Tips for Thermomechanical Data Storage, *Appl. Phys. Lett.* **83**(6), 1266 (2003).

[162] J. M. Kinaret, T. Nord, and S. Viefers, A Carbon-Nanotube-Based Nanorelay, *Appl. Phys. Lett.* **82**(8), 1287 (2003).

[163] W. B. Choi, S. Chae, E. Bae, J-W. Lee, B-H. Cheong, J-R. Kim, and J-J. Kim, Carbon-Nanotube-Based Nonvolatile Memory with Oxide–Nitride–Oxide Film and Nanoscale Channel, *Appl. Phys. Lett.* **82**(2), 275 (2003).

[164] J. B. Cui, R. Sordan, M. Burghard, and K. Kern, Carbon Nanotube Memory Devices of High Charge Storage Stability, *Appl. Phys. Lett.* **81**(17), 3260 (2002).

[165] A. Bachtold, P. Hadley, T. Nakanishi, and C. Dekker, Logic Circuits with Carbon Nanotubes, *AIP Conf. Proc.* **633**(1), 502 (2002).

[166] B. Zhao, R. Kozhuharova, T. Mühl, I. Mönch, H. Vinzelberg, M. Ritschel, A. Graff, M. Huhle, H. Lichte, and C. M. Schneider, Magnetic Systems with Carbon Nanotubes, *AIP Conf. Proc.* **633**(1), 583 (2002).

[167] Y-H. Kim and K. J. Chang, Electron Transport Through Quantum-Dot States of N-Type Carbon Nanotubes, *Appl. Phys. Lett.* **81**(12), 2264 (2002).

[168] A. Javey, M. Shim, and H. Dai, Electrical Properties and Devices of Large-Diameter Single-Walled Carbon Nanotubes, *Appl. Phys. Lett.* **80**(6), 1064 (2002).

[169] W. B. Choi, J. U. Chu, K. S. Jeong, E. J. Bae, J-W. Lee, J-J. Kim, and J-O. Lee, Ultrahigh-Density Nanotransistors by Using Selectively Grown Vertical Carbon Nanotubes, *Appl. Phys. Lett.* **79**(22), 3696 (2001).

[170] M. Freitag, M. Radosavljevic, Y. Zhou, A. T. Johnson, and W. F. Smith, Controlled Creation of a Carbon Nanotube Diode by a Scanned Gate, *Appl. Phys. Lett.* **79**(20), 3326 (2001).

[171] R. J. Luyken and F. Hofmann, Concept for a Highly Scalable Nonvolatile Nanotube Based Memory, *AIP Conf. Proc.* **591**(1), 581 (2001).

[172] C. Thelander, M. H. Magnusson, K. Deppert, L. Samuelson, P. R. Poulsen, J. Nygård, and J. Borggreen, Gold Nanoparticle Single-Electron Transistor with Carbon Nanotube Leads, *Appl. Phys. Lett.* **79**(13), 2106 (2001).

[173] B. Q. Wei, R. Vajtai, and P. M. Ajayan, Reliability and Current Carrying Capacity of Carbon Nanotubes, *Appl. Phys. Lett.* **79**(8), 1172 (2001).

[174] B. W. Alphenaar, K. Tsukagoshi, and M. Wagner, Spin Transport in Nanotubes, *J. Appl. Phys.* **89**(11), 6863 (2001).

[175] H. Watanabe, C. Manabe, T. Shigematsu, and M. Shimizu, Dual-Probe Scanning Tunneling Microscope: Measuring a Carbon Nanotube Ring Transistor, *Appl. Phys. Lett.* **78**(19), 2928 (2001).

[176] M. Krüger, M. R. Buitelaar, T. Nussbaumer, C. Schönenberger, and L. Forró, Electrochemical Carbon Nanotube Field-Effect Transistor, *Appl. Phys. Lett.* **78**(9), 1291 (2001).

[177] H. R. Huff, Silicon Wafers for the Mesoscopic Era, *AIP Conf. Proc.* **550**(1), 67 (2001).

[178] Y. Gao, Y. Bando, Z. Liu, D. Golberg, and H. Nakanishi, Temperature Measurement Using a Gallium-Filled Carbon Nanotube Nanothermometer, *Appl. Phys. Lett.* **83**(14), 2913 (2003).

[179] G. Z. Yue, Q. Qiu, Bo Gao, Y. Cheng, J. Zhang, H. Shimoda, S. Chang, J. P. Lu, and O. Zhou, Generation of Continuous and Pulsed Diagnostic Imaging X-Ray Radiation Using a Carbon-Nanotube-Based Field-Emission Cathode, *Appl. Phys. Lett.* **81**(2), 355 (2002).

[180] H. Sugie, M. Tanemura, V. Filip, K. Iwata, K. Takahashi, and F. Okuyama, Carbon Nanotubes as Electron Source in an X-Ray Tube, *Appl. Phys. Lett.* **78**(17), 2578 (2001).

[181] P. M. Ajayan and L. S. Schadler, Carbon Nanotube Filled Composites, Department of Materials Science and Engineering, Rensselaer Polytechnic Institute, Troy, NY, NanoteC'99 Conference, Sept. 8–10, 1999, University of Sussex at Brighton, U.K.

[182] A. N. Watkins, J. L. Ingram, J. D. Jordan, R. A. Wincheski, J. M. Smits, and P. A. Williams, Single Wall Carbon Nanotube-Based Structural Health Monitoring Sensing Materials, 2004 NSTI Nanotechnology Conference and Trade Show, Nanotech 2004, Mar. 7–11, 2004, Boston, MA.

[183] R. H. Cayton, Nanoparticle Composites for Coating Applications, 2004 NSTI Nanotechnology Conference and Trade Show, Nanotech 2004, Mar. 7–11, 2004, Boston, MA.

[184] T. Webster, Purdue University, Self-Assembling "Nanotubes" Offer Promise for Future Artificial Joints, Press Release, Apr. 9, 2004.

[185] A. L. Chun, J. G. Moralez, T. Webster, and H. Fenniri, Helical Rosette Nanotubes: A More Effective Orthopaedic Implant Material, *Nanotechnology* **15**(4), S234–S239 (2004).

[186] D. Loss and D. P. DiVincenzo, Quantum Computation with Quantum Dots, *Phys. Rev. A* **57**, 120 (1998).

[187] L. M. K. Vandersypen, R. Hanson, L. H. Willems van Beveren, J. M. Elzerman, J. S. Greidanus, S. De Franceschi, and L. P. Kouwenhoven, Quantum Computing with Electron Spins in Quantum Dots, in *Quantum Computing and Quantum Bits in Mesoscopic Systems*, A. J. Leggett, B. Ruggiero, and P. Silvestrini, Eds., Kluwer Academic/Plenum, New York, 2003.

[188] J. M. Elzerman, R. Hanson, J. S. Greidanus, L. H. Willems van Beveren, S. De Franceschi, L. M. K. Vandersypen, S. Tarucha, and L. P. Kouwenhoven, A Few-Electron Quantum Dot Circuit with Integrated Charge Read-out, *Phys. Rev. B* **67**, 161308(R) (2003).

[189] R. Hanson, B. Witkamp, L. M. K. Vandersypen, L. H. Willems van Beveren, J. M. Elzerman, and L. P. Kouwenhoven, Zeeman Energy and Spin Relaxation in a One-electron Quantum Dot, *Phys. Rev. Lett.* **91**, 196802 (2003).

[190] T. Fujisawa, D. G. Austing, Y. Tokura, Y. Hirayama, and S. Tarucha, Allowed and Forbidden Transitions in Artificial Hydrogen and Helium Atoms, *Nature* **419**, 278 (2002).

[191] L. Vandersypen, J. Elzerman, R. Hanson, L. W. van Beveren, J. Greidanus, J. Wever, B. Witkamp, and L. Kouwenhoven, Electron Spin Qubits in Quantum Dots, International Conference on Nanoelectronics, Lancaster University, Jan. 4–9, 2003, Lancaster, U.K.

[192] J. E. Mooij and L. P. Kouwenhoven, Electron Spins in Semiconductor Quantum Dots, Delft Spin Qubit Project, Kavli Institute of Nanoscience at Delft, Faculty of Applied Sciences at Delft University of Technology, Delft, http://qt.tn.tudelft.nl/research/ spinqubits/.

[193] The Nano-Optics Group, University of Rochester's Institute of Optics, Rochester, NY.

[194] S. M. Reimann and M. Manninen, Finite Fermion Systems, *Rev. Mod. Phys.* **74**, 1283ON (2002).

[195] P. A. Orellana, F. Domýnguez-Adame, I. Gómez, and M. L. Ladrón de Guevara, Transport Through a Quantum Wire with a Side Quantum-Dot Array, *Phys. Rev. B* **67** (2003).

[196] M. Victor, Self-Assembled Quantum Dots for Advanced Semiconductor Laser Diodes, K. U. Leuven IWETO-researchdatabase, Solid State Physics and Magnetism Section, Katholieke Universiteit Leuven, Leuven, Belgium.

[197] A. Zrenner et al., Coherent Properties of a Two-Level System Based on a Quantum-Dot Photodiode, *Nature* **418**, 612 (2002).

[198] Quantum Dot Corporation, Hayward, CA, http://www.qdots.com.

[199] Evident Technologies, Troy, NY, www.evidenttech.com.

[200] S. Chang, M. Zhou, and C. P. Grover, Information Coding and Retrieving Using Fluorescent Semiconductor Nanocrystals for Object identification, Institute for National Measurement Standards, National Research Council Canada, Ottawa, Ontario, Canada, Optical Society of America, Jan. 2004.

[201] A. P. Alivisatos, Perspectives on the Physical Chemistry of Semiconductor Nanocrystals, *J. Phys. Chem.* **100**, 13226–13239 (1996).

[202] M. Han, X. Gao, J. Z. Su, and S. Nie, Quantum-Dot-Tagged Microbeads for Multiplexed Optical Coding of Biomoleclues, *Nature Biotechnol.* **19**, 631–635 (2001).

[202a] K. R. Brown, D. A. Lidar, and K. B. Whaley, Quantum Computing with Quantum Dots on Quantum Linear Supports, *Phys. Rev. A* **65**, 012397 (2001).

[202b] X. Peng, M. C. Schlamp, A. V. Kadavanich, and A. P. Alivisatos, Epitaxial Growth of Highly Luminescent CdSe/CdS Core/Shell Nanocrystals with Photostability and Electronic Accessibility. *J. Am. Chem. Soc.* **119**, 7019 (1997).

[202c] D. Katz, T. Wizansky, O. Millo, E. Rothenberg, T. Mokari, and U. Banin, Size Dependent Tunneling and Optical Spectroscopy of CdSe Quantum Rods, *Phys. Rev. Lett.* **89**, 086801 (2002).

[202d] Y. A. Wang, J. J. Li, H. Chen, and X. Peng, Stabilization of Inorganic Nanocrystals by Organic Dendrons, *J. Am. Chem. Soc.* **124**, 2293 (2002).

[203] D. Lidke et al., *Nature Biotechnol.*, Feb. 2004 issue.

[204] N. Kane, Advanced Diamond Technologies, Chicago, IL, www.thindiamond.com.

[205] S-G. Liu, J. E. Dahl, and R. M. K. Carlson, Diamondoid Derivatization Chemistry—Towards Nanotechnology Application, 2004 NSTI Nanotechnology Conference and Trade Show, Nanotech 2004, Mar. 7–11, 2004, Boston, MA.

[206] P. M. Ajayan, L. S. Schadler, and P. V. Braun, *Nanocomposite Science and Technology*, Wiley, Hoboken, NJ, 2003.

[207] N. A. Kotov, Thin Film Nanocomposites: From Electronics and Photonics to Magnetic Materials and Tissue Engineering, Chemical Engineering, University of Michigan Department of Materials Science and Engineering, Ann Arbor, MI, msewww@umich.edu.

[208] M. Rieth and W. Schommers, Ed., *Handbook of Theoretical and Computational Nanotechnology*, American Scientific Publishers, Stevenson Ranch, CA.

[209] J-Y. Lee and A. J. Crosby, Combinatorial Investigation of Crazing in Thin Film Nanocomposites, Polymer Poster Symposium, Crosby Research Group, Dept. of Polymer Science and Engineering, University of Massachusetts, Amherst, MA, umass.edu, http://www.pse.umass.edu/crosby.

[210] A. Hult, Thin Film, KTH, Campus Valhallavägen, Besöksadress: Valhallavägen 79, Postadress: Kungliga Tekniska högskolan, Stockholm.

[211] S. Manne and I. A. Aksay, Thin Films and Nanolaminates Incorporating Organic/Inorganic Interfaces, *Curr. Opin. Solid State Mater. Sci.* **2**, 358–364 (1997).

[212] F. A. Mohamed, D. Chzran, J. C. Earthman, E. J. Lavernia, and J. W. Morris, Nano Highlight, Minimum Nanocrystalline Grain Size Obtainable During Cryomilling, NSF Nanoscale Science and Engineering Grantees Conference, Dec. 16–18, 2003.

[213] J. Freim and Y. Avniel, Capacitors Containing Nanocrystalline $BaTiO_3$ as Dielectric, LEW-16984, Nanomaterials Research Corp. work for Glenn Research Center NASA Glenn Research Center, Commercial Technology Office, Cleveland, OH.

[214] G. Poirier, The Self-Assembly Mechanism of Alkanethiols on Au(111), *Science*, **272**, 1145 (1996).

[215] J. Giles, Scientists Create Fifth Form of Carbon, *Nature*, Mar. 23, 2004.

[216] W. Wang and S. A. Asher, Incorporation of Quantum Dots in Colloids, *J. Am. Chem. Soc.* **123**, 12528 (2001).

[217] Staff, Duke University Engineers Fabricating Polymer "Nanobrushes," *Nanotechnology News*, Mar. 30, 2004.

[218] H. Lovy, Nano Is a Concept by Which We Measure Our Pain, Howard Lovy's NanoBot, Mar. 29, 2004.

[219] Staff, North Texas Researchers Study Buckyball Effects, *Star-Telegram*, Mar. 29, 2004.

[220] Staff, Type of Buckyball Shown to Cause Brain Damage in Fish, *EurekAlert*, Mar. 28, 2004.

[221] D. Normile, Coaxing Molecular Devices to Build Themselves, *Science* **290**, 1524–1525 (2000).

[222] K. E. Drexler, Molecular Engineering: An Approach to the Development of General Capabilities for Molecular Manipulation, *Proc. Natl. Acad. Sci.* **78,** 5275–5278 (1981).

[223] G. M. Whitesides and M. Boncheva, Beyond Molecules: Self-Assembly of Mesoscopic and Macroscopic Components, *Proc. Natl. Acad. Sci.* **99**, 4769–4774 (2002).

[224] K. E. Drexler, Building Molecular Machines, *Trends Biotechnol.* **17**, 5–7 (1999).

[225] F. Fabry, T. M. Gruenberger, J. Gonzalez-Aguilar, H. Okuno, E. Grivei, N. Probst, L. Fulcheri, G. Flamant, and J-C. Charlier, Continuous Mass Production of Carbon Nanotubes by 3-Phase AC Plasma Processing, 2004 NSTI Nanotechnology Conference and Trade Show, Nanotech 2004, Mar. 7–11, 2004, Boston, MA.

[226] L. Yowell, JSC Carbon Nanotube Project, NASA, Houston, TX, http://mmptdpublic.jsc.nasa.gov/jscnano//.

[227] T. Gokcen, Materials Processing, Center for Nanotechnology, NASA Ames Research Center, Moffett Field, CA, meyya@orbit.arc.nasa.gov.

[227a] Center for Nanotechnology, NASA Ames Research Center, meyya@orbit.arc.nasa.gov.

[228] Workshop on Purity and Dispersion Measurement Issues in Single Wall Carbon Nanotube Materials, Sponsored by NASA-JSC and NIST, May 27–29, 2003, NIST HQ, Gaithersburg, MD.

[229] A. E. Kaloyeros, E. T. Eisenbraun, J. Welch, and R. E. Geer, Exploiting Nanotechnology for Terahertz Interconnects, *Semiconductor International*, Jan. 1, 2003.

[230] E. Mena-Osteritz, 3D Self-Assembled Molecular Architectures, Dept. Organic Chemistry II, University of Ulm, Ulm, Germany, ECSCD-8 18-21 July 2004 (Segovia, Spain), elena.mena-osteritz@chemie.uni-ulm.de.

[231] Molecular Self-Assembly: Biomimetics as a Route to Novel Products and Processes, 2nd Nanoforum Summer School, University of Cambridge Nanoscience Centre, Sept. 5–10, 2004.

[232] J. Ouellette, *Exploiting Molecular Self-Assembly*, American Institute of Physics, Woodbury, NY, Dec. 2000.

[233] D. Frenkel, On Designer Colloidal “Atoms,” *Science* **296**, 65 (2002).

[234] B. J. Alder and T. E. Wainwright, Phase Transition for a Hard Sphere System, *J. Chem. Phys.* **27**, 1208 (1957).

[235] G. Ungar et al., Giant Supramolecular Liquid Crystal Lattice, *Science* **299**, 1208 (2003).

[236] J. M. Seddon, Lyotropic Phase Behaviour of Biological Amphiphiles, *Ber. Bunsenges Phys. Chem.* **100**, 380–393 (1996).

[236a] E. L. Thomas, D. M. Anderson, C. S. Henkee, and D. Hoffman, Periodic Area-Minimizing Surfaces In Block Copolymers, *Nature* **334**, 598 (1988).

[236b] I. W. Hamley, *The Physics of Block Copolymers*, Oxford University Press, Oxford, 1998.

[236c] M. A. Hillmyer et al., Macromolecules, *Science* **271**, 976 (1996).

[236d] A. M. Levelut and M. Impéror-Clerc, Structural Investigations on “Smectic D” and Related Mesophases, *Liquid Crystals* **24**, 105 (1998).

[237] E. Forsén, Organic Molecular Electronics—Biotronics, May 28, 2000, Technical University of Denmark, Department of Micro and Nanotechnology, http://www.mic.dtu.dk.

[238] C. P. Collier, E. W. Wong, M. Belohradsky, F. M. Raymo, J. F.Stoddart, P. J. Kuekes, R. S. Williams, and J. R. Heath, *Science* **285**, 391 (1999).

[239] J. S. Tans, A. R. M. Verschueren, and C. Dekker, *Nature* **393**, 49 (1998).

[240] National Science Foundation and IBM Research Press Releases, June 25, 2003.

[241] L. S. Penrose, Self-Reproducing Machines, *Sci. Am.* **200**(6), 105–114 (1959).

[242] H. Jacobson, On Models of Reproduction, *Am. Sc.* **46**, 255–284 (1958).

[243] J. Conner, Nanotechnology Pioneer Calms Fears of Runaway Replicators, News Release, Palo Alto, CA, June 9, 2004, Judy@foresight.org.

[244] A. Yazdani and C. M. Lieber, On Scanning Probe Microscopy, *Nature* **401**, 227 (1999).

[245] P. Clarke, EDA Pioneer Calls for Design, Manufacturing Links, *EE Times*, Nov. 24, 2003.

[246] I. Ionica, L. Montès, S. Ferraton, J. Zimmermann, V. Bouchiat, and L. Saminadayar, Silicon Nanostructures Patterned on SOI by AFM Lithography, 2004 NSTI Nanotechnology Conference and Trade Show, Nanotech 2004, Mar. 7–11, 2004, Boston, MA.

[247] C. Brown, Photonic Tech Comes to Light, *EE Times*, June 2, 2003.

[248] R. C. Johnson, Optical Computer Components Unveiled, *EE Times*, Feb. 16, 2004.

[248a] P. Singer, Optical Waveguides a Focus at Cornell, *Semiconductor Int.*, Mar. 1, 2004.

[248b] Yvonne Carts-Powell, Low-index Slot serves as Waveguide, *Laser Focus World*, Feb. 2004.

[249] E-MRS 2004 Spring Meeting, Symposium A2: Nanophotonic Materials, Palais de la Musique et des Congres, Strasbourg, France, May 24–28, 2004, Conference Proceedings.

[250] I. Birkby and C. Chai, Optical Switching Applications Using Nanotechnology, Online Magazine, azonano.com.

[251] M. Lipson, Cornell University, Cornell Nanophotonics Group, Department of ECE, Phillips Hall, Ithaca, NY.

[252] S. A. Maier, P. G. Kik, and H. A. Atwater, Observation of Coupled Plasmon-Polariton Modes in Au Nanoparticle Chain Waveguides of Different Lengths: Estimation of Waveguide Loss, *Appl. Phys. Lett.* **81**(9), 1714 (2002).

[253] S. Maier, California Institute of Technology, A. Polman, FOM-Institute AMOLF, Photonic Micro- and Nanostructures, Materials Research Society (MRS) Spring Meeting, Apr. 12, 2004, San Francisco, U.S., Materials Research Society, Warrendale PA, info@mrs.org.

[254] J. Cheng, Frontiers in Optics 2004, Photonics Division Themes, Feb. 5, 2004, Optical Society of America, www.osa.org/join/techgroups/pho/Photonics_ Presentation.ppt.

[255] Acronyms and Glossary, Optical Component and Systems Testing, Spirent Communications of Ottawa, Ottawa, Ontario, Canada.

[256] Staff, The Bragg Law, Dissemination of IT for the Promotion of Materials Science (DoITPoMS) Teaching and Learning Packages, Department of Materials Science and Metallurgy, University of Cambridge, Pembroke Street, Cambridge, U.K., doitpoms@msm.cam.ac.uk.

[257] Laser and Fiber Optics Glossary of Terms, California Eastern Laboratories, Santa Clara, Nov. 19, 2001, http://www.cel.com.

[258] S. A. Vittorio, MicroElectroMechanical Systems (MEMS), Oct. 2001, Cambridge Scientific Abstracts, Bethesda, MD, www.csa.com.

[259] Y. Vlasov, LEOS '03 Paper JWB3, Summer 2003 IEEE LEOS Summer Topicals Meeting Series—Holey Fibers and Photonic Crystals, Holey Fibers and Photonic Crystals Advance Program, July 14, 2003.

[260] A. Pham, Planar Lightguide Circuits: An Emerging Market for Refractive Index Profile Analysis, Application Note 053, EXFO, Industrial and Scientific Division, Vanier, Quebec, Canada, http://www.exfo.com.

[261] S. Suzuki, Y. Inoue, S. Mino, M. Ishii, I. Ogawa, R. Kasahara, Y. Doi, Y. Hashizume, and T. Kitagawa, Compactly Integrated 32-Channel AWG Multiplexer with Variable Optical Attenuators and Power Monitors Based on Multichip PLC Technique, Optical Fiber Communication Conference, Feb. 23–27, 2004, Los Angeles, CA, Optical Society of America.

[262] M. Notomi, *IEEE LEOS '03 Conference*, Tucson, AZ, Oct. 26–30, 2003.

[263] P. D. Trinh, S. Yegnanarayanan, F. Coppinger, and B. Jalali, Silicon-on-Insulator (SOI) Phased-Array Wavelength Multi/Demultiplexer with Extremely Low-Polarization Sensitivity, *IEEE Photon. Technol. Lett.* **9**(7), 1041–1135 (1997).

[264] J. C. Knigh, Photonic Crystal Fibers: Putting New Life into an Old Hat, Summer 2003 IEEE LEOS Summer Topicals Meeting Series—Holey Fibers and Photonic Crystals, Holey Fibers and Photonic Crystals Advance Program, Vancouver, BC, Canada, July 14, 2003.

[265] *Business Development & News*, Light and Nano: Quantum Mechanics vs. Classical Optics—Rice Physicists Show That Quantum Methods Can Predict Nanophotonic Behavior, News.nanoapex.com, Cleveland, OH, Oct. 16, 2003.

[266] V. M. Menon, W. Tong, C. Li, F. Xia, I. Glesk, P. R. Pruncnal, and S. R. Forrest, All-Optical Wavelength Conversion Using a Regrowth-Free Monolithically Integrated Sagnac Interferometer, *IEEE Photonics Tech. Lett.* **15**(2), 254–256 (2003).

[267] H. S. Wang, C. Li, and S. R. Forrest, *Photon. Technol. Lett.* **15**, 1189 (2003).

[268] O. Wada, High Spped Microelectro Mechanical Systems (MEMS), *IEEE LEOS '03 Conference*, Tucson, AZ, Oct. 26–30, 2003.

[269] J. Cheng et al., Long-Wavelength VCSEL Technology, *IEEE LEOS '03 Conference*, Tucson, AZ, Oct. 26–30, 2003.

[270] V. Jayaraman, M. Mehta, A. W. Jackson, Y. Okuno, J. Piprek, and J. E. Bowers, High-Power 1320-nm Wafer-Bonded VCSELs with Tunnel Junctions, *IEEE Photonics Technology Letters,* **15**(11), 1495–1497 (2003).

[271] B. Lemoff, Terabit/s Capacity, *Conference on Lasers and Electro-Optics* (CLEO '03), Baltimore, MD, June 2003.

[272] Y. Jeong et al., The Rising Power of Fiber Lasers, *IEEE LEOS'03 Conference,* Tucson, AZ, Oct. 2003.

[273] R. Selvas, J. Nilsson, J. Sahu, K. Yia-Jarkko, S. Alam, P. Turner, J. Moore, and A. Grudinin, High Power 977 nm Fiber Sources Based on Jacketed Air-clad Fibers, *Proc. OFC '2003,* Atlanta, GA Mar. 23–28, 2003.

[274] C. A. Codemard, C. Farrell, V. Phillppov, P. Dupriez, J. K. Sahu, and J. Nilsson, Narrow-linewidth Pulsed Fiber MOPA Source at 1535, *Conference Proceedings*, CLEO Europe 2005 Conference.

[275] M. D. Nielsen, N. Mortensen, J. R. Folkenberg, A. Petersson, and A. Bjarklev, Improved All-Silica Endlessly Single-Mode Photonic Crystal Fiber, in *Technical Digest Optical Fiber Communication Conference, OFC'03*, Vol. 2, Atlanta, GA, pp. 701–702, Mar. 2003. Note: Paper FI7.

[276] S. G. Leon-Saval, T. A. Birks, W. J. Wadsworth, P. St. J. Russell, M. W. Mason, Supercontinuum Generation in Submicro Fibre Waveguides, *Optics Express* **12**, 2864–2869 (2004).

[277] W. J. Wadsworth, Supercontinuum Generation of 400-1600 nm "White Light" Using 200 fs Pulses, *J. Optical Soc. Amer.* **B19**, 2148 (2002).

[278] D. G. Ouzounov, F. R. Ahmad, D. Müller, N. Venkataraman, M. T. Gallagher, M. G. Thomas, J. Silcox, K. W. Koch, and A. L. Gaeta., Generation of Megawatt Optical Solitons in Hollow-Core Photonic Band-Gap Fibers, *Science* **301**, 1702–1704 (2003).

[279] B. Eggleton, Microfluidics Applications with PCF's, FiO-03 (Frontiers in Optics Conference), Optical Society of America, Washington, DC 2003, www.osa.org.

[280] J. O'Brien, Photonic Crystal Devices, *IEEE LEOS '03 Conference*, Tucson, AZ, Oct. 26–30, 2003.

[281] Y. Lee, Polarization-Controlled Single Mode Photonic Crystal VCSEL, *IEEE LEOS '03 Conference*, Tucson, AZ, Oct. 26–30, 2003.

[282] S. J. Choi and P. D. Dapkus, Microdisk Lasers Vertically Coupled to Output Waveguides, *Photon.Technol. Lett.* **15**, 1330 (2003).

[283] I. Marcikic, H. de Riedmatten, W. Tittel, H. Zbiden, and N. Gisin, *Nature* **421**, 509 (2003).

[284] Nano-Optics Group, part of the Laboratory of Physical Chemistry at the Chemistry Department of the Swiss Federal Institute of Technology (ETH); Laboratorium für Physikalische Chemie, Zürich, Switzerland, http://nano-optics.ethz.ch/subjects/.

[285] T. Prasad, Photonic Crystals: Superprism Effect, Studies in Applied Physics at Rice University, Houston, TX, Nov. 2003.

[286] E. Smalley, Crystal Fiber Goes Distance, *Technol. Res. News*, Nov. 5/12, 2003.

[287] C. M. Smith, N. Venkataraman, M. T. Gallagher, D. Müller, J. A. West, N. F. Borrelli, D. C. Allan, and K. W. Koch, Low-Loss Hollow-Core Silica/Air Photonic Bandgap Fibre, *Nature*, **424**(6949) (2003).

[288] D. J. DiGiovanni, OFS Laboratories, Under the Hood: Photonic Crystal Fibers—A Source of Unlimited Capacity or a Bubble Ready to Burst? Communications Technology, June 2002, PBI Media's Broadband Group, Potomac, MD.

[289] E. Yablonovitch, Inhibited Spontaneous Emission in Solid-State Physics and Electronics, *Phys. Rev. Lett.* **58**, 2059–2062 (1987).

[290] J. O'Brien, W. Kuang, P-T. Lee, J. R. Cao, C. Kim, and W. Kim, Photonic Crystal Lasers, in H. S. Nalwa, Ed., *Encyclopedia of Nanoscience and Nanotechnology*, Vol. 8, American Scientific Publishers, pp. 1–12.

[291] T. Prasad, Superprism Phenomenon in Photonic Crystals: Guiding the Path of Light, Oct. 2003, Master's Thesis, Rice University, Houston, TX.

[292] R. M. Rosdale, Photonic Crystal, Free Definition, http://www.akademie-asp.de.

[293] G. Comtet, Nanosources of Photons and Quantum Computing with Nanocrystals, a New Chemical Approach, Laboratoire de Photophysique Moléculaire Bât 210, Université Paris, Orsay, France, NID Workshop, Feb. 4–6, 2004, Athens, Greece.

[294] P. Nordlander, N. J. Halas, E. Prodan, and C. Radloff, A Hybridization Model for the Plasmon Response of Complex Nanostructures, *Science*, Oct. 17, 2003.

[295] T. Prasad, V. Colvin, and D. Mittleman, Superprism Phenomenon in Three-Dimensional Macroporous Polymer Photonic Crystals, *Phys. Rev. B* **67**, 165103 (2003); also in Apr. 21, 2003, issue of the *Virtual Journal of Nanoscale Science and Technology*.

[296] T. W. Ebbesen, H. J. Lezec, H. F. Ghaemi, T. Thio, and P. A.Wolff, Extraordinary Optical Transmission Through Sub-Wavelength Hole Arrays, *Nature* **391**, 667 (1998).

[297] A. K. Sarychev, V. A. Podolskiy, A. M. Dykhe, and V. M. Shalaev Resonance Transmittance Through Metal Film with Subwavelength Holes, *IEEE J. Quantum Electron.* **38**, 956–963 (2002).

[298] Scientists Announce First 3-D Assembly of Magnetic and Semiconducting Nanoparticles, Press Release, National Science Foundation, *Nature*, June 26, 2003.

[299] D. Davidov, Specific Topics Related to Nanoscience and Nanotechnology, Racah Institute of Physics, Hebrew University of Jerusalem, Jerusalem, Israel, http://cond-mat.phys.huji.ac.il/davidov/.

[300] K. Constant, Photonic Band Gap Structures (PBG Structures), Iowa State University (Dept. of Material Science and Engineering), Ames Lab. (USDOE)/Office of Basic Energy Sciences.

[301] Y. Fink, The Many Colors of Photonic Bandgap Fibers: Theory, Materials, Processing, Characterization and Applications, Summer 2003 IEEE LEOS Summer Topicals Meeting Series—Holey Fibers and Photonic Crystals, Holey Fibers and Photonic Crystals Advance Program, Massachusetts Institute of Technology, Vancouver, Canada, July 14, 2003.

[302] M. Hatcher, Holey Fiber Delivers Industrial Laser, *Opto and Laser Europe Magazine*, Dec. 2002, p. 18.

[303] S. A. Maier, M. L. Brongersma, P. G. Kik, S. Meltzer, A. A. G. Requicha, and H. A. Atwater, Plasmonics—A Route to Nanoscale Optical Devices, *Adv. Mater.* **13**(19) (2001).

[303a] H. Atwater, Guiding Light, *SPIE's OE Magazine*, July 2002, p. 42ff.

[304] A. Driessen, Microresonators as Building Blocks for VLSI Photonics, University of Twente, MESA+, Enschede, The Netherlands, Abstracts of the Invited Lectures, Erice, Sicily, Italy, Oct. 18–25, 2003, http://w3.uniroma1.it/cattedra_michelotti/invited1.htm.

[305] Staff, Cornell-Developed Tools to Guide and Switch Light Could Lead to Photonic Microchips and Practical Home Fiber-Optic Lines, *Innovations-Report*, Feb. 16, 2004.

[305a] Staff, Tools to Guide and Switch Light for Photonic Microchips, *Newswise*, Feb. 11, 2004.

[306] E. Altewischer, M. P. van Exter, and J. P. Woerdman, Plasmon-Assisted Transmission of Entangled Photons, *Nature* **418** (2002).

[307] Asian Technology Information Program (ATIP), Quantum Computation in Europe, Report ATIP01.019r, Asian Technology Information Program, Minato-ku, Tokyo, May 2004.

[307a] Akiyama et al., Pattern-Effect-Free Amplification and Cross-Gain Modulation Achieved by Using Ultrafast Gain Nonlinearity in Quantum-Dot Semiconductor

Optical Amplifiers, Postdeadline Paper of ECOC 2003, Proc. 29th European Conference on Optical Commulation.

[308] Y. C. Xin, L. G. Vaughn, L. R. Dawson, A. Stintz, Y. Lin, L. F. Lester, and D. L. Huffaker, InAs Quantum-Dot GaAs-Based Lasers Grown on AlGaAsSb Metamorphic Buffers, *J. Appl. Phys.* **94**, 2133 (2003). Also, *Virtual J. Nanoscale Sci. Technol.*, July 28, 2003.

[309] Lucent Staff, Proof-of-Concept Device May Pave the way for New and Powerful Laser-on-a-Chip Applications: Bell Labs Researchers Build Novel Semiconductor Laser Using Photonic Crystal, Lucent Press Release, Oct. 31, 2003, Murray Hill, NJ.

[310] Complex Photonic Systems, University of Twente, Dept. Science and Technology, The Netherlands, cops@utwente.nl.

[311] E. Yablonovitch, Inhibited Spontaneous Emission in Solid-State Physics and Electronics, *Phys. Rev. Lett.* **58**(20), 2059–2062 (1987).

[312] S. John, Strong Localization of Photons in Certain Disordered Dielectric Superlattices, *Phys. Rev. Lett.* **58**, 2486–2489 (1987), issue 23—June 8, 1987.

[313] M. Deopura, C. Schuh, and Y. Fink, Tin Sulfide as a 1D Photonic Band-Gap Material for Visible and NIR, 2004 NSTI Nanotechnology Conference and Trade Show, Nanotech 2004, Mar. 7–11, 2004, Boston, MA.

[314] K. Busch, Band Structure of Dispersive and Absorptive Photonic Crystals, DFG Center for Functional Nanostructures (CFN), Universität Karlsruhe (TH), Karlsruhe, Germany, http://www.cfn.uni-karlsruhe.de/projects.html.

[315] S. W. Leonard, H. M. Van Driel, A. Birner, U. Gösele, and P. R. Villeneuve, Single-Mode Transmission in Two-Dimensional Macroporous Silicon Photonic Crystal Waveguides, *Opt. Lett.* **25**, 1550 (2000).

[316] A. Mekis, J. C. Chen, I. Kurland, S. Fan, P. Villeneuve, and J. D. Joannopoulos, High Transmission Through Sharp Bends in Photonic Crystal Waveguides, *Phys. Rev. Lett.* **77**, 3787 (1996).

[317] S. Fan, P. R. Villeneuve, J. D. Joannopoulos, and H. A. Haus, Channel Drop Tunneling Through Localized States, *Phys. Rev. Lett.* **80**, 960 (1998).

[318] K. Busch and S. John, Liquid-Crystal Photonic-Band-Gap Materials: The Tunable Electro-Magnetic Vacuum, *Phys. Rev. Lett.* **83**, 967 (1999).

[319] P. Halevi and F. Ramos-Mendietta, Tunable Photonic Crystals with Semiconducting Constituents, *Phys. Rev. Lett.* **85**, 1875 (2000).

[320] W. Y. Zhang, X. Y. Lei, Z. L. Wang, D. G. Zheng, W. Y. Tam, C. T. Chan, and P. Sheng, Robust Photonic Bandgap from Tunable Scatterers, *Phys. Rev. Lett.* **84**, 2853 (2000).

[321] A. Tip, A. Moroz, and J. M. Combes, Band Structure of Absorptive Photonic Crystals, *J. Phys. A* **33**, 6223 (2000).

[322] N. Aközbek and S. John, Optical Solitary Waves in Two- and Three-Dimensional Nonlinear Photonic Bandgap Structures, *Phys. Rev. E* **57**, 2287 (1998).

[323] D. Hermann, M. Frank, K. Busch, and P. Wölfle, Photonic Band Structure Computations, *Opt. Express* **8**, 167 (2001).

[324] O. Toader, S. John, and K. Busch, Optical Trapping, Field Enhancement and Laser Cooling in Photonic Crystals, *Opt. Express* **8**, 217 (2001).

[325] K. Busch and S. John, Photonic Band-Gap Formation in Certain Self-Organizing Systems, *Phys. Rev. E* **58**, 3896 (1998).

[326] Y. Y. Li et al., Polymer Replicas of Photonic Porous Silicon for Sensing and Drug Delivery Applications, *Science* **299**, 2045 (2003).

[327] S. Polarz and M. Antonietti, Preparation of Porous Silica Materials via Sol-Gel for the Precise Prediction of the Mesopore Size, *J. Chem. Soc. Chem. Commun.* **2593** (2002).

[328] M. Wirtz, M. Parker, Y. Kobayashi, and C. R. Martin, Template-Synthesized Nanotubes for Chemical Separations and Analysis, *Chem. Eur. J.* **16**, 3572 (2002).

[329] J. C. Hulteen and C. R. Martin, A General Template-Based Method for the Preparation of Nanomaterials, *J. Mater. Chem.* **7**, 1075 (1997).

[330] K. Moller and T. Bein, Inclusion Chemistry in Periodic Mesoporous Hosts, *Chem. Mater.* **10**, 2950 (1998).

[331] C. E. Reese, M. E. Baltusavich, J. P. Keim, and S. A. Asher, Development of an Intelligent Polymerized Crystalline Colloidal Array Colorimetric Reagent, *Anal. Chem.* **73,** 5038–5042 (2001).

[332] J. M. Kahn and K-P. Ho, Communications Technology: A Bottleneck for Optical Fibres, *Nature* **411**, 1007–1010 (2001).

[333] J. K. Ranka, R. S. Windeler, and A. J. Stentz, Optical Properties of High Delta Air-Silica Microstructured Optical Fibers, *Opt. Lett.* **25**, 796–798, (2000).

[334] B. J. Eggleton, C. Kerbage, P. Westbrook, R. S. Windeler, and A. Hale, Microstructured Optical Fiber Devices, *Opt. Express* **9**(13), 698–713 (2001).

[335] S. G. Johnson, M. Ibanescu, M. Skorobogitiy, O. Weisberg, T. D. Egeness, M. Soljacic, S. A. Jacobs, and J. D. Joannopoulos, Low-Loss Asymptotically Single-Mode Propagation in Large-Core OmniGuide Fibers, *Opt. Express* **9**(13), 748–779 (2001).

[336] J. C. Knight, T. A. Birks, P. S. J. Russell, and D. M. Atkin, All-Silica Singlemode Optical Fiber with Photonic Crystal Cladding, *Opt. Lett.* **21**, 1547–1549 (1996).

[337] R. F. Cregan, B. J. Managan, J. C. Knight, T. A. Birks, P. S. J. Russell, P. J. Roberts and D. C. Allan, Singlemode Photonic Bandgap Guidance of Light in Air, *Science* **285**, 1537–1539 (1999).

[337a] M. Gerken and D. A. B. Miller, Multilayer Thin-Film Structures with High Spatial Dispersion, *Appl. Opt.* **42**(7) (2003). Optical Society of America, *OCIS Codes:* 260.2030, 060.4230, 230.4170.

[338] N. J. Halas and T. R. Huser, *Plasmonics: Metallic Nanostructures and Their Optical Properties*, Conference, SPIE, The International Society of Optical Engineering, SPIE San Diego Symposium, Aug. 3–5, 2003.

[339] J. Steele, C. E. Moran, and N. Halas, Plasmonic Properties of Metallodielectric Periodic Structures Poster Titles, Oct. 8, 2003, Department of Electrical and Computer Engineering, Rice University.

[340] V. P. Drachev, A. K. Sarychev, et al., Plasmonic Nanophotonics and Optoelectronics, Photonics and Spectroscopy Lab, Purdue University, www.purdue.edu.

[341] V. M. Shalaev, *Nonlinear Optics of Random Media: Fractal Composites and Metal-Dielectric Films*, Springer Tracts in Modern Physics, Vol 158, Springer, Berlin, 2000.

[342] A. K. Sarychev and V. M. Shalaev, Electromagnetic Field Fluctuations and Optical Nonlinearities in Metal-Dielectric Composites, *Phys. Reps.* **335**, 275–371 (2000).

[343] V. M. Shalaev, Optical Properties of Fractal Composites, in V. M. Shalaev, Ed., *Optical Properties of Random Nanostructures*, Topics in Applied Physics Series, Springer, Berlin, 2002.

[344] J. Kalkman, C. Strohhofer, B. Gralak, and A. Polman, Surface Plasmon Polariton Modified Emission of Erbium in a Metallodielectric Grating, *Appl. Phys. Lett.* **83**, 30 (2003).

[345] D. J. Bergman and M. I. Stockman, Surface Plasmon Amplification by Stimulated Emission of Radiation: Quantum Generation of Coherent Surface Plasmons in Nanosystems, *Phys. Rev. Lett.* **90**, 027402 (2003).

[345a] T. Okamoto and I. Yamaguchi, Optical Absorption Study of the Surface Plasmon Resonance in Gold Nanoparticles Immobilized onto a Gold Substrate by Self-Assembly Technique, *J. Phys. Chem. B* **107**, 10321–10324 (2003).

[345b] K. Kaneko, H.-B. Sun, X.-M. Duan, and S. Kawata, Two-Photon Photoreduction of Metallic Nanoparticle Gratings in a Polymer Matrix, *Appl. Phys. Lett.* **83**, 1426–1428 (2003).

[345c] N. Hayazawa, T. Yano, H. Watanabe, Y. Inouye, and S. Kawata, Detection of an Individual Single-Wall Carbon Nanotube by Tip-Enhanced Near-Field Raman Spectroscopy, *Chem. Phys. Lett.* **376**, 174–180 (2003).

[345d] S. Kawata, H.-B. Sun, T. Tanaka, and K. Takada, Finer Features for Functional Microdevices, *Nature* **412**, 697–698 (2001).

[345e] S. Kawata, Ed., *Near Field Optics and Surface Plasmon Polaritons*, Springer-Verlag, New York, 2001.

[345f] K. Li, M. I. Stockman, and D. J. Bergman, Self-Similar Chain of Metal Nanospheres as an Efficient Nanolens, *Phys. Rev. Lett.* **91**, 227402 (2003).

[345g] C. Genet, M. P. van Exter, and J. P. Woerdman, Fano-Type Interpretation of Red Shifts and Red Tails in Hole Array Transmission Spectra, *Opt. Commun.* **225**, 331 (2003).

[345h] P. Lalanne, C. Sauvan, J. P. Hugonin, J. C. Rodier, and P. Chavel, Perturbative Approach for Surface Plasmon Effects on Flat Interfaces Periodically Corrugated by Subwavelength Apertures, *Phys. Rev. B* **68**, 125404 (2003).

[345i] S. Shinada, J. Hashizume, and F. Koyama, Surface Plasmon Resonance on Microaperture Vertical-Cavity Surface-Emitting Laser with Metal Grating, *Appl. Phys. Lett.* **83**, 836 (2003).

[345j] M. Sarrazin and J. P. Vigneron, Optical Properties of Tungsten Thin Films Perforated with a Bidimensional Array of Subwavelength Holes, *Phys. Rev. E* **68**, 016603 (2003).

[345k] I. R. Hooper and J. R. Sambles, Surface Plasmon Polaritons on Thin-Slab Metal Gratings, *Phys. Rev. B* **67**, 235404 (2003).

[345l] F. J. García-Vidal, H. J. Lezec, T. W. Ebbesen, and L. Martín-Moreno, Multiple Paths to Enhance Optical Transmission Through a Single Subwavelength Slit, *Phys. Rev. Lett.* **90**, 213901 (2003).

[345m] A. Moreau, G. Granet, F. I. Baida, and D. Van Labeke, Light Transmission by Subwavelength Square Coaxial Aperture Arrays in Metallic Films, *Opt. Express* **11**, 1131 (2003).

[345n] A. Nahata, R. A. Linke, T. Ishi, and K. Ohashi, Enhanced Nonlinear Optical Conversion from a Periodically Nanostructured Metal Film, *Opt. Lett.* **28**, 423 (2003).

[345o] N. Bonod, S. Enoch, P. F. Li, E. Popov, and M. Neviere, Resonant Optical Transmission Through Thin Metallic Films With and Without Holes, *Opt. Express* **11**, 482 (2003).

[345p] J. Seidel, S. Grafstrom, L. Eng, and L. Bischoff, Surface Plasmon Transmission Across Narrow Grooves in Thin Silver Films, *Appl. Phys. Lett.* **82**, 1368 (2003).

[345q] G. T. Shubeita, S. K. Sekatskii, G. Dietler, S. Takahashi, and A. V. Zayats, Near-Field Optical Microscopy of Strongly Scattering Microporous Metal/Polymer Membranes, *Opt. Commun.* **217**, 23 (2003).

[345r] M. Sarrazin, J-P. Vigneron, and J-M. Vigoureux, Role of Wood Anomalies in Optical Properties of Thin Metallic Films with a Bidimensional Array of Subwavelength Holes, *Phys. Rev. B* **67**, 085415 (2003).

[345s] S. A. Darmanyan, A. V. Zayats, Light Tunneling via Resonant Surface Plasmon Polariton States and the Enhanced Transmission of Periodically Nanostructured Metal Films: An Analytical Study, *Phys. Rev. B* **67**, 035424 (2003).

[345t] L. Savio, L. Vattuone, and M. Rocca, Surface Plasmon Dispersion on Sputtered and Nanostructured Ag(001), *Phys. Rev. B* **67**, 45406 (2003).

[346] U. Kreibig and M. Vollmer, *Optical Properties of Metal Clusters*, Springer, Berlin 1994.

[347] M. Brongersma, J. Hartman, H. Atwater (California Institute of Technology), Plasmonics: Electromagnetic Energy Transfer and Switching in Nanoparticle Chain—Arrays Below the Diffraction Limit, Session G29—Nanoparticles and Nanoparticle Systems IV: Optical Properties, 2000 March Meeting of the American Physical Society, Mar. 20–24, Minneapolis, MN.

[348] B. E. A. Saleh and M. C. Teich, *Fundamentals of Photonics*, Wiley, New York, 1991.

[349] A. Mekis, J. C. Chen, I. Kurland, S. Fan, P. R. Villeneuve, and J. D. Joannopoulos, High Transmission Through Sharp Bends in Photonic Crystal Waveguides, *Phys. Rev. Lett.* **77**(18), 3787–3790 (1996).

[350] J. Moosburger, M. Kamp, A. Forchel, S. Olivier, H. Benisty, C. Weisbuch, and U. Oesterle, Enhanced Transmission Through Photonic-Crystal-Based Bent Waveguides by Bend Engineering, *Appl. Phys. Lett.* **79**(22), 3579–3581 (2001).

[351] O. Painter, R. K. Lee, A. Scherer, A. Yariv, J. D. O'Brian, P. D. Dapkus, and I. Kim, Two-Dimensional Photonic Band-Gap Defect Mode Laser, *Science* **284**(5421), 1819–1821 (1999).

[352] H. Raether, *Surface Plasmons on Smooth and Rough Surfaces and on Gratings*, Springer, Berlin, 1988.

[353] J. Tominaga, C. Mihalcea, D. Büchel, H. Fukuda, T. Nakano, N. Atoda, H. Fuji, and T. Kikukawa, Local Plasmon Photonic Transistor, *Appl. Phys. Lett.* **78**(17), 2417–2419 (2001).

[354] G. P. Wiederrecht, R. Bachelot, A. Bouhelier, J. Hranisavljevic, and J. S. Im, Argonne National Laboratory, Chemistry Division (Argonne is operated by the University of Chicago for the U.S. Department of Energy's Office of Science), http://chemistry.anl.gov/Nanophotonics/.

[355] U. Kreibig and M. Vollmer, *Optical Properties of Metal Clusters*, Springer, Berlin, 1995.

[356] C. F. Bohren and D. R. Huffman, *Absorption and Scattering of Light by Small Particles*, Wiley, New York, 1983.

[357] T. A. Klar, M. Perner, S. Grosse, G. von Plessen, W. Spirkl, and J. Feldmann, Surface Plasmon Resonances in Single Metallic Particles, *Phys. Rev. Lett.* **80**(19), 4249–4252 (1998).

[358] M. Quinten, A. Leitner, J. R. Krenn, and F. R. Aussenegg, Electromagnetic Energy Transport via Linear Chains of Silver Nanoparticles, *Opt. Lett.* **23**(17), 1331–1333 (1998).

[359] M. L. Brongersma, J. W. Hartman, and H. A. Atwater, Electromagnetic Energy Transfer and Switching in Nanoparticle Chain Arrays below the Diffraction Limit, *Phys. Rev. B* **62**, R16356 (2000).

[360] H. G. Craighead and G. A. Niklasson, Characterization and Optical Properties of Arrays of Small Gold Particles, *Appl. Phys. Lett.* **44**(12), 1134–1136 (1984).

[361] J. P. Hoogenboom, D. L. J. Vossen, C. Faivre-Moskalenko, M. Dogterom, and A. van Blaaderen, Patterning Surfaces with Colloidal Particles Using Optical Tweezers, *Appl. Phys. Lett.* **80**(25), 4828–4830 (2002).

[362] R. A. McMillan, C. D. Paavola, J. Howard, S. L. Chan, N. J. Zaluzec, and J. D. Trent, Ordered Nanoparticle Arrays Formed on Engineered Protein Templates, *Nature Mat.* **1**, 247–252 (2002).

[363] T. Müller, K.-H. Heinig, and B. Schmidt, Template-Directed Self-Assembly of Buried Nanowires and the Pearling Instability, *Mat. Sci. Engr. C* **19**, 209–213 (2002).

[364] J. J. Penninkhof, A. Polman, L. A. Sweatlock, H. A. Atwater, A. Vredenberg, and B. J. Kooi, MeV Ion Beam Induced Anisotropic Plasmon Resonance of Silver Nanocrystals in Glass, *Appl. Phys. Lett.* (2003).

[365] D. S. Citrin, Coherent Transport of Excitons in Quantum-Dot Chains: Role of Retardation, *Opt. Lett.* **20**(8), 901–903 (1995).

[366] P. Poddar, T. Telem-Sharif, T. Fried, and G. Markovich, Dipolar Interactions in Two and Three-Dimensional Magnetic Nanoparticle Arrays, *Phys. Rev. B* **66**, 060403(R) (2002).

[367] A. Yariv, Y. Xu, R. K. Lee, and A. Scherer, Coupled-Resonator Optical Waveguide: A Proposal and Analysis, *Opt. Lett.* **24**(11), 711–713 (1999).

[368] B. E. A. Saleh and M. C. Teich, *Fundamentals of Photonics*, Wiley, New York, 1991.

[369] A. Mekis, J. C. Chen, I. Kurland, S. Fan, P. R. Villeneuve, and J. D. Joannopoulos, High Transmission Through Sharp Bends in Photonic Crystal Waveguides, *Phys. Rev. Lett.* **77**, 3787 (1996).

[370] M. L. Brongersma, J. W. Hartman, and H. A. Atwater, Electromagnetic Energy Transfer and Switching in Nanoparticle Chain-Arrays Below the Diffraction Limit, *Phys. Rev. B* **62**, R16 356 (2000).

[371] M. Quinten, A. Leitner, J. R. Krenn, and F. R. Aussenegg, Electromagnetic Energy Transport via Linear Chains of Silver Nanoparticles, *Opt. Lett.* **23**, 1331 (1998).

[372] S. A. Maier, M. L. Brongersma, and H. A. Atwater, Electromagnetic Energy Transport Along Arrays of Closely Spaced Metal Rods as an Analogue to Plasmonic Devices, *Appl. Phys. Lett.* **78**, 16 (2001).

[373] S. A. Maier, M. L. Brongersma, and H. A. Atwater, Energy Transport Along Yagi Arrays, *Mater. Res. Soc. (MRS) Symp. Proc.* **637**, E2.9 (2001).

[374] J. R. Krenn, A. Dereux, J. C.Weeber, E. Bourillot, Y. Lacroute, J. P. Goudonnet, G. Schider, W. Gotschy, A. Leitner, F. R. Aussenegg, and C. Girard, Squeezing the Optical Near-Field Zone by Plasmon Coupling of Metallic Nanoparticles, *Phys. Rev. Lett.* **82**, 2590–2593 (1999).

[375] C. Girard, C. Joachim, and S. Gauthier, The Physics of the Near-Field, *Rep. Prog. Phys.* **63**, 893–938 (2000).

[376] H. Furukawa and S. Kawata, Analysis of Image Formation in a Near-Field Scanning Optical Microscope: Effects of Multiple Scattering, *Opt. Commun.* **132** (1–2), 170–178 (1996).

[377] F. Ghebremichael, Nonlinear Optics, U.S. Air Force Academy, CO, http://www.usafa.af.mil/dfp/research/lorc/polymer/all_rev/node3.html.

[378] H. R. Khalesifard, *Introduction to Nonlinear Optics*, Institute for Advanced Studies in Basic Sciences, khalesi@iasbs.ac.ir.

[379] University of Alberta, ECE 671 Nonlinear Optics and Nanophotonics Description, Electrical and Computing Engineering, Electrical & Computer Engineering Research Facility (ECERF), Edmonton, Alberta, Canada, www.ece.ualberta.ca.

[380] E. M. Purcell, Spontaneous Emission Probabilities at Radio Frequencies, *Phys. Rev.* **69**, 681 (1946).

[381] S. M. Dutray, Correspondence Principle Approach to Cavity Losses, *Eur. J. Phys.* **18**, 194–198 (1997).

[382] A. Fiore, Quantum Dots and Microcavities: Electronic and Optical Confinement in Semiconductors, Ecole Polytechnique Fédérale de Lausanne, Lausanne, Switzerland, Microresonators as Building Blocks for VLSI Photonics, Abstracts of the Invited Lectures, Erice, Sicily, Italy, Oct. 18–25, 2003, http://w3.uniroma1.it/cattedra_michelotti/ invited1.htm.

[383] P. D. Dapkus, K. Djordjev, S-J. Choi, and S.-J. Choi, Active Photonic Integrated Circuit Components Based on Semiconductor Microdisk Resonators, IEEE LEOS '03 Conference, Tucson, AZ, Oct. 26–30, 2003.

[384] P. Absil, Semiconductor Optical Micro-ring resonators, University of Ghent, IMEC-Dept. of Information Technology (INTEC), Gent, Belgium, Microresonators as Building Blocks for VLSI Photonics, Abstracts of the Invited Lectures, Erice, Sicily, Italy, Oct. 18–25, 2003, http://w3.uniroma1.it/cattedra_michelotti/invited1.htm.

[385] J. Ctyroky, Modelling Guided-Wave Optical Microresonators, Institute of Radio Engineering and Electronics AS CR, Prague, Czech Republic, Microresonators as Building Blocks for VLSI Photonics, Abstracts of the Invited Lectures, Erice, Sicily, Italy, Oct. 18–25, 2003, http://w3.uniroma1.it/cattedra_michelotti/ invited1.htm.

[386] Staff, What Is Quantum Well Intermixing? Promotional Material, Intense Photonics Ltd., Blantyre, Glasgow, Scotland, http://www.intenseco.com/technology/.

[387] A. Griessner, D. Jaksch, and P. Zoller, Cavity Assisted Nondestructive Laser Cooling of Atomic Qubits, *J. Phys. B: At. Mol. Opt. Phys.* **37** (2004).

[388] P. Mataloni, Generation and Applications of Single Photon States and Entangled Photon Pairs, Università di Roma "La Sapienza," Dipartimento di Fisica, Rome, Italy, Microresonators as Building Blocks for VLSI Photonics, Abstracts of the Invited Lectures, Erice, Sicily, Italy, Oct. 18–25, 2003, http://w3.uniroma1.it/cattedra_michelotti/invited1.htm.

[388a] N. Fang, H. Lee, C. Sun, X. Zhang, Subdiffraction Limited Optical Imaging with a Silver Superlens, *Science* **308**, 534–537 (2005).

[388b] B. Dumé, Superlens Could Image Nanoscale with Light, 2 May 2005, online magazine, www.nanotechweb.org.

[389] G. E. Moore, Cramming More Components onto Integrated Circuits, *Electronics* **38**(8), 114–117 (1965).

[390] W. J. Gross, D. Vasileska, and D. K. Ferry, Ultrasmall MOSFETs: The Importance of the Full Coulomb Interaction on Device Characteristics, *IEEE Trans. Electron Devices*, **47**, 1831 (2000).

[391] J. R. Heath, P. J. Kuekes, G. S. Snider, and R. S. Williams, A Defect-tolerant Computer Architecture: Opportunities for Nanotechnology, *Science*, **280**, 1716–1721 (1998).

[391a] L. Deferm, Dive Intro 32 nm, *Semiconductor Internationl*, March 1, 2005.

[391b] Staff, Semiconductor International Special Report for July 28, 2005, *Semiconductor International Weekly*, July 28, 2005, siweekly@email.semiconductor.net.

[392] S. M. Rossnagel and H. Kim, From DVD to CVD to ALD for interconnects and Related Applications, *Proc. of the 2001 International Interconnect Technology Conference,* 2001, p. 3.

[393] P. Singer, NANOCMOS Project to Explore Limits of CMOS, *Semiconductor Int.*, Apr. 1, 2004.

[394] F. O. Heinz, F. M. Bufler, A. Schenk, and W. Fichtner, Quantum Transport Phenomena and Their Modeling, Integrated Systems Laboratory, ETH Zurich, Zurich, Switzerland, fw@iis.ee.ethz.ch.

[395] J. D. Meindl et al., Limits of Silicon Nanoelectronics, *Science* **293**, 2044 (2001).

[396] J. Lee et al., Bandgap Modulation of Carbon Nanotubes, *Nature* **415**, 1005 (2002).

[397] Staff, Manufacturing Upheaval, Responsible Nanotechnology, Mar. 28, 2004.

[398] Staff, Application-Specific ICs: Customer-Specific to Customer-Defined, Report No. IN0401746WHT, Sept. 2004, In-Stat/ MDR, Scottsdale, AZ, www.instat.com.

[399] W. Wang and S. A. Asher, Incorporation of Quantum Dots in Colloids, *J. Am. Chem. Soc.* **123**, 12528 (2001).

[400] Staff, Nanowires Span Silicon Contacts, *Technol. Rev.* Mar. 29, 2004.

[401] L. Peters, Emerging Transistor Structures, *Semiconductor International*, Mar. 1, 2002.

[402] J. H. Song et al., Fabrication of Metal Nanowires, *J. Am. Chem. Soc.* **123**, 10397 (2001).

[403] Staff, Nanowire Transistors are Turned on Their End, *Nature*, Mar. 25, 2004.

[404] A. Parker, Modeling of Resonant Tunneling Diodes, Macquarie University, Australia, http://www.elec.mq.edu.au.

[405] M. A. Reed and T. Lee, Eds., *Molecular Nanoelectronics*, American Scientific, 2003.

[406] Staff, The Foundations of Next-Generation Electronics: Condensed-Matter Physics at RLE, *RLE Currents* **10**(2) (1998). Quantum-Effect Devices Group of the Research Laboratory of Electronics, Massachusetts Institute of Technology, Cambridge, MA.

[407] M. A. Reed, Molecular—Scale Electronics, *Proc. IEEE* **87**(4), 652–658 (1999).

[408] J. Tour et al., Are Molecular Wires Conducting? *Science* **271**, 1705–1707 (1996).

[409] R. Amerson, R. J. Carter, W. B. Culbertson, P. Kuekes, and G. Snider, Teramac—Configurable Custom Computing, in *Proceedings of the 1995 IEEE Symposium on FPGA's for Custom Computing Machines*, pp. 32–38.

[410] R. Amerson and P. Kuekes, The Design of an Extremely Large MCM-C—A Case Study, *Int. J. Microcircuits Electronic Packaging*, **17**(4), 377–382.

[411] G. Snider, P. Kuekes, B. Culbertson, R. Carter, A. Berger, and R. Amerson, The Teramac Configurable Computer Engine, in *Proceedings of the 5th International Workshop, Field Programmable Logic*, 1995, Oxford, England, pp. 44–53.

[412] B. Culbertson, R. Amerson, R. Carter, P. Kuekes, and G. Snider, The Teramac Configurable Custom Computer, Field Programmable Gate Arrays (FPGAs) for Fast Board Development and Reconfigurable Computing, *Proc. SPIE* **2607** (1995).

[413] A. de Boer, Application of Large-Scale Programmable Logic for Artery Extraction Filtering Of 3D MRI Data, HP Laboratories Technical Report HPL-95-95, Hewlett-Packard Laboratories, Palo Alto, CA, Aug. 1995.

[414] R. Amerson, R. Carter, W. Culbertson, P. Kuekes, and G. Snider, Plasma: An FPGA for Million Gate Systems, ACM/SIGDA Fourth International Symposium on Field-Programmable Gate Arrays, Feb. 1996.

[415] B. Culbertson, R. Amerson, R. Carter, P. Kuekes, and G. Snider, Exploring Architectures for Volume Visualization on the Teramac Custom Computer, in *Proceedings of the 1996 IEEE Symposium on FPGA's for Custom Computing Machines*.

[416] B. Culbertson, R. Amerson, R. Carter, P. Kuekes, and G. Snider, The Teramac Custom Computer: Extending the Limits with Defect Tolerance, IEEE International Symposium on Defect and Fault Tolerance in VLSI Systems, Nov. 1996.

[417] U. Kanus, M. Meisner, W. Strasser, H. Pfister, A. Kaufman, R. Amerson, R. Carter, B. Culbertson, P. Kuekes, and G. Snider, Cube-4 Implementations on the Teramac Custom Computing Machine, in *Proceedings of the 11th Eurographics Hardware Workshop*, Aug. 1996, Poitiers, France, pp. 133–143.

[418] B. Culbertson, R. Amerson, R. Carter, P. Kuekes, G. Snider, Defect Tolerance on the Teramac Custom Computer, in *Proceedings of the 1997 IEEE Symposium on FPGA's for Custom Computing Machines*, pp. 116–123.

[419] J. Heath, P. Kuekes, G. Snider, and S. Williams, A Defect-Tolerant Computer Architecture: Opportunities for Nanotechnology, *Science* **280**, 1716–1721 (1998).

[420] C. H. Bennett and D. P. DiVincenzo, Quantum Information and Computation, *Nature*, **404**, 247–255 (2000).

[421] Staff, Quantum Computation: A Great Challenge with a Wonderful Promise, The Delft Spin Qubit Project, Kavli Institute of Nanoscience Delft, The Faculty of Applied Sciences at Delft University of Technology, Delft, http://qt.tn.tudelft.nl/research/qc/.

[422] M. Bayer, Quantum Dots and Quantum Information Processing, *Nature* **418**, 597 (2002).

[423] A. Rodriguez, F. Dominguez-Adame, I. Gomez, and P. A. Orellana, Dynamics of the Electron Transport in a Quantum Wire Coupled to a Quantum-Dot Array, *Phys. Rev. B* **67**, 085321 (2003).

[424] T. H. Oosterkamp et al., Microwave Spectroscopy of a Quantum Dot Molecule, *Nature* **395**, 873 (1998).

[425] F. Remacle and R. D. Levine, Quantum Dots, *Proc. Natl. Acad. Sci.* **97**, 553 (2000).

[426] D. Gammon, On the Optical Activity of Quantum Dots, *Nature* **405**, 899 (2000).

[427] W. Liang et al., Kondo Resonance in a Single-Molecule Transistor, *Nature* **417**, 725 (2002).

[428] C. Santori et al., Indistinguishable Photons from a Single-Photon Device, *Nature* **419**, 594 (2002).

[429] T. Tsutsui, Quantum Dots and Electroluminescence, *Nature* **420**, 752 (2002).

[430] J. Lee et al., Bandgap Modulation of Carbon Nanotubes by Encapsulated Metallofullerenes, *Nature* **415**, 1005 (2002).

[431] M. T. Woodside and P. L. McEuen, Scanned Probe Imaging of Single-Electron Charge States in Nanotube Quantum Dots, *Science* **296**, 1098 (2002).

[432] R. de Picciotto et al., Wonder Wire Puts Up No Resistance, *Nature* **411**, 51 (2001).

[433] R. McCullough and L. Ward, Carnegie Mellon University Announces "One-Step" Method to Make Polymer Nanowires, Public Release, Mar. 30, 2004, Carnegie Mellon University, Pittsburgh, PA, wardle@andrew.cmu.edu.

[434] K. Patch, Technology Research News, Glossary, TRN, Boston, MA, www.trnmag.com.

[435] S. Lloyd, D. DiVincenzo, U. Vazirani, G. Doolen, and B. Whaley, Theory Component of the Quantum Information Processing and Quantum Computing Roadmap—A

Quantum Information Science and Technology Roadmap—Part 1: Quantum Computation, Section 6.9, LA-UR-04-1777, Version 2.0, Apr. 2, 2004, produced for the Advanced Research and Development Activity (ARDA), http://qist.lanl.gov.

[436] P. Peercy, The Drive to Miniaturization, *Nature* **406**, 1023 (2000).

[437] G. Timp, *Nanotechnology*, Springer, New York, 1999, pp. 161–206.

[438] C. B. Murray, C. R. Kagan, and M. G. Bawendi, Self-Organization of CdSe Nanocrystallites into Three-Dimensional Quantum Dot Superlattices, *Science* **270**, 1335 (1995).

[439] C. A. Mirkin, R. L. Letsinger, R. C. Mucic, and J. J. Storhoff, A DNA-Based Method for Rationally Organizing Nanoparticles into Macroscopic Materials, *Nature* **382**, 607 (1996).

[440] A. P. Alivisatos et al., Organization of Nanocrystal Molecules Using DNA, *Nature* **382**, 609 (1996).

[441] J-M. Lehn and P. Ball, in *The New Chemistry*, N. Hall, Ed., Cambridge University Press, 2000, p. 347.

[442] A. Cataldo, Intel Fabricates Next-Gen 65-nm and EUV Masks, *EE Times*, Jan. 16, 2003.

[443] A. Hand, Nanoimprint Lithography: Plays Well with Others? *Semiconductor International*, Sept. 1, 2004.

[444] P. A. Houston, Quantum Wire Field Effect Transistor, Semiconductor Materials and Devices Group, *J. Phys. D: Appl. Phys.* **36**, 3027–3033 (2003). Also, www.shef.ac.uk/eee/smd.

[445] Y. Huang et al., On Nanowire Building Blocks, *Science* **294**, 1313 (2001).

[446] C. N. Fleming et al., On the Design of Nanoscale Materials, *J. Am. Chem. Soc.* **123**, 10336 (2001).

[447] C. Wright-Smith and C. M. Smith, Atomic Force Microscopy in Biology, *Scientist*, Jan. 22, 2001.

[448] G. A. Somorjai, in *The New Chemistry*, N. Hall, Ed., Cambridge University Press, 2000.

[449] H. G. Craighead, On Micro- and Nanoelectromechanical Systems, *Science* **290**, 1532 (2000).

[450] T. Kouh, D. Karabacak, D. H. Kim, and K. L. Ekinci, Ultimate Limits to Optical Displacement Detection in Nanoelectromechanical Systems, 2004 NSTI Nanotechnology Conference and Trade Show, Nanotech 2004, Mar. 7–11, 2004, Boston, MA.

[451] S. M. Blinder, Quantum Chemistry, Class Materials, Chemistry Class 461, Chapter 7: The Hydrogen Atom: Atomic Orbitals, Spring Term 2002, University of Michigan, Ann Arbor, MI, sblinder@umich.edu.

[452] Hilbert Spaces, Wikipedia, http://en2.wikipedia.org/wiki/Hilbert_space.

[453] J. P. Lowe, *Quantum Chemistry*, 2nd ed., Academic, New York, 1993; republished as *NIH Guide to Molecular Modeling*, Center for Molecular Modeling (CMM), Bethesda, MD.

[454] C. Levit, A New Pictorial Approach to Molecular Structure and Reactivity, White Paper, NASA Nanotechnology Publications, NASA Advanced Supercomputing Division, NAS Systems Division Office, gpost@mail.arc.nasa.gov.

[455] M. Sachs, Quantum Mechanics from General Relativity: Particle Probability from Interaction Weighting, *Ann. Fondation Louis de Broglie* **24** (1999).

[456] Classical and Quantum Description of a Particle, Promotional Material, Institute of Physics, Technical University of Budapest, Budapest, Hungary.

[457] Stanford University, Quantum Mechanics, Stanford Linear Accelerator Center and Virtual Visitor Center, Menlo Park, CA.

[458] NIH Guide to Molecular Modeling, Center for Molecular Modeling (CMM), part of the Division of Computational Bioscience, Center for Information Technology, National Institutes of Health, Bethesda, MD.

[459] J. P. Tollenaere and E. E. Moret, Hyper-Glossary of Terminology, Department of Medicinal Chemistry, Computational Medicinal Chemistry Group, Faculty of Pharmacy, Utrecht University, Utrecht, The Netherlands.

[460] N. C. Cohen, Ed. *Guidebook on Molecular Modeling in Drug Design*, Academic Press, 1996.

[461] W. J. Hehre, L. Radom, P. V. R. Schleyer, and J. A. Pople, *Ab Initio Molecular Orbital Theory*, Wiley, New York, 1986.

[462] Introduction to Path Integrals, University of Sidney, School of Mathematics and Statistics, Sydney, Australia.

[463] R. Jackiw, Field Theory: Why Have Some Physicists Abandoned It? Massachusetts Institute of Technology, Center for Theoretical Physics, Cambridge, MA; also *Proc. Natl. Acad Sci USA*. **95**(22), 12776–12778 (1998).

[464] D. Atkinson, E. Bergshoeff, and M. de Roo, Research Areas, Institute for Theoretical Physics, Groningen, The Netherlands.

[465] P. Steinbach, Y. S. Lee, S. A. Hassan, and J. Xiang, *Molecular Mechanics, The NIH Guide to Molecular Modeling*, Center for Molecular Modeling (CMM), Division of Computational Bioscience (DCB/CIT) of the National Institutes of Health, Bethesda, MD.

[466] D. K. Ferry, R. Akis, M. J. Gilbert, and G. Speyer, Modeling of Nanoelectronic and Quantum Devices, 2004 NSTI Nanotechnology Conference and Trade Show, Nanotech 2004, Mar. 7–11, 2004, Boston, MA.

[467] A. Round and R. Williamson, Scanning Probe Microscopy at Bristol, Physics Laboratory, University of Bristol, Bristol, U.K., http://spm.phy.bris.ac.uk/techniques/SNOM/.

[468] Staff, FEI Company Breaks the 1 Angstrom High Resolution Imaging Barrier, *Nanotechnology News*, Mar. 31, 2004.

[469] Promotional Material, FEI Company, Hillsboro, OR, http://www.feicompany.com.

[470] S. Pennycook et al., Microscope Brings Atoms into Focus, *Science*, Sept. 17, 2004.

[471] S. Pignolet-Brandom, Miscoscopy.info, Harvard, MA, info@microscopy.info.

[472] Summary Table of Surface Analytical Techniques, Charles Evans & Associates, Sunnyvale, CA, cea@eaglabs.com.

[473] L. Ladic and K. van der Wulp, Introduction to LSCM (on-line material), The University of British Columbia, Computer Science Department, Vancouver, B.C.

[474] F. M. Gilchrist, *The Activity of pH Membrane Disruptive Pseudopeptides and their Subcellular Fate in Mammalian Cells Cultured in vitro,* PhD thesis, The University of Aston in Birmingham, Birmingham, UK, 2002.

[474a] P. C. Cheng, T. H. Lin, W. L. Wu, and J. L. Wu, Eds., *Multidimensional Microscopy*, Springer, New York, 1994.

[475] E. Farhi, J. Goldstone, S. Gutmann, and M. Sipser, Quantum Computation by Adiabatic Evolution, Jan. 28, 2000.

[476] D. Aharonov and A. Ta-Shma, Adiabatic Quantum State Generation and Statistical Zero Knowledge, Jan. 3, 2003.

[477] D. Aharonov, W. van Dam, J. Kempe, Z. Landau, S. Lloyd, and O. Regev, On the Universality of Adiabatic Quantum Computation, 2003.

[478] A. M. Childs, R. C. Cleve, E. Deotto, E. Farhi, S. Gutmann, and D. A. Spielman, Exponential Algorithmic Speedup by a Quantum Walk, in *Proceedings of the 35th ACM Symposium on Theory of Computing (STOC 2003)*, ACM Press, New York, 2003, pp. 59–68.

[479] A. Ambainis, Quantum Walk Algorithm for Element Distinctness, Nov. 1, 2003.

[480] C. H. Bennett, E. Bernstein, G. Brassard, and U. Vazirani, Strengths and Weaknesses of Quantum Computing, *SIAM J. Computing* **26**, 1510–1523 (1997).

[481] W. van Dam, M. Mosca, and U. Vazirani, How Powerful Is Adiabatic Quantum Computation?, in *Proceedings of the 42nd Annual Symposium on the Foundations of Computer Science (FOCS'01)*, IEEE Computer Society Press, Los Alamitos, CA, 2001, pp. 279–287.

[482] S. Arora, R. Impagliazzo, and U. Vazirani, The Principle of Local Checkability and Relativizing Arguments in Complexity Theory, in *Proceedings of the 8th Annual Structure in Complexity Theory Conference*, IEEE Computer Society Press, Los Alamitos, CA, 1993.

[483] D. Aharonov and D. Gottesmann, Improved Threshold for Fault-Tolerant Quantum Computation, manuscript, 2002.

[484] A. M. Steane and B. Ibinson, Fault-Tolerant Logical Gate Networks for CSS Codes, Nov. 4, 2003.

[485] A. A. Razborov, An Upper Bound on the Threshold Quantum Decoherence Rate, Quantum Information and Computation, **4**(3), 222–228 (2004).

[486] A. Kitaev, Quantum Measurements and the Abelian Stabilizer Problem, in *Proceedings of the Electronic Colloquium on Computational Complexity* (ECCC-1996), 3(3), ECCC Report TR96-003 (1996).

[487] D. S. Abrams and S. Lloyd, A Quantum Algorithm Providing Exponential Speed Increase for Finding Eigenvalues and Eigenvectors, *Phys. Rev. Lett.* **83**, 5162–5165 (1999).

[488] J. Traub, and H. Wozniakowski, Path Integration on a Quantum Computer, *Quantum Information Proc.* **1**, 365–388 (2002).

[489] G. Vidal, Efficient Simulation of One-Dimensional Quantum Many-Body Systems, Oct. 14, 2003.

[490] J. S. Bell, On the Einstein-Podolski-Rosen Paradox, *Physics* 1, 195–200 (1964); reprinted in *Speakable and Unspeakable in Quantum Mechanics*, Cambridge University Press, Cambridge, 1987, pp. 14–21.

[491] J. S. Bell, On the Problem of Hidden Variables in Quantum Mechanics, *Rev. Modern Phys.* **38**, 447–452 (1966).

[492] R. Landauer, Irreversibility and Heat Generation in the Computing Process, *IBM J. Res. Devel.* **5**(3), 183–191 (1961).

[493] C. H. Bennett, Logical Reversibility of Computation, *IBM J. Res. Devel.* **17**(6), 525–530 (1973).

[494] P. Benioff, The Computer as a Physical System: A Microscopic Quantum Mechanical Hamiltonian Model of Computers as Represented by Turing Machines, *J. Statist. Phys.* **22**, 563–591 (1980).

[495] P. Benioff, Quantum Mechanical Models of Turing Machines That Dissipate no Energy, *Phys. Rev. Lett.* **48**, 1581–1585 (1982).

[496] R. P. Feynman, Simulating Physics with Computers, *Int. J. Theoret. Phys.* **21**, 467–488 (1982).

[497] D. Deutsch, Quantum Theory, the Church-Turing Principle and the Universal Quantum Computer, *Proc. Roy. Soc. London Ser. A Math. Phys. Sci. A* **400**(1818), 97–117 (1985).

[498] E. Bernstein and U. Vazirani, Quantum Complexity Theory, in *Proceedings of the 25th Annual ACM Symposium on Theory of Computing*, Association for Computing Machinery Press, New York, 1993, pp. 11–20.

[499] D. Deutsch and R. Josza, Rapid Solution of Problems by Quantum Computation, *Proc. Roy. Soc. London Ser. A Math. Phys. Sci. A* **439**, 553–558 (1992).

[500] D. Simon, On the Power of Quantum Computation, in *Proceedings of the 35th Annual Symposium on the Foundations of Computer Science (FOCS'94)*, IEEE Computer Society Press, Los Alamitos, CA, 1994, pp. 116–123.

[501] P. W. Shor, Algorithms for Quantum Computation: Discrete Logarithms and Factoring, in *Proceedings of the 35th Annual Symposium on the Foundations of Computer Science (FOCS'94)*, IEEE Computer Society Press, Los Alamitos, CA, 1994, pp. 124–134.

[502] L. Grover, A Fast Quantum Mechanical Algorithm for Database Search, in *Proceedings of the 28th Annual ACM Symposium on Theory of Computing*, Association for Computing Machinery Press, New York, 1999, pp. 212–219.

[503] P. W. Shor, Scheme for Reducing Decoherence in Quantum Computer Memory, *Phys. Rev. A* **52**, R2493–R2496 (1995).

[504] A. R. Calderbank, and P. W. Shor, Good Quantum Error-Correcting Codes Exist, *Phys. Rev. A* **54**, 1098–1105 (1996).

[505] A. M. Steane, Error Correcting Codes in Quantum Theory, *Phys. Rev. Lett.* **77**, 793–797 (1996).

[506] A. Y. Kitaev, Quantum Computations: Algorithms and Error Correction, Russian *Math. Surv.* **52**, 1191–1249 (1997).

[507] P. W. Shor, Fault-Tolerant Quantum Computation, in *Proceedings of the 37th Annual Symposium on the Foundations of Computer Science (FOCS'96)*, IEEE Computer Society Press, Los Alamitos, CA, 1996, pp. 56–67.

[508] D. Aharonov and M. Ben-Or, Fault Tolerant Quantum Computation with Constant Error, Nov. 14, 1996.

[509] E. Knill, R. Laflamme, and W. H. Zurek, Resilient Quantum Computation: Error Models and Thresholds, *Proc. Roy. Soc. London Ser. A Math. Phys. Sci. A* **454**, 365–384 (1998).

[510] D. Gottesmann, Stabilizer Codes and Quantum Error Correction, Ph.D. thesis, California Institute of Technology, 1997.

[511] J. Preskill, Reliable Quantum Computers, *Proc. Roy. Soc. London Ser. A Math. Phys. Sci. A* **454**, 385–410 (1998).

[512] S. Aaronson, Quantum Lower Bound for the Collision Problem, in *Proceedings of the 34th Annual ACM Symposium on Theory of Computing*, Association for Computing Machinery Press, New York, 2002, pp. 635–642.

[513] J. Watrous, On Quantum and Classical Space-Bounded Processes with Algebraic Transition Amplitudes, in *Proceedings of the 40th Annual Symposium on the Foundations of Computer Science (FOCS'99)*, IEEE Computer Society Press, Los Alamitos, CA, 1999, pp. 341–351.

[514] H. Buhrman, R. Cleve, and A. Wigderson, Quantum vs. Classical Communication and Computation, in *Proceedings of the 30th Annual ACM Symposium on Theory of Computing*, Association for Computing Machinery Press, New York, 1998, pp. 63–68.

[515] A. Ambainis, L. J. Schulman, A. Ta-Shma, U. Vazirani, and A. Wigderson, The Quantum Communication Complexity of Sampling, in *Proceedings of the 39th Annual Symposium on the Foundations of Computer Science (FOCS'98)*, IEEE Computer Society Press, Los Alamitos, CA, 1998, pp. 342–351.

[516] R. Raz, Exponential Separation of Quantum and Classical Communication Complexity, in *Proceedings of the 3st ACM Symposium on Theory of Computing (STOC 1999),* ACM Press, New York, 2001, pp. 358–367.

[517] Z. Bar-Yossef, T. S. Jayram, and I. Kerenidis, Exponential Separation of Quantum and Classical One-Way Communication Complexity, 36th Annual ACM Symposium on Theory of Computing (STOC 2004), Chicago, IL, June 13–15, 2004.

[518] A. S. Holevo, Bounds for the Quantity of Information Transmitted by a Quantum Communication Channel, *Problems of Information Transmission* **9**(3), 177–183 (1973).

[519] W. K. Wooters and W. H. Zurek, A Single Quantum Cannot Be Cloned, *Nature* **299**, 802–803 (1982).

[520] C. H. Bennett and S. J. Wiesner, Communication via One- and Two-Particle Operators on Einstein-Podolsky-Rosen States, *Phys. Rev. Lett.* **69**, 2881–2884 (1992).

[521] C. H. Bennett, G. Brassard, C. Crepeau, R. Jozsa, A. Peres, and W. Wootters, Teleporting an Unknown Quantum State via Dual Classical and Einstein-Podolsky-Rosen Channels, *Phys. Rev. Lett.* **70**, 1895–1899 (1993).

[522] B. Schumacher, M. Westmoreland, and W. K. Wootters, Limitation on the Amount of Accessible Information in a Quantum Channel, *Phys. Rev. Lett.* **76**, 3452–3455 (1997).

[523] M. Sasaki, K. Kato, M. Izutsu, and O. Hirota, Quantum Channels Showing Superadditivity in Classical Capacity, *Phys. Rev. A* **58**, 146–158 (1998).

[524] A. S. Holevo, On Capacity of a Quantum Communications Channel, *Problems of Information Transmission* **15**(4), 247–253 (1979).

[525] B. Schumacher and M. Westmoreland, Sending Classical Information via Noisy Quantum Channels, *Phys. Rev. A* **56**, 131–138 (1997).

[526] A. S. Holevo, The Capacity of the Quantum Channel with General Signal States, *IEEE Trans. Inf. Theory* **IT-44**(#1), 269–273 (1998).

[527] C. H. Bennett, H. J. Bernstein, S. Popescu, and B. Schumacher, Concentrating Partial Entanglement by Local Operations, *Phys. Rev. A* **53**, 2046–2052 (1996).

[528] C. H. Bennett, G. Brassard, S. Popescu, B. Schumacher, J. A. Smolin, and W. K. Wootters, Purification of Noisy Entanglement and Faithful Teleportation via Noisy Channels, *Phys. Rev. Lett.* **76**, 722–725 (1996).

[529] C. H. Bennett, D. P. DiVincenzo, J. A. Smolin, and W. K. Wootters, Mixed-State Entanglement and Quantum Error Correction, *Phys. Rev.* A **54**, 3824–3851 (1996).

[530] C. H. Bennett and G. Brassard, Quantum Cryptography: Public Key Distribution and Coin Tossing, in *Proceedings of the IEEE International Conference on Computers Systems and Signal Processing*, IEEE, New York, 1984, pp. 175–179.

[531] S. Hallgren, Polynomial-Time Quantum Algorithms for Pell's Equation and the Principal Ideal Problem, in *Proceedings of the 34th Annual ACM Symposium on Theory of Computing*, Association for Computing Machinery Press, New York, 2002, pp. 653–658.

[532] W. van Dam and S. Hallgren, Efficient Quantum Algorithms for Shifted Quadratic Character Problems, Nov. 15, 2000.

[533] L. Ip, Solving Shift Problems and Hidden Coset Problem Using the Fourier Transform, May 7, 2002.

[534] W. van Dam and G. Seroussi, Efficient Quantum Algorithms for Estimating Gauss Sums, July 23, 2002.

[535] O. Regev, Quantum Computation and Lattice Problems, in *Proceedings of the 43rd Annual Symposium on the Foundations of Computer Science (FOCS'02)*, IEEE Computer Society Press, Los Alamitos, CA, 2002, pp. 520–530.

[536] M. Grigni, L. Schulman, M. Vazirani, and U. Vazirani, Quantum Mechanical Algorithms for the Nonabelian Hidden Subgroup Problem, in *Proceedings of the 33rd ACM Symposium on Theory of Computing (STOC 2001)*, ACM Press, New York, 2001, pp. 68–74.

[537] M. Ettinger, P. Hoyer, and E. Knill, The Quantum Query Complexity of the Hidden Subgroup Problem Is Polynomial, Jan. 12, 2004.

[538] A. Kuperberg, Subexponential-Time Quantum Algorithm for the Dihedral Hidden Subgroup Problem, Feb. 14, 2003.

[539] F. Magniez, M. Santha, and M. Szegedy, Quantum Algorithm for Detecting Triangles, 2003.

[540] W. van Dam and U. Vazirani, Limits on Quantum Adiabatic Optimization, 5th Workshop on Quantum Information Processing (QIP 2002), IBM T. J. Watson Research Center, Yorktown Heights, NY, January 14–17, 2002.

[541] B. Reichardt, The Quantum Adiabatic Optimization Algorithm and Local Minima, 36th Annual ACM Symposium on Theory of Computing (STOC 2004), Chicago, IL, June 13–15, 2004.

[542] E. Farhi, J. Goldstone, S. Gutman, B. Reichardt, and U. Vazirani, Tunneling in Quantum Adiabatic Optimization, manuscript in preparation, 2004.

[543] E. Farhi and S. Gutmann, Quantum Mechanical Square Root Speedup in a Structured Search Problem, Nov. 18, 1997.

[544] J. Watrous, Quantum Simulations of Classical Random Walks and Undirected Graph Connectivity, *J. Computer System Sci.* **62**(2), 376–391 (2001). A preliminary version appeared in *Proceedings of the 14th Annual IEEE Conference on Computational Complexity*, 1999, pp. 180–187.

[545] A. Ambainis, D. Aharonov, J. Kempe, and U. V. Vazirani, Quantum Walks on Graphs, in *Proceedings of the 33rd ACM Symposium on Theory of Computing (STOC 2001)*, ACM Press, New York, 2001, pp. 50–59.

[546] H.-K. Lo and H. F. Chau, Unconditional Security of Quantum Key Distribution Over Arbitrarily Long Distances, *Science* **283**, 2050–2056 (1999).

[547] D. A. Lidar and K. B. Whaley, Decoherence-Free Subspaces and Subsystems in Irreversible Quantum Dynamics, in Vol. 622, F. Benatti and R. Floreanini, Eds., *Springer Lecture Notes in Physics*, Springer-Verlag, Berlin, 2003, pp. 83120.

[548] D. Bacon, K. R. Brown, and K. B. Whaley, Coherence-Preserving Quantum Bits, *Phys. Rev. Lett.* **87**, 247902 (2001).

[549] M. Freedman, A. Kitaev, M. J. Larsen, and Z. Wang, Topological Quantum Computation, *Bull. Am. Math. Soc.* **40**, 31–38 (2003).

[550] D. Gottesman, A. Y. Kitaev, and J. Preskill, Encoding a Qubit in an Oscillator, *Phys. Rev. A* **64**, 012310 (2001).

[551] D. Gottesman, An Introduction to Quantum Error Correction, in *Quantum Computation: A Grand Mathematical Challenge for the Twenty-First Century and the Millennium*, S. Lomonaco, Jr., Ed., American Mathematical Society, Providence, RI, 2002, pp. 221–235.

[552] A. Y. Kitaev and J. Watrous, Parallelization, Amplification, and Exponential Time Simulation of Quantum Interactive Proof Systems, in *Proceedings of the 32nd ACM Symposium on Theory of Computing (STOC 2000)*, ACM Press, New York, 2000, pp. 608–617.

[553] J. Watrous, Limits on the Power of Quantum Statistical Zero-Knowledge, in *Proceedings of the 43rd Annual IEEE Symposium on Foundations of Computer Science (FOCS'02)*, IEEE Computer Society Press, Los Alamitos, CA, 2002) X2, pp. 459–468.

[554] H. Kobayashi and K. Matsumoto, Quantum Multiprover Interactive Proof Systems with Limited Prior Entanglement, *J. Computer Syst. Sci.* **66**(3), 429–450 (2003).

[555] S. Somaroo, C. H. Tseng, T. Havel, R. Laflamme, and D. G. Cory, Quantum Simulation of a Quantum Computer, *Phys. Rev. Lett.* **82**, 5381–5384 (1999).

[556] C. H. I Tseng, S. S. Somaroo, Y. S. Sharf, E. Knill, R. Laflamme, T. F. Havel, and D. G. Cory, Quantum Simulation of a Three-Body Interaction Hamiltonian on an NMR Quantum Computer, *Phys. Rev. A* **61**, 12302–12308 (2000).

[557] L. Viola, E. M. Fortunato, S. Lloyd, C.-H. Tseng, and D. G. Cory, Stochastic Resonance and Nonlinear Response by NMR Spectroscopy, *Phys. Rev. Lett.* **84**, 5466–5470 (2000).

[558] Y. Weinstein, S. Lloyd, J. V. Emerson, and D. G. Cory, Experimental Implementation of the Quantum Baker's Map, *Phys. Rev. Lett.* **89**, 157902 (2002).

[559] J. Emerson, Y. S. Weinstein, S. Lloyd, and D. G. Cory, Fidelity Decay as an Efficient Indicator of Quantum Chaos, *Phys. Rev. Lett.* **89**, 284102 (2002).

[560] G. Teklemariam, E. M. Fortunato, M. A. Pravia, T. F. Havel, and D. G. Cory, Experimental Investigations of Decoherence on a Quantum Information Processor, *Chaos, Solitons, and Fractals* **16**, 457–465 (2002).

[561] W. Zhang and D. G. Cory, First Direct Measurement of the Spin Diffusion Rate in a Homogenous Solid, *Phys. Rev. Lett.* **80**, 1324–1327 (1998).

[562] G. S. Boutis, D. Greenbaum, H. Cho, D. G. Cory, and C. Ramanathan, Spin Diffusion of Correlated Two-Spin States in a Dielectric Crystal, *Phys. Rev. Lett.* **92**, 137201 (2004).

[563] G. Vidal, J. I. Latorre, E. Rico, and A. Y. Kitaev, Entanglement In Quantum Critical Phenomena, *Phys. Rev. Lett.* **90**, 227902 (2003).

[564] E. Knill, R. Laflamme, and G. J. Milburn, Efficient Linear Optics Quantum Computation, *Nature* **409**, 46–52 (2001).

[565] M. A. Nielsen, Conditions for a Class of Entanglement Transformations, *Phys. Rev. Lett.* **83**(2), 436–439 (1999).

[566] R. Raussendorf and H. J. Briegel, A One-Way Quantum Computer, *Phys. Rev. Lett.* **86**, 5188 (2001).

[567] R. Schack and C. M. Caves, Classical Model for Bulk-Ensemble NMR Quantum Computation, Apr. 30, 1999.

[568] E. Knill and R. Laflamme, On the Power of One Bit of Quantum Information, *Phys. Rev. Lett.* **81**, 5672–5675 (1998).

[569] D. Poulin, R. Blume-Kohout, R. Laflamme, and H. Ollivier, Exponential Speed-Up with a Single Bit of Quantum Information: Testing the Quantum Butterfly Effect, Oct. 6, 2003.

[570] G. Brassard, Quantum Communication Complexity: A Survey, *Foundations Phys.* **33**(11), 1593–1616 (2003).

[571] P. M. B. Vitanyi, Quantum Kolmogorov Complexity Based on Classical Descriptions, *IEEE Trans. Information Theory* **47**(6), 2464–2479 (2001).

[572] P. Gacs, Quantum Algorithmic Entropy, *J. Phys. A: Math. Gen.* **34**(35), 6859–6880 (2001).

[573] A. S. Holevo, Problems in the Mathematical Theory of Quantum Communication Channels, *Repts. Math. Phys.* **12**(2), 273–278 (1977).

[574] I. Devetak and A. Winter, Distilling Common Randomness from Bipartite Quantum States, in *Proceedings of the IEEE International Symposium on Information Theory*, IEEE, Piscataway, NJ, 2003, p. 403.

[575] M. A. Nielsen and I. L. Chuang, *Quantum Computation and Quantum Information*, Cambridge University Press, Cambridge, 2000, Sec. 8.2.

[576] P. W. Shor, Equivalence of Additivity Questions in Quantum Information Theory, May 7, 2003.

[577] P. W. Shor, Capacities of Quantum Channels and How to Find Them, *Math. Programming* **97**(1–2), 311–335 (2003).

[578] C. H. Bennett, P. W. Shor, J. A. Smolin, and A. V. Thapliyal, Entanglement-Assisted Classical Capacity of Noisy Quantum Channels, *Phys. Rev. Lett.* **83**, 3081 (1999).

[579] C. H. Bennett, P. W. Shor, J. A. Smolin, and A. V. Thapliyal, Entanglement-Assisted Capacity of a Quantum Channel and the Reverse Shannon Theorem, *IEEE Trans. Information Theory* **48**, 2637–2655 (2002).

[580] H.-K. Lo, Classical-Communication Cost in Distributed Quantum-Information Processing: A Generalization of Quantum-Communication Complexity, *Phys. Rev. A* **62**, 012313 (2000).

[581] C. H. Bennett, D. P. DiVincenzo, J. A. Smolin, B. M. Terhal, and W. K. Wootters, Remote State Preparation, *Phys. Rev. Lett.* **87**, 077902 (2001).

[582] P. Hayden, R. Jozsa, and A. Winter, Trading Quantum for Classical Resources in Quantum Data Compression, *J. Math. Phys.* **43**(9), 4404–4444 (2002).

[583] A. Winter, and S. Massar, Compression of Quantum Measurement Operations, *Phys. Rev. A* **64**, 012311 (2001).

[584] A. Einstein, B. Podolsky, and N. Rosen, Can Quantum-Mechanical Description of Physical Reality Be Considered Complete? *Phys. Rev.* **47**, 777 (1935).

[585] E. Schrodinger, Die gegenwartige Situation der Quantenmechanik, *Naturw* **23**, 807, 823, 844 (1935).

[586] J. S. Bell, On the Einstein-Podolsky-Rosen Paradox, *Physics* **1**, 195–200 (1964).

[587] R. Jozsa, and N. Linden, On the Role of Entanglement in Quantum Computational Speed-Up, *Proc. Roy. Soc. London Ser. A* **459**(2036), 2011–2032 (2003).

[588] P. W. Shor, J. A. Smolin, and B. M. Terhal, Nonadditivity of Bipartite Distillable Entanglement Follows from Conjecture on Bound Entangled Werner States, *Phys. Rev. Lett.* **86**, 2681–2684 (2001).

[589] P. Horodecki, Separability Criterion and Inseparable Mixed States with Positive Partial Transposition, *Phys. Lett. A* **232**(5), 333–339 (1997).

[590] G. Vidal and J. I. Cirac, Irreversibility in Asymptotic Manipulations of Entanglement, *Phys. Rev. Lett.* **86**, 5803–5806 (2001).

[591] V. Vedral, M. B. Plenio, M. A. Rippin, and P. L. Knight, Quantifying Entanglement, *Phys. Rev. Lett.* **78**, 2275–2279 (1997).

[592] V. Vedral, The Role of Relative Entropy in Quantum Information Theory, *Rev. Modern Phys.* **74**, 197 (2002).

[593] C. H. Bennett, D. P. DiVincenzo, C. A. Fuchs, T. Mor, E. Rains, P. W. Shor, J. A. Smolin, and W. K. Wootters, Quantum Nonlocality Without Entanglement, *Phys. Rev. A* **59**, 1070–1091 (1999).

[594] G. Vidal, Entanglement Monotones, *J. Modern Opt.* **47**, 355 (2000).

[595] G. Vidal and R. Tarrach, Robustness of Entanglement, *Phys. Rev. A* **59**(1), 141–155 (1999).

[596] A. Peres, Separability Criterion for Density Matrices, *Phys. Rev. Lett.* **77**, 1413 (1996).

[597] M. Horodecki, P. Horodecki, and R. Horodecki, Separability of Mixed States: Necessary and Sufficient Conditions, *Phys. Lett. A* **223**, 1 (1996).

[598] C. H. Bennett, D. P. DiVincenzo, T. Mor, P. W. Shor, J. A. Smolin, and B. M. Terhal, Unextendible Product Bases and Bound Entanglement, *Phys. Rev. Lett.* **82**, 5385 (1999).

[599] D. P. DiVincenzo, T. Mor, P. W. Shor, J. A. Smolin, and B. M. Terhal, Unextendible Product Bases, Uncompletable Product Bases, and Bound Entanglement, *Commun. Math. Phys.* **238**, 379–410 (2003).

[600] M. Lewenstein, B. Krauss, J. I. Cirac, and P. Horodecki, Optimization of Entanglement Witnesses, *Phys. Rev. A* **62**, 052310 (2000).

[601] E. M. Rains, Rigorous Treatment of Distillable Entanglement, *Phys. Rev. A* **60**, 173 (1999).

[602] B. M. Terhal, A Family of Indecomposable Positive Linear Maps Based on Entangled Quantum States, *Linear Algebra Applications* **323**, 61–73 (2000).

[603] A. K. Ekert, C. M. Alves, D. K. L. Oi, M. Horodecki, P. Horodecki, and L. C. Kwek, Direct Estimations of Linear and Nonlinear Functionals of a Quantum State, *Phys. Rev. Lett.* **88**, 217901 (2002).

[604] D. Gottesman and I. L. Chuang, Demonstrating the Viability of Universal Quantum Computation Using Teleportation and Single Qubit Operations, *Nature* **402**, 390–393 (1999).

[605] S. Wiesner, Conjugate Coding, *SIGACT News* **15**, 78–88 (1983).

[606] C. H. Bennett, Quantum Cryptography Using Any Two Nonorthogonal States, *Phys. Rev. Lett.* **68**, 3121–3124 (1992).

[607] A. K. Ekert, Quantum Cryptography Based on Bell's Theorem, *Phys. Rev. Lett.* **67**, 661–663 (1991).

[608] A. K. Ekert, J. G. Rarity, P. R. Tapster, and G. M. Palma, Practical Quantum Cryptography Based on Two-Photon Interferometry, *Phys. Rev. Lett.* **69**, 1293–1295 (1992).

[609] B. Huttner and A. K. Ekert, Information Gain in Quantum Eavesdropping, *J. Modern Opt.* **41**, 2455–2466 (1994).

[610] D. Deutsch, A. K. Ekert, R. Jozsa, C. Macchiavello, S. Popescu, and A. Sanpera, Quantum Privacy Amplification and the Security of Quantum Cryptography Over Noisy Channels, *Phys. Rev. Lett.* **77**, 2818–2821 (1996).

[611] D. Mayers, Unconditionally Secure Quantum Bit Commitment Is Impossible, *Phys. Rev. Lett.* **78**, 3414–3417 (1997).

[612] P. W. Shor and J. Preskill, Simple Proof of Security of the BB84 Quantum Key Distribution Protocol, *Phys. Rev. Lett.* **85**, 441–444 (2000).

[613] K. Tamaki, M. Koashi, and N. Imoto, Unconditionally Secure Key Distribution Based on Two Nonorthogonal States, *Phys. Rev. Lett.* **90**, 167904 (2003).

[614] D. Mayers, Unconditionally Secure Quantum Bit Commitment Is Impossible, *Phys. Rev. Lett.* **78**, 3414–3417 (1997).

[615] R. W. Spekkens and T. Rudolph, Degrees of Concealment and Bindingness in Quantum Bit Commitment Protocols, *Phys. Rev. A* **65**, 012310 (2002).

[616] A. Y. Kitaev, Quantum Coin Tossing, MSRI lecture, http://www.msri.org/publications/ln/msri/2002/qip/kitaev/1/.

[617] R. Cleve, D. Gottesman, and H.-K. Lo, How to Share a Quantum Secret, *Phys. Rev. Lett.* **83**, 648–651 (1999).

[618] D. P. DiVincenzo, D. W. Leung, and B. M. Terhal, Quantum Data Hiding, *IEEE Trans. Inf. Theory* **48**, 580–598 (2002).

[619] T. Eggeling and R. F. Werner, Hiding Classical Data in Multipartite Quantum States, *Phys. Rev. Lett.* **89**, 097905 (2002).

[620] D. P. DiVincenzo, P. Hayden, and B. M. Terhal, Hiding Quantum Data, *Found. Phys.* **33**, 11, 1629–1647 (2003).

[621] H. Buhrman, R. Cleve, J. Watrous, and R. de Wolf, Quantum Fingerprinting, *Phys. Rev. Lett.* **87**(16), 167902 (2001).

[622] C. Crepeau, D. Gottesman, and A. Smith, Secure Multiparty Quantum Computing, *Proceedings of the 34th ACM Symposium on Theory of Computing*, ACM Press, New York, 2001, pp. 643–652.

[623] H. Barnum, C. Crepeau, D. Gottesman, A. Smith, and A. Tapp, Authentication of Quantum Messages, in *Proceedings of the 43rd Annual Symposium on Foundations of Computer Science*, IEEE Computer Society Press, Los Alamitos, CA, 2002, pp. 449–458.

[624] A. Ambainis, M. Mosca, A. Tapp, and R. de Wolf, Private Quantum Channels, in *Proceedings of the 41st Annual Symposium on Foundations of Computer Science*, IEEE Computer Society Press, Los Alamitos, CA, 2000, pp. 547–553.

[625] M. Mosca, A. Tapp, and R. de Wolf, Private Quantum Channels and the Cost of Randomizing Quantum Information, Mar. 24, 2000.

[626] P. O. Boykin and V. Roychowdhury, Optimal Encryption of Quantum Bits, *Phys. Rev. A* **67**(4), 042317 (2003).

[627] P. Hayden, D. W. Leung, P. W. Shor, and A. Winter, Randomizing Quantum States: Constructions and Applications, Nov. 13, 2003.

[628] D. W. Leung, Quantum Vernam Cipher, *Quantum Information and Computation* **2**(1), 14–34 (2002).

[629] D. Gottesman and I. Chuang, Quantum Digital Signatures, May 8, 2001.

[630] I. Kerenidis and R. de Wolf, Quantum Symmetrically-Private Information Retrieval, July 10, 2003.

[631] R. Cleve and J. Watrous, Fast Parallel Circuits for the Quantum Fourier Transform, *Proceedings of the 41st Annual Symposium on Foundations of Computer Science*, IEEE Computer Society Press, Los Alamitos, CA, 2000, pp. 526–536.

[632] L. Hales and S. Hallgren, An Improved Quantum Fourier Transform Algorithm and Applications, in *Proceedings of the 41st Annual Symposium on Foundations of Computer Science*, IEEE Computer Society Press, Los Alamitos, CA, 2000, pp. 515–525.

[633] D. P. DiVincenzo, D. Bacon, J. Kempe, G. Burkard, and K.B. Whaley, Universal Quantum Computation with the Exchange Interaction, *Nature* **408**, 339–342, (2000).

[634] S. Lloyd, A Potentially Realizable Quantum Computer, *Science* **261**, 1569–1571 (1993).

[635] G. P. Berman, G. D. Doolen, D. I. Kamenev, and V. I. Tsifrinovich, Perturbation Theory for Quantum Computation with a Large Number of Qubits, *Phys. Rev. A* **65**, 012321 (2002).

[636] G. P. Berman, G. D. Doolen, G. V. Lopez, and V. I. Tsifrinovich, A Quantum Full Adder for a Scalable Nuclear Spin Quantum Computer, *Computer Phys. Commun.* **146**(3), 324–330 (2002).

[637] G. P. Berman, F. Borgonovi, H. S. Goan, S. A. Gurvitz, and V. I. Tsifrinovich, Single-Spin Measurement and Decoherence in Magnetic-Resonance Force Microscopy, *Phys. Rev. B* **67**, 094425 (2003).

[638] W. H. Zurek, Decoherence, Einselection, and the Quantum Origins of the Classical, *Rev. Modern Phys.* **75**, 715–775 (2003).

[639] K. Blum, *Density Matrix Theory and Applications*, 2nd Ed., Plenum Press, New York, 1996.

[640] A. Bohm and K. Kraus, *States, Effects and Operations: Fundamental Notions of Quantum Theory*, Springer-Verlag, Berlin, 1983.

[641] J. Preskill, Lecture notes for Caltech graduate course Quantum Computation Physics 219/Computer Science 219, http://www.theory.caltech.edu/people/preskill/-ph229/#lecture.

[642] A. Shirman and G. Schön, Dephasing and Renormalization in Quantum Two-Level Systems, in *Proceedings of NATO ARW Workshop on Quantum Noise in Mesoscopic Physics,* Y. V. Nazarov, Ed., Kluwer Academic, Dordrecht, The Netherlands, 2002.

[643] A. M. Steane, Quantum Computing and Error Correction, in *Decoherence and Its Implications in Quantum Computation and Information Transfer,* A. Gonis and P. E. A. Turchi, Eds., IOS Press, Amsterdam, 2001, pp. 284–298.

[644] M. Freedman, A. Kitaev, M. J. Larsen, and Z. Wang, Topological Quantum Computation, *Bull. Am. Math. Soc.* **40**, 31 (2003).

[645] L. Viola and E. Knill, Robust Dynamical Decoupling of Quantum Systems with Bounded Controls, *Phys. Rev. Lett.* **90**, 037901 (2003).

[646] M. S. Byrd and D. A. Lidar, Combined Error Correction Techniques for Quantum Computing Architectures, *J. Modern Opt.* **50**, 1285–1297 (2003).

[647] L. Fedichkin, A. Fedorov, and V. Privman, Measures of Decoherence, in *Proceedings of the 2003 International Society for Optical Engineering (SPIE) Conference on Quantum Information and Computation*, E. Donkor, A. R. Pirich, and H. E. Brandt, Eds., SPIE, Bellingham, WA, 2003, pp. 243–254.

[648] D. Loss, and D. P. Divincenzo, Exact Born Approximation for the Spin-Boson Model, Apr. 10, 2003.

[649] G. A. Fiete and E. J. Heller, Semiclassical Theory of Coherence and Decoherence, *Phys. Rev. A* **68**, 022112 (2003).

[650] R. W. Simmonds, K. M. Lang, D. A. Hite, D. P. Pappas, and J. M. Martinis, Decoherence in Josephson Qubits from Junction Resonances, Feb. 18, 2004.

[651] D. Vion, A. Aassime, A. Cottet, P. Joyez, H. Pothier, C. Urbina, D. Esteve, and M. H. Devoret, Manipulating the Quantum State of an Electrical Circuit, *Science* **296**, 886–889 (2002).

[652] P. C. Chen, C. Piermarocchi, and L. J. Sham, Control of Exciton Dynamics in Nanodots for Quantum Operations, *Phys. Rev. Lett.* **87**, 067401 (2001).

[653] E. Myrgren and K. B. Whaley, Implementing a Quantum Algorithm with Exchange-Coupled Quantum Dots: A Feasibility Study, *Quantum Inf.* Process. **2**(5), 1 (2003); J. M. Raimond, M. Brune, and S. Haroche, Manipulating Quantum Entanglement with Atoms and Photons in a Cavity, *Rev. Modern Phys.* **73**, 565–582 (2001).

[654] D. Zollman, W. Axmann, B. Grabhorn, C. Regehr, and P. Donovan, Visual Quantum Mechanics, Contemporary Physics Course, Spring 1994, The University of Oklahoma, Department of Physics & Astronomy, Norman, OK.

[655] S. Shellswell, Everyscience Glossary, Worcester, Oxford, http://www.everyscience.com/Chemistry/Glossary/S.php.

[656] Technical Glossary, Applied Materials, Santa Clara, CA.

[657] Glossary, National Physical Laboratory (NPL), UK's National Measurement Laboratory, Teddington, Middlesex, U.K., enquiry@npl.co.uk.

[658] Fibercore Ltd., Promotional Materials, Fibercore House, Southampton, UK. info@fibercore.com.

[659] MATTER, Department of Engineering, The University of Liverpool, Liverpool, U.K., http://www.matter.org.uk/default.htm. MATTER has been set up as a nonprofit consortium of U.K. materials science departments in 1993 to develop and help integrate computer-based learning materials into mainstream teaching.

[660] T. Laude, Nanostructures of Layered Materials, June 2001, National Institute of Material Science (NIMS), Tsukuba, Japan, t.laude@teijin.co.jp, http://www.umi-nokai.net/ nanotube/intro.htm.

[661] *Encyclopedia Nanotech* (also known as *The Nanotech Dictionary*), Nanoword.net, an online distributor of publications, http://www.nanoword.net/library/def/bldefindex.htm.

[662] A. A. Lucas et al., Carbon and the Structure of Fullerenes, *Rev. Mod. Phys.* **74**, 1 (2002).

[663] The Shodor Education Foundation, in cooperation with the Department of Chemistry, University of North Carolina at Chapel Hill, Durham, NC, rpanoff@shodor.org.

[664] Board on Physics and Astronomy (BPA), *Frontiers in High Energy Density Physics: The X-Games of Contemporary Science*, National Academy Press, Washington, DC, 2003.

[665] Steering Committee for Cyberinfrastructure Research and Development in the Atmospheric Sciences (CyRDAS) (http://www.cyrdas.org), Division of Atmospheric Sciences of the National Science Foundation, Arlington, VA. http://www.geo.nsf.gov/atm/.

[666] H. Abelson, Thomas F. Knight, Gerald Jay Sussman, et al., *Amorphous Computing Manifesto*, MIT, Cambridge, MA, 1996.

[667] Promotional Material, Autonomic Computing Glossary, IBM Corporation, White Plains, NY, http://www.ibm.com.

[668] D. Minoli, *A Networking Approach to Grid Computing*, Wiley, Hoboken, NJ, 2005.

[669] R. Bajcsy, National Science Foundation, House Science Committee Hearing on *Beyond Silicon Computing: Quantum and Molecular Computing*, Sept. 12, 2000.

[670] National Institute of Standards and Technology, *Pervasive Computing 2001*, May 1–2, 2001, Gaithersburg, MD, http://www.nist.gov/pc2001/.

[671] J. West, *The Quantum Computer, An Introduction*, Computer Science Department, California Institute of Technology, Pasadena, CA, 2000.

[672] D. Deutsch, Quantum Theory, The Church-Touring Principle and The Universal Quantum Computer, *Proc. Roy. Soc. London Ser. A* **400**, 96–117 (1985).

[673] D. Deutsch and A. Ekert, Quantum Computation, *Physics World*, March 1998.

[674] R. P. Feynman, Simulating Physics with Computers, *Int. J. Theor. Phys.* **21**, 467 (1982).

[675] M. Weiser, Hot Topics: Ubiquitous Computing, *IEEE Computer*, Oct. 1993.

[676] B. J. Rhodes, et al., Wearable Computing Meets Ubiquitous Computing: Reaping the Best of Both Worlds, in *Proceedings of the Third International Symposium on Wearable Computers (ISWC '99)*, San Francisco, CA, Oct. 18–19, 1999, pp. 141–149.

[677] *Mathematical Challenges from Theoretical/Computational Chemistry*, Commission on Physical Sciences, Mathematics, and Applications, National Research Council, National Academy of Sciences, Washington, DC, 1995.

[678] Introduction to Confocal Microscopy, Promotional Material, Nikon USA; also MWW Group, East Rutherford, NJ.

[679] K. W. Dunn and E. Wang, Introduction to Confocal Microscopy, Department of Medicine, Indiana University, School of Medicine, Indianapolis, IN.

[680] S. W. Paddock, E. J. Hazen, and P. J. DeVries, Introduction to Confocal Microscopy, Laboratory of Molecular Biology, Howard Hughes Medical Institute, University of Wisconsin, Madison, WI.

[681] J. B. Pawley, Introduction to Confocal Microscopy, Department of Zoology, University of Wisconsin, Madison, WI.

[682] M. Parry-Hill, T. J. Fellers, and M. W. Davidson, Introduction to Confocal Microscopy, National High Magnetic Field Laboratory, Florida State University, Tallahassee, FL.

[683] *JCrystal* (a computer program for creating, editing, displaying and deploying crystal shapes), JCrystalSoft, Steffen Weber, Livermore, CA, 2003.

[684] F. Axel, F. Denoyer, and J. P. Gazeau, Eds., *From Quasicrystals to More Complex Systems*. Springer, Heidelberg, New York, 2000.

[685] F. Gähler, P. Kramer, H.-R. Trebin, and K. Urban, Eds., Proceedings of the 7th International Conference on Quasicrystals, *Mater. Sci. Eng. A* **294–296**, 1–912 (2000).

[686] Reciprocal Space Images of Aperiodic Crystals, in *International Tables for Crystallography*, Vol. B, U. Shmueli, Ed., Kluwer Academic Publishers, Dordrecht, 2001, pp. 486–518.

[687] E. Rochat, High Power Optical Amplifiers for Coherent Inter-Satellite Communications, Thesis, Feb. 2000, University of Neuchatel, Switzerland.

[688] Glossary, University of Wisconsin-Madison, Materials Research Science and Engineering Center (MRSEC), Madison, WI, http://mrsec.wisc.edu.

[689] Cambridge Scientific Abstracts, Bethesda, MD, www.csa.com.

[690] *Browse Dictionary*, McGraw-Hill Encyclopedia of Science and Technology Online, Access Science, McGraw-Hill, New York.

[691] S. M. Cronenwett, T. H. Oosterkamp, and L. P. Kouwenhoven, A Tunable Kondo Effect in Quantum Dots, *Science* **281**, 540 (1998).

[692] E. T. Gendlin and J. Lemke, A Critique of Relativity and Localization, *Math. Modeling* **4**, 61–72, (1983).

[693] J. Conner, Foresight Institute, Judy@foresight.org.

[694] C. Sealy, *Materials Today*, Elsevier Publishers.

[695] Nanotechnology Now, On-line Nanotechnology Portal, http://nanotech-now.com/about_us.htm.

[696] International Union of Pure and Applied Chemistry (IUPAC), *Compendium of Chemical Terminology: Recommendations*, compiled by A. D. McNaught and A. Wilkinson, Blackwell Science, 1997.

[697] Glossary, University of Nebraska at Lincoln, Center for Materials Research and Analysis, Lincoln, NE.

[698] H. F. Bettinger et al., Structure and Properties of Carbon Nanotubes, *J. Am. Chem. Soc.* **123**, 12849 (2001).

[699] G. M. Whitesides and J. C. Love, The Art of Building Small, *Sci. Am.* **285**(3), 39–47 (2001).

[700] P. J. Moyer and M. Paesler, *Near-Field Optics: Theory, Instrumentation, and Applications*, Wiley-Interscience, Hoboken, NJ, 1996.

[701] A. D. Sokal, Transgressing the Boundaries: Towards a Transformative Hermeneutics of Quantum Gravity, Department of Physics, New York University, New York, sokal@nyu.edu.

[702] J. P. Hornak, The Basics of NMR, Rochester Institute of Technology, Magnetic Resonance Laboratory, Center for Imaging Science, Rochester, NY, jphsch@rit.edu.

[703] L. Fortnow and S. Homer, A Short History of Computational Complexity, NEC Research Institute, Princeton, NJ, Nov. 14, 2002, Computer Science Department, Boston University, Boston, MA.

[704] Quantum Mechanics, Educational/Promotional Material, Stanford Linear Accelerator Center and Virtual Visitor Center, Stanford University, Menlo Park, CA.

[705] Quantum Glossary, Stanford Linear Accelerator Center and Virtual Visitor Center, Stanford University, Menlo Park, CA.

[706] C. H. Bennett, F. Bessette, G. Brassard, L. Salvail, and J. Smolin, Experimental Quantum Cryptography, *J. Cryptology* **5**, 3–28 (1992).

[707] W. McCarthy, *Hacking Matter: Levitating Chairs, Quantum Mirages, and the Infinite Weirdness of Programmable Atoms*, Basic Books, Feb. 2003.

[708] Wenger, Promotional Material, Kansas City, http://www.wenger.com.

[709] *Photonics Magazine*, *Photonics Dictionary*, Laurin Publishing Co., Pittsfield, MA, www.Photonics.com.

[710] Comprehensive Glossary, Electrical and Computer Engineering Department, University of Colorado, Boulder, CO.

[711] Quantum Glossary, About, Inc., New York. www.physics.about.com.

[712] L. H. Dubois and R. G. Nuzzo, Synthesis, Structure, and Properties of Model Organic Surfaces, *Annu. Rev. Phys. Chem.* **43**, 437–463 (1992).

[713] G. Alber, Technische Universitat-Darmstadt, Institut für Angewandte Physik, Experimentelle Licht- und Teilchenoptik, Darmstadt, Germany, email dekanat@ physik.tu-darmstadt.de.

INDEX

Nanotechnology Applications to Telecommunications and Networking, By Daniel Minoli